T0181736

Herbert Amann
Joachim Escher

Analysis I

Translated from the German by Gary Brookfield

Birkhäuser Verlag
Basel · Boston · Berlin

Authors:

Herbert Amann
Institut für Mathematik
Universität Zürich
Winterthurerstr. 190
CH-8057 Zürich
e-mail: amann@math.unizh.ch

Joachim Escher
Institut für Angewandte Mathematik
Universität Hannover
Welfengarten 1
D-30167 Hannover
e-mail: escher@ifam.uni-hannover.de

Originally published in German under the same title by Birkhäuser Verlag, Switzerland

2000 Mathematical Subject Classification 26-01, 26Axx; 03-01, 30-01, 40-01, 54-01

A CIP catalogue record for this book is available from the
Library of Congress, Washington D.C., USA

Bibliografische Information Der Deutschen Bibliothek
Die Deutsche Bibliothek verzeichnet diese Publikation in der Deutschen Nationalbibliografie;
detaillierte bibliografische Daten sind im Internet über <http://dnb.ddb.de> abrufbar.

ISBN 3-7643-7153-6 Birkhäuser Verlag, Basel – Boston – Berlin

Part of Springer Science+Business Media
Cover design: Micha Lotrovsky, 4106 Therwil, Switzerland
Printed on acid-free paper produced from chlorine-free pulp. TCF ∞

ISBN 3-7643-7153-6

9 8 7 6 5 4 3 2 1

www.birkhauser.ch

Preface

Logical thinking, the analysis of complex relationships, the recognition of underlying simple structures which are common to a multitude of problems — these are the skills which are needed to do mathematics, and their development is the main goal of mathematics education.

Of course, these skills cannot be learned 'in a vacuum'. Only a continuous struggle with concrete problems and a striving for deep understanding leads to success. A good measure of abstraction is needed to allow one to concentrate on the essential, without being distracted by appearances and irrelevancies.

The present book strives for clarity and transparency. Right from the beginning, it requires from the reader a willingness to deal with abstract concepts, as well as a considerable measure of self-initiative. For these efforts, the reader will be richly rewarded in his or her mathematical thinking abilities, and will possess the foundation needed for a deeper penetration into mathematics and its applications.

This book is the first volume of a three volume introduction to analysis. It developed from courses that the authors have taught over the last twenty six years at the Universities of Bochum, Kiel, Zurich, Basel and Kassel. Since we hope that this book will be used also for self-study and supplementary reading, we have included far more material than can be covered in a three semester sequence. This allows us to provide a wide overview of the subject and to present the many beautiful and important applications of the theory. We also demonstrate that mathematics possesses, not only elegance and inner beauty, but also provides efficient methods for the solution of concrete problems.

Analysis itself begins in Chapter II. In the first chapter we discuss quite thoroughly the construction of number systems and present the fundamentals of linear algebra. This chapter is particularly suited for self-study and provides practice in the logical deduction of theorems from simple hypotheses. Here, the key is to focus on the essential in a given situation, and to avoid making unjustified assumptions. An experienced instructor can easily choose suitable material from this chapter to make up a course, or can use this foundational material as its need arises in the study of later sections.

In this book, we have tried to lay a solid foundation for analysis on which the reader will be able to build in later forays into modern mathematics. Thus most concepts and definitions are presented, right from the beginning, in their general form — the form which is used in later investigations and in applications. This way the reader needs to learn each concept only once, and then with this basis, can progress directly to more advanced mathematics.

We refrain from providing here a detailed description of the contents of the three volumes and instead refer the reader to the introductions to each chapter, and to the detailed table of contents. We also wish to direct the reader's attention to the numerous exercises which appear at the end of each section. Doing these exercises is an absolute necessity for a thorough understanding of the material, and serves also as an effective check on the reader's mathematical progress.

In the writing of this first volume, we have profited from the constructive criticism of numerous colleagues and students. In particular, we would like to thank Peter Gabriel, Patrick Guidotti, Stephan Maier, Sandro Merino, Frank Weber, Bea Wollenmann, Bruno Scarpellini and, not the least, our students, who, by their positive reactions and later successes, encouraged our particular method of teaching analysis.

From Peter Gabriel we received support 'beyond the call of duty'. He wrote the appendix 'Introduction to Mathematical Logic' and unselfishly allowed it to be included in this book. For this we owe him special thanks.

As usual, a large part of the work necessary for the success of this book was done 'behind the scenes'. Of inestimable value are the contributions of our 'typesetting perfectionist' who spent innumerable hours in front of the computer screen and participated in many intense discussions about grammatical subtleties. The typesetting and layout of this book are entirely due to her, and she has earned our warmest thanks.

We also wish to thank Andreas who supplied us with latest versions of TEX[1] and stood ready to help with software and hardware problems.

Finally, we thank Thomas Hintermann for the encouragement to make our lectures accessible to a larger audience, and both Thomas Hintermann and Birkhäuser Verlag for a very pleasant collaboration.

Zurich and Kassel, June 1998 H. Amann and J. Escher

[1]The text was typeset using LATEX. For the graphs, CorelDRAW! and Maple were also used.

Preface to the second edition

In this new edition we have eliminated the errors and imprecise language that have been brought to our attention by attentive readers. Particularly valuable were the comments and suggestions of our colleagues H. Crauel and A. Ilchmann. All have our heartfelt thanks.

Zurich and Hannover, March 2002 H. Amann and J. Escher

Preface to the English translation

It is our pleasure to thank Gary Brookfield for his work in translating this book into English. As well as being able to preserve the 'spirit' of the German text, he also helped improve the mathematical content by pointing out inaccuracies in the original version and suggesting simpler and more lucid proofs in some places.

Zurich and Hannover, May 2004 H. Amann und J. Escher

Contents

Chapter V Sequences of Functions

Chapter I

Foundations

Most of this first chapter is about numbers — natural numbers, integers, real numbers and complex numbers. Without a clear understanding of these numbers, a deep investigation of mathematics is not possible. This makes a thorough discussion of number systems absolutely necessary.

To that end we have chosen to present a constructive formulation of these number systems. Starting with the Peano axioms for the natural numbers, we construct successively the integers, the rational numbers, the real numbers and finally, the complex numbers. At each step, we are guided by a desire to solve certain 'naturally' occurring equations. These constructions are relatively long and require considerable stamina from the reader, but those readers who persevere will be rewarded with considerable practice in mathematical thinking.

Even before we can talk about the natural numbers, the simplest of all number systems, we must consider some of the fundamentals of set theory. Here the main goal is to develop a precise mathematical language. The axiomatic foundations of logic and set theory are beyond the scope of this book.

The reader may well be familiar with some of the material in Sections 1–4. Even so, we have deliberately avoided appealing to the reader's intuitions and previous experience, and have instead chosen a relatively abstract framework for our presentation. In particular, we have been strict about avoiding any concepts that are not already precisely defined, and using claims that are not previously proved. It is important that, right from the beginning, students learn to work with definitions and derive theorems from them without introducing spurious additional assumptions.

The transition from the simplest number system, the natural numbers, to the most complicated number system, the complex numbers, is paralleled by a corresponding increasing complexity in the algebra needed. Therefore, in Sections 7–8 we discuss fairly thoroughly the most important concepts of algebra. Here again we have chosen an abstract approach with the goal that beginning students become familiar with certain mathematical structures which appear in later chapters of this book and, in fact, throughout mathematics.

A deeper understanding of these concepts is the goal of (linear) algebra and, in the corresponding literature, the reader will find many other applications. The goal of algebra is to derive rules which hold in systems satisfying certain small sets of axioms. The discovery that these axioms hold in complex problems of analysis will enable us to recognize underlying unity in diverse situations and to maintain an overview of an otherwise unwieldy area of mathematics. In addition, the reader should see early on that mathematics is a whole — it is not made up of disjoint research areas, isolated from each other.

Since the beginner usually studies linear algebra in parallel with an introduction to analysis, we have restricted our discussion of algebra to the essentials. In the choice of the concepts to present we have been guided by the needs of later chapters. This is particularly true about the material in Section 12, namely vector spaces and algebras. These we will meet frequently, for example, in the form of function algebras, as we penetrate further into analysis.

The somewhat 'dry' material of this first chapter is made more palatable by the inclusion of many applications. Since, as already mentioned, we want to train the reader to use only what has previously been proved, we are limited at first to very simple 'internal' examples. In later sections this becomes less of a restriction, as, for example, the discussion of the interpolation problems in Section 12 shows.

We remind the reader that this book is intended to be used either as a textbook for a course on analysis, or for self study. For this reason, in this first chapter, we are more thorough and cover more material than is possible in lectures. We encourage the reader to work through these 'foundations' with diligence. In the first reading, the proofs of Theorems 5.3, 9.1, 9.2 and 10.4 can be skipped. At a later time, when the reader is more comfortable with proofs, these gaps should filled.

1 Fundamentals of Logic

To make complicated mathematical relationships clear it is convenient to use the notation of symbolic logic. Symbolic logic is about **statements** which one can meaningfully claim to be true or false. That is, each statement has the *truth value* 'true' (T) or 'false' (F). There are no other possibilities, and no statement can be both true and false.

Examples of statements are 'It is raining', 'There are clouds in the sky', and 'All readers of this book find it to be excellent'. On the other hand, 'This sentence is false' is not a statement. Indeed, if the sentence were true, then it says that it is false, and if it is false, it follows that the sentence is true.

Any statement A has a **negation** $\neg A$ ('not A') defined by $\neg A$ is true if A is false, and $\neg A$ is false if A is true. We can represent this relationship in a *truth table*:

A	T	F
$\neg A$	F	T

Of course, in normal language 'not A' can be expressed in many ways. For example, if A is the statement 'There are clouds in the sky', then $\neg A$ could be expressed as 'There are no clouds in the sky'. The negation of the statement 'All readers of this book find it to be excellent' is 'There is at least one reader of this book who finds that it is not excellent' (but not 'No readers of this book find it to be excellent').

Two statements, A and B, can be combined using **conjunction** $\wedge$ and **disjunction** $\vee$ to make new statements. The statement $A \wedge B$ ('A and B') is true if both A and B are true, and is false in all other cases. The statement $A \vee B$ ('A or B') is false when both A and B are false, and is true in all other cases. The following truth table makes the definitions clear:

A	B	$A \wedge B$	$A \vee B$
T	T	T	T
T	F	F	T
F	T	F	T
F	F	F	F

Note that the 'or' of disjunction has the meaning 'and/or', that is, 'A or B' is true if A is true, if B is true, or if both A and B are true.

If $E(x)$ is an expression which becomes a statement when x is replaced by an object (member, thing) of a specified class (collection, universe) of objects, then E is a **property**. The sentence 'x has property E' means '$E(x)$ is true'. If x belongs to a class X, that is, x is an **element** of X, then we write $x \in X$, otherwise[1] $x \notin X$.

[1] It is usual when abbreviating statements with symbols (such as $\in$, $=$, etc.) to denote their negations using the corresponding slashed symbol ($\notin$, $\neq$, etc.).

Then

$$\{\, x \in X \ ; \ E(x) \,\}$$

is the class of all elements x of the collection X which have property E. If X is the class of all readers of this book and $E(x)$ is the statement 'x wears glasses', then $\{\, x \in X \ ; \ E(x) \,\}$ is the class of all readers of this book who wear glasses.

We write $\exists$ for the **quantifier** 'there exists'. The expression

$$\exists\, x \in X : E(x)$$

has the meaning 'There is (at least) one object x in (the class) X which has property E'. We write $\exists!\, x \in X : E(x)$ when exactly one such object exists.

We use the symbol $\forall$ for the quantifier 'for all'. Once again, in normal language statements containing $\forall$ can be expressed in various ways. For example,

$$\forall\, x \in X : E(x) \tag{1.1}$$

means that 'For each (object) x in (the class) X, the statement $E(x)$ is true', or 'Every x in X has the property E'. The statement (1.1) can also be written as

$$E(x)\ , \qquad \forall\, x \in X\ , \tag{1.2}$$

that is, 'Property E is true for all x in X'. In a statement such as (1.2) we usually leave out the quantifier $\forall$ and write simply

$$E(x)\ , \qquad x \in X\ . \tag{1.3}$$

Finally, we use the symbol $:=$ to mean 'is defined by'. Thus

$$a := b\ ,$$

means that the object (or symbol) a is defined by the object (or expression) b. One says also 'a is a new name for b' or 'a stands for b '. Of course $a = b$ means that objects a and b are equal, that is, a and b are simply different representations of the same object (statement, etc.).

1.1 Examples Let A and B be statements, X and Y classes of objects, and E a property. Then, using truth tables or other methods, one can easily verify the following statements:

(a) $\neg\neg A := \neg(\neg A) = A$.

(b) $\neg(A \wedge B) = (\neg A) \vee (\neg B)$.

(c) $\neg(A \vee B) = (\neg A) \wedge (\neg B)$.

(d) $\neg\bigl(\forall\, x \in X : E(x)\bigr) = \bigl(\exists\, x \in X : \neg E(x)\bigr)$. Example: The negation of the statement 'Every reader of this book wears glasses' is 'At least one reader of this book does not wear glasses'.

(e) $\neg(\exists x \in X : E(x)) = (\forall x \in X : \neg E(x))$. Example: The negation of the statement 'There is a bald man in London' is 'No man in London is bald'.

(f) $\neg(\forall x \in X : (\exists y \in Y : E(x,y))) = (\exists x \in X : (\forall y \in Y : \neg E(x,y)))$.
Example: The negation of the statement 'Each reader of this book finds at least one sentence in Chapter I which is trivial' is 'At least one reader of this book finds every sentence of Chapter I nontrivial'.

(g) $\neg(\exists x \in X : (\forall y \in Y : E(x,y))) = (\forall x \in X : (\exists y \in Y : \neg E(x,y)))$.
Example: The negation of the statement 'There is a Londoner who is a friend of every New Yorker' is 'For each Londoner there is at least one New Yorker who is not his/her friend'. ■[2]

1.2 Remarks **(a)** For clarity, in the above examples, we have been careful to include all possible parentheses. This practice is to be recommended for complicated statements. On the other hand, statements are often easier to understand without parentheses and even without the membership symbol $\in$, so long as no ambiguity arises. In all cases, it is the order of the quantifiers that is significant. Thus '$\forall x\ \exists y : E(x,y)$' and '$\exists y\ \forall x : E(x,y)$' are different statements: In the first case, for all x there is some y such that $E(x,y)$ is true. Thus y depends on x, that is, for each x one has to find a (possibly) different y such that $E(x,y)$ is true. In the second case it suffices to find a fixed y such that the statement $E(x,y)$ is true for all x. For example, if $E(x,y)$ is the statement 'Reader x of this book finds the mathematical concept y to be trivial', then the first statement is 'Each reader of this book finds at least one mathematical concept to be trivial'. The second statement is 'There is a mathematical concept which every reader of this book finds to be trivial'.

(b) Using the quantifiers $\exists$ and $\forall$, negation becomes a purely 'mechanical' process in which the symbols $\exists$ and $\forall$ (as well as $\wedge$ and $\vee$) are interchanged (without changing the order) and statements which appear are negated (see Examples 1.1). For example, the negation of the statement '$\forall x\ \exists y\ \forall z : E(x,y,z)$' is '$\exists x\ \forall y\ \exists z : \neg E(x,y,z)$'. ■

Let A and B be statements. Then one can define a new statement, the **implication** $A \Rightarrow B$, ('A implies B') as follows:

$$(A \Rightarrow B) := (\neg A) \vee B\ . \tag{1.4}$$

Thus $A \Rightarrow B$ is false if A is true and B is false, and is true in all other cases (see Examples 1.1(a), (c)). In other words, $A \Rightarrow B$ is true when A and B are both true, or when A is false (independent of whether B is true or false). This means that a true statement cannot imply a false statement, and also that a false

[2]We use a black square to indicate the end of a list of examples or remarks, or the end of a proof.

statement implies any statement — true or false. It is common to express $A \Rightarrow B$ as 'To prove B it **suffices** to prove A', or 'B is **necessary** for A to be true', in other words, A is a **sufficient condition** for B, and B is a **necessary condition** for A.

The **equivalence** $A \Leftrightarrow B$ ('A and B are equivalent') of the statements A and B is defined by

$$(A \Leftrightarrow B) := (A \Rightarrow B) \wedge (B \Rightarrow A) .$$

Thus the statements A and B are equivalent when both $A \Rightarrow B$ and its **converse** $B \Rightarrow A$ are true, or when A is a **necessary and sufficient** condition for B (or vice versa). Another common way of expressing this equivalence is to say 'A is true **if and only if** B is true'.

A fundamental observation is that

$$(A \Rightarrow B) \Leftrightarrow (\neg B \Rightarrow \neg A) . \tag{1.5}$$

This follows directly from (1.4) and Example 1.1(a). The statement $\neg B \Rightarrow \neg A$ is called the **contrapositive** of the statement $A \Rightarrow B$.

If, for example, A is the statement 'There are clouds in the sky' and B is the statement 'It is raining', then $B \Rightarrow A$ is the statement 'If it is raining, then there are clouds in the sky'. Its contrapositive is, 'If there are no clouds in the sky, then it is not raining'.

If $B \Rightarrow A$ is true it does not, in general, follow that $\neg B \Rightarrow \neg A$ is true! Even when 'it is not raining', it is possible that 'there are clouds in the sky'.

To define a statement A so that it is true whenever the statement B is true, we write

$$A :\Leftrightarrow B$$

and say 'A is true, by definition, if B is true'.

In mathematics a true statement is often called a proposition, theorem, lemma or corollary.[3] Especially common are propositions of the form $A \Rightarrow B$. Since this statement is automatically true if A is false, the only interesting case is when A is true. Thus to prove that $A \Rightarrow B$ is true, one supposes that A is true and then shows that B is true.

The proof can proceed directly or 'by contradiction'. In the first case, one can use the fact (which the reader can easily check) that

$$(A \Rightarrow C) \wedge (C \Rightarrow B) \Rightarrow (A \Rightarrow B) . \tag{1.6}$$

If the statements $A \Rightarrow C$ and $C \Rightarrow B$ are already known to be true, then, by (1.6), $A \Rightarrow B$ is also true. If $A \Rightarrow C$ and $C \Rightarrow B$ are not known to be true and the

[3]All theorems, lemmas and corollaries are propositions. A theorem is a particularly important proposition. A lemma is a proposition which precedes a theorem and is needed for its proof. A corollary is a proposition which follows directly from a theorem.

implications $A \Rightarrow C$ and $C \Rightarrow B$ can be similarly decomposed, this procedure can be used to show $A \Rightarrow C$ and $C \Rightarrow B$ are true.

For a proof by contradiction one supposes that B is false, that is, $\neg B$ is true. Then one proves, using also the assumption that A is true, a statement C which is already known to be false. It follows from this 'contradiction' that $\neg B$ cannot be true, and hence that B is true.

Instead of $A \Rightarrow B$, it is often easier to prove its contrapositive $\neg B \Rightarrow \neg A$. According to (1.5) these statements are equivalent, that is, one is true if and only if the other is true.

At this point, we prefer not to provide examples of the above concepts since they would be necessarily rather contrived. Instead the reader is encouraged to identify these structures in the proofs in following section (see, in particular, the proof of Proposition 2.6).

The preceding discussion is incomplete in that we have neither defined the word 'statement' nor explained how to tell whether a statement is true or false. A further difficultly lies in our use of the English language, which, like most languages, contains many sentences whose meaning is ambiguous. Such sentences cannot be considered to be statements in the sense of this section.

For a more solid understanding of the rules of deduction, one needs mathematical logic. This provides a formal language in which the only statements appearing are those which can be derived from a given system of 'axioms' by means of well defined constructions. These axioms are 'unprovable' statements which are recognized as fundamental universal truths.

We do not wish to go further here into such formal systems. Instead, interested readers are directed to the appendix, 'Introduction to Mathematical Logic', which contains a more precise presentation of these ideas.

Exercises

1 "The Simpsons are coming to visit this evening," announced Maud Flanders. "The whole family — Homer, Marge and their three kids, Bart, Lisa and Maggie?" asked Ned Flanders dismayed. Maud, who never misses a chance to stimulate her husband's logical thinking, replied, "I'll explain it this way: If Homer comes then he will bring Marge too. At least one of the two children, Maggie and Lisa, are coming. Either Marge or Bart is coming, but not both. Either both Bart and Lisa are coming or neither is coming. And if Maggie comes, then Lisa and Homer are coming too. So now you know who is visiting this evening."

Who is coming to visit?

2 In the library of Count Dracula no two books contain exactly the same number of words. The number of books is greater than the total number of words in all the books. These statements suffice to determine the content of at least one book in Count Dracula's library. What is in this book?

2 Sets

Even though the reader is probably familiar with basic set theory, we review in this section some of the relevant concepts and notation.

Elementary Facts

If X and Y are sets, then $X \subseteq Y$ ('X is a **subset** of Y' or 'X is contained in Y') means that each element of X is also an element of Y, that is, $\forall x \in X : x \in Y$. Sometimes it is convenient to write $Y \supseteq X$ ('Y contains X') instead of $X \subseteq Y$. Equality of sets is defined by

$$X = Y :\Longleftrightarrow (X \subseteq Y) \wedge (Y \subseteq X) .$$

The statements

$$\begin{aligned} &X \subseteq X && \text{(reflexivity)} \\ &(X \subseteq Y) \wedge (Y \subseteq Z) \Rightarrow (X \subseteq Z) && \text{(transitivity)} \end{aligned}$$

are obvious. If $X \subseteq Y$ and $X \neq Y$, then X is called a **proper subset** of Y. We denote this relationship by $X \subset Y$ or $Y \supset X$ and say 'X is properly contained in Y'.

If X is a set and E is a property then $\{\, x \in X \ ; \ E(x) \,\}$ is the subset of X consisting of all elements x of X such that $E(x)$ is true. The set

$$\emptyset_X := \{\, x \in X \ ; \ x \neq x \,\}$$

is the **empty subset** of X.

2.1 Remarks **(a)** Let E be a property. Then

$$x \in \emptyset_X \Rightarrow E(x)$$

is true for each $x \in X$ ('The empty set possesses every property').

Proof From (1.4) we have

$$(x \in \emptyset_X \Rightarrow E(x)) = \neg(x \in \emptyset_X) \vee E(x) .$$

The negation $\neg(x \in \emptyset_X)$ is true for each $x \in X$. ■

(b) If X and Y are sets, then $\emptyset_X = \emptyset_Y$, that is, there is exactly one **empty set**. This set is denoted $\emptyset$ and is a subset of any set.

Proof From (a) we get $x \in \emptyset_X \Rightarrow x \in \emptyset_Y$, hence $\emptyset_X \subseteq \emptyset_Y$. By symmetry, $\emptyset_Y \subseteq \emptyset_X$, and so $\emptyset_X = \emptyset_Y$. ■

The set containing the single element x is denoted $\{x\}$. Similarly, the set consisting of the elements a, b, ..., $*$, $\odot$ is written $\{a, b, \ldots, *, \odot\}$.

The Power Set

If X is a set, then so is its **power set** $\mathcal{P}(X)$. The elements of $\mathcal{P}(X)$ are the subsets of X. Sometimes the power set is written 2^X for reasons which are made clear in Section 3 and in Exercise 3.6. The following are clearly true:

$$\emptyset \in \mathcal{P}(X)\ , \quad X \in \mathcal{P}(X)\ .$$
$$x \in X \Longleftrightarrow \{x\} \in \mathcal{P}(X)\ .$$
$$Y \subseteq X \Longleftrightarrow \ Y \in \mathcal{P}(X)\ .$$

In particular, $\mathcal{P}(X)$ is never empty.

2.2 Examples (a) $\mathcal{P}(\emptyset) = \{\emptyset\}$, $\mathcal{P}(\{\emptyset\}) = \{\emptyset, \{\emptyset\}\}$.

(b) $\mathcal{P}(\{*, \odot\}) = \{\emptyset, \{*\}, \{\odot\}, \{*, \odot\}\}$. ■

Complement, Intersection and Union

Let A and B be subsets of a set X. Then

$$A \backslash B := \{\, x \in X\ ;\ (x \in A) \wedge (x \notin B)\,\}$$

is the (relative) **complement of** B **in** A. When the set X is clear from context, we write also

$$A^c := X \backslash A$$

and call A^c the **complement** of A.

The set

$$A \cap B := \{\, x \in X\ ;\ (x \in A) \wedge (x \in B)\,\}$$

is called the **intersection** of A and B. If $A \cap B = \emptyset$, that is, if A and B have no element in common, then A and B are **disjoint**. Clearly, $A \backslash B = A \cap B^c$. The set

$$A \cup B := \{\, x \in X\ ;\ (x \in A) \vee (x \in B)\,\}$$

is called the **union** of A and B.

2.3 Remark It is useful to represent graphically the relationships between sets using **Venn diagrams**. Each set is represented by a region of the plane enclosed by a curve.

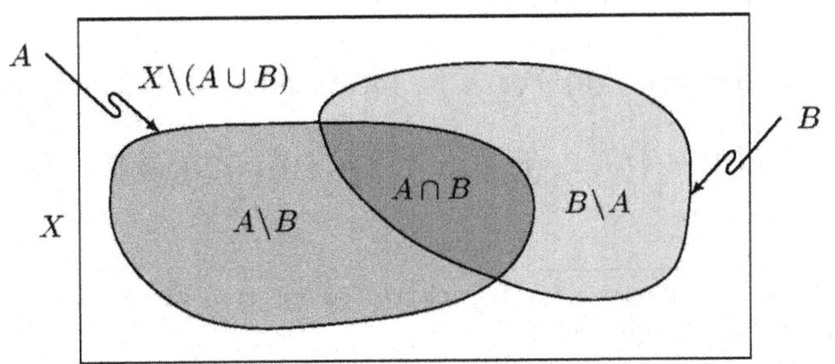

Such diagrams cannot be used to prove theorems, but, by providing intuition about the possible relationships between sets, they do suggest what statements about sets might be provable. ∎

In the following proposition we collect together some simple algebraic properties of the intersection and union operations.

2.4 Proposition *Let X, Y and Z be subsets of a set.*

(i) $X \cup Y = Y \cup X$, $X \cap Y = Y \cap X$. (commutativity)

(ii) $X \cup (Y \cup Z) = (X \cup Y) \cup Z$, $X \cap (Y \cap Z) = (X \cap Y) \cap Z$. (associativity)

(iii) $X \cup (Y \cap Z) = (X \cup Y) \cap (X \cup Z)$,
$X \cap (Y \cup Z) = (X \cap Y) \cup (X \cap Z)$. (distributivity)

(iv) $X \subseteq Y \Longleftrightarrow X \cup Y = Y \Longleftrightarrow X \cap Y = X$.

Proof These follow directly from the definitions.[1] ∎

Products

From two objects a and b we can form a new object, the **ordered pair** (a, b). Equality of two ordered pairs (a, b) and (a', b') is defined by

$$(a, b) = (a', b') :\Longleftrightarrow (a = a') \wedge (b = b') .$$

The objects a and b are called the first and second **components** of the ordered pair (a, b). For $x = (a, b)$, we also define

$$\mathrm{pr}_1(x) := a , \quad \mathrm{pr}_2(x) := b ,$$

and, for $j = 1, 2$ (that is, for $j \in \{1, 2\}$), we call $\mathrm{pr}_j(x)$ the jth **projection** of x.

If X and Y are sets, then the **(Cartesian) product** $X \times Y$ of X and Y is the set of all ordered pairs (x, y) with $x \in X$ and $y \in Y$.

2.5 Example and Remark **(a)** For $X := \{a, b\}$ and $Y := \{*, \odot, \Box\}$ we have

$$X \times Y = \big\{(a, *), (b, *), (a, \odot), (b, \odot), (a, \Box), (b, \Box)\big\} .$$

---

[1] By this and similar statements ('This is clear', 'Trivial' etc.) we mean, of course, that the reader should prove the claim his/herself!

(b) As in Remark 2.3, it is useful to have a graphical representation of the product $X \times Y$. In this diagram the sets X and Y are represented by lines, and $X \times Y$ by the rectangle. Once again we stress that such diagrams cannot be used to prove theorems, but serve only to help the intuition. ■

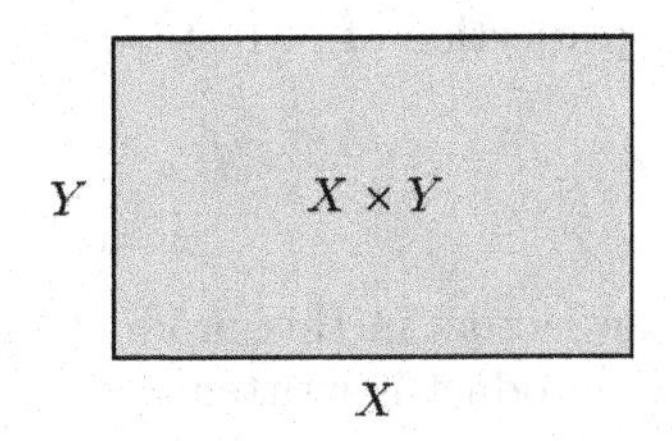

We provide a complete proof for the following Proposition 2.6(i) so that the reader may become familiar with the ways that proofs are constructed and written.

2.6 Proposition *Let X and Y be sets.*

(i) $X \times Y = \emptyset \Leftrightarrow (X = \emptyset) \vee (Y = \emptyset)$.

(ii) *In general*: $X \times Y \neq Y \times X$.

Proof (i) We have two statements to prove, namely

$$X \times Y = \emptyset \Rightarrow (X = \emptyset) \vee (Y = \emptyset)$$

and its converse. The corresponding parts of the proof are labelled using the symbols '$\Rightarrow$' and '$\Leftarrow$'.

'$\Rightarrow$' This part of the proof is done by contradiction. Suppose that $X \times Y = \emptyset$ and that the statement $(X = \emptyset) \vee (Y = \emptyset)$ is false. Then, by Example 1.1(c), the statement $(X \neq \emptyset) \wedge (Y \neq \emptyset)$ is true and so there are elements $x \in X$ and $y \in Y$. But then $(x, y) \in X \times Y$, contradicting $X \times Y = \emptyset$. Thus $X \times Y = \emptyset$ implies $(X = \emptyset) \vee (Y = \emptyset)$.

'$\Leftarrow$' We prove the contrapositive of the statement

$$(X = \emptyset) \vee (Y = \emptyset) \Rightarrow X \times Y = \emptyset \ .$$

Suppose that $X \times Y \neq \emptyset$. Then there is some $(x, y) \in X \times Y$ with $x \in X$ and $y \in Y$. Consequently we have $(X \neq \emptyset) \wedge (Y \neq \emptyset) = \neg\bigl((X = \emptyset) \vee (Y = \emptyset)\bigr)$.

(ii) See Exercise 4. ■

The product of three sets X, Y and Z is defined by

$$X \times Y \times Z := (X \times Y) \times Z \ .$$

This construction can be repeated[2] to define the product of n sets:

$$X_1 \times \cdots \times X_n := (X_1 \times \cdots \times X_{n-1}) \times X_n \ .$$

For x in $X_1 \times \cdots \times X_n$ we write $(x_1, \ldots, x_n)$ instead of $\bigl(\cdots((x_1, x_2), x_3), \ldots, x_n\bigr)$ and call x_j the jth **component** of x for $1 \leq j \leq n$. The element x_j is also $\mathrm{pr}_j(x)$,

[2]See Proposition 5.11.

the jth **projection** of x. Instead of $X_1 \times \cdots \times X_n$ we can also write

$$\prod_{j=1}^{n} X_j \, .$$

If all the factors in this product are the same, that is, $X_j = X$ for $j = 1, \ldots, n$, then the product is written X^n.

Families of Sets

Let A be a nonempty set and, for each $\alpha \in \mathsf{A}$, let A_α be a set. Then $\{ A_\alpha \; ; \; \alpha \in \mathsf{A} \}$ is called a **family of sets** and A is an **index set** for this family. Note that we do not require that $A_\alpha \neq A_\beta$ whenever the indices α and β are different, nor do we require that A_α is nonempty for each index. Note also that a family of sets is never empty.

Let X be a set and $\mathcal{A} := \{ A_\alpha \; ; \; \alpha \in \mathsf{A}\}$ a family of subsets of X. Generalizing the above concepts we define the **intersection** and the **union** of this family by

$$\bigcap_\alpha A_\alpha := \{ x \in X \; ; \; \forall \alpha \in \mathsf{A} : x \in A_\alpha \}$$

and

$$\bigcup_\alpha A_\alpha := \{ x \in X \; ; \; \exists \alpha \in \mathsf{A} : x \in A_\alpha \}$$

respectively. Note that $\bigcap_\alpha A_\alpha$ and $\bigcup_\alpha A_\alpha$ are subsets of X. Instead of $\bigcap_\alpha A_\alpha$, we sometimes write $\bigcap_{\alpha \in \mathsf{A}} A_\alpha$, or $\bigcap_\alpha \{ x \in X \; ; \; x \in A_\alpha \}$, or $\bigcap_{A \in \mathcal{A}} A$, or simply $\bigcap \mathcal{A}$. If $\mathcal{A}$ is a finite family of sets, then it can be indexed with finitely many natural numbers[3] $\{0, 1, \ldots, n\}$: $\mathcal{A} = \{ A_j \; ; \; j = 0, \ldots, n \}$. Then we also write $\bigcup_{j=0}^{n} A_j$ or $A_0 \cup \cdots \cup A_n$ for $\bigcup \mathcal{A}$.

The following proposition generalizes Proposition 2.4 to families of sets.

2.7 Proposition *Let $\{ A_\alpha \; ; \; \alpha \in \mathsf{A} \}$ and $\{ B_\beta \; ; \; \beta \in \mathsf{B} \}$ be families of subsets of a set X.*

(i) $\left(\bigcap_\alpha A_\alpha\right) \cap \left(\bigcap_\beta B_\beta\right) = \bigcap_{(\alpha,\beta)} A_\alpha \cap B_\beta$.
$\left(\bigcup_\alpha A_\alpha\right) \cup \left(\bigcup_\beta B_\beta\right) = \bigcup_{(\alpha,\beta)} A_\alpha \cup B_\beta$. (associativity)

(ii) $\left(\bigcap_\alpha A_\alpha\right) \cup \left(\bigcap_\beta B_\beta\right) = \bigcap_{(\alpha,\beta)} A_\alpha \cup B_\beta$.
$\left(\bigcup_\alpha A_\alpha\right) \cap \left(\bigcup_\beta B_\beta\right) = \bigcup_{(\alpha,\beta)} A_\alpha \cap B_\beta$. (distributivity)

(iii) $\left(\bigcap_\alpha A_\alpha\right)^c = \bigcup_\alpha A_\alpha^c$.
$\left(\bigcup_\alpha A_\alpha\right)^c = \bigcap_\alpha A_\alpha^c$. (de Morgan's laws)

Here (α, β) runs through the index set $\mathsf{A} \times \mathsf{B}$.

[3] See Section 5.

Proof These follow easily from the definitions. For (iii), see also Examples 1.1. ■

2.8 Remark The attentive reader will have noticed that we have not explained what a set is. Indeed the word 'set', as well as the word 'element', are undefined concepts of mathematics. Hence one needs **axioms**, that is, rules that are assumed to be true without proof, which say how these concepts are to be used. Statements about sets in this and following sections which are not provided with proofs can be considered to be axioms. For example, the statement 'The power set of a set is a set' is such an axiom. In this book we cannot discuss the axiomatic foundations of set theory — except perhaps in a few remarks in Section 5. Instead, we direct the interested reader to the relevant literature. Short and understandable presentations of the axiomatic foundations of set theory can be found, for example, in [Dug66], [Ebb77], [FP85] and [Hal74]. Even so, the subject requires a certain mathematical maturity and is not recommended for beginners.

We emphasize that the question of what sets and elements 'are' is unimportant. What matters are the rules with which one deals with these undefined concepts. ■

Exercises

1 Let X, Y and Z be sets. Prove the *transitivity of inclusion*, that is,

$$(X \subseteq Y) \wedge (Y \subseteq Z) \Rightarrow X \subseteq Z .$$

2 Verify the claims of Proposition 2.4.

3 Provide a complete proof of Proposition 2.7.

4 Let X and Y be nonempty sets. Show that $X \times Y = Y \times X \Leftrightarrow X = Y$.

5 Let A and B be subsets of a set X. Determine the following sets:

(a) $(A^c)^c$.

(b) $A \cap A^c$.

(c) $A \cup A^c$.

(d) $(A^c \cup B) \cap (A \cap B^c)$.

(e) $(A^c \cup B) \cup (A \cap B^c)$.

(f) $(A^c \cup B^c) \cap (A \cup B)$.

(g) $(A^c \cup B^c) \cap (A \cap B)$.

6 Let X be a set. Prove

$$\bigcup_{A \in \mathcal{P}(X)} A = X \quad \text{and} \quad \bigcap_{A \in \mathcal{P}(X)} A = \emptyset .$$

7 Let X and A be subsets of a set U and let Y and B be subsets of a set V. Prove the following:

(a) If $A \times B \neq \emptyset$, then $A \times B \subseteq X \times Y \Longleftrightarrow (A \subseteq X) \wedge (B \subseteq Y)$.

(b) $(X \times Y) \cup (A \times Y) = (X \cup A) \times Y$.

(c) $(X \times Y) \cap (A \times B) = (X \cap A) \times (Y \cap B)$.

(d) $(X \times Y) \backslash (A \times B) = \bigl((X \backslash A) \times Y\bigr) \cup \bigl(X \times (Y \backslash B)\bigr)$.

8 Let $\{ A_\alpha \ ; \ \alpha \in \mathsf{A} \}$ and $\{ B_\beta \ ; \ \beta \in \mathsf{B} \}$ be families of subsets of a set. Prove the following:

(a) $\bigl(\bigcap_\alpha A_\alpha\bigr) \times \bigl(\bigcap_\beta B_\beta\bigr) = \bigcap_{(\alpha,\beta)} A_\alpha \times B_\beta$.

(b) $\bigl(\bigcup_\alpha A_\alpha\bigr) \times \bigl(\bigcup_\beta B_\beta\bigr) = \bigcup_{(\alpha,\beta)} A_\alpha \times B_\beta$.

3 Functions

Functions are of fundamental importance for all mathematics. Of course, this concept has undergone many changes on the way to its modern meaning. An important step in its development was the removal of any connection to arithmetic, algorithmic or geometric ideas. This lead (neglecting certain formal hair-splitting discussed in Remark 3.1) to the set theoretical definition which we present below.

In this section X, Y, U and V are arbitrary sets.

A **function** or **map** f **from** X **to** Y is a rule which, for *each* element of X, specifies *exactly one* element of Y. We write

$$f : X \to Y \qquad \text{or} \qquad X \to Y\ , \quad x \mapsto f(x)\ ,$$

and sometimes also $f : X \to Y,\ x \mapsto f(x)$. Here $f(x) \in Y$ is the **value** of f at x. The set X is called the **domain** of f and is denoted $\mathrm{dom}(f)$, and Y is the **codomain** of f. Finally

$$\mathrm{im}(f) := \{\, y \in Y\ ;\ \exists x \in X : y = f(x) \,\}$$

is called the **image** of f.

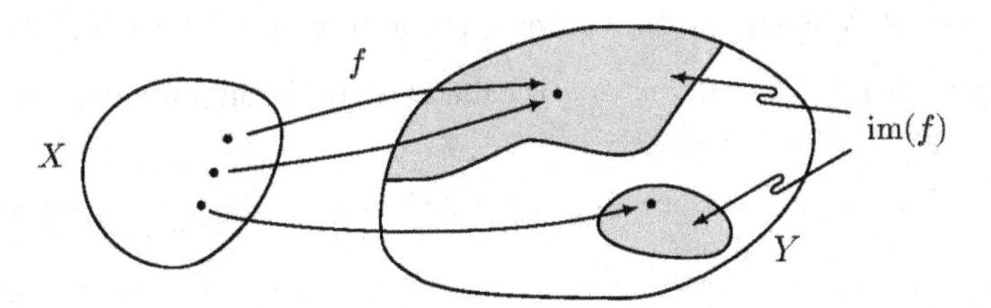

If $f : X \to Y$ is a function, then

$$\mathrm{graph}(f) := \{\, (x,y) \in X \times Y\ ;\ y = f(x) \,\} = \{\, (x, f(x)) \in X \times Y\ ;\ x \in X \,\}$$

is called the **graph** of f. Clearly, the graph of a function is a subset of the Cartesian product $X \times Y$. In the following diagrams of subsets G and H of $X \times Y$, G is the graph of a function from X to Y, whereas H is not the graph of such a function.

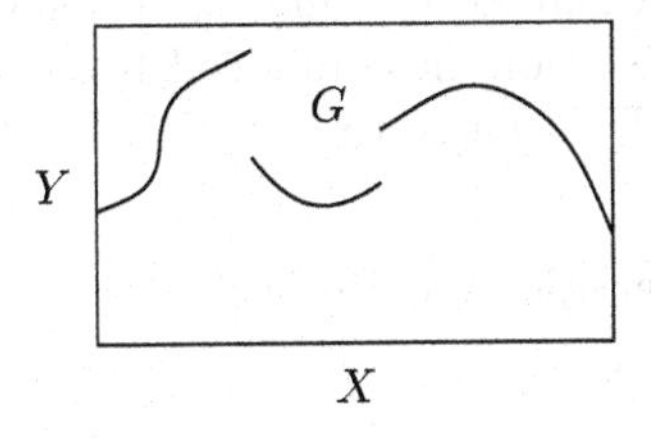

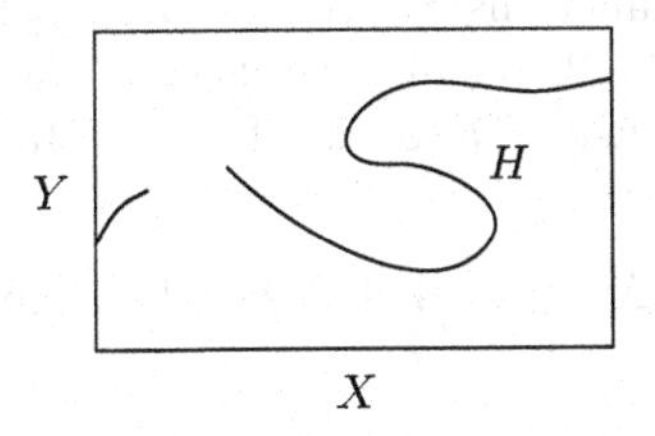

3.1 Remark Let G be a subset of $X \times Y$ having the property that, for each $x \in X$, there is exactly one $y \in Y$ with $(x,y) \in G$. Then we can define a function $f : X \to Y$ using the rule that, for each $x \in X$, $f(x) := y$ where $y \in Y$ is the unique element such that $(x,y) \in G$. Clearly $\mathrm{graph}(f) = G$. This observation motivates the following *definition*: A function $X \to Y$ is an ordered triple (X,G,Y) with $G \subseteq X \times Y$ such that, for each $x \in X$, there is exactly one $y \in Y$ with $(x,y) \in G$. This definition avoids the useful but imprecise expression 'rule' and uses only set theoretical concepts (see however Remark 2.8). ■

Simple Examples

Notice that we have not excluded $X = \emptyset$ and $Y = \emptyset$. If X is empty, then there is exactly one function from X to Y, namely the **empty function** $\emptyset : \emptyset \to Y$. If $Y = \emptyset$ but $X \neq \emptyset$, then there are no functions from X to Y. Two functions $f : X \to Y$ and $g : U \to V$ are **equal**, in symbols $f = g$, if

$$X = U \ , \quad Y = V \quad \text{and} \quad f(x) = g(x) \ , \qquad x \in X \ .$$

Thus, for two functions to be equal, they must have the same domain, codomain and rule. If one of these conditions fails, then the functions are distinct.

3.2 Examples **(a)** The function $\mathrm{id}_X : X \to X$, $x \mapsto x$ is the **identity function** (of X). If the set X is clear from context, we often write id for id_X.

(b) If $X \subseteq Y$, then $i : X \to Y$, $x \mapsto x$ is called the **inclusion** (embedding, injection) **of X into Y**. Note that $i = \mathrm{id}_X \iff X = Y$.

(c) If X and Y are nonempty and $b \in Y$, then $X \to Y$, $x \mapsto b$ is a **constant function**.

(d) If $f : X \to Y$ and $A \subseteq X$, then $f \,|\, A : A \to Y$, $x \mapsto f(x)$ is the **restriction of f to A**. Clearly $f \,|\, A = f \iff A = X$.

(e) Let $A \subseteq X$ and $g : A \to Y$. Then any function $f : X \to Y$ with $f \,|\, A = g$ is called an **extension of** g, written $f \supseteq g$. For example, with the notation of (b) we have $\mathrm{id}_Y \supseteq i$. (The set theoretical notation $f \supseteq g$ follows naturally from Remark 3.1.)

(f) Let $f : X \to Y$ be a function with $\mathrm{im}(f) \subseteq U \subseteq Y \subseteq V$. Then there are 'induced' functions $f_1 : X \to U$ and $f_2 : X \to V$ defined by $f_j(x) := f(x)$ for $x \in X$ and $j = 1, 2$. Usually we use the same symbol f for these induced functions and hence consider f to be a function from X to U, from X to Y or from X to V as needed.

(g) Let $X \neq \emptyset$ and $A \subseteq X$. Then the **characteristic function** of A is

$$\chi_A : X \to \{0,1\} \ , \quad x \mapsto \begin{cases} 1 \ , & x \in A \ , \\ 0 \ , & x \in A^c \ . \end{cases}$$

(h) If $X_1, \ldots, X_n$ are nonempty sets, then the projections

$$\mathrm{pr}_k : \prod_{j=1}^{n} X_j \to X_k \ , \quad x = (x_1, \ldots, x_n) \mapsto x_k \ , \qquad k = 1, \ldots, n \ ,$$

are functions. ∎

Composition of Functions

Let $f : X \to Y$ and $g : Y \to V$ be two functions. Then we define a new function $g \circ f$, the **composition** of f and g (more precisely, 'f followed by g'), by

$$g \circ f : X \to V \ , \quad x \mapsto g\bigl(f(x)\bigr) \ .$$

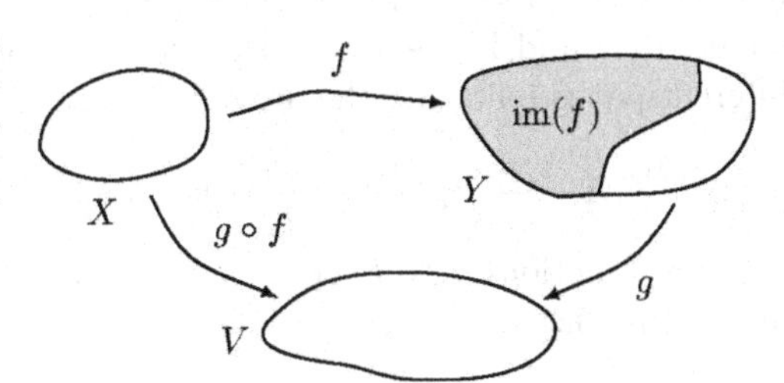

3.3 Proposition *Let $f : X \to Y$, $g : Y \to U$ and $h : U \to V$ be functions. Then the compositions $(h \circ g) \circ f$ and $h \circ (g \circ f) : X \to V$ are well defined and*

$$(h \circ g) \circ f = h \circ (g \circ f) \tag{3.1}$$

(associativity of composition).

Proof This follows directly from the definition. ∎

In view of this proposition, it is unnecessary to use parentheses when composing three functions. The function (3.1) can be written simply as $h \circ g \circ f$. This notational simplification also applies to compositions of more than three functions. See Examples 4.9(a) and 5.10.

Commutative Diagrams

It is frequently useful to represent compositions of functions in a diagram. In such a diagram we write $X \xrightarrow{f} Y$ in place of $f : X \to Y$. The **diagram**

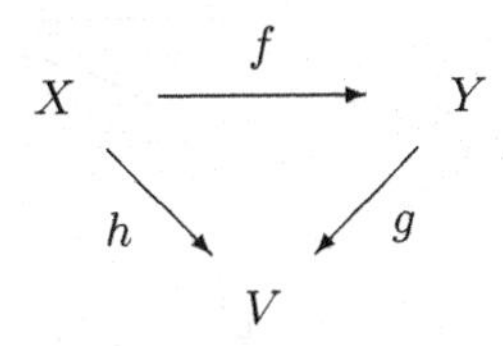

is **commutative** if $h = g \circ f$.

Similarly the diagram

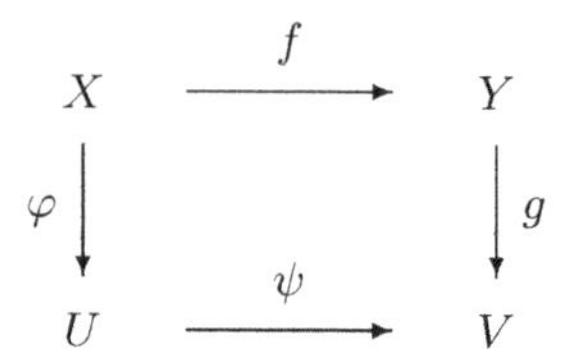

is **commutative** if $g \circ f = \psi \circ \varphi$. Occasionally one has complicated diagrams with many 'arrows', that is, functions. Such diagrams are **commutative** if the following is true: If X and Y are sets in the diagram and one can get from X to Y via two different paths following the arrows, for example,

$$X \xrightarrow{f_1} A_1 \xrightarrow{f_2} A_2 \xrightarrow{f_3} \cdots \xrightarrow{f_n} Y \quad \text{and} \quad X \xrightarrow{g_1} B_1 \xrightarrow{g_2} B_2 \xrightarrow{g_3} \cdots \xrightarrow{g_m} Y \ ,$$

then the functions $f_n \circ f_{n-1} \circ \cdots \circ f_1$ and $g_m \circ g_{m-1} \circ \cdots \circ g_1$ are equal. For example, the diagram

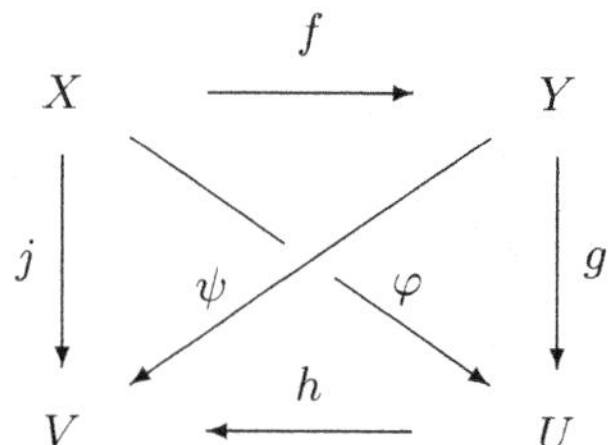

is commutative if $\varphi = g \circ f$, $\psi = h \circ g$ and $j = h \circ g \circ f = h \circ \varphi = \psi \circ f$, which is the associativity statement of Proposition 3.3.

Injections, Surjections and Bijections

Let $f \colon X \to Y$ be a function. Then f is **surjective** if $\mathrm{im}(f) = Y$, **injective** if $f(x) = f(y)$ implies $x = y$ for all $x, y \in X$, and **bijective** if f is both injective and surjective. One says also that f is a **surjection**, **injection** or **bijection** respectively. The expressions 'onto' and 'one-to-one' are often used to mean 'surjective' and 'injective'.

3.4 Examples **(a)** The functions graphed below illustrate these properties:

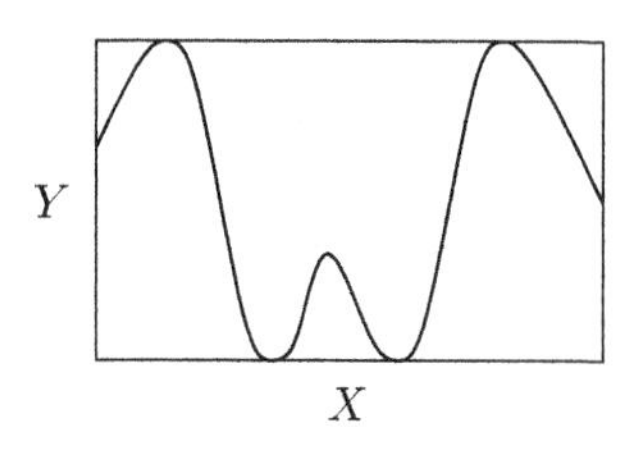

Surjective, not injective

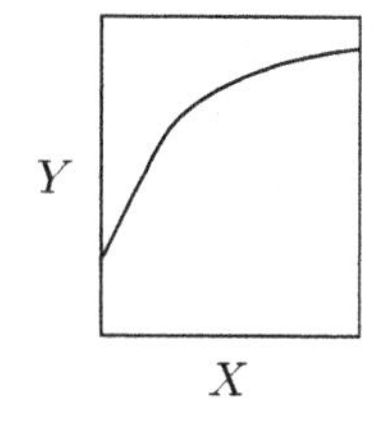

Injective, not surjective

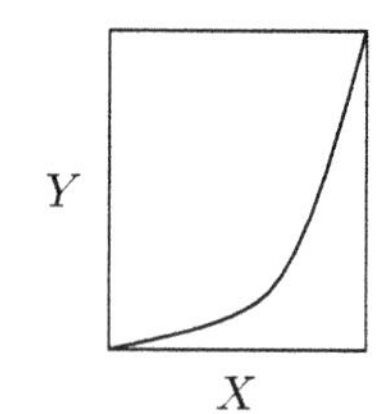

Bijective

(b) Let $X_1, \ldots, X_n$ be nonempty sets. Then for each $k \in \{1, \ldots, n\}$ the kth projection $\mathrm{pr}_k : \prod_{j=1}^n X_j \to X_k$ is surjective, but not, in general, injective. ∎

3.5 Proposition *Let $f : X \to Y$ be a function. Then f is bijective if and only if there is a function $g : Y \to X$ such that $g \circ f = \mathrm{id}_X$ and $f \circ g = \mathrm{id}_Y$. In this case, g is uniquely determined by f.*

Proof (i) '$\Rightarrow$' Suppose that $f : X \to Y$ is bijective. Since f is surjective, for each $y \in Y$ there is some $x \in X$ with $y = f(x)$. Since f is injective, this x is uniquely determined by y. This defines a function $g : Y \to X$ with the desired properties.

(ii) '$\Leftarrow$' From $f \circ g = \mathrm{id}_Y$ it follows immediately that f is surjective. Now let $x, y \in X$ and $f(x) = f(y)$. Then we have $x = g\big(f(x)\big) = g\big(f(y)\big) = y$. Hence f is injective.

(iii) If $h : Y \to X$ with $h \circ f = \mathrm{id}_X$ and $f \circ h = \mathrm{id}_Y$, then, from Proposition 3.3, we have

$$g = g \circ \mathrm{id}_Y = g \circ (f \circ h) = (g \circ f) \circ h = \mathrm{id}_X \circ h = h \; .$$

Thus g is uniquely determined by f. ∎

Inverse Functions

Proposition 3.5 motivates the following definition: Let $f : X \to Y$ be bijective. Then the **inverse function** f^{-1} of f is the unique function $f^{-1} : Y \to X$ such that $f \circ f^{-1} = \mathrm{id}_Y$ and $f^{-1} \circ f = \mathrm{id}_X$.

The proof of the following proposition is left as an exercise (see Exercises 1 and 3).

3.6 Proposition *Let $f : X \to Y$ and $g : Y \to V$ be bijective. Then $g \circ f : X \to V$ is bijective and*

$$(g \circ f)^{-1} = f^{-1} \circ g^{-1} \; .$$

Let $f : X \to Y$ be a function and $A \subseteq X$. Then

$$f(A) := \big\{\, f(a) \in Y \; ; \; a \in A \,\big\}$$

is called the **image of A under f**. For each $C \subseteq Y$,

$$f^{-1}(C) := \big\{\, x \in X \; ; \; f(x) \in C \,\big\}$$

is called the **preimage of C under f**.

3.7 Example Let $f: X \to Y$ be the function whose graph is below.

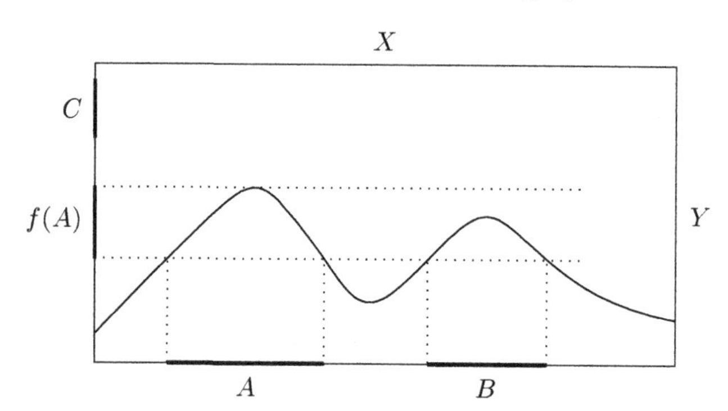

Then $f^{-1}(C) = \emptyset$ and $f^{-1}\big(f(A)\big) = A \cup B$, and, in particular, $f^{-1}\big(f(A)\big) \supset A$. ∎

Set Valued Functions

Let $f: X \to Y$ be a function. Then, using the above definitions, we have two 'induced' set valued functions,

$$f: \mathcal{P}(X) \to \mathcal{P}(Y) , \quad A \mapsto f(A) \quad \text{and} \quad f^{-1}: \mathcal{P}(Y) \to \mathcal{P}(X) , \quad B \mapsto f^{-1}(B) .$$

Using the same symbol f for two different functions leads to no confusion since the intent is always clear from context.

If $f: X \to Y$ is bijective, then $f^{-1}: Y \to X$ exists and $\{f^{-1}(y)\} = f^{-1}(\{y\})$ for all $y \in Y$. In this equation, and in general, the context makes clear which version of f^{-1} is meant. If f is not bijective, then only the set valued function f^{-1} is defined, so no confusion is possible. In either case, we write $f^{-1}(y)$ for $f^{-1}(\{y\})$ and call $f^{-1}(y) \subseteq X$ the **fiber** of f at y. The fiber $f^{-1}(y)$ is simply the solution set $\{\, x \in X \; ; \; f(x) = y \,\}$ of the equation $f(x) = y$. This could, of course, be empty.

3.8 Proposition *The following hold for the set valued functions induced from f:*

(i) $A \subseteq B \subseteq X \Rightarrow f(A) \subseteq f(B)$.

(ii) $A_\alpha \subseteq X \quad \forall \alpha \in \mathsf{A} \Rightarrow f\big(\bigcup_\alpha A_\alpha\big) = \bigcup_\alpha f(A_\alpha)$.

(iii) $A_\alpha \subseteq X \quad \forall \alpha \in \mathsf{A} \Rightarrow f\big(\bigcap_\alpha A_\alpha\big) \subseteq \bigcap_\alpha f(A_\alpha)$.

(iv) $A \subseteq X \Rightarrow f(A^c) \supseteq f(X) \backslash f(A)$.

(i′) $A' \subseteq B' \subseteq Y \Rightarrow f^{-1}(A') \subseteq f^{-1}(B')$.

(ii′) $A'_\alpha \subseteq Y \quad \forall \alpha \in \mathsf{A} \Rightarrow f^{-1}\big(\bigcup_\alpha A'_\alpha\big) = \bigcup_\alpha f^{-1}(A'_\alpha)$.

(iii′) $A'_\alpha \subseteq Y \quad \forall \alpha \in \mathsf{A} \Rightarrow f^{-1}\big(\bigcap_\alpha A'_\alpha\big) = \bigcap_\alpha f^{-1}(A'_\alpha)$.

(iv′) $A' \subseteq Y \Rightarrow f^{-1}(A'^c) = \big[f^{-1}(A')\big]^c$.

If $g: Y \to V$ is another function, then $(g \circ f)^{-1} = f^{-1} \circ g^{-1}$.

The easy proofs of these claims are left to the reader.

In short, Proposition 3.8(i′)–(iv′) says that the function $f^{-1}: \mathcal{P}(Y) \to \mathcal{P}(X)$ respects all set operations. The same is not true, in general, of the induced function $f: \mathcal{P}(X) \to \mathcal{P}(Y)$ as can be seen in (iii) and (iv).

Finally, we denote the **set of all functions** from X to Y by $\mathrm{Funct}(X, Y)$. Because of Remark 3.1, $\mathrm{Funct}(X, Y)$ is a subset of $\mathcal{P}(X \times Y)$. For $\mathrm{Funct}(X, Y)$ we write also Y^X. This is consistent with the notation X^n for the nth Cartesian product of the set X with itself, since this coincides with the set of all functions from $\{1, 2, \ldots, n\}$ to X. If $U \subseteq Y \subseteq V$, then

$$\mathrm{Funct}(X, U) \subseteq \mathrm{Funct}(X, Y) \subseteq \mathrm{Funct}(X, V) , \tag{3.2}$$

where we have used the conventions of Example 3.2(f).

Exercises

1 Prove Proposition 3.6.

2 Prove Proposition 3.8 and show that the given inclusions are, in general, proper.

3 Let $f: X \to Y$ and $g: Y \to V$ be functions. Show the following:

(a) If f and g are injective (surjective), then so is $g \circ f$.

(b) $f$is injective $\Longleftrightarrow \exists\, h: Y \to X$ such that $h \circ f = \mathrm{id}_X$.

(c) f is surjective $\Longleftrightarrow \exists\, h: Y \to X$ such that $f \circ h = \mathrm{id}_Y$.

4 Let $f: X \to Y$ be a function. Show that the following are equivalent:

(a) f is injective.

(b) $f^{-1}(f(A)) = A, \ \ A \subseteq X$.

(c) $f(A \cap B) = f(A) \cap f(B), \ \ A, B \subseteq X$.

5 Determine the fibers of the projections pr_k.

6 Prove that, for each nonempty set X, the function

$$\mathcal{P}(X) \to \{0, 1\}^X , \quad A \mapsto \chi_A$$

is bijective.

7 Let $f: X \to Y$ be a function and $i: A \to X$ the inclusion of a subset $A \subseteq X$ in X. Show the following:

(a) $f \,|\, A = f \circ i$.

(b) $(f \,|\, A)^{-1}(B) = A \cap f^{-1}(B), \ \ B \subseteq Y$.

4 Relations and Operations

In order to describe relationships between elements of a set X it is useful to have a simple set theoretical meaning for the word 'relation': A (binary) **relation** on X is simply a subset $R \subseteq X \times X$. Instead of $(x, y) \in R$, we usually write xRy or $x \underset{R}{\sim} y$.

A relation R on X is **reflexive** if xRx for all $x \in X$, that is, if R contains the **diagonal**

$$\Delta_X := \{ (x,x) \; ; \; x \in X \} .$$

It is **transitive** if

$$(xRy) \wedge (yRz) \Rightarrow xRz .$$

If

$$xRy \Rightarrow yRx$$

holds, then R is **symmetric**.

Let Y be a nonempty subset of X and R a relation on X. Then the set $R_Y := (Y \times Y) \cap R$ is a relation on Y called the **restriction** of R to Y. Obviously $xR_Y y$ if and only if $x, y \in Y$ and xRy. Usually we write R instead of R_Y when the context makes clear the set involved.

Equivalence Relations

A relation on X which is reflexive, transitive and symmetric is called an **equivalence relation** on X and is usually denoted $\sim$. For each $x \in X$, the set

$$[x] := \{ y \in X \; ; \; y \sim x \}$$

is the **equivalence class** of (or, containing) x, and each $y \in [x]$ is a **representative** of this equivalence class. Finally,

$$X/\!\sim \; := \{ [x] \; ; \; x \in X \} ,$$

'X modulo $\sim$', is the set of all equivalence classes of X. Clearly $X/\!\sim$ is a subset of $\mathcal{P}(X)$.

A **partition** of a set X is a subset $\mathcal{A} \subseteq \mathcal{P}(X) \backslash \{\emptyset\}$ with the property that, for each $x \in X$, there is a unique $A \in \mathcal{A}$ such that $x \in A$. That is, $\mathcal{A}$ consists of pairwise disjoint subsets of X whose union is X.

4.1 Proposition *Let $\sim$ be an equivalence relation on X. Then $X/\!\sim$ is a partition of X.*

Proof Since $x \in [x]$ for all $x \in X$, we have $X = \bigcup_{x \in X} [x]$. Now suppose that $z \in [x] \cap [y]$. Then $z \sim x$ and $z \sim y$, and hence $x \sim y$. This shows that $[x] = [y]$. Hence two equivalence classes are either identical or disjoint. ■

It follows immediately from the definition that the function

$$p := p_X : X \to X/{\sim} \ , \quad x \mapsto [x]$$

is a well defined surjection, the (canonical) **quotient function** from X to $X/{\sim}$.

4.2 Examples **(a)** Let X be the set of inhabitants of London. Define a relation on X by $x \sim y :\Longleftrightarrow$ (x and y have the same parents). This is clearly an equivalence relation, and two inhabitants of London belong to the same equivalence class if and only if they are siblings.

(b) The 'smallest' equivalence relation on a set X is the diagonal Δ_X, that is, the equality relation.

(c) Let $f : X \to Y$ be a function. Then

$$x \sim y :\Longleftrightarrow f(x) = f(y)$$

is an equivalence relation on X. The equivalence class of $x \in X$ is $[x] = f^{-1}(f(x))$. Moreover, there is a unique function $\widetilde{f}$ such that the diagram

$$\begin{array}{ccc} X & \xrightarrow{\quad f \quad} & Y \\ & {\scriptstyle p}\searrow \quad \nearrow{\scriptstyle \widetilde{f}} & \\ & X/{\sim} & \end{array}$$

is commutative. The function $\widetilde{f}$ is injective and $\mathrm{im}(\widetilde{f}) = \mathrm{im}(f)$. In particular, $\widetilde{f}$ is bijective if f is surjective.

(d) If $\sim$ is an equivalence relation on a set X and Y is a nonempty subset of X, then the restriction of $\sim$ to Y is an equivalence relation on Y. ■

Order Relations

A relation $\leq$ on X is a **partial order** on X if it is reflexive, transitive and **antisymmetric**, that is,

$$(x \leq y) \wedge (y \leq x) \Rightarrow x = y \ .$$

If $\leq$ is a partial order on X, then the pair $(X, \leq)$ is called a **partially ordered set**. If the partial order is clear from context, we write simply X for $(X, \leq)$ and say X is a partially ordered set. If, in addition,

$$\forall\, x, y \in X : (x \leq y) \vee (y \leq x) \ ,$$

then $\leq$ is called a **total order** on X and $(X, \leq)$ is a **totally ordered set**.

4.3 Remarks **(a)** The following notation is useful:

$$
\begin{aligned}
x \geq y &:\Longleftrightarrow y \leq x \ , \\
x < y &:\Longleftrightarrow (x \leq y) \wedge (x \neq y) \ , \\
x > y &:\Longleftrightarrow y < x \ .
\end{aligned}
$$

(b) If X is totally ordered, then, for each pair of elements $x, y \in X$, exactly one of the following is true:

$$x < y \ , \quad x = y \ , \quad x > y \ .$$

If X is partially ordered but not totally ordered, then there are at least two elements $x, y \in X$ which are incomparable, meaning that neither $x \leq y$ nor $y \leq x$ is true. ■

4.4 Examples **(a)** Let $(X, \leq)$ be a partially ordered set and Y a subset of X. Then the restriction of $\leq$ to Y is a partial order.

(b) $\big(\mathcal{P}(X), \subseteq\big)$ is a partially ordered set and $\subseteq$ is called the **inclusion order** on $\mathcal{P}(X)$. In general, $\big(\mathcal{P}(X), \subseteq\big)$ is not totally ordered.

(c) Let X be a set and $(Y, \leq)$ a partially ordered set. Then

$$f \leq g :\Longleftrightarrow f(x) \leq g(x) \ , \quad x \in X \ ,$$

defines a partial order on $\mathrm{Funct}(X, Y)$. The set $\mathrm{Funct}(X, Y)$ is not, in general, totally ordered, even if Y is totally ordered. ■

Convention Unless otherwise stated, $\mathcal{P}(X)$, and by restriction, any subset of $\mathcal{P}(X)$, is considered to be a partially ordered set with the inclusion order as described above.

Let $(X, \leq)$ be a partially ordered set and A a nonempty subset of X. An element $s \in X$ is an **upper bound** of A if $a \leq s$ for all $a \in A$. Similarly, s is a **lower bound** of A if $a \geq s$ for all $a \in A$. The subset A is **bounded above** if it has an upper bound, **bounded below** if it has a lower bound, and simply **bounded** if it is bounded above and below.

An element $m \in X$ is the **maximum**, $\max(A)$, of A if $m \in A$ and m is an upper bound of A. An element $m \in X$ is the **minimum**, $\min(A)$, of A if $m \in A$ and m is a lower bound of A. Note that A has at most one minimum and at most one maximum.

Let A be a subset of a partially ordered set X which is bounded above. If the set of all upper bounds of A has a minimum, then this element is called the **least upper bound** of A or **supremum** of A and is written $\sup(A)$, that is,

$$\sup(A) := \min\{\, s \in X \ ;\ s \text{ is an upper bound of } A \,\} .$$

Similarly, for a nonempty subset A of X which is bounded below we define

$$\inf(A) := \max\{\, s \in X \ ;\ s \text{ is a lower bound of } A \,\} ,$$

and call $\inf(A)$, if this element exists, the **greatest lower bound** of A or **infimum** of A. If A has two elements, $A = \{a, b\}$, we often use the notation $a \vee b := \sup(A)$ and $a \wedge b := \inf(A)$.

4.5 Remarks **(a)** It should be emphasized that a set which is bounded above (or below) does not necessarily have a least upper (or greatest lower) bound (see Example 10.3).

(b) If $\sup(A)$ and $\inf(A)$ exist, then, in general, $\sup(A) \notin A$ and $\inf(A) \notin A$.

(c) If $\sup(A)$ exists and $\sup(A) \in A$, then $\sup(A) = \max(A)$. Similarly, if $\inf(A)$ exists and $\inf(A) \in A$, then $\inf(A) = \min(A)$.

(d) If $\max(A)$ exists then $\sup(A) = \max(A)$. Similarly, if $\min(A)$ exists then $\inf(A) = \min(A)$. ∎

4.6 Examples **(a)** Let $\mathcal{A}$ be a nonempty subset of $\mathcal{P}(X)$. Then

$$\sup(\mathcal{A}) = \bigcup \mathcal{A} , \quad \inf(\mathcal{A}) = \bigcap \mathcal{A} .$$

(b) Let X be a set with at least two elements and $\mathcal{X} := \mathcal{P}(X) \backslash \{\emptyset\}$ with the inclusion order. Suppose further that A and B are nonempty disjoint subsets of X and $\mathcal{A} := \{A, B\}$. Then $\mathcal{A} \subseteq \mathcal{X}$ and $\sup(\mathcal{A}) = A \cup B$, but $\mathcal{A}$ has no maximum, and $\mathcal{A}$ is not bounded below. In particular, $\inf(\mathcal{A})$ does not exist. ∎

Let $X := (X, \leq)$ and $Y := (Y, \leq)$ be partially ordered sets and $f : X \to Y$ a function. (Here we use the same symbol $\leq$ for the partial orders on both X and Y.) Then f is called **increasing** (or **decreasing**) if $x \leq y$ implies $f(x) \leq f(y)$ (or $f(x) \geq f(y)$). We say that$f$ is **strictly increasing** (or **strictly decreasing**) if $x < y$ implies that $f(x) < f(y)$ (or $f(x) > f(y)$). Finally f is called (**strictly**) **monotone** if f is (strictly) increasing or (strictly) decreasing.

Let X be an arbitrary set and $Y := (Y, \leq)$ a partially ordered set. A function $f : X \to Y$ is called **bounded, bounded above** or **bounded below** if the same is true of its image $\operatorname{im}(f) = f(X)$ in Y. If X is also a partially ordered set, then f is called **bounded on bounded sets** if, for each bounded subset A of X, the restriction $f \,|\, A$ is bounded.

4.7 Examples **(a)** Let X and Y be sets and $f \in Y^X$. Proposition 3.8 says that the induced functions $f : \mathcal{P}(X) \to \mathcal{P}(Y)$ and $f^{-1} : \mathcal{P}(Y) \to \mathcal{P}(X)$ are increasing.

(b) Let X be a set with at least two elements and $\mathcal{X} := \mathcal{P}(X) \backslash \{X\}$ with the inclusion order. Then the identity function $\mathcal{X} \to \mathcal{X}$, $A \mapsto A$ is bounded on bounded sets but not bounded. ■

Operations

A function $\circledast : X \times X \to X$ is often called an **operation** on X. In this case we write $x \circledast y$ instead of $\circledast(x, y)$. For nonempty subsets A and B of X we write $A \circledast B$ for the image of $A \times B$ under $\circledast$, that is,

$$A \circledast B = \{ a \circledast b \; ; \; a \in A, \; b \in B \} . \tag{4.1}$$

If $A = \{a\}$, we write $a \circledast B$ instead of $A \circledast B$. Similarly $A \circledast b = \{ a \circledast b \; ; \; a \in A \}$. A nonempty subset A of X is **closed under the operation** $\circledast$, if $A \circledast A \subseteq A$, that is, if the image of $A \times A$ under the function $\circledast$ is contained in A.

4.8 Examples **(a)** Let X be a set. Then composition $\circ$ of functions is an operation on $\mathrm{Funct}(X, X)$.

(b) $\cup$ and $\cap$ are operations on $\mathcal{P}(X)$. ■

An operation $\circledast$ on X is **associative** if

$$x \circledast (y \circledast z) = (x \circledast y) \circledast z \; , \qquad x, y, z \in X \; , \tag{4.2}$$

and $\circledast$ is **commutative** if $x \circledast y = y \circledast x$ for $x, y \in X$. If $\circledast$ is associative then the parentheses in (4.2) are unnecessary and we write simply $x \circledast y \circledast z$.

4.9 Examples **(a)** By Proposition 3.3, composition is an associative operation on $\mathrm{Funct}(X, X)$. It may not be commutative (see Exercise 3).

(b) $\cup$ and $\cap$ are associative and commutative on $\mathcal{P}(X)$. ■

Let $\circledast$ be an operation on the set X. An element $e \in X$ such that

$$e \circledast x = x \circledast e = x \; , \qquad x \in X \; ,$$

is called an **identity element** of X (with respect to the operation $\circledast$).

4.10 Examples **(a)** id_X is an identity element in $\mathrm{Funct}(X, X)$ with respect to composition.

(b) $\emptyset$ is an identity element of $\mathcal{P}(X)$ with respect to $\cup$. X is an identity element of $\mathcal{P}(X)$ with respect to $\cap$.

(c) Clearly $\mathcal{X} := \mathcal{P}(X)\backslash\{\emptyset\}$ contains no identity element with respect to $\cup$ whenever X has more than one element. ■

The following proposition shows that an identity element is unique if it exists at all.

4.11 Proposition *There is at most one identity element with respect to a given operation.*

Proof Let e and e' be identity elements with respect to an operation $\circledast$ on a set X. Then, directly from the definition, we have $e = e \circledast e' = e'$. ■

4.12 Example Let $\circledast$ be an operation on a set Y and X a nonempty set. Then we define the operation on $\text{Funct}(X, Y)$ **induced from** $\circledast$ by

$$(f \odot g)(x) := f(x) \circledast g(x) , \qquad x \in X .$$

It is clear that $\odot$ is associative or commutative whenever the same is true of $\circledast$. If Y has an identity element e with respect to $\circledast$, then the constant function

$$X \to Y , \qquad x \mapsto e$$

is the identity element of $\text{Funct}(X, Y)$ with respect to $\odot$. Henceforth we will use the same symbol $\circledast$ for the operation on Y and for the induced operation on $\text{Funct}(X, Y)$. From the context it will be clear which function the symbol represents. We will soon see that this simple and natural construction is extremely useful. Important applications can be found in Examples 7.2(d), 8.2(b), 12.3(e) and 12.11(a), as well as in Remark 8.14(b). ■

Exercises

1 Let $\sim$ and $\dot\sim$ be equivalence relations on the sets X and Y respectively. Suppose that a function $f \in Y^X$ is such that $x \sim y$ implies $f(x) \,\dot\sim\, f(y)$ for all $x, y \in X$. Prove that there is a unique function f_* such that the diagram below is commutative.

$$\begin{array}{ccc} X & \xrightarrow{\ f\ } & Y \\ {\scriptstyle p_X}\downarrow & & \downarrow{\scriptstyle p_Y} \\ X/\!\sim & \xrightarrow{\ f_*\ } & Y/\dot\sim \end{array}$$

2 Verify that the function f of Example 4.7(b) is not bounded.

3 Show that composition $\circ$ is not, in general, a commutative operation on $\text{Funct}(X, X)$.

4 An operation $\circledast$ on a set X is called *anticommutative* if it satisfies the following:

(i) There is a right identity element $r := r_X$, that is, $\exists r \in X : x \circledast r = x, \ \ x \in X$.

(ii) $x \circledast y = r \Longleftrightarrow (x \circledast y) \circledast (y \circledast x) = r \Longleftrightarrow x = y$ for all $x, y \in X$.

Show that, whenever X has more than one element, an anticommutative operation $\circledast$ on X is not commutative and has no identity element.

5 Let $\circledast$ and $\circledcirc$ be anticommutative operations on X and Y respectively. Further, let $f : X \to Y$ satisfy

$$f(r_X) = r_Y \ , \quad f(x \circledast y) = f(x) \circledcirc f(y) \ , \qquad x, y \in X \ .$$

Prove the following:

(a) $x \sim y :\Longleftrightarrow f(x \circledast y) = r_Y$ defines an equivalence relation on X.

(b) The function

$$\widetilde{f} : X/{\sim} \ \to Y \ , \quad [x] \mapsto f(x)$$

is well defined and injective. If, in addition, f is surjective, then $\widetilde{f}$ is bijective.

6 Let $(X, \leq)$ be a partially ordered set with nonempty subsets A, B, C and D. Suppose that A and B are bounded above and C and D are bounded below. Assuming that the relevant suprema and infima exist, prove the following:

(a) $\sup(A \cup B) = \sup\{\sup(A), \sup(B)\}$, $\inf(C \cup D) = \inf\{\inf(C), \inf(D)\}$.

(b) If $A \subseteq B$ and $C \subseteq D$, then

$$\sup(A) \leq \sup(B) \quad \text{and} \quad \inf(C) \geq \inf(D) \ .$$

(c) If $A \cap B$ and $C \cap D$ are nonempty, then

$$\sup(A \cap B) \leq \inf\{\sup(A), \sup(B)\} \ , \quad \inf(C \cap D) \geq \sup\{\inf(C), \inf(D)\} \ .$$

(d) In (a), the claim that $\sup(A \cup B) = \sup\{\sup(A), \sup(B)\}$ cannot be strengthened to $\sup(A \cup B) = \max\{\sup(A), \sup(B)\}$.
(Hint: Consider the power set of a nonempty set.)

7 Let R be a relation on X and S a relation on Y. Define a relation $R \times S$ on $X \times Y$ by

$$(x, y)(R \times S)(u, v) :\Longleftrightarrow (xRu) \wedge (ySv)$$

for $(x, y), (u, v) \in X \times Y$. Prove that, if R and S are equivalence relations, then so is $R \times S$.

8 Show by example that the partially ordered set $(\mathcal{P}(X), \subseteq)$ may not be totally ordered.

9 Let $\mathcal{A}$ be a nonempty subset of $\mathcal{P}(X)$. Show that $\sup(\mathcal{A}) = \bigcup \mathcal{A}$ and $\inf(\mathcal{A}) = \bigcap \mathcal{A}$ (see Example 4.6(a)).

5 The Natural Numbers

In 1888, R. Dedekind published the book 'Was sind and was sollen die Zahlen?' (What are the numbers and what should they be?) [Ded95] about the set theoretical foundation of the natural number system. It is a milestone in the development of this subject, and indeed one of the high points of the history of mathematics.

Starting in this section with a simple and 'natural' axiom system for the natural numbers, we will construct in later sections the integers, the rational numbers, the real numbers and finally the complex numbers. This *constructive* approach has the advantage over the *axiomatic* formulation of the real numbers of D. Hilbert 1899 (see [Hil23]), that the entire structure of mathematics can be built up from a few foundation stones coming from mathematical logic and axiomatic set theory.

The Peano Axioms

We define the natural numbers using a system of axioms due to G. Peano which formalizes the idea that, given any natural number, there is always a next largest natural number.

The **natural numbers** consist of a set $\mathbb{N}$, a distinguished element $0 \in \mathbb{N}$, and a function $\nu \colon \mathbb{N} \to \mathbb{N}^\times := \mathbb{N} \backslash \{0\}$ with following properties:

(N_0) ν is injective.

(N_1) If a subset N of $\mathbb{N}$ contains 0 and if $\nu(n) \in N$ for all $n \in N$, then $N = \mathbb{N}$.

5.1 Remarks **(a)** For $n \in \mathbb{N}$, the element $\nu(n)$ is called the **successor** of n, and ν is called the **successor function**. The element 0 is the only natural number which is not a successor of a natural number, that is, the function $\nu : \mathbb{N} \to \mathbb{N}^\times$ is surjective (and, with (N_0), bijective).

Proof Let

$$N := \{\, n \in \mathbb{N} \ ; \ \exists n' \in \mathbb{N} : \nu(n') = n \,\} \cup \{0\} = \operatorname{im}(\nu) \cup \{0\} .$$

For $n \in N$ we have $\nu(n) \in \operatorname{im}(\nu) \subseteq N$. Since also $0 \in N$, (N_1) implies that $N = \mathbb{N}$. From this it follows immediately that $\operatorname{im}(\nu) = \mathbb{N}^\times$. ■

(b) Instead of $0, \nu(0), \nu\bigl(\nu(0)\bigr), \nu\bigl(\nu(\nu(0))\bigr), \ldots$ one usually writes $0, 1, 2, 3, \ldots$

(c) Some authors prefer to start the natural numbers with 1 rather than with 0. This is, of course, without mathematical significance.

(d) Axiom (N_1) is one form of the **principle of induction**. We discuss this important principle more thoroughly in Proposition 5.7 and Examples 5.8. ■

5.2 Remarks **(a)** We will later see that everything one learns in school about the arithmetic of numbers can be deduced from the Peano axioms. Even so, for a mathematician, two important questions arise: (1) Does there exist a system $(\mathbb{N}, 0, \nu)$ in which the Peano axioms hold? That is, is there a **model** for the natural numbers? (2) If so, how many models are there? We briefly consider these questions here.

To simplify our discussion we introduce the following concept: A set M is called an *infinite system*, if there is an injective function $f : M \to M$ such that $f(M) \subset M$. Clearly the natural numbers, if they exist, form an infinite system. The significance of such systems is seen in the following theorem proved by R. Dedekind: *Any infinite system contains a model* $(\mathbb{N}, 0, \nu)$ *for the natural numbers.*

Thus the question of the existence of the natural numbers can be reduced to the question of the existence of infinite systems. Dedekind gave a proof of the existence of such systems which implicitly uses the 'comprehension axiom' introduced by G. Frege in 1893: *For each property E of sets, the set*

$$M_E := \{\, x \;;\; x \textit{ is a set which satisfies } E \,\}$$

exists. In 1901 B. Russell recognized that this axiom leads to contradictions, so called *antinomies*. Russell chose for E the property 'x is a set and x is not an element of itself'. Then the comprehension axiom ensures the existence of the *set*

$$M := \bigl\{\, x \;;\; (x \text{ is a set}) \wedge (x \notin x) \,\bigr\} \;.$$

This clearly leads to the contradiction

$$M \in M \Longleftrightarrow M \notin M \;.$$

It is no surprise that such antinomies shook the foundations of the set theory. Closer inspection showed that such problems in set theory arise only when one considers sets which are 'too big'. To avoid Russell's antinomy one can distinguish two types of collections of objects: *classes* and *sets*. Sets are special 'small' classes. If a class is a set, then it can be described axiomatically. The comprehension axiom then becomes: *For each property E of sets, the class*

$$M_E := \{\, x \;;\; x \textit{ is a set which satisfies } E \,\}$$

exists. Then $M = \bigl\{\, x \;;\; (x \text{ is a set}) \wedge (x \notin x) \,\bigr\}$ is a class and not a set, and Russell's contradiction no longer occurs.

One needs, in addition, a separate axiom which implies the fact, which we have already used many times, that *For each set X and property E of sets,*

$$\bigl\{\, x \;;\; (x \in X) \wedge E(x) \,\bigr\} =: \bigl\{\, x \in X \;;\; E(x) \,\bigr\} \textit{ is a set.}$$

For a more complete discussion of these questions we have to refer the reader to the literature (for example, [FP85]).

Dedekind's investigation showed that, to prove the existence of the natural numbers in the framework of axiomatic set theory, one needs the *Infinity Axiom*: *An inductive set exists.* Here an **inductive set** is a set N which contains $\emptyset$ and such that for all $z \in N$, $z \cup \{z\}$ is also in N. Consider the set

$$\mathbb{N} := \bigcap \{\, m \;;\; m \text{ is an inductive set} \,\} \;,$$

and the function $\nu : \mathbb{N} \to \mathbb{N}$ defined by $\nu(z) := z \cup \{z\}$. Finally, set $0 := \emptyset$. It can be shown that $\mathbb{N}$ is itself an inductive set and that $(\mathbb{N}, 0, \nu)$ satisfies the Peano axioms. Thus $(\mathbb{N}, 0, \nu)$ is a model for the natural numbers.

Now let $(\mathbb{N}', 0', \nu')$ be some other model of the natural numbers. Then, in the framework of set theory, it can be shown that there is a bijection $\varphi : \mathbb{N} \to \mathbb{N}'$ such that $\varphi(0) = 0'$ and $\varphi \circ \nu = \nu' \circ \varphi$, that is, φ is an **isomorphism** from $(\mathbb{N}, 0, \nu)$ to $(\mathbb{N}', 0', \nu')$. Thus, *the natural numbers are unique up to isomorphism.* It is thus meaningful to speak of *the* natural numbers. For proofs and details, see [FP85].

(b) In the previous remark we have limited our discussion to the *von Neumann-Bernays-Gödel* (NBG) axiom system in which the concept of classes is central. This concept can, in fact, be completely avoided. For example, the equally popular *Zermelo-Fraenkel set theory with the axiom of choice* (ZFC) does not require this concept. Fortunately, it can be shown that both axiom systems are equivalent in the sense that in both systems the same statements about sets are provable. ■

The Arithmetic of Natural Numbers

Starting from the Peano axioms we can deduce all of the usual rules of the arithmetic of the natural numbers.

5.3 Theorem *There are operations* **addition** $+$, **multiplication** $\cdot$ *and a partial order* $\leq$ *on* $\mathbb{N}$ *which are uniquely determined by the following conditions:*

(i) *Addition is associative, commutative and has the identity element* 0.

(ii) *Multiplication is associative, commutative and has* $1 := \nu(0)$ *as its identity element.*

(iii) *The distributive law holds:*

$$(\ell + m) \cdot n = \ell \cdot n + m \cdot n , \qquad \ell, m, n \in \mathbb{N} .$$

(iv) $0 \cdot n = 0$ *and* $\nu(n) = n + 1$ *for* $n \in \mathbb{N}$.

(v) $\mathbb{N}$ *is totally ordered by* $\leq$ *and* $0 = \min(\mathbb{N})$.

(vi) *For* $n \in \mathbb{N}$ *there is no* $k \in \mathbb{N}$ *with* $n < k < n + 1$.

(vii) *For all* $m, n \in \mathbb{N}$,

$$m \leq n \iff \exists\, d \in \mathbb{N} \ : m + d = n ,$$
$$m < n \iff \exists\, d \in \mathbb{N}^\times : m + d = n .$$

The element d *is unique and is called the* **difference** *of* n *and* m, *in symbols:* $d := n - m$.

(viii) *For all* $m, n \in \mathbb{N}$,

$$m \leq n \iff m + \ell \leq n + \ell , \quad \ell \in \mathbb{N} ,$$
$$m < n \iff m + \ell < n + \ell , \quad \ell \in \mathbb{N} .$$

(ix) *For all* $m, n \in \mathbb{N}^\times$, $m \cdot n \in \mathbb{N}^\times$.

(x) *For all* $m, n \in \mathbb{N}$,

$$m \le n \iff m \cdot \ell \le n \cdot \ell \ , \quad \ell \in \mathbb{N}^\times \ ,$$
$$m < n \iff m \cdot \ell < n \cdot \ell \ , \quad \ell \in \mathbb{N}^\times \ .$$

Proof We show only the existence and uniqueness of an operation $+$ on $\mathbb{N}$, such that (i) and

$$n + \nu(m) = \nu(n+m) \ , \qquad n, m \in \mathbb{N} \ , \tag{5.1}$$

are satisfied. For the remaining claims we recommend the book [Lan30]. The proofs are elementary. The main difficulty for beginners is to avoid using facts from ordinary arithmetic *before* they are derived from the Peano axioms. In particular, at the beginning, 0 and 1 are simply certain distinguished elements of a set $\mathbb{N}$, and have nothing to do with the *numbers* 0 and 1 as we usually think of them.

(a) Suppose first that $\circledast$ is a commutative operation on $\mathbb{N}$ such that

$$0 \circledast 0 = 0 \ , \quad n \circledast 1 = \nu(n) \quad \text{and} \quad n \circledast \nu(m) = \nu(n \circledast m) \ , \qquad n, m \in \mathbb{N} \ . \tag{5.2}$$

Consider the set

$$N := \{ n \in \mathbb{N} \ ; \ 0 \circledast n = n \} \ .$$

Clearly 0 is in N. If n is in N then $0 \circledast n = n$, and hence, from (5.2),

$$0 \circledast \nu(n) = \nu(0 \circledast n) = \nu(n).$$

Thus $\nu(n)$ is also in N. From (N_1) we then have $N = \mathbb{N}$, that is,

$$0 \circledast n = n \ , \qquad n \in \mathbb{N} \ . \tag{5.3}$$

(b) Suppose that $\circledcirc$ is another commutative operation on $\mathbb{N}$ which also satisfies (5.2), that is,

$$0 \circledcirc 0 = 0 \ , \quad n \circledcirc 1 = \nu(n) \quad \text{and} \quad n \circledcirc \nu(m) = \nu(n \circledcirc m) \ , \qquad n, m \in \mathbb{N} \ . \tag{5.4}$$

For an arbitrary, but fixed, $n \in \mathbb{N}$, set

$$M := \{ m \in \mathbb{N} \ ; \ m \circledast n = m \circledcirc n \} \ .$$

Just as in (a), it follows from (5.4) that $0 \circledcirc n = n$. From (5.3) we get $0 \circledast n = n = 0 \circledcirc n$, that is, $0 \in M$. Now suppose that m is in M. Then $m \circledast n = m \circledcirc n$ and hence, from (5.2) and (5.4),

$$\nu(m) \circledast n = n \circledast \nu(m) = \nu(m \circledast n) = \nu(m \circledcirc n) = n \circledcirc \nu(m) = \nu(m) \circledcirc n \ .$$

Thus $\nu(m)$ is also in M. The axiom (N_1) implies that $M = \mathbb{N}$. Since $n \in \mathbb{N}$ was arbitrary, we have shown that $m \circledast n = m \circledcirc n$ for all $m, n \in \mathbb{N}$. Consequently there is at most one commutative operation $\circledast : \mathbb{N} \times \mathbb{N} \to \mathbb{N}$ which satisfies (5.2).

(c) We construct next an operation on $\mathbb{N}$ with the property (5.1). Define

$$N := \{ n \in \mathbb{N} \; ; \; \exists \varphi_n : \mathbb{N} \to \mathbb{N} \text{ with } \varphi_n(0) = \nu(n) \text{ and } \varphi_n(\nu(m)) = \nu(\varphi_n(m)) \;\; \forall m \in \mathbb{N} \} . \tag{5.5}$$

Setting $\varphi_0 := \nu$ we see that $0 \in N$. Let $n \in N$. Then there is a function $\varphi_n : \mathbb{N} \to \mathbb{N}$ such that $\varphi_n(0) = \nu(n)$ and $\varphi_n(\nu(m)) = \nu(\varphi_n(m))$ for all $m \in \mathbb{N}$. Define

$$\psi : \mathbb{N} \to \mathbb{N} , \quad m \mapsto \nu(\varphi_n(m)) .$$

Then $\psi(0) = \nu(\varphi_n(0)) = \nu(\nu(n))$ and also

$$\psi(\nu(m)) = \nu(\varphi_n(\nu(m))) = \nu(\nu(\varphi_n(m))) = \nu(\psi(m)) , \qquad m \in \mathbb{N} .$$

Thus we have shown that $n \in N$ implies $\nu(n) \in N$. Once again, (N_1) implies $N = \mathbb{N}$.

We show further that, for each $n \in \mathbb{N}$, the function φ_n in (5.5) is unique. For $n \in \mathbb{N}$, suppose that $\psi_n : \mathbb{N} \to \mathbb{N}$ is a function such that

$$\psi_n(0) = \nu(n) \quad \text{and} \quad \psi_n(\nu(m)) = \nu(\psi_n(m)) , \qquad m \in \mathbb{N} ,$$

and define

$$M_n := \{ m \in \mathbb{N} \; ; \; \varphi_n(m) = \psi_n(m) \} .$$

From $\varphi_n(0) = \nu(n) = \psi_n(0)$ we deduce that $0 \in M_n$. If $m \in M_n$, then it follows that $\varphi_n(\nu(m)) = \nu(\varphi_n(m)) = \nu(\psi_n(m)) = \psi_n(\nu(m))$. Thus $\nu(m)$ is also in M_n. The axiom (N_1) implies that $M_n = \mathbb{N}$, which means that $\varphi_n = \psi_n$.

We have therefore shown that for each $n \in \mathbb{N}$ there is exactly one function

$$\varphi_n : \mathbb{N} \to \mathbb{N} \text{ such that } \varphi_n(0) = \nu(n) \text{ and } \varphi_n(\nu(m)) = \nu(\varphi_n(m)) , \qquad m \in \mathbb{N} .$$

Now we define

$$+ : \mathbb{N} \times \mathbb{N} \to \mathbb{N} , \quad (n,m) \mapsto n + m := \begin{cases} n , & m = 0 , \\ \varphi_n(m') , & m = \nu(m') . \end{cases} \tag{5.6}$$

Because of Remark 5.1(a), $+$ is a well defined operation on $\mathbb{N}$ which satisfies (5.1). Also

$$\left. \begin{aligned} & n + 0 = n , \\ & n + 1 = n + \nu(0) = \varphi_n(0) = \nu(n) = \nu(n+0) , \end{aligned} \right. \qquad n \in \mathbb{N} ,$$

and

$$n + \nu(m) = \varphi_n(m) = \varphi_n(\nu(m')) = \nu(\varphi_n(m')) = \nu(n+m)$$

for all $n \in \mathbb{N}$, $m \in \mathbb{N}^\times$ and $m' := \nu^{-1}(m)$. Thus we have shown the existence of an operation $+$ on $\mathbb{N}$ which satisfies (5.1). We have already shown that $n + 0 = n$ for all $n \in \mathbb{N}$. Together with (5.3) this implies that 0 is the identity element for $+$.

(d) We verify the associativity of addition. Let $\ell, m \in \mathbb{N}$ be arbitrary and set

$$N := \{ n \in \mathbb{N} \; ; \; (\ell + m) + n = \ell + (m + n) \} .$$

Clearly $0 \in N$ and, by (5.1), for all $n \in N$ we have

$$(\ell + m) + \nu(n) = \nu((\ell + m) + n) = \nu(\ell + (m + n)) = \ell + \nu(m + n) = \ell + (m + \nu(n)) .$$

Hence $n \in N$ implies $\nu(n) \in N$. Using axiom (N_1) we conclude that $N = \mathbb{N}$.

(e) To prove the commutativity of addition, we consider first the set

$$N := \{ n \in \mathbb{N} \ ; \ n + 1 = 1 + n \} .$$

This set certainly contains 0. For $n \in N$ it follows from (5.1) that

$$\nu(n) + 1 = \nu\big(\nu(n)\big) = \nu(n+1) = \nu(1+n) = 1 + \nu(n) .$$

Thus $\nu(n) \in N$, and (N_1) implies $N = \mathbb{N}$. Hence we know that

$$n + 1 = 1 + n , \qquad n \in \mathbb{N} . \tag{5.7}$$

Now we fix $n \in \mathbb{N}$ and define

$$M := \big\{ m \in \mathbb{N} \ ; \ m + n = n + m \big\} .$$

Once again $0 \in M$. For $m \in M$ we have from (d) and (5.7) that

$$\begin{aligned} \nu(m) + n &= (m+1) + n = m + (1+n) = m + (n+1) \\ &= (m+n) + 1 = \nu(m+n) = \nu(n+m) = n + \nu(m) , \end{aligned}$$

where in the last step we have used (5.1) again. Thus $\nu(m)$ is in M, and from (N_1) we have $M = \mathbb{N}$. Since $n \in \mathbb{N}$ was arbitrary, we have shown that $n + m = m + n$ for all $m, n \in \mathbb{N}$. ■

Henceforth we use, without further comment, all of the familiar facts about the arithmetic of natural numbers learned in school. For practice, the reader is encouraged to prove a few of these, for example, $1 + 1 = 2$, $2 \cdot 2 = 4$ and $3 \cdot 4 = 12$.

As usual, we write mn for $m \cdot n$, and make the convention that 'multiplication takes precedence over addition', that is, $mn + k$ means $(m \cdot n) + k$ (and not $m(n + k)$). Finally, the elements of $\mathbb{N}^\times$ are called the **positive natural numbers**.

The Division Algorithm

A simple consequence of Theorem 5.3(x) is the following **cancellation rule**:

$$\textit{If } m, n \in \mathbb{N} \textit{ and } k \in \mathbb{N}^\times \textit{ satisfy } mk = nk, \textit{ then } m = n. \tag{5.8}$$

We call $m \in \mathbb{N}^\times$ a **divisor** of $n \in \mathbb{N}$ if there is some $k \in \mathbb{N}$ such that $mk = n$. If m is a divisor of n we write $m \mid n$ ('m divides n'). The unique natural number k is called the **quotient** of n by m and is written $\frac{n}{m}$ or n/m. If m and n are two positive natural numbers then it is often true that m does not divide n or vice versa. The following proposition, called the **division algorithm**, clarifies the general situation.

5.4 Proposition *For each $m \in \mathbb{N}^\times$ and $n \in \mathbb{N}$, there are unique $l, k \in \mathbb{N}$ such that*

$$n = km + \ell \quad \text{and} \quad \ell < m .$$

Proof (a) We verify first the existence statement. Fix $m \in \mathbb{N}^\times$ and set

$$N := \{ n \in \mathbb{N} \; ; \; \exists\, k, \ell \in \mathbb{N} : n = km + \ell, \;\; \ell < m \} .$$

Our goal is to prove that $N = \mathbb{N}$. Clearly 0 is in N because $0 = 0 \cdot m + 0$ by Theorem 5.3(i) and (iv). Now suppose that $n \in N$. Then there are $k, \ell \in \mathbb{N}$ with $n = km + \ell$ and $\ell < m$ from which follows $n + 1 = km + (\ell + 1)$. If $\ell + 1 < m$ then $n + 1$ is in N. On the other hand, if $\ell + 1 = m$, then, by Theorem 5.3(iii), we have $n + 1 = (k + 1)m$ and so $n + 1$ is again in N. Thus we have shown that $0 \in N$ and that $n \in N$ implies $n + 1 \in N$. By induction, that is, by (N_1), we conclude that $N = \mathbb{N}$.

(b) To prove uniqueness we suppose that there are $m \in \mathbb{N}^\times$ and $k, k', \ell, \ell' \in \mathbb{N}$ such that

$$km + \ell = k'm + \ell' \quad \text{and} \quad \ell < m , \quad \ell' < m . \tag{5.9}$$

We can also assume that $\ell \leq \ell'$, since the case $\ell' \leq \ell$ would follow by symmetry. From $\ell \leq \ell'$ and (5.9) we have $k'm + \ell' = km + \ell \leq km + \ell'$ and hence $k'm \leq km$, by Theorem 5.3(viii). Then, from Theorem 5.3(x), we get $k' \leq k$.

On the other hand, from $\ell' < m$ we have the inequalities

$$km \leq km + \ell = k'm + \ell' < k'm + m = (k' + 1)m .$$

Here we have used (viii) and (iii) of Theorem 5.3. From (x) of the same theorem it follows that $k < k' + 1$. Together with $k' \leq k$ we find that $k' \leq k < k' + 1$, which, because of Theorem 5.3(vi), is possible only if $k = k'$. From $k = k'$, (5.9) and the uniqueness claim of Theorem 5.3(vii) it follows that $\ell = \ell'$. ∎

In the above proof we have made all references to Theorem 5.3 explicit. In future discussion we will use these rules without reference.

The Induction Principle

We have already made considerable use of the induction axiom (N_1). It is frequently convenient to use this axiom in an alternative form, the **well ordering principle**.

5.5 Proposition *The natural numbers $\mathbb{N}$ are* **well ordered**, *that is, each nonempty subset of $\mathbb{N}$ has a minimum.*

Proof We prove this claim by contradiction. Suppose that $A \subseteq \mathbb{N}$ is nonempty and has no minimum. Set

$$B := \{\, n \in \mathbb{N} \ ; \ n \text{ is lower bound of } A \,\} \ .$$

Clearly $0 \in B$. Now suppose that $n \in B$. Since A has no minimum, n cannot be in A, so we have, in fact, $a > n$ for all $a \in A$. This implies that $a \geq n+1$ for all $a \in A$, that is, $n+1 \in B$. Because of the induction axiom (N_1) we have $B = \mathbb{N}$. But this implies that $A = \emptyset$ because, if $m \in A$, then $m \in \mathbb{N} = B$ which means that m is a lower bound and, hence a minimum element, of A, which is not possible. We have therefore found the desired contradiction: $A \neq \emptyset$ and $A = \emptyset$. ■

For an example of the use of the well ordering principle, we discuss the prime factorization of natural numbers. We say that a natural number $p \in \mathbb{N}$ is **prime** if $p \geq 2$ and p has no divisors except 1 and p.

5.6 Proposition *Except for* 0 *and* 1, *every natural number is a product of finitely many prime numbers, its* **prime factors**. *Here 'products' with only one factor are allowed. This* **prime factorization** *is, up to the order of the factors, unique.*

Proof Suppose that the claim is false. By Proposition 5.5 there is a smallest natural number n_0 which cannot be factored into prime numbers. In particular, n_0 cannot be a prime number, so there are $n, m \in \mathbb{N}$ with $n_0 = n \cdot m$ and $n, m > 1$. This implies $n < n_0$ and $m < n_0$. From the minimality of n_0 it follows that n and m are each products of finitely many prime numbers, and hence $n_0 = n \cdot m$ is also such a product. This contradicts our assumption, so we have we shown the existence of a prime factorization for any natural number greater than 1.

To prove the uniqueness of prime factorizations we suppose, to the contrary, that there is a number with two different prime factorizations. Let p be the least such number with prime factorizations $p = p_0 p_1 \cdots p_k = q_0 q_1 \cdots q_n$. We have $p_i \neq q_j$ for all i and j, since any common factor could be divided out to give a smaller natural number p' with two different prime factorizations, in contradiction to the choice of p.

We can suppose that $p_0 \leq p_1 \leq \cdots \leq p_k$ and $q_0 \leq q_1 \leq \cdots \leq q_n$ as well as $p_0 < q_0$. Set $q := p_0 q_1 \cdots q_n$. Then $p_0 \,|\, q$ and $p_0 \,|\, p$, hence $p_0 \,|\, (p-q)$. Consequently we have the prime factorization

$$p - q = p_0 r_1 \cdots r_\ell$$

for some prime numbers $r_1, \ldots, r_\ell$. Because $p - q = (q_0 - p_0) q_1 \cdots q_n$, the number $p - q$ is positive. Write $q_0 - p_0$ as a product of prime numbers: $q_0 - p_0 = t_0 \cdots t_s$. Then

$$p - q = t_0 \cdots t_s q_1 \cdots q_n$$

is a second prime factorization of $p - q$. It is clear that p_0 does not divide $q_0 - p_0$. Hence we have two prime factorizations of $p - q$, only one of which contains p_0. Because $0 < p - q < p$, this contradicts the minimality of p. ∎

All of the proofs in this section have depended on the induction axiom (N_1) either directly, or via the well ordering principle. This axiom is used so frequently in mathematics that it is worthwhile formalizing 'proof by induction'. For each $n \in \mathbb{N}$, let $\mathcal{A}(n)$ be a statement. To prove *by induction on* n that $\mathcal{A}(n)$ is true for each $n \in \mathbb{N}$, one uses the following procedure:

(a) *Prove that* $\mathcal{A}(0)$ *is true.*

(b) *This step has two parts:*

(α) *Induction hypothesis*: Suppose that $\mathcal{A}(n)$ is true for some $n \in \mathbb{N}$.

(β) *Induction step* $(n \to n+1)$: Prove that $\mathcal{A}(n+1)$ follows from (α) and other previously proved statements.

If (a) and (b) can be done, then $\mathcal{A}(n)$ is true for all $n \in \mathbb{N}$. To see this, let

$$N := \{\, n \in \mathbb{N} \ ; \ \mathcal{A}(n) \text{ is true} \,\} \ .$$

Then (a) implies that $0 \in N$, and from (b) we have that $n \in N$ implies $n + 1 \in N$ for all $n \in \mathbb{N}$. It follows from (N_1) that $N = \mathbb{N}$.

In many applications it is useful to start the induction with some number other than 0. This leads to a slight generalization of the above method.

5.7 Proposition (induction principle) *Let* $n_0 \in \mathbb{N}$ *and, for each* $n \geq n_0$, *let* $\mathcal{A}(n)$ *be a statement. If*

(i) $\mathcal{A}(n_0)$ *is true, and*

(ii) *for each* $n \geq n_0$, $\mathcal{A}(n+1)$ *can be proved from the assumption that* $\mathcal{A}(n)$ *is true,*

then $\mathcal{A}(n)$ *is true for all* $n \geq n_0$.

Proof Set $N := \{\, n \in \mathbb{N} \ ; \ \mathcal{A}(n + n_0) \text{ is true} \,\}$. Then $N = \mathbb{N}$ follows from (N_1) as above. ∎

For $m \in \mathbb{N}$ and $n \in \mathbb{N}^\times$, we write

$$m^n := \underbrace{m \cdot m \cdot \cdots \cdot m}_{n \text{ times}} \ .$$

Using this notation we can give some simple applications of the induction principle.

5.8 Examples **(a)** For $n \in \mathbb{N}^\times$, we have $1+3+5+\cdots+(2n-1)=n^2$.

Proof (By induction) We can start the induction with $n_0 = 1$ since $1 = 1 \cdot 1 = 1^2$. The induction hypothesis is

Suppose that for some $n \in \mathbb{N}$ we have $1+3+5+\cdots+(2n-1)=n^2$.

The induction step proceeds as follows:

$$\begin{aligned} 1+3+5+\cdots+\bigl(2(n+1)-1\bigr) &= 1+3+5+\cdots+(2n+1) \\ &= 1+3+5+\cdots+(2n-1)+(2n+1) \\ &= n^2+2n+1 \ . \end{aligned}$$

Here we have used the induction hypothesis in the last step. Since $n^2+2n+1=(n+1)^2$, which follows easily from the distributive law (Theorem 5.3(iii)), we have completed the induction step and hence proved the claim. ■

(b) For all $n \in \mathbb{N}$ with $n \geq 5$, we have $2^n > n^2$.

Proof We start the induction with $n_0 = 5$ since $32 = 2^5 > 5^2 = 25$. The induction hypothesis is

$$\text{Suppose, that for some } n \in \mathbb{N} \text{ with } n \geq 5, \text{ we have } 2^n > n^2 \ . \tag{5.10}$$

The induction step can be done as follows: From (5.10) we have

$$2^{n+1} = 2 \cdot 2^n > 2 \cdot n^2 = n^2 + n \cdot n \ . \tag{5.11}$$

Since $n \geq 5$, we have also $n \cdot n \geq 5n > 2n+1$. Together with (5.11), this implies

$$2^{n+1} > n^2+2n+1=(n+1)^2.$$

This completes the induction step and we have proved the claim. ■

We formulate one more version of the induction principle which allows one to assume that all of the statements $\mathcal{A}(k)$ for $n_0 \leq k \leq n$ are true in proving the induction step $n \to n+1$.

5.9 Proposition *Let $n_0 \in \mathbb{N}$, and for each $n \geq n_0$, let $\mathcal{A}(n)$ be a statement. If*

(i) *$\mathcal{A}(n_0)$ is true, and*

(ii) *for each $n \geq n_0$, $\mathcal{A}(n+1)$ can be proved from the assumption that $\mathcal{A}(k)$ is true for all $n_0 \leq k \leq n$,*

then $\mathcal{A}(n)$ is true for all $n \geq n_0$.

Proof Set

$$N := \bigl\{ n \in \mathbb{N} \ ; \ n \geq n_0 \text{ and } \mathcal{A}(n) \text{ is false} \bigr\}$$

and suppose that $N \neq \emptyset$. By the well ordering principle (Proposition 5.5), N has a minimum element, $m := \min(N)$, which, by (i), satisfies $m > n_0$. Thus there is a unique $n \in \mathbb{N}$ with $n+1=m$. Further, it follows from our choice of m that $\mathcal{A}(k)$ is true for all $k \in \mathbb{N}$ such that $n_0 \leq k \leq n$. Then (ii) implies that $\mathcal{A}(n+1)=\mathcal{A}(m)$ is true, a contradiction. ■

5.10 Example Let $\circledast$ be an associative operation on a set X. Then the value of any valid expression involving $\circledast$, elements of X and parentheses, is independent of the placement of the parentheses. For example,

$$(a_1 \circledast a_2) \circledast (a_3 \circledast a_4) = ((a_1 \circledast a_2) \circledast a_3) \circledast a_4 = a_1 \circledast (a_2 \circledast (a_3 \circledast a_4)) .$$

Proof In this proof, K_n always stands for some 'expression of length n', that is, an expression consisting of n elements $a_1, \ldots, a_n \in X$, $n-1$ operation symbols and an arbitrary number of (correctly nested) parentheses, for example,

$$K_7 := \bigl((a_1 \circledast a_2) \circledast (a_3 \circledast a_4)\bigr) \circledast \bigl((a_5 \circledast (a_6 \circledast a_7))\bigr) .$$

We will prove by induction on n that

$$K_n = \bigl(\cdots(a_1 \circledast a_2) \circledast a_3) \circledast \cdots\bigr) \circledast a_{n-1}\bigr) \circledast a_n , \qquad n \in \mathbb{N} .$$

For $n = 3$, the claim is true by definition of associativity. Our induction hypothesis is

$$K_k = \bigl(\cdots(a_1 \circledast a_2) \circledast a_3) \circledast \cdots\bigr) \circledast a_{k-1}\bigr) \circledast a_k$$
for all expressions K_k of length $k \in \mathbb{N}$ with $3 \le k \le n$.

Now let K_{n+1} have length $n+1$. Then there are $\ell, m \in \mathbb{N}^\times$ such that $\ell + m = n+1$ and expressions K_ℓ and K_m such that $K_{n+1} = K_\ell \circledast K_m$. Now we have two cases:

Case 1: $m = 1$. Then $\ell = n$, $K_m = a_{n+1}$, and by the induction hypothesis,

$$K_\ell = \bigl(\cdots(a_1 \circledast a_2) \circledast a_3)\cdots\bigr) \circledast a_n .$$

Consequently,

$$K_{n+1} = \bigl(\bigl(\cdots(a_1 \circledast a_2) \circledast a_3)\cdots\bigr) \circledast a_n\bigr) \circledast a_{n+1} .$$

Case 2: $m > 1$. By the induction hypothesis, K_m can be written in the form $K_m = K_{m-1} \circledast a_{n+1}$, and so

$$K_{n+1} = K_\ell \circledast (K_{m-1} \circledast a_{n+1}) = (K_\ell \circledast K_{m-1}) \circledast a_{n+1} .$$

But $K_\ell \circledast K_{m-1}$ is an expression of length n, so, by the induction hypothesis again,

$$K_\ell \circledast K_{m-1} = \bigl(\cdots(a_1 \circledast a_2) \circledast a_3)\cdots\bigr) \circledast a_n .$$

This implies

$$K_{n+1} = \bigl(\bigl(\cdots(a_1 \circledast a_2) \circledast a_3)\cdots\bigr) \circledast a_n\bigr) \circledast a_{n+1} ,$$

completing the induction step. ■

Recursive Definitions

We come now to a further application of induction: *recursive definitions*. Its significance will be made clear in the examples at the end of this section.

5.11 Proposition *Let X be a nonempty set and $a \in X$. For each $n \in \mathbb{N}^\times$, let $V_n : X^n \to X$ be a function. Then there is a unique function $f : \mathbb{N} \to X$ with the following properties:*

(i) $f(0) = a$.

(ii) $f(n+1) = V_{n+1}\big(f(0), f(1), \ldots, f(n)\big), \quad n \in \mathbb{N}$.

Proof (a) We show first, using induction, that there can be at most one such function. Let $f, g : \mathbb{N} \to X$ be such that $f(0) = g(0) = a$ and

$$\begin{aligned} f(n+1) &= V_{n+1}\big(f(0), \ldots, f(n)\big) , \\ g(n+1) &= V_{n+1}\big(g(0), \ldots, g(n)\big) , \end{aligned} \qquad n \in \mathbb{N} . \tag{5.12}$$

We want to show that $f = g$, that is, $f(n) = g(n)$ for all $n \in \mathbb{N}$. The condition $f(0) = g(0)$ $(= a)$ starts the induction. For the induction hypothesis we assume that $f(k) = g(k)$ for $0 \le k \le n$. From (5.12) it follows that $f(n+1) = g(n+1)$. From Proposition 5.9 we have that $f(n) = g(n)$ for all $n \in \mathbb{N}$, that is, $f = g$.

(b) We turn to the existence of the function f. We first claim that, for each $n \in \mathbb{N}$, there is a function $f_n : \{0, 1, \ldots, n\} \to X$ such that

$$\begin{aligned} f_n(0) &= a , \\ f_n(k) &= f_k(k) , \\ f_n(k+1) &= V_{k+1}\big(f_n(0), \ldots, f_n(k)\big) , \end{aligned} \qquad 0 \le k < n .$$

Once again, the proof of this claim uses induction. Clearly the claim is true for $n = 0$ since there are no $k \in \mathbb{N}$ with $0 \le k < 0$. To do the induction step $n \to n+1$, define

$$f_{n+1}(k) := \begin{cases} f_n(k) , & 0 \le k \le n , \\ V_{n+1}\big(f_n(0), \ldots, f_n(n)\big) , & k = n+1 . \end{cases}$$

By the induction hypothesis,

$$f_{n+1}(k) = f_n(k) = f_k(k) , \qquad k \in \mathbb{N} , \quad 0 \le k \le n , \tag{5.13}$$

and, together with (5.13), we have

$$\begin{aligned} f_{n+1}(k+1) = f_n(k+1) &= V_{k+1}\big(f_n(0), \ldots, f_n(k)\big) \\ &= V_{k+1}\big(f_{n+1}(0), \ldots, f_{n+1}(k)\big) \end{aligned}$$

for $0 < k+1 \le n$, and hence

$$f_{n+1}(n+1) = V_{n+1}\big(f_n(0), \ldots, f_n(n)\big) = V_{n+1}\big(f_{n+1}(0), \ldots, f_{n+1}(n)\big) .$$

This completes the induction step and proves the existence of the functions f_n for all $n \in \mathbb{N}$.

(c) After these preliminary steps we finally define $f : \mathbb{N} \to X$ by

$$f : \mathbb{N} \to X \ , \quad f(n) := \begin{cases} a \ , & n = 0 \ , \\ f_n(n) \ , & n \in \mathbb{N}^\times \ . \end{cases}$$

Because of the properties of the functions f_n proved in (b), we have

$$\begin{aligned} f(n+1) = f_{n+1}(n+1) &= V_{n+1}\big(f_{n+1}(0), \ldots, f_{n+1}(n)\big) \\ &= V_{n+1}\big(f_0(0), \ldots, f_n(n)\big) \\ &= V_{n+1}\big(f(0), \ldots, f(n)\big) \ . \end{aligned}$$

This completes the proof. ■

5.12 Example Let $\odot$ be an associative operation on a set X and $x_k \in X$ for all $k \in \mathbb{N}$. For each $n \in \mathbb{N}$, define

$$\bigodot_{k=0}^{n} x_k := x_0 \odot x_1 \odot \cdots \odot x_n \ . \tag{5.14}$$

This definition is not complete unless we explain the meaning of the three dots on the right. This is accomplished most easily using a recursive definition. Thus, for $n \in \mathbb{N}^\times$, let

$$V_n : X^n \to X \ , \quad (y_0, \ldots, y_{n-1}) \mapsto y_{n-1} \odot x_n \ .$$

By Proposition 5.11, there is a unique function $f : \mathbb{N} \to X$ such that $f(0) = x_0$ and

$$f(n) = V_n\big(f(0), \ldots, f(n-1)\big) = f(n-1) \odot x_n \ , \qquad n \in \mathbb{N}^\times \ .$$

Now define $\bigodot_{k=0}^{n} x_k := f(n)$ for $n \in \mathbb{N}$. From this definition we get the recursion rules

$$\bigodot_{k=0}^{0} x_k = x_0 \ , \qquad \bigodot_{k=0}^{n} x_k = \bigodot_{k=0}^{n-1} x_k \odot x_n \ , \quad n \in \mathbb{N}^\times \ ,$$

which justify the notation of (5.14). ■

When we use the symbol $+$ or $\cdot$ for an associative operation on a set X, then we will call $+$ an *addition* and $\cdot$ a *multiplication* on X. For *sums* and *products* we use the usual notation

$$\sum_{k=0}^{n} x_k := x_0 + x_1 + \cdots + x_n \quad \text{and} \quad \prod_{k=0}^{n} x_k := x_0 \cdot x_1 \cdot \cdots \cdot x_n \ .$$

Note that the order is significant since the operation may not be commutative.

5.13 Remarks (a) Sums and products are independent of the choice of index, that is,

$$\sum_{k=0}^{n} x_k = \sum_{j=0}^{n} x_j \quad \text{and} \quad \prod_{k=0}^{n} x_k = \prod_{j=0}^{n} x_j \ .$$

If addition $+$ and multiplication $\cdot$ are commutative, we have

$$\sum_{k=0}^{n} x_k = \sum_{j=0}^{n} x_{\sigma(j)} \quad \text{and} \quad \prod_{k=0}^{n} x_k = \prod_{j=0}^{n} x_{\sigma(j)}$$

for any **permutation** σ of the numbers $0, \ldots, n$, that is, for any bijective function $\sigma \colon \{0, \ldots, n\} \to \{0, \ldots, n\}$.

(b) Let $+$ and $\cdot$ be associative and commutative operations on X which satisfy the **distributive law** $(x + y) \cdot z = x \cdot z + y \cdot z$ for $x, y, z \in X$. Then the following hold:

(α) $\displaystyle\sum_{k=0}^{n} a_k + \sum_{k=0}^{n} b_k = \sum_{k=0}^{n} (a_k + b_k)$.

(β) $\displaystyle\prod_{k=0}^{n} a_k \cdot \prod_{k=0}^{n} b_k = \prod_{k=0}^{n} (a_k \cdot b_k)$.

(γ) $\displaystyle\sum_{j=0}^{m} a_j \cdot \sum_{k=0}^{n} b_k = \sum_{\substack{0 \le j \le m \\ 0 \le k \le n}} (a_j \cdot b_k)$.

The right hand side of (γ) is the sum of the terms $a_j \cdot b_k$ for all possible $0 \le j \le m$ and $0 \le k \le n$. These rules can be proved using induction — a job we leave to the reader. ■

5.14 Examples (a) For a further use of a recursive definition consider a nonempty set X and an associative operation $\circledast$ on X with identity element e. For $a \in X$ define

$$a^0 := e \ , \quad a^{n+1} := a^n \circledast a \ , \qquad n \in \mathbb{N} \ .$$

From Proposition 5.11 it follows that a^n, the nth **power of** a, is defined for all $n \in \mathbb{N}$. Clearly $a^1 = a$ as well as

$$e^n = e \ , \quad a^n \circledast a^m = a^{n+m} \ , \quad (a^n)^m = a^{nm} \ , \qquad n, m \in \mathbb{N} \ . \tag{5.15}$$

If a and b **commute**, that is, $a \circledast b = b \circledast a$, then

$$a^n \circledast b^n = (a \circledast b)^n \ , \qquad n \in \mathbb{N} \ .$$

If the operation is commutative and written using **additive notation**, then the identity element is denoted by 0_X or simply 0 when there is no chance of confusion.

In the commutative case we define recursively

$$0 \cdot a := 0_X , \quad (n+1) \cdot a := (n \cdot a) + a , \qquad n \in \mathbb{N} , \quad a \in X ,$$

and call $n \cdot a$, n **times** a. Then

$$n \cdot a = \sum_{k=1}^{n} a := \underbrace{a + \cdots + a}_{n \text{ terms}} , \qquad n \in \mathbb{N}^{\times} ,$$

and the rules in (5.15) become

$$n \cdot 0_X = 0_X , \quad n \cdot a + m \cdot a = (n+m) \cdot a , \quad m \cdot (n \cdot a) = (mn) \cdot a$$

and

$$n \cdot a + n \cdot b = n \cdot (a+b)$$

for $a, b \in X$ and $m, n \in \mathbb{N}$.

Once again we leave the simple proofs of these statements to the reader.

(b) Define a function $\mathbb{N} \to \mathbb{N}$, $n \mapsto n!$, the **factorial function**, recursively by

$$0! := 1 , \quad (n+1)! := (n+1)n! , \qquad n \in \mathbb{N} .$$

It is not difficult to see that $n! = \prod_{k=1}^{n} k$ for $n \in \mathbb{N}^{\times}$. Note that the factorial function grows very quickly:

$$0! = 1 , \quad 1! = 1 , \quad 2! = 2 , \quad 3! = 6 , \quad 4! = 24 , \quad \ldots , \quad 10! > 3,628,000 \quad \ldots ,$$

$$100! > 9 \cdot 10^{157} , \quad \ldots , \quad 1,000! > 4 \cdot 10^{2,567} , \quad \ldots , \quad 10,000! > 2 \cdot 10^{35,659} , \quad \ldots$$

In Chapter VI we derive a formula which can be used to estimate this rapid growth. ■

Exercises

1 Provide complete proofs for the rules in Remark 5.13(b) and the rules of exponents in Example 5.14(a).

2 Verify the following equalities using induction:

(a) $\sum_{k=0}^{n} k = n(n+1)/2$, $n \in \mathbb{N}$.

(b) $\sum_{k=0}^{n} k^2 = n(n+1)(2n+1)/6$, $n \in \mathbb{N}$.

3 Verify the following inequalities using induction:

(a) For all $n \geq 2$, we have $n+1 < 2^n$.

(b) If $a \in \mathbb{N}$ with $a \geq 3$, then $a^n > n^2$ for all $n \in \mathbb{N}$.

4 Let A be a set with n elements. Show that $\mathcal{P}(A)$ has 2^n elements.

5 (a) Show that $m!\,(n-m)!$ divides $n!$ for all $m, n \in \mathbb{N}$ with $m \leq n$. (Hint: $(n+1)! = n!\,(n+1-m) + n!\,m$.)

(b) For $m, n \in \mathbb{N}$, the **binomial coefficient** $\binom{n}{m} \in \mathbb{N}$ is defined by

$$\binom{n}{m} := \begin{cases} \frac{n!}{m!\,(n-m)!}\ , & m \leq n\ , \\ 0\ , & m > n\ . \end{cases}$$

Prove the following:

(i) $\binom{n}{m} = \binom{n}{n-m}$.

(ii) $\binom{n}{m-1} + \binom{n}{m} = \binom{n+1}{m}$, $1 \leq m \leq n$.

(iii) $\sum_{k=0}^{n} \binom{n}{k} = 2^n$.

(iv) $\sum_{k=0}^{m} \binom{n+k}{n} = \binom{n+m+1}{n+1}$.

Remark The formula (ii) makes calculating small binomial coefficients easy when they are written down in the form of a **Pascal triangle**. In this triangle, the symmetry (i) and the equation (iv) are easy to see.

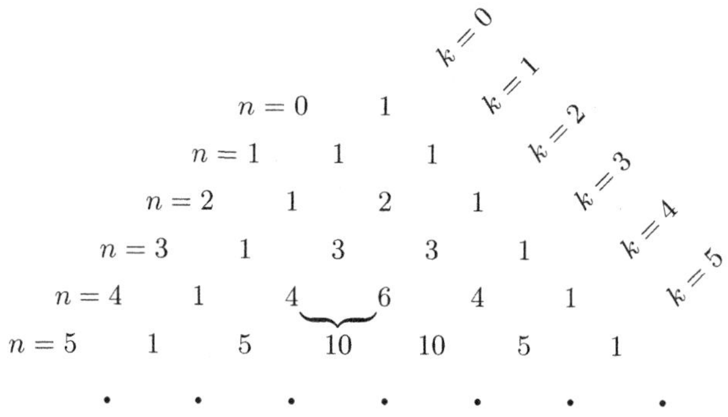

$k=0$, $k=1$, $k=2$, $k=3$, $k=4$, $k=5$

$n=0$ 1

$n=1$ 1 1

$n=2$ 1 2 1

$n=3$ 1 3 3 1

$n=4$ 1 4 6 4 1

$n=5$ 1 5 10 10 5 1

. . . .

6 Simplify the sum

$$S(m,n) := \sum_{k=0}^{n} \left[\binom{m+n+k}{k} 2^{n+1-k} - \binom{m+n+k+1}{k} 2^{n-k} \right]$$

for $m, n \in \mathbb{N}$. (Hint: For $1 \leq j < \ell$ we have $\binom{\ell}{j} - \binom{\ell}{j-1} = \binom{\ell+1}{j} - 2\binom{\ell}{j-1}$.)

7 Let $p \in \mathbb{N}$ with $p > 1$. Prove that p is a prime number if and only if, for all $m, n \in \mathbb{N}$,

$$p \,|\, mn \Rightarrow (p \,|\, m \text{ or } p \,|\, n)\ .$$

8 (a) Let $n \in \mathbb{N}^{\times}$. Show that none of the n consecutive numbers

$$(n+1)!+2,(n+1)!+3,\ldots,(n+1)!+(n+1)$$

is prime. Hence there are arbitrarily large gaps in the set of prime numbers.

(b) Show that there is no greatest prime number.
(Hint: Suppose that there is a greatest prime number and let $\{p_0,\ldots,p_m\}$ be the set of all prime numbers. Consider $q := p_0 \cdot \cdots \cdot p_m + 1$.)

9 The famous American mathematician M.I. Stake has finally found a mathematical proof of Thomas Jefferson's assertion that 'all men are created equal':

Proposition *If M is a finite set of men and $a, b \in M$, then a and b are equal.*

Proof We prove the claim by induction on the number of men in M:

(a) If M contains exactly one man, then the claim is obviously true.

(b) Induction step: Suppose that the claim is true for all sets of n men. Let M be a set containing $n+1$ men and let a and b be two men in M. We will show that a and b are equal. Let $M_a = M \setminus \{a\}$ and $M_b = M \setminus \{b\}$. These sets contain n men each. Let c be in the intersection of M_a and M_b. Since $a, c \in M_b$, the induction hypothesis implies that a and c are equal. Similarly, since $b, c \in M_a$, we have that b and c are equal. The claim then follows from the transitivity of equality. ■

What is wrong with this proposition?

10 Show that 7 divides $1 + 2^{(2^n)} + 2^{(2^{n+1})}$ for all $n \in \mathbb{N}$.

11 Fix some $g \in \mathbb{N}$ with $g \geq 2$. Show that each $n \in \mathbb{N}^{\times}$ can be written in the form

$$n = \sum_{j=0}^{\ell} y_j g^j \tag{5.16}$$

where $y_k \in \{0,\ldots,g-1\}$ for $k \in \{0,\ldots,\ell\}$ and $y_\ell > 0$. Show further that the expression (5.16) is unique, that is, if $n = \sum_{j=0}^{m} z_j g^j$ with $z_k \in \{0,\ldots,g-1\}$ for $k \in \{0,\ldots,m\}$ and $z_m > 0$, then $\ell = m$ and $y_k = z_k$ for $k \in \{0,\ldots,\ell\}$.

6 Countability

In the previous section we saw that 'infinite sets' are necessary for the construction of the natural numbers. However, the bijection $\mathbb{N} \to 2\mathbb{N}$, $n \mapsto 2n$, which suggests that there are exactly as many even numbers as natural numbers, encourages caution in dealing with infinity. How can there be room for the odd numbers $1, 3, 5, \ldots$ in $\mathbb{N}$? In this section we consider the concept of infinity again, and, in particular, we show that there is more than one kind of infinity.

A set X is called **finite**, if X is empty or if there are $n \in \mathbb{N}^\times$ and a bijection from $\{1, \ldots, n\}$ to X. If a set is not finite, it is called **infinite**.

6.1 Examples (a) The set $\mathbb{N}$ is infinite.

Proof Suppose, to the contrary, that $\mathbb{N}$ is finite. Since $\mathbb{N}$ is nonempty, there is a bijection φ from $\mathbb{N}$ to $\{1, \ldots, m\}$ for some $m \in \mathbb{N}^\times$. Thus $\psi := \varphi \,|\, \{1, \ldots, m\}$ is an injection from $\{1, \ldots, m\}$ to itself, and so, by Exercise 1, a bijection. Since $\varphi(m+1) \in \{1, \ldots, m\}$ there is, in particular, some $n \in \{1, \ldots, m\}$ such that $\varphi(n) = \psi(n) = \varphi(m+1)$. But this contradicts the injectivity of φ. ■

(b) It is not difficult to see that any infinite system as in Remark 5.2(a) is an infinite set (see Exercise 2). ■

The above discussion suggests that the 'size' of a finite set X can be determined by counting, that is, with a bijection from $\{1, \ldots, n\}$ to X. For infinite sets, of course, this idea will not work. Nonetheless it is very useful to define $\mathrm{Num}(X)$ for both infinite and finite sets by

$$\mathrm{Num}(X) := \begin{cases} 0\,, & X = \emptyset\,, \\ n\,, & n \in \mathbb{N}^\times \text{ and a bijection from } \{1, \ldots, n\} \text{ to } X \text{ exists}\,, \\ \infty\,, & X \text{ is infinite}\,.^1 \end{cases}$$

If X is finite with $\mathrm{Num}(X) = n \in \mathbb{N}$, then we say that X has n elements or that X is an n element set.

6.2 Remark If $m, n \in \mathbb{N}^\times$ and φ and ψ are bijections from X to $\{1, \ldots, m\}$ and $\{1, \ldots, n\}$ respectively, then $\varphi \circ \psi^{-1}$ is a bijection from $\{1, \ldots, n\}$ to $\{1, \ldots, m\}$, and it follows from Exercise 2 that $m = n$. Thus the above definition makes sense, that is, $\mathrm{Num}(X)$ is well defined. ■

---

[1] The symbol ∞ ('infinity') is not a natural number. It is nonetheless useful to (partially) extend addition and multiplication on $\mathbb{N}$ to $\bar{\mathbb{N}} := \mathbb{N} \cup \{\infty\}$ using the conventions $n + \infty := \infty + n := \infty$ for all $n \in \bar{\mathbb{N}}$, and $n \cdot \infty := \infty \cdot n := \infty$ for $n \in \mathbb{N}^\times \cup \{\infty\}$. Further, we define $n < \infty$ for all $n \in \mathbb{N}$.

Permutations

Let X be a finite set. A bijective function from X to itself is called a **permutation** of X. (Note that, by Exercise 1, an injective function from X to itself is necessarily bijective too.) We denote the set of all permutations of X by S_X.

6.3 Proposition *If X is an n element set, then* $\mathrm{Num}(\mathsf{S}_X) = n!$. *That is, there are $n!$ permutations of an n element set.*

Proof We consider first the case when $X = \emptyset$. Then there is a unique function $\emptyset : \emptyset \to \emptyset$. This is function is bijective[2] so the claim is true this case.

We prove the case $n \in \mathbb{N}^\times$ by induction. Since $\mathsf{S}_X = \{\mathrm{id}_X\}$ for any one element set X, we can start the induction with $n_0 = 1$. The induction hypothesis is that for each n element set X, we have $\mathrm{Num}(\mathsf{S}_X) = n!$.

Now let $Y = \{a_1, \ldots, a_{n+1}\}$ be an $(n+1)$ element set. In view of the induction hypothesis, there are, for each $j \in \{1, \ldots, n+1\}$, exactly $n!$ permutations of Y which send a_j to a_1. So in total (see Exercise 5) there are $(n+1)n! = (n+1)!$ permutations of Y. ∎

Equinumerous Sets

Two sets X and Y are called **equinumerous** or **equipotent**, written $X \sim Y$, if there is a bijection from X to Y. If M is a set of sets then $\sim$ is clearly an equivalence relation on M (see Proposition 3.6).

A set X is called **countably infinite** if $X \sim \mathbb{N}$, and we say X is **countable** if $X \sim \mathbb{N}$ or X is finite. Finally, X is **uncountable** if X is not countable.

6.4 Remark If $X \sim \mathbb{N}$ then it follows from Example 6.1(a) that X is not finite. Thus a set cannot be both finite and countably infinite. ∎

Of course, the set of natural numbers is countably infinite. More interesting is the observation that proper subsets of countably infinite sets can themselves be countably infinite, as the example of the set of even natural numbers $2\mathbb{N} = \{\, 2n \ ; \ n \in \mathbb{N} \,\}$ shows. In the other direction, we will meet, in the next section, countably infinite sets which properly contain $\mathbb{N}$.

Before we investigate further the properties of countable sets, we show the existence of uncountable sets. To that end we prove the following fundamental result due to G. Cantor.

---

[2]This is vacuously true since none of the conditions in the definition of bijective is ever tested. The real intention here is not to make $n = 0$ a special case, thus avoiding cumbersome case distinctions in upcoming proofs.

6.5 Theorem *There is no surjection from a set X to $\mathcal{P}(X)$.*

Proof For a function $\varphi\colon X \to \mathcal{P}(X)$, consider the subset

$$A := \{\, x \in X \ ; \ x \notin \varphi(x) \,\}$$

of X. We show that A is not in the image of φ. Indeed if $y \in X$ with $\varphi(y) = A$, then either $y \in A$ and hence $y \notin \varphi(y) = A$, a contradiction, or $y \notin A = \varphi(y)$ and so $y \in A$ which is also a contradiction. This shows that φ is not surjective. ■

An immediate consequence of this theorem is the existence of uncountable sets.

6.6 Corollary *$\mathcal{P}(\mathbb{N})$ is uncountable.*

Countable Sets

We now return to countable sets and prove some seemingly obvious propositions:

6.7 Proposition *Any subset of a countable set is countable.*

Proof (a) Let X be a countable set and $A \subseteq X$. We are done if A is finite (see Exercise 9), so we can assume that A is infinite, in which case X must be countably infinite. That is, there are a bijection φ from X to $\mathbb{N}$ and a bijection $\psi := \varphi \,|\, A$ from A to $\varphi(A)$. Therefore we can assume, without loss of generality, that $X = \mathbb{N}$ and A is an infinite subset of $\mathbb{N}$.

(b) We define recursively a function $\alpha : \mathbb{N} \to A$ by

$$\alpha(0) := \min(A) \ , \quad \alpha(n+1) := \min\{\, m \in A \ ; \ m > \alpha(n) \,\} \ .$$

Because of Proposition 5.5 and the supposition that $\operatorname{Num}(A) = \infty$, $\alpha : \mathbb{N} \to A$ is well defined. It is clear that

$$\alpha(n+1) > \alpha(n) \ , \quad \alpha(n+1) \geq \alpha(n) + 1 \ , \qquad n \in \mathbb{N} \ . \tag{6.1}$$

(c) We have $\alpha(n+k) > \alpha(n)$ for $n \in \mathbb{N}$ and $k \in \mathbb{N}^\times$. This follows easily from the first inequality of (6.1) by induction on k. In particular, α is injective.

(d) We verify the surjectivity of α. First we prove by induction that

$$\alpha(m) \geq m \ , \qquad m \in \mathbb{N} \ . \tag{6.2}$$

For $m = 0$, this is certainly true. The induction step $m \to m+1$ follows from the second inequality of (6.1) and the induction hypothesis,

$$\alpha(m+1) \geq \alpha(m) + 1 \geq m + 1 \ .$$

Now let $n_0 \in A$ be given. We need to find some $m_0 \in \mathbb{N}$ such that $\alpha(m_0) = n_0$. Consider the set $B := \{ m \in \mathbb{N} \ ; \ \alpha(m) \geq n_0 \}$. Because of (6.2), B is not empty. So there exists, by Proposition 5.5, some $m_0 := \min(B)$. If $m_0 = 0$, then

$$\min(A) = \alpha(0) \geq n_0 \geq \min(A) ,$$

and hence $n_0 = \alpha(0)$. So we can suppose that $n_0 > \min(A)$ and so $m_0 \in \mathbb{N}^\times$. But then $\alpha(m_0 - 1) < n_0 \leq \alpha(m_0)$ and, by the definition of α, we have $\alpha(m_0) = n_0$. ■

6.8 Proposition *A countable union of countable sets is countable.*

Proof For each $n \in \mathbb{N}$, let X_n be a countable set. By Proposition 6.7, we can assume that the X_n are countably infinite and pairwise disjoint. Thus we have $X_n = \{x_{n,k} \ ; \ k \in \mathbb{N}\}$ with $x_{n,k} \neq x_{n,j}$ for $k \neq j$, that is, $x_{n,k}$ is the image of $k \in \mathbb{N}$ under a bijection from $\mathbb{N}$ to X_n. Now we order the elements of $X := \bigcup_{n=0}^\infty X_n$ as indicated by the arrows in the 'infinite matrix' below. This induces a bijection from X to $\mathbb{N}$.

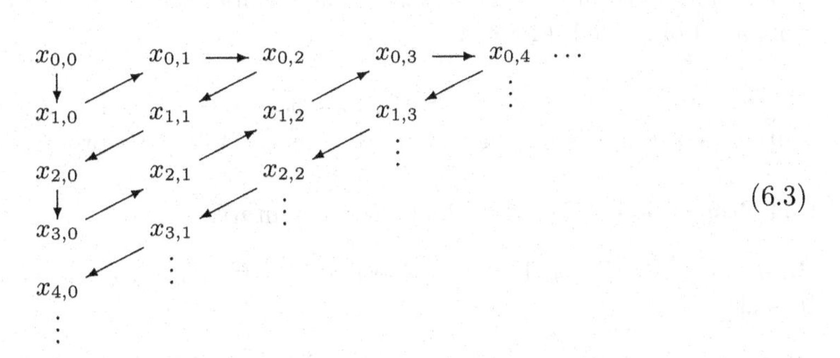

(6.3)

We leave to the reader the task of defining this bijection explicitly. ■

6.9 Proposition *A finite product of countable sets is countable.*

Proof Let X_j, $j = 0, 1, \ldots, n$ be countable sets, and $X := \prod_{j=0}^n X_j$. By definition $X = \left(\prod_{j=0}^{n-1} X_j\right) \times X_n$, so it suffices to consider the case $n = 1$. Thus we suppose $X := X_0 \times X_1$ with X_0 and X_1 countably infinite. Write $X_0 = \{ y_k \ ; \ k \in \mathbb{N}\}$ and $X_1 = \{ z_k \ ; \ k \in \mathbb{N}\}$, and set $x_{j,k} := (y_j, z_k)$ for $j, k \in \mathbb{N}$. Using this notation we have $X = \{ x_{j,k} \ ; \ j, k \in \mathbb{N}\}$ and so we can use (6.3) again to define a bijection from X to $\mathbb{N}$. ■

Infinite Products

Proposition 6.9 is no longer correct if we allow 'infinite products' of countable sets. To make this claim more precise, we need to explain first what an 'infinite product' is. Suppose that $\{ X_\alpha \ ; \ \alpha \in \mathsf{A}\}$ is a family of subsets of a fixed set. Then the **Cartesian product** $\prod_{\alpha \in \mathsf{A}} X_\alpha$ is defined to be the set of all functions

$\varphi\colon \mathsf{A} \to \bigcup_{\alpha\in\mathsf{A}} X_\alpha$ such that $\varphi(\alpha) \in X_\alpha$ for each $\alpha \in \mathsf{A}$. In place of φ one often writes $\{\, x_\alpha \;;\; \alpha \in \mathsf{A} \,\}$, where, of course, $x_\alpha := \varphi(\alpha)$.

In the special case that $\mathsf{A} = \{1,\ldots,n\}$ for some $n \in \mathbb{N}^\times$, $\prod_{\alpha\in\mathsf{A}} X_\alpha$ is clearly identical to the product $\prod_{k=1}^n X_k$ which was introduced in Section 2. If $X_\alpha = X$ for each $\alpha \in \mathsf{A}$, then we write $X^{\mathsf{A}} := \prod_{\alpha\in\mathsf{A}} X_\alpha$.

6.10 Remark It is clear that $\prod_{\alpha\in\mathsf{A}} X_\alpha = \emptyset$ if one (or more) of the X_α is empty. On the other hand, even if $X_\alpha \neq \emptyset$ for each $\alpha \in \mathsf{A}$, it is not possible to prove that $\prod_{\alpha\in\mathsf{A}} X_\alpha$ is nonempty using the axioms of set theory we have seen so far. To do that one needs to know that a function $\varphi\colon \mathsf{A} \to \bigcup_{\alpha\in\mathsf{A}} X_\alpha$ exists such that $\varphi(\alpha) \in X_\alpha$ for each $\alpha \in \mathsf{A}$, that is, a rule which chooses a single element from each set X_α. To ensure that such a function exists one needs the **axiom of choice**, which we formulate as follows: For any family of sets $\{\, X_\alpha \;;\; \alpha \in \mathsf{A} \,\}$,

$$\prod_{\alpha\in\mathsf{A}} X_\alpha \neq \emptyset \iff (X_\alpha \neq \emptyset \quad \forall\, \alpha \in \mathsf{A}) \ .$$

In the following we will use this naturally appearing axiom without comment. Readers who are interested in the foundations of mathematics are directed to the literature, for example, [Ebb77] and [FP85]. ■

Surprisingly, in contrast to Proposition 6.9, countably infinite products of *finite* sets are, in general, not countable, as the following proposition shows.

6.11 Proposition *The set $\{0,1\}^{\mathbb{N}}$ is uncountable.*

Proof Let $A \in \mathcal{P}(\mathbb{N})$. Then the characteristic function χ_A is an element of $\{0,1\}^{\mathbb{N}}$. It is clear that the function

$$\mathcal{P}(\mathbb{N}) \to \{0,1\}^{\mathbb{N}} \ , \quad A \mapsto \chi_A \tag{6.4}$$

is injective. For $\varphi \in \{0,1\}^{\mathbb{N}}$, let $A(\varphi) := \varphi^{-1}(1) \in \mathcal{P}(\mathbb{N})$. Then $\chi_{A(\varphi)} = \varphi$. This shows that the function (6.4) is surjective. (See also Exercise 3.6.) Thus $\{0,1\}^{\mathbb{N}}$ and $\mathcal{P}(\mathbb{N})$ are equinumerous and the claim follows from Corollary 6.6. ■

6.12 Corollary *The sets $\{0,1\}^{\mathbb{N}}$ and $\mathcal{P}(\mathbb{N})$ are equinumerous.*

Exercises

1 Let $n \in \mathbb{N}^\times$. Prove that any injective function from $\{1,\ldots,n\}$ to itself is bijective. (Hint: Use induction on n. Let $f\colon \{1,\ldots,n+1\} \to \{1,\ldots,n+1\}$ be an injective function and $k := f(n+1)$. Consider the functions

$$g(j) := \begin{cases} n+1 \ , & j = k \ , \\ k \ , & j = n+1 \ , \\ j & \text{otherwise} \ , \end{cases}$$

together with $h := g \circ f$ and $h \,|\, \{1,\ldots,n\}$.)

2 Prove the following:

(a) Let $m, n \in \mathbb{N}^\times$. Then there is a bijective function from $\{1, \ldots, m\}$ to $\{1, \ldots, n\}$ if and only if $m = n$.

(b) If M is an infinite system, then $\operatorname{Num}(M) = \infty$ (Hint: Exercise 1).

3 Show that the number of m element subsets of an n element set is $\binom{n}{m}$. (Hint: Let N be an n element set and M an m element subset of N. From Proposition 6.3 deduce that there are $m!\,(n-m)!$ bijections from $\{1, \ldots, n\}$ to N such that $\{1, \ldots, m\}$ goes to M.)

4 Let M and N be finite sets. How many injective functions are there from M to N?

5 Let $X_0, \ldots, X_m$ be finite sets. Show that $X := \bigcup_{j=0}^m X_j$ is also finite and that

$$\operatorname{Num}(X) \leq \sum_{j=0}^{m} \operatorname{Num}(X_j) \ .$$

When do we get equality?

6 Let $X_0, \ldots, X_m$ be finite sets. Prove that $X := \prod_{j=0}^m X_j$ is also finite and that

$$\operatorname{Num}(X) = \prod_{j=0}^{m} \operatorname{Num}(X_j) \ .$$

7 Show that a nonempty set X is countable if and only if there is a surjection from $\mathbb{N}$ to X.

8 Let X be a countable set. Show that the set of all finite subsets of X is countable. (Hint: Consider the functions $X^n \to \mathcal{E}_n(X),\ (x_1, \ldots, x_n) \mapsto \{x_1, \ldots, x_n\}$ where $\mathcal{E}_n(X)$ is the set of all subsets with at most n elements.)

9 Show that any subset of a finite set is finite.

7 Groups and Homomorphisms

In Theorem 5.3 we defined the difference $n - m$ of two natural numbers m and n when $m \leq n$. We defined also the quotient n/m of two natural numbers m and n when m is a divisor of n. In both cases, the given restrictions on m and n are needed to ensure that the difference and the quotient are once again natural numbers. If we want to define the 'difference' $n - m$ or the 'quotient' n/m of arbitrary natural numbers m and n, then we have to leave the realm of natural numbers. In Sections 9–11 we will construct new kinds of numbers and so extend the set of natural numbers to larger number systems in which these operations can be used (almost) without restriction.

Of course these new number systems must be constructed so that the usual rules of addition and multiplication hold. For this purpose, it is extremely useful to investigate these rules themselves, independent of any connection to a particular number system. Such an investigation also provides further practice in the logical deduction of propositions from definitions and axioms.

A thorough discussion of the questions appearing here and in the following sections is algebra rather than analysis, and so our presentation is relatively short and we prove only a few of the most important theorems. Our goal is to be able to recognize general algebraic structures which appear over and over again in various disguises. The derivation of a large number of arithmetic rules from a small number of axioms will allow us to bring order to an otherwise huge mass of formulas and results, and to keep our attention on the essential. The propositions that we derive from the axioms are true whenever the axioms are true, independent of the context in which they hold. Things that have been proved once, do not need to be proved again for each special case.

In this and the following sections we give only a few concrete examples of the new concepts. We are primarily interested in providing a language and hope that the reader will recognize in later sections the usefulness of this language and will see also the mathematical content behind the formalism.

Groups

Groups are systems consisting of one set, one operation and three axioms. Since they have such a simple algebraic structure, they occur everywhere in mathematics.

A pair $(G, \odot)$ consisting of a nonempty set G and an operation $\odot$ is called a **group** if the following holds:

(G_1) $\odot$ is associative.

(G_2) $\odot$ has an identity element e.

(G_3) Each $g \in G$ has an **inverse** $h \in G$ such that $g \odot h = h \odot g = e$.

A group $(G, \odot)$ is called **commutative** or **Abelian** if $\odot$ is a commutative operation on G. If the operation is clear from the context, we often write simply G for $(G, \odot)$.

7.1 Remarks Let $G = (G, \odot)$ be a group.

(a) By Proposition 4.11, the identity element e is unique.

(b) Each $g \in G$ has a unique inverse which we denote (temporarily) by $g^\flat$. In particular $e^\flat = e$.

Proof In view of (G_3), only the uniqueness needs to be proved. Suppose that h and k are inverses of $g \in G$, that is, $g \odot h = h \odot g = e$ and $g \odot k = k \odot g = e$. Then

$$h = h \odot e = h \odot (g \odot k) = (h \odot g) \odot k = e \odot k = k\ ,$$

which shows the uniqueness.

Since $e \odot e = e$ the second claim is clear. ■

(c) For each pair $a, b \in G$, there is a unique $x \in G$ such that $a \odot x = b$ and a unique $y \in G$ such that $y \odot a = b$. That is, the 'equations' $a \odot x = b$ and $y \odot a = b$ have unique solutions.

Proof Let $a, b \in G$ be given. If we set $x := a^\flat \odot b$ and $y := b \odot a^\flat$, then $a \odot x = b$ and $y \odot a = b$. This proves the existence statement. To verify the uniqueness of the solution of the first equation, suppose that $x, z \in G$ are such that $a \odot x = b$ and $a \odot z = b$. Then

$$x = (a^\flat \odot a) \odot x = a^\flat \odot (a \odot x) = a^\flat \odot b = a^\flat \odot (a \odot z) = (a^\flat \odot a) \odot z = z\ .$$

A similar argument for the equation $y \odot a = b$ completes the proof. ■

(d) For each $g \in G$, we have $(g^\flat)^\flat = g$.

Proof Directly from the definition of the inverse we get the equations

$$\begin{aligned} g \odot g^\flat &= g^\flat \odot g = e\ , \\ (g^\flat)^\flat \odot g^\flat &= g^\flat \odot (g^\flat)^\flat = e\ , \end{aligned}$$

which, together with (c), imply that $g = (g^\flat)^\flat$. ■

(e) Let H be a nonempty set with an associative operation $\circledast$ and identity element e. If every element $h \in H$ has a **left inverse** $\overline{h}$ such that $\overline{h} \circledast h = e$, then $(H, \circledast)$ is a group and $\overline{h} = h^\flat$. Similarly, if every element $h \in H$ has a **right inverse** $\underline{h}$ such that $h \circledast \underline{h} = e$, then $(H, \circledast)$ is a group and $\underline{h} = h^\flat$.

Proof Suppose h is in H, $\overline{h}$ is a left inverse of h, and $\overline{\overline{h}}$ is a left inverse of $\overline{h}$. Then $\overline{\overline{h}} \circledast \overline{h} = e$ and so

$$h = e \circledast h = (\overline{\overline{h}} \circledast \overline{h}) \circledast h = \overline{\overline{h}} \circledast (\overline{h} \circledast h) = \overline{\overline{h}} \circledast e = \overline{\overline{h}}\ ,$$

from which $h \circledast \overline{h} = e$ follows. Therefore $\overline{h}$ is also a right inverse of h, and thereby an inverse of h. Similarly one shows that, if every element has a right inverse, then each right inverse is also a left inverse. ■

(f) For arbitrary group elements g and h, $(g \odot h)^\flat = h^\flat \odot g^\flat$.

Proof Since $(h^\flat \odot g^\flat) \odot (g \odot h) = h^\flat \odot (g^\flat \odot g) \odot h = h^\flat \odot e \odot h = h^\flat \odot h = e$, the claim follows from (e). ■

In order to show that an axiom system is free of contradictions, it suffices to exhibit some mathematical system which satisfies the axioms. In the case of the group axioms (G_1)–(G_3), this is quite easy to do, as the following examples show.

7.2 Examples **(a)** Let $G := \{e\}$ be a one element set. Then $\{G, \odot\}$ is an Abelian group, the **trivial group**, with the (only possible) operation $e \odot e = e$.

(b) Let $G := \{a, b\}$ be a set with operation $\odot$ defined by the table on the right. Then $(G, \odot)$ is an Abelian group.

$\odot$	a	b
a	a	b
b	b	a

(c) Let X be a nonempty set and S_X the set of all bijections from X to itself. Then $\mathsf{S}_X := (\mathsf{S}_X, \circ)$ is a group with identity element id_X when $\circ$ denotes the composition of functions. Further, the inverse function f^{-1} is the inverse of $f \in \mathsf{S}_X$ in the group. In view of Exercise 4.3, S_X is, in general, not commutative. When X is finite, the elements of S_X are called permutations (see Section 6) and S_X is called the **permutation group** of X.

(d) Let X be a nonempty set and $(G, \odot)$ a group. With the induced operation $\odot$ as in Example 4.12, $(G^X, \odot)$ is a group. The inverse of $f \in G^X$ is the function

$$f^\flat : X \to G , \quad x \mapsto \bigl(f(x)\bigr)^\flat .$$

In particular, for $m \geq 2$, G^m with the operation

$$(g_1, \ldots, g_m) \odot (h_1, \ldots, h_m) = (g_1 \odot h_1, \ldots, g_m \odot h_m)$$

is a group.

(e) Let $G_1, \ldots, G_m$ be groups. Then $G_1 \times \cdots \times G_m$ with operation defined analogously to (d) is a group called the **direct product** of $G_1, \ldots, G_m$. ■

Subgroups

Let $G = (G, \odot)$ be a group and H a nonempty subset of G which is closed under the operation $\odot$, that is,

(SG_1) $H \odot H \subseteq H$.

If, in addition,

(SG_2) $h^\flat \in H$ for all $h \in H$,

then $H := (H, \odot)$ is itself a group and is called a **subgroup** of G. Here we use the same symbol $\odot$ for the restriction of the operation to H. Since H is nonempty, there is some $h \in H$ and so, from (SG_1) and (SG_2), $e = h^\flat \odot h$ is also in H.

7.3 Examples Let $G = (G, \odot)$ be a group.

(a) The trivial subgroup $\{e\}$ and G itself are subgroups of G, the smallest and largest subgroups with respect to inclusion (see Example 4.4.(b)).

(b) If $H_\alpha,\ \alpha \in \mathsf{A}$ are subgroups of G, then $\bigcap_\alpha H_\alpha$ is also a subgroup of G. ∎

Cosets

Let N be a subgroup of G and $g \in G$. Then $g \odot N$ is the **left coset** and $N \odot g$ is the **right coset** of $g \in G$ with respect to N. If we define

$$g \sim h :\Longleftrightarrow g \in h \odot N \ , \tag{7.1}$$

then $\sim$ is an equivalence relation on G: Indeed $\sim$ is reflexive because $e \in N$. If $g \in h \odot N$ and $h \in k \odot N$, then

$$g \in (k \odot N) \odot N = k \odot (N \odot N) = k \odot N \ ,$$

since, of course,

$$N \odot N = N \ . \tag{7.2}$$

Thus $\sim$ is transitive. If $g \in h \odot N$, then there is some $n \in N$ with $g = h \odot n$. Then it follows from (SG_2) that $h = g \odot n^\flat \in g \odot N$. Thus $\sim$ is also symmetric and (7.1) defines an equivalence relation on G. For the equivalence classes $[\cdot]$ with respect to $\sim$, we have

$$[g] = g \odot N \ , \qquad g \in G \ . \tag{7.3}$$

For this reason, we denote $G/\!\sim$ by G/N, and call G/N the **set of left cosets of** G **modulo** N.

Of particular importance are subgroups N such that

$$g \odot N = N \odot g \ , \qquad g \in G \ . \tag{7.4}$$

Such a subgroup is called a **normal** subgroup of G. In this case one calls $g \odot N$ the **coset of** g **modulo** N since each left coset is a right coset and vice versa.

For a normal subgroup N of G it follows from (7.2), (7.4) and the associativity of the operation, that

$$(g \odot N) \odot (h \odot N) = g \odot (N \odot h) \odot N = (g \odot h) \odot N \ , \qquad g, h, \in G \ .$$

This shows that there is a well defined operation on G/N, induced from $\odot$, such that

$$(G/N) \times (G/N) \to G/N \ , \quad (g \odot N, h \odot N) \mapsto (g \odot h) \odot N \ . \tag{7.5}$$

We will use the same symbol $\odot$ for this induced operation.

7.4 Proposition *Let G be a group and N a normal subgroup of G. Then G/N with the induced operation is a group, the* **quotient group of G modulo N.**

Proof The reader can easily check that the induced operation is associative. Since $(e \odot N) \odot (g \odot N) = (e \odot g) \odot N = g \odot N$, the identity element of G/N is $N = e \odot N$. Since also

$$(g^\flat \odot N) \odot (g \odot N) = (g^\flat \odot g) \odot N = e \odot N = N$$

the claim follows from Remark 7.1(e). ■

7.5 Remarks **(a)** In the notation of (7.3), $[e] = N$ is the identity element of G/N and $[g]^\flat = [g^\flat]$ is the inverse of $[g] \in G/N$. Because of (7.3) and (7.5) we have

$$[g] \odot [h] = [g \odot h] , \qquad g, h \in G .$$

In other words, to combine two cosets with the operation $\odot$, one can choose a representative of each coset, combine these elements using $\odot$ and then take the coset which contains the resulting element. Since the operation on G/N is well defined, the final result is independent of the particular choice of representatives.

(b) Any subgroup N of an Abelian group G is normal and so G/N is a group. Of course, G/N is also Abelian. ■

Homomorphisms

Among functions between groups, those which preserve the group structure are of particular interest.

Let $G = (G, \odot)$ and $G' = (G', \circledast)$ be groups. A function $\varphi \colon G \to G'$ is called a **(group) homomorphism** if

$$\varphi(g \odot h) = \varphi(g) \circledast \varphi(h) , \qquad g, h \in G .$$

A homomorphism from G to itself is called a **(group) endomorphism**.

7.6 Remarks **(a)** Let e and e' be the identity elements of G and G' respectively, and let $\varphi \colon G \to G'$ be a homomorphism. Then

$$\varphi(e) = e' \quad \text{and} \quad \bigl(\varphi(g)\bigr)^\flat = \varphi(g^\flat) , \qquad g \in G .$$

Proof From $e' \circledast \varphi(e) = \varphi(e) = \varphi(e \odot e) = \varphi(e) \circledast \varphi(e)$ and Remark 7.1(c) it follows that $\varphi(e) = e'$. Suppose $g \in G$. Then $e' = \varphi(e) = \varphi(g^\flat \odot g) = \varphi(g^\flat) \circledast \varphi(g)$ and, similarly, $e' = \varphi(g) \circledast \varphi(g^\flat)$. Thus, from Remark 7.1(b), we get $\bigl(\varphi(g)\bigr)^\flat = \varphi(g^\flat)$. ■

(b) Let $\varphi\colon G \to G'$ be a homomorphism. The **kernel** of φ, $\ker(\varphi)$, defined by

$$\ker(\varphi) := \varphi^{-1}(e') = \{\, g \in G \; ; \; \varphi(g) = e' \,\} \; ,$$

is a normal subgroup of G.

Proof For all $g, h \in \ker(\varphi)$ we have

$$\varphi(g \odot h) = \varphi(g) \circledast \varphi(h) = e' \circledast e' = e' \; .$$

Thus (SG_1) is satisfied. Because $\varphi(g^\flat) = \bigl(\varphi(g)\bigr)^\flat = (e')^\flat = e'$, (SG_2) also holds, and so $\ker(\varphi)$ is a subgroup of G. Let $h \in g \odot \ker(\varphi)$. Then there is some $n \in G$ such that $\varphi(n) = e'$ and $h = g \odot n$. For $m := g \odot n \odot g^\flat$, we have

$$\varphi(m) = \varphi(g) \circledast \varphi(n) \circledast \varphi(g^\flat) = \varphi(g) \circledast \varphi(g^\flat) = e' \; ,$$

and hence $m \in \ker(\varphi)$. Since $m \odot g = g \odot n = h$, this implies that $h \in \ker(\varphi) \odot g$. Similarly one can show $\ker(\varphi) \odot g \subseteq g \odot \ker(\varphi)$, and so $\ker(\varphi)$ is a normal subgroup of G. ■

(c) Let $\varphi\colon G \to G'$ be a homomorphism and $N := \ker(\varphi)$. Then

$$g \odot N = \varphi^{-1}\bigl(\varphi(g)\bigr) \; , \qquad g \in G \; ,$$

and so

$$g \sim h \Longleftrightarrow \varphi(g) = \varphi(h) \; , \qquad g, h \in G \; ,$$

where $\sim$ denotes the equivalence relation (7.1).

Proof For $h \in g \odot N$ we have

$$\varphi(h) \in \varphi(g \odot N) = \varphi(g) \circledast \varphi(N) = \varphi(g) \circledast \{e'\} = \{\varphi(g)\} \; ,$$

and so $h \in \varphi^{-1}\bigl(\varphi(g)\bigr)$. Conversely if $h \in \varphi^{-1}\bigl(\varphi(g)\bigr)$, that is, $\varphi(h) = \varphi(g)$, then

$$\varphi(g^\flat \odot h) = \varphi(g^\flat) \circledast \varphi(h) = \bigl(\varphi(g)\bigr)^\flat \circledast \varphi(g) = e' \; ,$$

which means that $g^\flat \odot h \in N$ and hence $h \in g \odot N$. ■

(d) A homomorphism is injective if and only if its kernel is trivial, that is, $\ker(\varphi) = \{e\}$.

Proof This follows directly from (c). ■

(e) The image $\operatorname{im}(\varphi)$ of a homomorphism $\varphi\colon G \to G'$ is a subgroup of G'. ■

7.7 Examples **(a)** The constant function $G \to G'$, $g \mapsto e'$ is a homomorphism, the **trivial** homomorphism.

(b) The identity function $\mathrm{id}_G : G \to G$ is an endomorphism.

(c) Compositions of homomorphisms (endomorphisms) are homomorphisms (endomorphisms).

(d) Let N be a normal subgroup of G. Then the quotient function

$$p\colon G \to G/N \ , \quad g \mapsto g \odot N$$

is a surjective homomorphism, the **quotient homomorphism**, with $\ker(p) = N$.

Proof Since N is a normal subgroup of G, the quotient group G/N is well defined. Because of (7.1) and Proposition 4.1, the quotient function p is well defined, and Remark 7.5(a) shows that p is a homomorphism. Since N is the identity element of G/N, $\ker(p) = N$.

(e) If $\varphi\colon G \to G'$ is a bijective homomorphism, then so is $\varphi^{-1}\colon G' \to G$. ∎

Isomorphisms

A homomorphism $\varphi\colon G \to G'$ is called a (**group**) **isomorphism** from G to G' if φ is bijective. In this circumstance, we say that the groups G and G' are **isomorphic** and write $G \cong G'$. An isomorphism from G to itself, that is, a bijective endomorphism, is called a (**group**) **automorphism** of G.

7.8 Examples **(a)** The identity function $\mathrm{id}_G\colon G \to G$ is an automorphism. If φ and ψ are automorphisms of G, then so are $\varphi \circ \psi$ and φ^{-1}. It follows easily from this that the set of all automorphisms of a group G, with composition as operation, forms a group, the **automorphism group** of G. This is a subgroup of the permutation group S_G.

(b) For each $a \in G$, the function $g \mapsto a \odot g \odot a^\flat$ is an automorphism of G.

(c) Let $\varphi\colon G \to G'$ be a homomorphism. Then there is a unique injective homomorphism $\widetilde{\varphi}\colon G/\ker(\varphi) \to G'$ such that the diagram

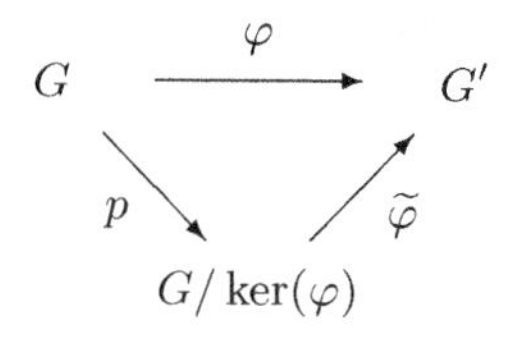

is commutative. If φ is surjective, then $\widetilde{\varphi}$ is an isomorphism.

Proof It follows from Remark 7.6(c) and Example 4.2(c) that there is a unique injective function $\widetilde{\varphi}$ which makes the diagram commutative, and that $\mathrm{im}(\varphi) = \mathrm{im}(\widetilde{\varphi})$. It is easy to check that $\widetilde{\varphi}$ is a homomorphism. ∎

(d) Let $(G, \odot)$ be a group, G' a nonempty set, and $\varphi\colon G \to G'$ a bijection from G to G'. Define an operation $\circledast$ on G' by

$$g' \circledast h' := \varphi^{-1}(g') \odot \varphi^{-1}(h') \ , \qquad g', h' \in G' \ .$$

Then $(G', \circledast)$ is a group and φ is an isomorphism from G to G'. The operation $\circledast$ is called the **operation on G' induced from $\odot$ via φ** .

(e) If $G = \{e\}$ and $G' = \{e'\}$ are trivial groups, then G and G' are isomorphic.

(f) Let G be the group of Example 7.2(b) and G' the group produced when the symbols a and b are interchanged in the table. Then G and G' are isomorphic. More precisely, the operation on G' is induced from the operation G via the function $\varphi\colon \{a,b\} \to \{a,b\}$ defined by $\varphi(a) := b$ and $\varphi(b) := a$.

(g) Let X and Y be nonempty sets and $\varphi\colon X \to Y$ a bijective function. Then

$$\widehat{\varphi}\colon \mathsf{S}_X \to \mathsf{S}_Y\ , \quad f \mapsto \varphi \circ f \circ \varphi^{-1}$$

is an isomorphism from the permutation group S_X to the permutation group S_Y. ■

If φ is an isomorphism from the group $(G, \odot)$ to the group $(G', \circledast)$, then even though the groups may differ in the labeling of their elements, they have identical group structure. For example, if g and h are two elements of G, then to calculate $g \odot h$ one can just as well calculate $\varphi(g) \circledast \varphi(h)$ in G', and then $g \odot h$ is the image of $\varphi(g) \circledast \varphi(h)$ under the inverse isomorphism φ^{-1}. In practice it may be much easier to work with $(G', \circledast)$ than with $(G, \odot)$. (See, in particular, Sections 9 and 10.)

From the viewpoint of group theory, isomorphic groups are essentially identical. In fact, isomorphism $\cong$ is an equivalence relation on any set $\mathcal{G}$ of groups, as is easy to verify. Hence $\mathcal{G}$ can be partitioned using $\cong$ into equivalence classes, called **isomorphism classes**. It suffices then to investigate the set $\mathcal{G}/\!\cong$ of isomorphism classes rather than $\mathcal{G}$ itself. In other words, one 'identifies' (makes identical) isomorphic groups. This is the sense in which one speaks of *the* trivial group, since, by Example 7.8(e), any two trivial groups are isomorphic. Similarly, there is (up to isomorphism) only *one* group of order[1] two, that is, with exactly two elements (see Example 7.8(f)). If $n \in \mathbb{N}^\times$, then, by Example 7.8(g), there is only one permutation group S_X with $\mathrm{Num}(X) = n$ to consider, for example, *the* **permutation group** (or the **symmetric group**) **of order** $n!$,

$$\mathsf{S}_n := \mathsf{S}_{\{1,\ldots,n\}}\ ,$$

that is, the permutation group on the set $\{1,\ldots,n\}$. (See Proposition 6.3.)

> Convention In the following, we usually denote the operation in a group G by $\cdot$, and, instead of $x \cdot y$, write simply xy for $x, y \in G$. With this 'multiplicative' notation, the operation is called (**group**) **multiplication**, and for $x^\flat$ we write x^{-1} ('x inverse'). If the group is Abelian, it is common to use 'additive notation' meaning that the group operation is written $+$ and is called **addition**, and the inverse $x^\flat$ of x is written $-x$ ('negative x').

The reader is again reminded that notation is not important, it is the axioms that matter. The same symbol can have completely different meanings in different

---

[1] The **order** of a finite group is the number of its elements.

contexts, even when the same axioms apply. The use of familiar symbols, such as $+$ or $\cdot$, should not lead the reader to think that the familiar context is intended. One has to be clear about which axioms are in play and use only those rules which follow from them.

That a single symbol can have various context-dependent meanings may seem illogical and confusing to the beginner. Nonetheless it makes possible an elegant and concise presentation of complex ideas, and avoids overwhelming the reader with a multitude of different symbols.

Exercises

1 Let N be a subgroup of a finite group G. Show that $\operatorname{Num}(G) = \operatorname{Num}(N) \cdot \operatorname{Num}(G/N)$ so, in particular, the order of a subgroup divides the order of the group.

2 Verify the claims in Examples 7.2(c) and (d).

3 Prove the claim in Example 7.3(b) and show that the intersection of a set of normal subgroups is also normal.

4 Prove Remark 7.6(e). Is $\operatorname{im}(\varphi)$ a normal subgroup of G'?

5 Let $\varphi\colon G \to G'$ be a homomorphism and N' a normal subgroup of G'. Show that $\varphi^{-1}(N')$ is a normal subgroup of G.

6 Let G be a group and X a nonempty set. Then G **acts (from the left) on** X if there is a function

$$G \times X \to X\ , \quad (g, x) \mapsto g \cdot x$$

such that the following hold:

(GA_1) $e \cdot x = x$ for all $x \in X$.

(GA_2) $g \cdot (h \cdot x) = (gh) \cdot x$ for all $g, h \in G$ and $x \in X$.

(a) For each $g \in G$, show that $x \mapsto g \cdot x$ is a bijection on X with inverse $x \mapsto g^{-1} \cdot x$.

(b) For $x \in X$, $G \cdot x$ is called the **orbit** of x (under the action of G). Show that the relation 'y is in the orbit of x' is an equivalence relation on X.

(c) Show that if H is a subgroup of G, then $(h, g) \mapsto h \cdot g$ and $(h, g) \mapsto hgh^{-1}$ define actions of H on G.

(d) Show that

$$\mathsf{S}_m \times \mathbb{N}^m \to \mathbb{N}^m\ , \quad (\sigma, \alpha) \mapsto \sigma \cdot \alpha := (\alpha_{\sigma(1)}, \ldots, \alpha_{\sigma(m)})$$

defines an action of S_m on $\mathbb{N}^m$.

7 Let $G = (G, \odot)$ be a finite group of order m with identity element e. Show that for each $g \in G$, there is a least natural number $k > 0$ such that

$$g^k := \bigodot_{j=1}^{k} g = e\ .$$

Show that $g^m = e$ for all $g \in G$. (Hint: Exercise 1.)

8 The tables below define three operations on the set $G = \{e, a, b, c\}$.

$\odot$	e	a	b	c
e	e	a	b	c
a	a	e	c	b
b	b	c	e	a
c	c	b	a	e

$\circledast$	e	a	b	c
e	e	a	b	c
a	a	e	c	b
b	b	c	a	e
c	c	b	e	a

$\oplus$	e	a	b	c
e	e	a	b	c
a	a	b	c	e
b	b	c	e	a
c	c	e	a	b

(a) Verify that $(G, \circledast)$ and $(G, \oplus)$ are isomorphic groups.

(b) Show that the groups $(G, \odot)$ and $(G, \circledast)$ are not isomorphic.

(c) Determine all other possible group structures on G. Sort these groups into isomorphism classes.

9 Show that S_3 is not Abelian.

10 Let G and H be groups, and let

$$p \colon G \times H \to G\ , \quad (g, h) \mapsto g$$

be the projection onto the first factor. Show that p is a surjective homomorphism. Set $H' := \ker(p)$. Show that $(G \times H)/H'$ and G are isomorphic groups.

11 Let G be a set with an operation $\odot$ and identity element. For $g \in G$, define the function $Lg \colon G \to G,\ h \mapsto g \odot h$, called **left translation** by g. Suppose that

$$L := \{\, Lg\ ;\ g \in G \,\} \subseteq \mathsf{S}_G\ ,$$

that is, each Lg is bijective. Prove that

$$(G, \odot) \text{ is a group} \iff L \text{ is a subgroup of } \mathsf{S}_G\ .$$

8 Rings, Fields and Polynomials

In this section we consider sets on which two operations are defined. Here we assume that, with respect to one of the operations, the set forms an Abelian group and that the two operations satisfy an appropriate 'distributive law'. This leads to the concepts of 'rings' and 'fields', which formalize the rules of arithmetic. As particularly important examples of rings we consider power series rings and polynomial rings in one (and many) indeterminates and derive some of their fundamental properties. Polynomial functions are relatively easy to work with and are important in analysis because 'complicated functions can be approximated arbitrarily well by polynomials', a claim that we will make more precise later.

Rings

A triple $(R, +, \cdot)$ consisting of a nonempty set R and operations, **addition** $+$ and **multiplication** $\cdot$, is called a **ring** if:

(R_1) $(R, +)$ is an Abelian group.

(R_2) Multiplication is associative.

(R_3) The **distributive law** holds:

$$(a+b)\cdot c = a\cdot c + b\cdot c\ , \quad c\cdot(a+b) = c\cdot a + c\cdot b\ , \qquad a,b,c \in R\ .$$

Here we make the usual convention that multiplication takes precedence over addition. For example, $a \cdot b + c$ means $(a \cdot b) + c$ (the multiplication $d := a \cdot b$ is done first and the addition $d + c$ second) and not $a \cdot (b + c)$. Also we usually write ab for $a \cdot b$.

A ring is called **commutative** if multiplication is commutative. In this case, the distributive law (R_3) reduces to

$$(a+b)c = ac + bc\ , \qquad a,b,c \in R\ . \tag{8.1}$$

If there is an identity element with respect to multiplication, then it is written 1_R or simply 1, and is called the **unity** (or **multiplicative identity**) of R, and we say $(R, +, \cdot)$ is a **ring with unity**. When the addition and multiplication operations are clear from context, we write simply R instead of $(R, +, \cdot)$.

8.1 Remarks Let $R := (R, +, \cdot)$ be a ring.

(a) The identity element of the **additive group** $(R, +)$ of a ring R is, as in Example 5.14, denoted by 0_R, or simply 0, and is called the **zero** (or **additive identity**) of the ring R. In view of Proposition 4.11, 0_R and also 1_R, if it exists, are unique.

(b) From Remark 7.1(d) it follows that $-(-a) = a$ for each $a \in R$.

(c) For each pair $a, b \in R$, there is, by Remark 7.1(c), a unique solution $x \in R$ of the equation $a + x = b$, namely $x = b + (-a) =: b - a$ ('b minus a'), the **difference** of a and b.

(d) For all $a \in R$, we have $0a = a0 = 0$ and $-0 = 0$. If $a \neq 0$ and there is some $b \neq 0$ with $ab = 0$ or $ba = 0$, then a is called a **zero divisor** of R. If R is commutative and has no zero divisors, that is, $ab = 0$ implies $a = 0$ or $b = 0$, then R is called a **domain**.

Proof Since $0 = 0 + 0$, we have $a0 = a(0 + 0) = a0 + a0$. It then follows from (c) and the equation $a0 + 0 = a0$ that $a0 = 0$. Similarly one can show that $0a = 0$. The second claim also follows from (c). ■

(e) For all $a, b \in R$, we have $a(-b) = (-a)b = -(ab) =: -ab$ and $(-a)(-b) = ab$.

Proof From $0 = b + (-b)$ and (d) we get $0 = a0 = ab + a(-b)$. Hence, just as above, $a(-b) = -ab$. Similarly one can show that $(-a)b = -ab$. Using this fact twice we get

$$(-a)(-b) = -\big(a(-b)\big) = -(-ab) = ab \ ,$$

in which the last equality follows from (b). ■

(f) If R is a ring with unity then $(-1)a = -a$ for all $a \in R$.

Proof This is a special case of (e). ■

(g) In view of Example 5.14(a), $n \cdot a = na$ is well defined for all $n \in \mathbb{N}$ and $a \in R$ and the rules of this example hold. In particular, $0_{\mathbb{N}} \cdot a := 0_R$. From (d) we also have $0_R \cdot a := 0_R$, and so dropping the subscripts from $0_{\mathbb{N}}$ and 0_R leads to no ambiguity. Similarly, if R is a ring with unity, then $1_{\mathbb{N}} \cdot a = 1_R \cdot a = a$. ■

8.2 Examples **(a)** The **trivial ring** has exactly one element 0 and is itself denoted by 0. A ring with more than one element is **nontrivial**. The trivial ring is clearly commutative and has a unity element. If R is a ring with unity, then it follows from $1_R \cdot a = a$ for each $a \in R$, that R is trivial if and only if $1_R = 0_R$.

(b) Let $R := (R, +, \cdot)$ be a ring and X a nonempty set. Then R^X is a ring with the operations

$$(f + g)(x) := f(x) + g(x) \ , \quad (fg)(x) := f(x)g(x) \ , \qquad x \in X \ , \quad f, g \in R^X \ .$$

If R is a commutative ring (a ring with unity), then so is $R^X := (R^X, +, \cdot)$ (see Example 4.12). In particular, for $m \geq 2$, the **direct product** R^m of the ring R with the operations

$$(a_1, \ldots, a_m) + (b_1, \ldots, b_m) = (a_1 + b_1, \ldots, a_m + b_m)$$

and

$$(a_1, \ldots, a_m)(b_1, \ldots, b_m) = (a_1 b_1, \ldots, a_m b_m)$$

is a ring called the **product ring**. If R is a nontrivial ring with unity and X has at least two elements, then R^X has zero divisors.

Proof For the first claim, see Example 4.12. For the second claim, suppose that $x, y \in X$ are such that $x \neq y$, and $f, g \in R^X$ satisfy $f(x) = 1$ and $f(x') = 0$ for all $x' \in X \backslash \{x\}$ as well as $g(y) = 1$ and $g(y') = 0$ for all $y' \in X \backslash \{y\}$. Then $fg = 0$. ∎

(c) Suppose R is a ring and S is a nonempty subset of R that satisfies the following:
(SR_1) S is a subgroup of $(R, +)$.
(SR_2) $S \cdot S \subseteq S$.
Then S is itself a ring, a **subring** of R, and R is called an **overring** of S. Clearly, $0 = \{0\}$ and R are subrings of R. Even if R is a ring with unity, the same may not be true of S (see (e)). Even so, if $1_R \in S$, then 1_R is the unity of S. Of course, if R is commutative then so is S. The converse is not true in general.

(d) Intersections of subrings are subrings.

(e) Let R be a nontrivial ring with unity and S the set of all $g \in R^{\mathbb{N}}$ with $g(n) = 0$ for **almost all**, that is, for all but finitely many $n \in \mathbb{N}$. Then S is a subring of $R^{\mathbb{N}}$ without unity. (Why?)

(f) Let X be a set. For subsets A and B of X define their **symmetric difference** $A \triangle B$ by

$$A \triangle B := (A \cup B) \backslash (A \cap B) = (A \backslash B) \cup (B \backslash A) .$$

Then $\big(\mathcal{P}(X), \triangle, \cap\big)$ is a commutative ring with unity. ∎

Let R and R' be rings. A **(ring) homomorphism** is a function $\varphi \colon R \to R'$ which is compatible with the ring operations, that is,

$$\varphi(a + b) = \varphi(a) + \varphi(b) , \quad \varphi(ab) = \varphi(a)\varphi(b) , \qquad a, b \in R . \tag{8.2}$$

If, in addition, φ is bijective, then φ is called a **(ring) isomorphism** and R and R' are **isomorphic**.

A homomorphism φ from R to itself is a **(ring) endomorphism**. If φ is an isomorphism, then it is a **(ring) automorphism**.[1]

8.3 Remarks **(a)** A ring homomorphism $\varphi \colon R \to R'$ is, in particular, a group homomorphism from $(R, +)$ to $(R', +)$. The **kernel**, $\ker(\varphi)$, of φ is defined to be the kernel of this group homomorphism, that is,

$$\ker(\varphi) = \big\{\, a \in R \,;\; \varphi(a) = 0 \,\big\} = \varphi^{-1}(0) .$$

(b) The **zero function** $R \to R'$, $a \mapsto 0_{R'}$ is a homomorphism with $\ker(\varphi) = R$.

(c) Let R and R' be rings with unity and $\varphi \colon R \to R'$ a homomorphism. As (b) shows, it does not follow that $\varphi(1_R) = 1_{R'}$. This can be seen as a consequence of the fact that, with respect to multiplication, a ring is not a group. ∎

[1] We will use the words 'homomorphism', 'isomorphism', 'endomorphism', etc. when it is clear from the context what type of homomorphism — group, ring (and later field, vector space or algebra) — is intended.

The Binomial Theorem

We next show that the ring axioms (R_1)–(R_3) have other important consequences beyond the rules in Remark 8.1.

8.4 Theorem (binomial theorem) *Let a and b be two commuting elements (that is, $ab = ba$) of a ring R with unity. Then, for all $n \in \mathbb{N}$,*

$$(a+b)^n = \sum_{k=0}^{n} \binom{n}{k} a^k b^{n-k} . \tag{8.3}$$

Proof First we note that, by Examples 5.14, Remark 8.1(g) and Exercise 5.5, both sides of (8.3) are well defined, and that the claim is true for $n = 0$. If (8.3) holds for some $n \in \mathbb{N}$, then

$$\begin{aligned}
(a+b)^{n+1} &= (a+b)^n (a+b) = \Bigl(\sum_{k=0}^{n} \binom{n}{k} a^k b^{n-k}\Bigr)(a+b) \\
&= \sum_{k=0}^{n} \binom{n}{k} a^{k+1} b^{n-k} + \sum_{k=0}^{n} \binom{n}{k} a^k b^{n+1-k} \\
&= a^{n+1} + \sum_{k=0}^{n-1} \binom{n}{k} a^{k+1} b^{n-k} + \sum_{k=1}^{n} \binom{n}{k} a^k b^{n+1-k} + b^{n+1} \\
&= a^{n+1} + \sum_{k=1}^{n} \Bigl\{ \binom{n}{k-1} + \binom{n}{k} \Bigr\} a^k b^{n+1-k} + b^{n+1} .
\end{aligned}$$

From Exercise 5.5 we have $\binom{n}{k-1} + \binom{n}{k} = \binom{n+1}{k}$, and so

$$(a+b)^{n+1} = a^{n+1} + \sum_{k=1}^{n} \binom{n+1}{k} a^k b^{n+1-k} + b^{n+1} .$$

The claim then follows from the induction principle of Proposition 5.7. ■

The Multinomial Theorem

We want to generalize the binomial theorem so that, on the left side of (8.3), sums with more than two terms are allowed. To make this formula as simple as possible, it is useful to introduce the following notation:

For $m \in \mathbb{N}$ with $m \geq 2$, an element $\alpha = (\alpha_1, \ldots, \alpha_m) \in \mathbb{N}^m$ is called a **multi-index (of order** m**)**. The **length** $|\alpha|$ **of a multi-index** $\alpha \in \mathbb{N}^m$ is defined by

$$|\alpha| := \sum_{j=1}^{m} \alpha_j .$$

Set also

$$\alpha! := \prod_{j=1}^{m} (\alpha_j)! ,$$

and define the **natural (partial) order** on $\mathbb{N}^m$ by

$$\alpha \le \beta :\Longleftrightarrow (\alpha_j \le \beta_j,\ 1 \le j \le m) .$$

Finally, let R be a commutative ring with unity and $m \in \mathbb{N}$ with $m \ge 2$. Then we set

$$a^\alpha := \prod_{j=1}^{m} (a_j)^{\alpha_j}$$

for $a = (a_1, \ldots, a_m) \in R^m$ and $\alpha = (\alpha_1, \ldots, \alpha_m) \in \mathbb{N}^m$.

8.5 Theorem (multinomial theorem) *Let R be a commutative ring with unity. Then for all $m \ge 2$,*

$$\Bigl(\sum_{j=1}^{m} a_j\Bigr)^k = \sum_{|\alpha|=k} \frac{k!}{\alpha!} a^\alpha , \qquad a = (a_1, \ldots, a_m) \in R^m , \quad k \in \mathbb{N} . \tag{8.4}$$

Here $\sum_{|\alpha|=k}$ is the sum over all multi-indices of length k in $\mathbb{N}^m$.

Proof We begin by proving, by induction on m, that

$$k!/\alpha! \in \mathbb{N}^\times \qquad \text{for } k \in \mathbb{N} \text{ and } \alpha \in \mathbb{N}^m \text{ with } |\alpha| = k . \tag{8.5}$$

We consider first the case $m = 2$. Let $\alpha \in \mathbb{N}^2$ be an arbitrary multi-index of length k. Then $\alpha = (\ell, k - \ell)$ for some $\ell \in \mathbb{N}$ with $0 \le \ell \le k$, and so, by Exercise 5.5(b),

$$\frac{k!}{\alpha!} = \frac{k!}{\ell!\,(k-\ell)!} = \binom{k}{\ell} \in \mathbb{N}^\times .$$

Now suppose that (8.5) is true for some $m \ge 2$. Let $\alpha \in \mathbb{N}^{m+1}$ be arbitrary with $|\alpha| = k$. Set $\alpha' := (\alpha_2, \ldots, \alpha_{m+1}) \in \mathbb{N}^m$. It follows from the induction hypothesis and Exercise 5.5(a) that

$$\frac{k!}{\alpha!} = \frac{(k - \alpha_1)!}{\alpha'!} \binom{k}{\alpha_1} \in \mathbb{N}^\times . \tag{8.6}$$

This completes the induction and the proof of (8.5).

To prove (8.4) we again use induction on m. The case $m = 2$ is the binomial theorem. Thus we suppose that $a = (a_1, \ldots, a_m, a_{m+1}) \in R^{m+1}$ for $m \ge 2$ and

$k \in \mathbb{N}$ are given. We set $b := \sum_{j=2}^{m+1} a_j$ and calculate using Theorem 8.4 and the induction hypothesis as follows:

$$
\begin{aligned}
\Bigl(\sum_{j=1}^{m+1} a_j\Bigr)^k = (a_1+b)^k &= \sum_{\alpha_1=0}^{k} \binom{k}{\alpha_1} a_1^{\alpha_1} b^{k-\alpha_1} \\
&= \sum_{\alpha_1=0}^{k} \binom{k}{\alpha_1} a_1^{\alpha_1} \sum_{|\alpha'|=k-\alpha_1} \frac{(k-\alpha_1)!}{\alpha'!} a_2^{\alpha_2} \cdot \cdots \cdot a_{m+1}^{\alpha_{m+1}} \\
&= \sum_{\alpha_1=0}^{k} \sum_{|\alpha'|=k-\alpha_1} \frac{(k-\alpha_1)!}{\alpha'!} \binom{k}{\alpha_1} a_1^{\alpha_1} \cdot \cdots \cdot a_{m+1}^{\alpha_{m+1}} \\
&= \sum_{|\alpha|=k} \frac{k!}{\alpha!} a^\alpha ,
\end{aligned}
$$

where in the last step we have used (8.6). This completes the induction and the proof of the theorem. ∎

8.6 Remarks (a) The **multinomial coefficients** are defined by[2]

$$
\binom{k}{\alpha} := \frac{k!}{\alpha!\,(k-|\alpha|)!} , \qquad k \in \mathbb{N} , \quad \alpha \in \mathbb{N}^m , \quad |\alpha| \le k .
$$

Then $\binom{k}{\alpha} \in \mathbb{N}^\times$ and, if R is a commutative ring with unity,

$$
(1 + a_1 + \cdots + a_m)^k = \sum_{|\alpha| \le k} \binom{k}{\alpha} a^\alpha , \qquad a = (a_1, \ldots, a_m) \in R^m , \quad k \in \mathbb{N} .
$$

Proof If $\beta := (\alpha_1, \ldots, \alpha_m, k - |\alpha|) \in \mathbb{N}^{m+1}$, then we have $|\beta| = k$ for all $|\alpha| \le k$ and $\binom{k}{\alpha} = k!/\beta!$. The claim now follows from Theorem 8.5. ∎

(b) Clearly Theorem 8.5 and (a) are also true if $a_1, \ldots, a_m$ are pairwise commuting elements of an arbitrary ring with unity. ∎

Fields

A ring R has especially nice properties when $R\backslash\{0\}$ forms a group with respect to multiplication. Such rings are called fields. Specifically, K is a **field** when the following are satisfied:

[2]We use the same symbol $(\ )$ for multinomial coefficients and binomial coefficients. This should cause no misunderstanding, since, for a multinomial coefficient $\binom{k}{\alpha}$, we have $\alpha \in \mathbb{N}^m$ with $m \ge 2$, and for a binomial coefficient $\binom{k}{\ell}$, ℓ is always a natural number.

(F_1) K is a commutative ring with unity.

(F_2) $0 \neq 1$.

(F_3) $K^\times := K\backslash\{0\}$ is an Abelian group with respect to multiplication.

The Abelian group $K^\times = (K^\times, \cdot)$ is called the **multiplicative group** of K.

Of course, a field has all the properties that we have shown to occur in rings. Since $K^\times$ is an Abelian group, we get as well the following important rules from Remarks 7.1.

8.7 Remarks Let K be a field.

(a) For all $a \in K^\times$, $(a^{-1})^{-1} = a$.

(b) A field has no zero divisors.

Proof Suppose that $ab = 0$. If $a \neq 0$ then multiplication of $ab = 0$ by a^{-1} yields $b = 0$. ■

(c) Let $a \in K^\times$ and $b \in K$. Then there is a unique $x \in K$ with $ax = b$, namely the **quotient** $\frac{b}{a} := b/a := ba^{-1}$ ('b over a').

(d) For $a, c \in K$ and $b, d \in K^\times$, we have the following:[3]

(i) $\dfrac{a}{b} = \dfrac{c}{d} \Longleftrightarrow ad = bc$.

(ii) $\dfrac{a}{b} \pm \dfrac{c}{d} = \dfrac{ad \pm bc}{bd}$.

(iii) $\dfrac{a}{b} \cdot \dfrac{c}{d} = \dfrac{ac}{bd}$.

(iv) $\dfrac{a}{b} \Big/ \dfrac{c}{d} = \dfrac{ad}{bc}, \quad c \neq 0$.

Proof The first three claims are proved by multiplying both sides of the equation by bd and then using the rule that $bdx = bdy$ implies $x = y$. Rule (iv) is an easy consequence of (i). ■

(e) In view of (c), for $a, b \in K^\times$ the equation $ax = b$ has a unique solution. On the other hand, by Remark 8.1(d), any $x \in K$ is a solution of the equation $0x = 0$. This is because 0 has no multiplicative inverse. Indeed, the existence of 0^{-1} would imply $0 \cdot 0^{-1} = 1$ and then, since $0 \cdot 0^{-1} = 0$, we would have $0 = 1$, contradicting (F_2). This illustrates the special role of zero with respect to multiplication which finds expression in the definition of $K^\times$ and in the familiar idea that 'division by zero is not allowed'.

[3] Using the symbols $\pm$ and $\mp$ one can write two equations as if they were one: For one of these equations, the upper symbol ($+$ or $-$) is used throughout, and for the other, the lower symbol is used throughout.

(f) Let K' be a field and $\varphi\colon K \to K'$ a homomorphism with $\varphi \neq 0$. Then

$$\varphi(1_K) = 1_{K'} \quad \text{and} \quad \varphi(a^{-1}) = \varphi(a)^{-1}\ , \qquad a \in K^\times\ .$$

Proof Since φ is a group homomorphism from $K^\times$ to $K'^\times$, this follows from Remark 7.6(a). ∎

When we use the words 'homomorphism', 'isomorphism', etc. in connection with fields, we mean, of course, 'ring homomorphism', 'ring isomorphism', etc. and not group homomorphism.

The following example shows that fields do, in fact, exist and therefore that the axioms (F_1)–(F_3) do not lead to contradictions.

8.8 Example Define addition $+$ and multiplication $\cdot$ on $\{0,1\}$ using the tables below.

+	0	1
0	0	1
1	1	0

·	0	1
0	0	0
1	0	1

Then one can verify that $\mathbb{F}_2 := (\{0,1\},+,\cdot)$ is a field. Indeed, up to isomorphism, $\mathbb{F}_2$ is the only field with two elements. ∎

Ordered Fields

The rings and fields which are important in analysis usually have an order structure in addition to their algebraic structure. Of course, to prove interesting theorems, one expects that these two structures should be compatible in some way. Thus, a ring R with an order $\leq$ is called an **ordered ring** if the following holds:[4]

(OR_0) $(R,\leq)$ is totally ordered.

(OR_1) $x < y \Rightarrow x + z < y + z,\ \ z \in R$.

(OR_2) $x, y > 0 \Rightarrow xy > 0$.

Of course, an element $x \in R$ is called **positive** if $x > 0$ and **negative** if $x < 0$. We gather in the next proposition some simple properties of ordered fields.

8.9 Proposition *Let K be an ordered field and $x, y, a, b \in K$.*

(i) $x > y \Longleftrightarrow x - y > 0$.

(ii) *If $x > y$ and $a > b$, then $x + a > y + b$.*

(iii) *If $a > 0$ and $x > y$, then $ax > ay$.*

(iv) *If $x > 0$, then $-x < 0$. If $x < 0$, then $-x > 0$.*

[4] Here, and in the following, we write $a, b, \ldots, w > 0$ for $a > 0,\ b > 0,\ \ldots,\ w > 0$.

(v) *Let* $x > 0$. *If* $y > 0$, *then* $xy > 0$. *If* $y < 0$, *then* $xy < 0$.

(vi) *If* $a < 0$ *and* $x > y$, *then* $ax < ay$.

(vii) $x^2 > 0$ *for all* $x \in K^\times$. *In particular*, $1 > 0$.

(viii) *If* $x > 0$, *then* $x^{-1} > 0$.

(ix) *If* $x > y > 0$, *then* $0 < x^{-1} < y^{-1}$ *and* $xy^{-1} > 1$.

Proof All of these claims are easy consequences of the axioms (OR_1) and (OR_2). We verify only that (ix) follows from (i), (viii) and (OR_2), and leave the remaining proofs to the reader.

If $x > y > 0$, then $x - y > 0$, $x^{-1} > 0$ and $y^{-1} > 0$. From (OR_2) we get

$$0 < (x-y)x^{-1}y^{-1} = y^{-1} - x^{-1} ,$$

which implies $x^{-1} < y^{-1}$, and

$$0 < (x-y)y^{-1} = xy^{-1} - 1 ,$$

which implies $xy^{-1} > 1$. ■

The claims (ii) and (vii) of Proposition 8.9 imply that the field $\mathbb{F}_2$ of Example 8.8 cannot be ordered since otherwise we would have $0 = 1 + 1 > 0$. In the next section we show that ordered fields do exist.

For an ordered field K, the **absolute value function**, $|\cdot| : K \to K$ and the **sign function**, $\mathrm{sign}(\cdot) : K \to K$ are defined by

$$|x| := \begin{cases} x , & x > 0 , \\ 0 , & x = 0 , \\ -x , & x < 0 , \end{cases} \qquad \mathrm{sign}\, x := \begin{cases} 1 , & x > 0 , \\ 0 , & x = 0 , \\ -1 , & x < 0 . \end{cases}$$

8.10 Proposition *Let* K *be an ordered field and* $x, y, a, \varepsilon \in K$ *with* $\varepsilon > 0$.

(i) $x = |x| \,\mathrm{sign}(x)$, $|x| = x \,\mathrm{sign}(x)$.

(ii) $|x| = |-x|$, $x \le |x|$.

(iii) $|xy| = |x|\,|y|$.

(iv) $|x| \ge 0$ *and* $\bigl(|x| = 0 \Longleftrightarrow x = 0\bigr)$.

(v) $|x - a| < \varepsilon \Longleftrightarrow a - \varepsilon < x < a + \varepsilon$.

(vi) $|x + y| \le |x| + |y|$ (triangle inequality).

Proof The first four claims follow immediately from the definitions. From (vi) and (ii) of Proposition 8.9 we have

$$|x - a| < \varepsilon \Longleftrightarrow -\varepsilon < x - a < \varepsilon \Longleftrightarrow a - \varepsilon < x < a + \varepsilon ,$$

which proves (v). To verify (vi), we first suppose that $x+y \geq 0$. Then it follows from (ii) that $|x+y| = x+y \leq |x|+|y|$. If $x+y<0$, then $-(x+y)>0$, and hence

$$|x+y| = |-(x+y)| = |(-x)+(-y)| \leq |-x|+|-y| = |x|+|y| ,$$

which completes the proof. ■

8.11 Corollary (reversed triangle inequality) *In any ordered field K we have*

$$|x-y| \geq \big||x|-|y|\big| , \qquad x,y \in K .$$

Proof The triangle inequality applied to the equation $x=(x-y)+y$ yields $|x| \leq |x-y|+|y|$, that is, $|x|-|y| \leq |x-y|$. Interchanging x and y in this inequality gives $|y|-|x| \leq |y-x| = |x-y|$. ■

Formal Power Series

Let R be a nontrivial ring with unity. On the set $R^{\mathbb{N}} = \mathrm{Funct}(\mathbb{N}, R)$ define addition by

$$(p+q)_n := p_n + q_n , \qquad n \in \mathbb{N} , \tag{8.7}$$

and multiplication by **convolution**,

$$(pq)_n := (p \cdot q)_n := \sum_{j=0}^{n} p_j q_{n-j} = p_0 q_n + p_1 q_{n-1} + \cdots + p_n q_0 \tag{8.8}$$

for $n \in \mathbb{N}$. Here p_n denotes the value of $p \in R^{\mathbb{N}}$ at $n \in \mathbb{N}$ and is called the nth **coefficient** of p. In this situation an element $p \in R^{\mathbb{N}}$ is called a **formal power series over** R, and we set $R[\![X]\!] := (R^{\mathbb{N}}, +, \cdot)$. The following proposition shows that $R[\![X]\!]$ is a ring. Note that this ring is not the same as the function ring $R^{\mathbb{N}}$ introduced in Example 8.2(b).

8.12 Proposition *$R[\![X]\!]$ is a ring with unity, the* **formal power series ring over** *R. If R is commutative, then so is $R[\![X]\!]$.*

Proof Because of (8.7) and Example 7.2(d), $\big(R[\![X]\!], +\big)$ is an Abelian group.

We show next that (R_2) holds. If $p,q,r \in R[\![X]\!]$, then

$$\big((pq)r\big)_n = \sum_{j=0}^{n} (pq)_j r_{n-j} = \sum_{j=0}^{n} \sum_{k=0}^{j} p_k q_{j-k} r_{n-j} \tag{8.9}$$

for all $n \in \mathbb{N}$. The double sum in (8.9) is done over all pairs (j,k) corresponding to the dots in the diagram on the right. Since addition is an associative and commutative operation, the summation can be done in any order. In particular, the summation can be changed from 'column first' to 'row first', in which case the right side of (8.9) becomes

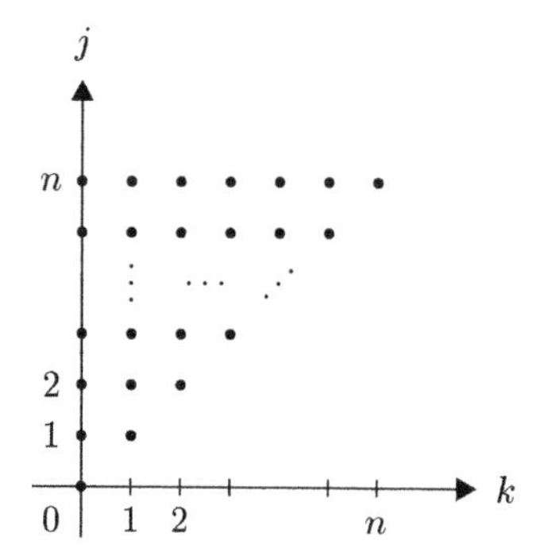

$$\sum_{k=0}^{n}\sum_{j=k}^{n} p_k q_{j-k} r_{n-j} = \sum_{k=0}^{n} p_k \sum_{\ell=0}^{n-k} q_\ell r_{n-k-\ell} = \sum_{k=0}^{n} p_k (qr)_{n-k} = \bigl(p(qr)\bigr)_n ,$$

where we have set $\ell := j - k$.

The validity of (R_3) is clear, as well as the fact that the formal power series p with $p_0 = 1$ and $p_n = 0$ for $n \in \mathbb{N}^\times$ is the unity element of $R[\![X]\!]$. The last claim is trivial. ■

We write X for the power series

$$X_n := \begin{cases} 1 , & n = 1 , \\ 0 & \text{otherwise} . \end{cases}$$

Then for X^m we have (see Example 5.14(a))

$$X_n^m := \begin{cases} 1 , & n = m , \\ 0 , & n \neq m , \end{cases} \qquad m, n \in \mathbb{N} . \tag{8.10}$$

In particular, X^0 is the unity element of $R[\![X]\!]$.

For $a \in R$, we denote by aX^0 the **constant** power series,

$$aX_n^0 := \begin{cases} a , & n = 0 , \\ 0 , & n > 0 , \end{cases}$$

and by RX^0 the set of all constant power series. From (8.7) and (8.8), it is clear that RX^0 is a subring of $R[\![X]\!]$ containing the unity element, and that the function

$$R \to RX^0 , \quad a \mapsto aX^0 \tag{8.11}$$

is an isomorphism. In the following we will usually *identify* R with RX^0, that is, we will write a for the constant power series aX^0 and consider R to be a subring of $R[\![X]\!]$. Note that (8.8) also implies

$$(ap)_n = ap_n , \qquad n \in \mathbb{N} , \quad a \in R , \qquad p \in R[\![X]\!] . \tag{8.12}$$

Polynomials

A **polynomial over** R is a formal power series $p \in R[\![X]\!]$ such that $\{ n \; ; \; p_n \neq 0 \}$ is finite, in other words, $p_n = 0$ 'almost everywhere'. It is easy to see that the set of all polynomials in $R[\![X]\!]$ is a subring of $R[\![X]\!]$ containing the unity element. This subring is denoted by $R[X]$ and called the **polynomial ring over** R.

If p is a polynomial, then there is some $n \in \mathbb{N}$ such that $p_k = 0$ for $k > n$. From (8.10) and (8.12) it follows that p can be written in the form

$$p = \sum_k p_k X^k = \sum_{k=0}^{n} p_k X^k = p_0 + p_1 X + p_2 X^2 + \cdots + p_n X^n \tag{8.13}$$

where $p_0, \ldots, p_n \in R$. Of course, it is possible that $p_k = 0$ for some (or all) $k \leq n$. When polynomials are written as in (8.13), the rules (8.7) and (8.8) take the form

$$\sum_k p_k X^k + \sum_k q_k X^k = \sum_k (p_k + q_k) X^k \tag{8.14}$$

and

$$\Bigl(\sum_k p_k X^k\Bigr)\Bigl(\sum_j q_j X^j\Bigr) = \sum_n \Bigl(\sum_{j=0}^{n} p_j q_{n-j}\Bigr) X^n \; . \tag{8.15}$$

Note that (8.15) can be obtained by applying the distributive law and the rule

$$(aX^j)(bX^k) = abX^{j+k} \; , \qquad a, b \in R \; , \quad j, k \in \mathbb{N} \; ,$$

to the left side of the equation.

As a simple application of the fact that $R[X]$ is a ring, we prove the following *addition theorem for binomial coefficients* which generalizes formula (ii) of Exercise 5.5.

8.13 Proposition For all $\ell, m, n \in \mathbb{N}$,

$$\binom{m+n}{\ell} = \sum_{k=0}^{\ell} \binom{m}{k}\binom{n}{\ell-k} = \sum_{k=0}^{\ell} \binom{m}{\ell-k}\binom{n}{k} \; .$$

Proof For $1 + X \in R[X]$ it follows from (5.15) that

$$(1+X)^m (1+X)^n = (1+X)^{m+n} \; . \tag{8.16}$$

Since X commutes with $1 = 1X^0 = X^0$, that is, $X1 = 1X$, the binomial theorem (8.4) implies

$$(1+X)^j = \sum_{i=0}^{j} \binom{j}{i} X^i \; , \qquad j \in \mathbb{N} \; . \tag{8.17}$$

Thus, from (8.15), we get

$$(1+X)^m(1+X)^n = \Big(\sum_{k=0}^{m}\binom{m}{k}X^k\Big)\Big(\sum_{j=0}^{n}\binom{n}{j}X^j\Big) = \sum_{\ell}\Big(\sum_{k=0}^{\ell}\binom{m}{k}\binom{n}{\ell-k}\Big)X^\ell ,$$

and then, with (8.16) and (8.17), it follows that

$$\sum_{\ell}\Big(\sum_{k=0}^{\ell}\binom{m}{k}\binom{n}{\ell-k}\Big)X^\ell = \sum_{\ell}\binom{m+n}{\ell}X^\ell ,$$

taking into account that $\binom{\ell}{k} = 0$ for $k > \ell$. The claim can now be obtained by matching the coefficients of X^ℓ on both sides of the equal sign.[5] ■

If $p = \sum_k p_k X^k \neq 0$ is a polynomial, then there is, by Proposition 5.5, a smallest $m \in \mathbb{N}$ such that $p_k = 0$ for $k > m$. The number m is called the **degree** of p, written $\deg(p)$, and p_m is called the **leading coefficient** of p. By convention, the degree of the **zero polynomial**, $p = 0$, is $-\infty$ ('negative infinity') for which the following relations hold:[6]

$$-\infty < k\ , \quad k \in \mathbb{N}\ , \qquad -\infty + k = k + (-\infty) = -\infty\ , \quad k \in \mathbb{N} \cup \{-\infty\}\ . \tag{8.18}$$

For $k + (-\infty)$ we write also $k - \infty$.

It is clear that

$$\deg(p+q) \leq \max\big(\deg(p), \deg(q)\big)\ , \quad \deg(pq) \leq \deg(p) + \deg(q) \tag{8.19}$$

for all $p, q \in R[X]$. If R has no zero divisors, in particular, if R is a field, then we have

$$\deg(pq) = \deg(p) + \deg(q)\ . \tag{8.20}$$

It is also convenient to write an arbitrary element $p \in R[\![X]\!]$ in the form

$$p = \sum_k p_k X^k\ , \tag{8.21}$$

which explains the name 'formal power series'. Since 'infinite sums' have no meaning in $R[\![X]\!]$, this should be considered only as an alternative way of writing the function $p \in R^{\mathbb{N}}$. That is, X^k is simply a placeholder used to indicate that the function p has the value $p_k \in R$ at $k \in \mathbb{N}$. Even so, the relations (8.14)–(8.15) can be used to calculate with such infinite sums.

---

[5]Here we use the fact that two polynomials, that is, two functions from $\mathbb{N}$ to R, are equal if and only if their coefficients match up.

[6]The conventions in (8.18) are chosen so that rules such as (8.19) hold for also zero polynomials. Of course, $-\infty$ is not a natural number, nor can it be an element of some Abelian group which contains the natural numbers. (Why not?)

Polynomial Functions

Let $p = \sum_{k=0}^{n} p_k X^k$ be a polynomial over R. Then we define the **value of** p **at** $x \in R$ by

$$p(x) := \sum_{k=0}^{n} p_k x^k \in R \; .$$

This defines a function

$$\underline{p} \colon R \to R \; , \quad x \mapsto p(x) \; ,$$

the **polynomial function,** $\underline{p} \in R^R$, corresponding to $p \in R[X]$.

8.14 Remarks **(a)** The polynomial function corresponding to the constant polynomial a is the constant function $(x \mapsto a) \in R^R$. The polynomial function corresponding to X is the identity function $\mathrm{id}_R \in R^R$.

(b) Let R be commutative. Then for all $p, q \in R[X]$,

$$(p+q)(x) = p(x) + q(x) \; , \quad (pq)(x) = p(x)q(x) \; , \qquad x \in R \; ,$$

that is, the function

$$R[X] \to R^R \; , \quad p \mapsto \underline{p} \tag{8.22}$$

is a homomorphism when R^R has the ring structure of Example 8.2(b). Moreover this homomorphism takes 1 to 1.

Proof The simple verification is left to the reader. ■

(c) If R is a nontrivial finite ring, then the function (8.22) is not injective. The rings which are important in analysis are infinite and for such rings the function (8.22) is injective.

Proof For the first claim, we note that, since R has at least two elements, the set $R[X] = R^{\mathbb{N}}$ is, by Propositions 6.7 and 6.11, uncountable. Since R^R is a finite set, there can be no injective function from $R[X]$ to R^R. The second claim is proved in Remark 8.19(d). ■

(d) Let M be a ring with unity. Suppose that there is a function $R \times M \to M$ which we denote by $(a, m) \mapsto am$. Then we can define the value of $p = \sum_{k=0}^{n} p_k X^k$ at $m \in M$ by

$$p(m) := \sum_{k=0}^{n} p_k m^k \; .$$

A trivial, but important, case is when R is a subring of M. Then any $p \in R[X]$ can be considered also as an element of $M[X]$ and hence $R[X] \subseteq M[X]$. In Remark 12.12 we will return to this general situation.

(e) Let $p = \sum_k p_k X^k \in R[\![X]\!]$ be a formal power series. Then a definition of the form $p(x) := \sum_k p_k x^k$ for $x \in R$ is meaningless since 'infinite sums' are, in general, undefined in R. Even so, in Section II.9 we will meet certain formal power series which have the property that for certain $x \in R$ the value of p at x, $p(x) := \sum_k p_k x^k \in R$, makes sense.

(f) For an efficient calculation of $p(x)$, note that p can be written in the form

$$p = \Big(\big(\cdots\big((p_n X + p_{n-1})X + p_{n-2}\big)\cdots\big)X + p_1\Big)X + p_0$$

(which can easily be proved using induction). This suggests an 'iterative process' for evaluating $p(x)$: Calculate $x_n, x_{n-1}, \ldots, x_0$ using

$$x_n := x \ , \quad x_{k-1} := p_k x_k + p_{k-1} \ , \qquad k = n, n-1, \ldots, 1 \ ,$$

and then set $p(x) = x_0$. This 'algorithm' is easy to program and requires only n multiplications and n additions. A 'direct' calculation, on the other hand, requires $2n - 1$ multiplications and n additions. ■

Division of Polynomials

For polynomials over a field K, we now prove an important version of the division algorithm of Proposition 5.4.

8.15 Proposition *Let K be a field and $p, q \in K[X]$ with $q \neq 0$. Then there are unique polynomials r and s such that*

$$p = sq + r \quad \text{and} \quad \deg(r) < \deg(q) \ . \tag{8.23}$$

Proof (a) Existence: If $\deg(p) < \deg(q)$, then $s := 0$ and $r := p$ satisfy (8.23). So we can assume that $n := \deg(p) \geq \deg(q) =: m$. Thus we have

$$p = \sum_{k=0}^{n} p_k X^k \ , \quad q = \sum_{j=0}^{m} q_j X^j \ , \qquad p_n \neq 0 \ , \quad q_m \neq 0 \ .$$

Set $s_{(1)} := p_n q_m^{-1} X^{n-m} \in K[X]$. Then $p_{(1)} := p - s_{(1)} q$ is a polynomial such that $\deg(p_{(1)}) < \deg(p)$. If $\deg(p_{(1)}) < m$, then $s := s_{(1)}$ and $r := p_{(1)}$ satisfy (8.23). Otherwise we apply the above argument to $p_{(1)}$ in place of p. Repeating as necessary, after a finite number of steps we find polynomials r and s which satisfy (8.23).

(b) Uniqueness: Suppose that $s_{(1)}$ and $r_{(1)}$ are other polynomials with the property that $p = s_{(1)} q + r_{(1)}$ and $\deg(r_{(1)}) < \deg(q)$. Then $(s_{(1)} - s)q = r - r_{(1)}$. If $s_{(1)} - s \neq 0$, then from (8.20) we would get

$$\deg(r - r_{(1)}) = \deg\big((s_{(1)} - s)q\big) = \deg(s_{(1)} - s) + \deg(q) > \deg(q) \ ,$$

which, because $\deg(r - r_{(1)}) \leq \max\big(\deg(r), \deg(r_{(1)})\big) < \deg(q)$, is not possible. Thus $s_{(1)} = s$ and also $r_{(1)} = r$. ■

Note that the above proof is 'constructive', that is, the polynomials r and s can be calculated using the method described in (a).

As a first application of Proposition 8.15 we prove that a polynomial can be 'expanded about' any $a \in K$.

8.16 Proposition *Let K be a field, $p \in K[X]$ a polynomial of degree $n \in \mathbb{N}$ and $a \in K$. Then there are unique $b_0, b_1, \ldots, b_n \in K$ such that*

$$p = \sum_{k=0}^{n} b_k (X-a)^k = b_0 + b_1(X-a) + b_2(X-a)^2 + \cdots + b_n(X-a)^n . \tag{8.24}$$

In particular, $b_n \neq 0$.

Proof Since $\deg(X-a) = 1$, it follows from Proposition 8.15 that there are unique $p_{(1)} \in K[X]$ and $b_0 \in K$ such that $p = (X-a)p_{(1)} + b_0$. From (8.20) we have that $\deg(p_{(1)}) = \deg(p) - 1$ so the claim can then be proved by induction. ■

Linear Factors

A direct consequence of Proposition (8.16) is the following factorization theorem.

8.17 Theorem *Let K be a field and $p \in K[X]$ with $\deg(p) \geq 1$. If $a \in K$* is a **zero** *of p, that is, if $p(a) = 0$, then $X - a \in K[X]$ divides p, that is, $p = (X-a)q$ for some unique $q \in K[X]$ with $\deg(q) = \deg(p) - 1$.*

Proof Evaluating both sides of (8.24) at a gives $0 = p(a) = b_0$, and so

$$p = \sum_{k=1}^{n} b_k (X-a)^k = \Bigl(\sum_{j=0}^{n-1} b_{j+1}(X-a)^j\Bigr)(X-a) ,$$

which proves the claim. ■

8.18 Corollary *A nonconstant polynomial of degree m over a field has at most m zeros.*

8.19 Remarks Let K be a field.

(a) In general, a nonconstant polynomial may have no zeros. For example, if K is an ordered field, then by Proposition 8.9(ii) and (vii), the polynomial $X^2 + 1$ has no zeros.

(b) Let $p \in K[X]$ with $\deg(p) = m \geq 1$. If $a_1, \ldots, a_n \in K$ are all the zeros of p, then p can be written uniquely in the form

$$p = q \prod_{j=1}^{n} (X - a_j)^{m(j)}$$

where $q \in K[X]$ has no zeros and $m(j) \in \mathbb{N}^\times$. Here $m(j)$ is called the **multiplicity of the zero** a_j of p. The zero a_j is **simple** if $m(j) = 1$. In addition, $\sum_{j=1}^n m(j) \le m$.

Proof This follows from Theorem 8.17 by induction. ■

(c) If p and q are polynomials over K of degree $\le n$ such that $p(a_i) = q(a_i)$ for some distinct $a_1, a_2, \ldots, a_{n+1} \in K$, then $p = q$ (*identity theorem for polynomials*).

Proof From (8.19) we have $\deg(p-q) \le n$. Since $p - q$ has $n+1$ zeros, the claim follows from Corollary 8.18. ■

(d) If K is an infinite field, that is, if the set K is infinite, then the homomorphism (8.22) is injective.[7]

Proof If $p, q \in K[X]$ are such that $\underline{p} = \underline{q}$, then $p(x) = q(x)$ for all $x \in K$. Since K is infinite, $p = q$ follows from (c). ■

Polynomials in Several Indeterminates

To complete this section, we extend the above results to the case of formal power series and polynomials in m indeterminates. In analogy to the $m = 1$ cases, namely $R[\![X]\!]$ and $R[X]$, for $m \in \mathbb{N}^\times$, we define addition and multiplication on the set $R^{(\mathbb{N}^m)} = \mathrm{Funct}(\mathbb{N}^m, R)$ by

$$(p+q)_\alpha := p_\alpha + q_\alpha , \qquad \alpha \in \mathbb{N}^m , \tag{8.25}$$

and

$$(pq)_\alpha := \sum_{\beta \le \alpha} p_\beta q_{\alpha-\beta} , \qquad \alpha \in \mathbb{N}^m . \tag{8.26}$$

In (8.26), the sum is over all multi-indices $\beta \in \mathbb{N}^m$ with $\beta \le \alpha$. In this situation, $p \in R^{(\mathbb{N}^m)}$ is called a **formal power series in** m **indeterminates over** R. We set

$$R[\![X_1, \ldots, X_m]\!] := \bigl(R^{(\mathbb{N}^m)}, +, \cdot\bigr) ,$$

where $+$ and $\cdot$ are as in (8.25) and (8.26).

A formal power series $p \in R[\![X_1, \ldots, X_m]\!]$ is called a **polynomial in** m **indeterminates over** R if $p_\alpha = 0$ for almost all $\alpha \in \mathbb{N}^m$. The set of all such polynomials is written $R[X_1, \ldots, X_m]$.

Set $X := (X_1, \ldots, X_m)$ and, for $\alpha \in \mathbb{N}^m$, denote by X^α the formal power series (that is, the function $\mathbb{N}^m \to R$) such that

$$X^\alpha_\beta := \begin{cases} 1 , & \beta = \alpha , \\ 0 , & \beta \ne \alpha , \end{cases} \qquad \beta \in \mathbb{N}^m .$$

Then each $p \in R[\![X_1, \ldots, X_m]\!]$ can be written uniquely in the form

$$p = \sum_{\alpha \in \mathbb{N}^m} p_\alpha X^\alpha .$$

[7] For finite fields this statement is false. See Remark 8.14(c) and Exercise 16.

The rules (8.25) and (8.26) become

$$\sum_{\alpha\in\mathbb{N}^m} p_\alpha X^\alpha + \sum_{\alpha\in\mathbb{N}^m} q_\alpha X^\alpha = \sum_{\alpha\in\mathbb{N}^m} (p_\alpha + q_\alpha) X^\alpha \tag{8.27}$$

and

$$\Bigl(\sum_{\alpha\in\mathbb{N}^m} p_\alpha X^\alpha\Bigr)\Bigl(\sum_{\beta\in\mathbb{N}^m} q_\beta X^\beta\Bigr) = \sum_{\alpha\in\mathbb{N}^m}\Bigl(\sum_{\beta\le\alpha} p_\beta q_{\alpha-\beta}\Bigr) X^\alpha \ . \tag{8.28}$$

Once again (8.27) and (8.28) can be obtained by using the distributive law and the rule

$$aX^\alpha bX^\beta = abX^{\alpha+\beta} \ , \qquad a,b\in R \ , \quad \alpha,\beta\in\mathbb{N}^m \ .$$

The **degree** of a polynomial

$$p = \sum_{\alpha\in\mathbb{N}^m} p_\alpha X^\alpha \in R[X_1,\ldots,X_m] \tag{8.29}$$

is defined by[8]

$$\deg(p) := \max\{\,|\alpha|\in\mathbb{N} \ ; \ p_\alpha \ne 0\,\} \ .$$

A polynomial of the form $p_\alpha X^\alpha$ with $\alpha\in\mathbb{N}^m$ is called a **monomial**. The polynomial (8.29) is **homogeneous of degree** k if $p_\alpha = 0$ whenever $|\alpha|\ne k$. Every homogeneous polynomial of degree $k\in\mathbb{N}$ has the form

$$\sum_{|\alpha|=k} p_\alpha X^\alpha \ , \qquad p_\alpha\in R \ .$$

Polynomials of degree ≤ 0 are called **constant**, polynomials of degree 1 are called **linear**, and polynomials of degree 2 are called **quadratic**.

8.20 Remarks (a) $R[\![X_1,\ldots,X_m]\!]$ is a ring with unity $X^0 = X^{(0,0,\ldots,0)}$, that is, X^0 is the function $\mathbb{N}^m\to R$ which has the value 1 at $(0,0,\ldots,0)$ and is zero otherwise. If R is commutative then so is $R[\![X_1,\ldots,X_m]\!]$. The polynomial ring in the indeterminates $X_1,\ldots,X_n$, that is, $R[X_1,\ldots,X_n]$, is a subring of $R[\![X_1,\ldots,X_m]\!]$. R is isomorphic to the subring $RX^0 := \{\,aX^0 \ ; \ a\in R\,\}$ of $R[X_1,\ldots,X_n]$. By means of this isomorphism we identify R and RX^0, and hence we consider R to be a subring of $R[X_1,\ldots,X_n]$ and write a for aX^0.

(b) Let R be a *commutative* ring and $p\in R[X_1,\ldots,X_m]$. Then we define the **value** of p at $x := (x_1,\ldots,x_m)\in R^m$ by

$$p(x) := \sum_{\alpha\in\mathbb{N}^m} p_\alpha x^\alpha \in R \ ,$$

[8]We use the conventions that $\max(\emptyset) = -\infty$ and $\min(\emptyset) = \infty$.

and the corresponding **polynomial function** (in m variables) by

$$\underline{p}\colon R^m \to R , \quad x \mapsto p(x) .$$

The function

$$R[X_1,\dots,X_m] \to R^{(R^m)} , \quad p \mapsto \underline{p} \tag{8.30}$$

is a homomorphism when $R^{(R^m)}$ is given the ring structure of Example 8.2(b).

(c) Let K be an infinite field. Then the homomorphism (8.30) is injective.

Proof Let $p \in K[X_1,\dots,X_m]$. Then, by Remark 7.6(d), it suffices to show that p is zero if $\underline{p}(x) = 0$ for all $x = (x_1,\dots,x_m) \in K^m$. Clearly $p = \sum_\alpha p_\alpha X^\alpha$ can be written in the form

$$p = \sum_{j=0}^{n} q_j X_m^j \tag{8.31}$$

for suitable $n \in \mathbb{N}$ and $q_j \in K[X_1,\dots,X_{m-1}]$. This suggests a proof by induction on the number of indeterminates: For $m = 1$, the claim is true by Remark 8.19(d). We suppose next that the claim is true for $1 \le k \le m-1$. Using (8.31), set

$$p_{(x')} := \sum_{j=0}^{n} \underline{q}_j(x_1,\dots,x_{m-1}) X^j \in K[X] , \qquad x' := (x_1,\dots,x_{m-1}) \in K^{m-1} .$$

Because $\underline{p}(x) = 0$ for $x \in K^m$, we have $\underline{p_{(x')}}(\xi) = 0$ for each $\xi \in K$ and fixed $x' \in K^{m-1}$. Remark 8.19(d) implies that $p_{(x')} = 0$, that is, $\underline{q}_j(x_1,\dots,x_{m-1}) = 0$ for all $0 \le j \le n$. Since $x' \in K^{m-1}$ was arbitrary, we have, by induction, that $q_j(X_1,\dots,X_{m-1}) = 0$ for all $j = 0,\dots,n$. This, of course, implies $p = 0$. ■

Convention Let K be an infinite field and $m \in \mathbb{N}^\times$. Then we identify the polynomial ring $K[X_1,\dots,X_m]$ with its image in $K^{(K^m)}$ under the homomorphism (8.30). In other words, we identify the polynomial $p \in K[X_1,\dots,X_m]$ with the polynomial function

$$K^m \to K , \quad x \mapsto p(x) .$$

Hence $K[X_1,\dots,X_m]$ is a subring of $K^{(K^m)}$, which we call the **polynomial ring in m indeterminates over R.**

Exercises

1 Let a and b be commuting elements of a ring with unity and $n \in \mathbb{N}$. Prove the following:

(a) $a^{n+1} - b^{n+1} = (a-b)\sum_{j=0}^{n} a^j b^{n-j}$.

(b) $a^{n+1} - 1 = (a-1)\sum_{j=0}^{n} a^j$.

Remark $\sum_{j=0}^{n} a^j$ is called a **finite geometric series** in R.

2 For a ring R with unity, show that $(1-X)\sum_k X^k = (\sum_k X^k)(1-X) = 1$ in $R[\![X]\!]$.
Remark $\sum_k X^k$ is called a **geometric series**.

3 Show that a polynomial ring in one indeterminate over a field has no zero divisors.

4 Show that a finite field cannot be ordered.

5 Prove Remarks 8.20(a) and (b).

6 Let R be a ring with unity. A subring I is called an **ideal** of R if $RI = IR = I$. An ideal is **proper** if it is a proper subset of R. Show the following:
(a) An ideal I is proper if and only if $1 \notin I$.
(b) A field K has exactly two ideals: $\{0\}$ and K.
(c) If $\varphi : R \to R'$ is a ring homomorphism, then $\ker(\varphi)$ is an ideal of R.
(d) The intersection of a set of ideals is an ideal.
(e) Let I be an ideal of R and let R/I be the quotient group $(R,+)/I$. Define an operation on R/I by

$$R/I \times R/I \to R/I \ , \quad (a+I, b+I) \mapsto ab + I \ .$$

Show that, with this operation as multiplication, R/I is a ring and the quotient homomorphism $p\colon R \to R/I$ is a ring homomorphism.
Remark R/I is called the **quotient ring of R modulo I**, and, for $a \in R$, $a+I$ is the **coset** of a **modulo** I. Instead of $a \in b+I$, we often write $a \equiv b \pmod{I}$ ('a is **congruent to** b **modulo** I').

7 Let R be a commutative ring with unity and $m \in \mathbb{N}$ with $m \geq 2$. Let

$$\mathsf{S}_m \times \mathbb{N}^m \to \mathbb{N}^m \ , \quad (\sigma, \alpha) \mapsto \sigma \cdot \alpha$$

be the action of the symmetric group S_m on $\mathbb{N}^m$ as in Exercise 7.6(d). Show the following:
(a) The equation

$$\sigma \cdot \sum_\alpha a_\alpha X^\alpha := \sum_\alpha a_\alpha X^{\sigma\cdot\alpha}$$

defines an action

$$\mathsf{S}_m \times R[X_1,\ldots,X_m] \to R[X_1,\ldots,X_m] \ , \quad (\sigma, p) \mapsto \sigma \cdot p$$

of S_m on the polynomial ring $R[X_1,\ldots,X_m]$.
(b) For each $\sigma \in \mathsf{S}_m$, $p \mapsto \sigma \cdot p$ is an automorphism of $R[X_1,\ldots,X_m]$.
(c) Determine the orbits $\mathsf{S}_3 \cdot p$ in the following cases:
(i) $p := X_1$.
(ii) $p := X_1^2$.
(iii) $p := X_1^2 X_2 X_3^3$.
(d) A polynomial $p \in R[X_1,\ldots,X_m]$ is called **symmetric** if $\mathsf{S}_m \cdot p = \{p\}$, that is, when it is fixed by all permutations. Show that p is symmetric if and only if it has the form

$$p = \sum_{[\alpha]\in\mathbb{N}^m/\mathsf{S}_m} a_{[\alpha]} \bigl(\sum_{\beta\in[\alpha]} X^\beta\bigr)$$

where $a_{[\alpha]} \in R$ for all $[\alpha] \in \mathbb{N}^m/\mathsf{S}_m$.

(e) Determine all symmetric polynomials in 3 indeterminates of degree ≤ 3.

(f) Show that the **elementary symmetric functions**

$$\begin{aligned}
s_1 &:= \textstyle\sum_{1\leq j\leq m} X_j \\
s_2 &:= \textstyle\sum_{1\leq j<k\leq m} X_j \cdot X_k \\
&\vdots \\
s_k &:= \textstyle\sum_{1\leq j_1<j_2<\cdots<j_k\leq m} X_{j_1} \cdot X_{j_2} \cdot \cdots \cdot X_{j_k} \\
&\vdots \\
s_m &:= X_1 X_2 \cdots X_m
\end{aligned}$$

are symmetric polynomials.

(g) Show that the polynomial

$$(X - X_1)(X - X_2)\cdots(X - X_m) \in R[X_1,\ldots,X_m][X]$$

in one indeterminate X over the ring $R[X_1,\ldots,X_m]$ satisfies

$$(X - X_1)(X - X_2)\cdots(X - X_m) = \textstyle\sum_{k=0}^{m}(-1)^k s_k X^{m-k}$$

where $s_0 := 1 \in R$.

8 Let R be a commutative ring with unity. For $r \in R$, define the power series $p[r] \in R[\![X]\!]$ by $p[r] := \sum_k r^k X^k$. Show the following:

(a) $\displaystyle (p[1])^m = \sum_k \binom{m+k-1}{k} X^k,\quad m \in \mathbb{N}^\times.$

(b) $\displaystyle \prod_{j=1}^{m} p[a_j] = \sum_k \Bigl(\sum_{|\alpha|=k} a^\alpha\Bigr) X^k,\quad a := (a_1,\ldots,a_m) \in R^m,\quad m \in \mathbb{N} \text{ with } m \geq 2.$

(c) $\displaystyle \sum_{\substack{\alpha\in\mathbb{N}^m \\ |\alpha|=k}} 1 = \binom{m+k-1}{k}.$

(d) $\displaystyle \sum_{\substack{\alpha\in\mathbb{N}^m \\ |\alpha|\leq k}} 1 = \binom{m+k}{k}.$

9 Verify that, for an arbitrary set X, $\bigl(\mathcal{P}(X),\triangle,\cap\bigr)$ is a commutative ring with unity (see Example 8.2(f)).

10 Let K be an ordered field and $a,b,c,d \in K$.

(a) Prove the inequality $\displaystyle \frac{|a+b|}{1+|a+b|} \leq \frac{|a|}{1+|a|} + \frac{|b|}{1+|b|}.$

(b) Show that, if $b > 0$, $d > 0$ and $\dfrac{a}{b} < \dfrac{c}{d}$, then $\dfrac{a}{b} < \dfrac{a+c}{b+d} < \dfrac{c}{d}.$

(c) Show that, if $a,b \in K^\times$, then $\left|\dfrac{a}{b} + \dfrac{b}{a}\right| \geq 2.$

11 Show that, in any ordered field K, we have

$$\begin{aligned}\sup\{a,b\} = \max\{a,b\} &= \frac{a+b+|a-b|}{2}\ , \\ \inf\{a,b\} = \min\{a,b\} &= \frac{a+b-|a-b|}{2}\ ,\end{aligned} \qquad a,b \in K\ .$$

12 Let R be an ordered ring and $a, b \in R$ such that $a \geq 0$ and $b \geq 0$. Suppose that there is some $n \in \mathbb{N}^\times$ such that $a^n = b^n$. Show that $a = b$.

13 Prove the statements in Examples 8.2(d) and (e).

14 Let K be a field. For $p = \sum_{k=0}^n p_k X^k \in K[X]$, set

$$Dp := \sum_{k=1}^n k p_k X^{k-1} \in K[X]\ ,$$

if $n \in \mathbb{N}^\times$, and $Dp = 0$, if p is constant. Prove that

$$D(pq) = pDq + qDp\ , \qquad p, q \in K[X]\ .$$

15 Find $r, s \in K[X]$ with $\deg(r) < 3$ such that

$$X^5 - 3X^4 + 4X^3 = s(X^3 - X^2 + X - 1) + r\ .$$

16 Let K be a finite field. Show that the homomorphism

$$K[X] \to K^K\ , \quad p \mapsto \underline{p}$$

from Remark 8.14(b) is not, in general, injective. (Hint: $p := X^2 - X \in \mathbb{F}_2[X]$.)

9 The Rational Numbers

After the algebraic investigations of the previous two sections, we return to our original question about the extension of the natural numbers to larger number systems. We want such extensions to preserve the usual commutivity, associativity and distributive laws of the natural numbers. As well, arbitrary differences and (almost arbitrary) quotients of elements should exist. In view of Remarks 8.1 and 8.7, these desired properties characterize fields, so a more precise goal is to 'embed' $\mathbb{N}$ in a field such the restriction of the field operations to $\mathbb{N}$ coincide with the usual addition and multiplication of natural numbers as seen in Theorem 5.3. Since $\mathbb{N}$ has a total order which is compatible with the operations $+$ and $\cdot$, we expect that this order structure should also extend to the entire field. Theorem 5.3 shows that the rules for calculating with the natural numbers, at least, do not contradict the rules that occur in ordered fields. We will see in this section that our question has an essentially unique answer. To show this we first embed $\mathbb{N}$ in the ring of integers, and then extend this ring to the field of rational numbers.

The Integers

From Theorem 5.3 we see that $\mathbb{N} = (\mathbb{N}, +, \cdot)$ is 'almost' a commutative ring with unity. The only property missing is the existence of an additive inverse $-n$ for each $n \in \mathbb{N}$.

Suppose that Z is a ring which contains $\mathbb{N}$, and that the ring operations on Z restrict to the usual operations on $\mathbb{N}$. Then for all $(m, n) \in \mathbb{N}^2$ the difference $m - n$ is a well defined element of Z, and

$$m - n = m' - n' \Longleftrightarrow m + n' = m' + n \ , \qquad (m', n') \in \mathbb{N}^2 \ . \tag{9.1}$$

For the sum of two such elements we have

$$(m - n) + (m' - n') = (m + m') - (n + n') \ , \tag{9.2}$$

and for their product

$$(m - n) \cdot (m' - n') = (mm' + nn') - (mn' + m'n) \ . \tag{9.3}$$

Note that the additions and multiplications in parentheses on the right side of each equation can be carried out completely within $\mathbb{N}$. This observation suggests *defining* addition and multiplication on $(m, n) \in \mathbb{N}^2$ using (9.2) and (9.3). In doing so we should not overlook (9.1) which indicates that two different pairs of natural numbers may correspond to a single element of the ring we are constructing. The following theorem shows the success of this strategy.

9.1 Theorem *There is a smallest domain (commutative ring without zero divisors) with unity, $\mathbb{Z}$, such that $\mathbb{N} \subseteq \mathbb{Z}$ and the ring operations on $\mathbb{Z}$ restrict to the usual operations on $\mathbb{N}$. This ring is unique up to isomorphism and is called the* **ring of integers**.

Proof We outline only the most important steps in the proof and leave to the reader the easy verifications that the operations are well defined and that the ring axioms (R_1)–(R_3) are satisfied.

Define an equivalence relation on $\mathbb{N}^2$ by

$$(m,n) \sim (m',n') :\Longleftrightarrow m+n' = m'+n \ ,$$

and set $\mathbb{Z} := \mathbb{N}^2/{\sim}$. Define addition and multiplication on $\mathbb{Z}$ by

$$[(m,n)] + [(m',n')] := [(m+m', n+n')]$$

and

$$[(m,n)] \cdot [(m',n')] := [(mm'+nn', mn'+m'n)] \ .$$

The rules of arithmetic in $\mathbb{N}$ from Theorem 5.3 imply that $\mathbb{Z} := (\mathbb{Z},+,\cdot)$ is a commutative ring without zero divisors. The zero and unity of $\mathbb{Z}$ are the equivalence classes $[(0,0)]$ and $[(1,0)]$ respectively.

The function

$$\mathbb{N} \to \mathbb{Z} \ , \quad m \mapsto [(m,0)] \tag{9.4}$$

is injective and compatible with the addition and multiplication operations in $\mathbb{N}$ and $\mathbb{Z}$. Consequently, we can identify $\mathbb{N}$ with its image under (9.4). Then $\mathbb{N} \subseteq \mathbb{Z}$, and the operations on $\mathbb{Z}$ restrict to the usual operations on $\mathbb{N}$.

Now let $R \supseteq \mathbb{N}$ be some commutative ring with unity and without zero divisors, such that the operations on R restrict to the usual operations on $\mathbb{N}$. Since $\mathbb{Z}$, by construction is clearly minimal, there is a unique injective homomorphism $\varphi : \mathbb{Z} \to R$ with $\varphi \,|\, \mathbb{N} =$ (inclusion of $\mathbb{N}$ in R). This implies the claimed uniqueness up to isomorphism. ■

In the following we do not distinguish different isomorphic copies of $\mathbb{Z}$ and speak of **the** (unique) **ring of integers**. (Another approach: Fix once and for all a particular representative of the isomorphism class of $\mathbb{Z}$ and call it **the** ring of integers.) The elements of $\mathbb{Z}$ are the **integers**, and $-\mathbb{N}^\times := \{\, -n \ ; \ n \in \mathbb{N}^\times \,\}$ is the set of **negative integers**. Clearly $\mathbb{Z} = \mathbb{N}^\times \cup \{0\} \cup (-\mathbb{N}^\times) = \mathbb{N} \cup (-\mathbb{N}^\times)$ as disjoint unions.

The Rational Numbers

In the ring $\mathbb{Z}$, we can now form arbitrary differences $m-n$, but, in general, the quotient of two integers m/n remains undefined, even if $n \neq 0$. For example, the equation $2x = 1$ has no solution in $\mathbb{Z}$ since, if $2(m-n) = 1$ with $m,n \in \mathbb{N}$, then $2m = 2n+1$, contradicting Proposition 5.4. To overcome this 'defect' we will construct a field K which contains $\mathbb{Z}$ as a subring. Of course, we choose K 'as small as possible'.

Following the pattern established for the extension of $\mathbb{N}$ to $\mathbb{Z}$, we suppose first that K is a such field. Then, for $a,c \in \mathbb{Z}$ and $b,d \in \mathbb{Z}^\times := \mathbb{Z}\backslash\{0\}$, we have relation (i) of Remark 8.7(d). This suggests that we introduce 'fractions' first as pairs of integers and define operations on these pairs so that the rules of Remark 8.7(d) hold. The following theorem shows that this idea works.

9.2 Theorem *There is, up to isomorphism, a unique smallest field $\mathbb{Q}$, which contains $\mathbb{Z}$ as a subring.*

Proof Once again we give only the most important steps in the proof and leave the verifications to the reader.

Define an equivalence relation on $\mathbb{Z} \times \mathbb{Z}^{\times}$ by

$$(a,b) \sim (a',b') :\Longleftrightarrow ab' = a'b ,$$

and set $\mathbb{Q} := (\mathbb{Z} \times \mathbb{Z}^{\times})/{\sim}$. Define addition and multiplication on $\mathbb{Q}$ by

$$\bigl[(a,b)\bigr] + \bigl[(a',b')\bigr] := \bigl[(ab' + a'b, bb')\bigr]$$

and

$$\bigl[(a,b)\bigr] \cdot \bigl[(a',b')\bigr] := \bigl[(aa', bb')\bigr] .$$

With these operations $\mathbb{Q} := (\mathbb{Q}, +, \cdot)$ is a field.

The function

$$\mathbb{Z} \to \mathbb{Q} , \quad z \mapsto \bigl[(z,1)\bigr] \tag{9.5}$$

is an injective ring homomorphism, and so we can identify $\mathbb{Z}$ with its image under (9.5) in $\mathbb{Q}$. Thus $\mathbb{Z}$ is a subring of $\mathbb{Q}$.

Let Q be a field which contains $\mathbb{Z}$ as a subring. By construction, $\mathbb{Q}$ is minimal and so there is a unique injective homomorphism $\varphi : \mathbb{Q} \to Q$ such that $\varphi | \mathbb{Z} =$ (inclusion of $\mathbb{Z}$ in Q). This implies the claimed uniqueness of $\mathbb{Q}$ up to isomorphism. ■

The elements of $\mathbb{Q}$ are called **rational numbers**. (Again, we do not distinguish isomorphic copies of $\mathbb{Q}$.)

9.3 Remarks **(a)** It is not hard to see that

$$r \in \mathbb{Q} \Longleftrightarrow \exists\, (p,q) \in \mathbb{Z} \times \mathbb{N}^{\times} \text{ with } r = p/q .$$

By Proposition 5.5, $\mathbb{N}$ is well ordered, and so, for a fixed $r \in \mathbb{Q}$, the set

$$\Bigl\{ q \in \mathbb{N}^{\times} \ ; \ \exists\, p \in \mathbb{Z} \text{ with } \frac{p}{q} = r \Bigr\}$$

has a unique minimum $q_0 := q_0(r)$. With $p_0 := p_0(r) := r q_0(r)$ we get a unique representation $r = p_0/q_0$ of r in **lowest terms**.

(b) In the construction of $\mathbb{Q}$ as an 'extension field' of $\mathbb{Z}$ in Theorem 9.2, no use is made of the fact that the elements of $\mathbb{Z}$ are 'numbers'. All that was necessary was that $\mathbb{Z}$ be a domain. So this proof shows that any domain R is a subring of a unique (up to isomorphism) minimal field Q. This field is called the **quotient field of** R.

(c) Let K be a field. Then the polynomial ring $K[X]$ is a domain (see Exercise 8.3). The corresponding quotient field, $K(X)$, is called the **field of rational functions over** K. Consequently a **rational function over** K is a quotient of two polynomials over K,

$$r = p/q\ , \qquad p,q \in K[X]\ , \quad q \neq 0\ ,$$

with the condition that, if $p', q' \in K[X]$, then $p'/q' = p/q = r$ if $pq' = p'q$. ■

9.4 Proposition *$\mathbb{Z}$ and $\mathbb{Q}$ are countably infinite.*

Proof Since $\mathbb{N} \subseteq \mathbb{Z} \subseteq \mathbb{Q}$, Example 6.1(a) shows that $\mathbb{Z}$ and $\mathbb{Q}$ are infinite. It is not difficult to see that

$$\varphi : \mathbb{N} \to \mathbb{Z}\ , \quad \varphi(n) := \begin{cases} n/2\ , & n \text{ even}\ , \\ -(n+1)/2\ , & n \text{ odd}\ , \end{cases}$$

is a bijection, and hence $\mathbb{Z}$ is countable. In view of Proposition 6.9, $\mathbb{Z} \times \mathbb{N}^\times$ is also countable. Expressing each element of $\mathbb{Q}$ in lowest terms as in Remark 9.3(a), one sees that there is a bijection from $\mathbb{Q}$ to a certain subset of $\mathbb{Z} \times \mathbb{N}^\times$. It then follows from Proposition 6.7 that $\mathbb{Q}$ is countable. ■

We define an order on $\mathbb{Q}$ by

$$\frac{m}{n} \le \frac{m'}{n'} :\Longleftrightarrow m'n - mn' \in \mathbb{N}\ , \qquad m, m' \in \mathbb{Z}\ , \quad n, n' \in \mathbb{N}^\times\ .$$

One can easily check that $\le$ is well defined.

9.5 Theorem *$\mathbb{Q} := (\mathbb{Q}, \le)$ is an ordered field and the order on $\mathbb{Q}$ restricts to the usual order on $\mathbb{N}$.*

Proof The simple verifications are left as an exercise. ■

Even though $\mathbb{Z}$ is not a field, the order on $\mathbb{Q}$ restricts to a total order on $\mathbb{Z}$ for which Proposition 8.9(i)–(vii) hold. In contrast to $\mathbb{N}$, neither $\mathbb{Z}$ nor $\mathbb{Q}$ is well ordered by $\le$.[1] For example, neither $\mathbb{Z}$, nor the set of even integers

$$2\mathbb{Z} = \{\, 2n\ ;\ n \in \mathbb{Z} \,\}\ ,$$

nor the set of odd integers

$$2\mathbb{Z} + 1 = \{\, 2n + 1\ ;\ n \in \mathbb{Z} \,\}$$

has a minimum — a fact which the Peano axioms, Theorem 5.3(vii) and Proposition 8.9(iv) make clear.

[1]However, it is possible to construct another order $\prec$ on $\mathbb{Q}$ so that $(\mathbb{Q}, \prec)$ is well ordered. See Exercise 9.

Rational Zeros of Polynomials

With the construction of the field $\mathbb{Q}$ we have found a number system in which familiar school arithmetic can be used without restriction. In particular, in $\mathbb{Q}$ we can now solve (uniquely) any equation of the form $ax = b$ with arbitrary $a, b \in \mathbb{Q}$ and $a \neq 0$.

What about solutions of equations of the form $x^n = b$ with $b \in \mathbb{Q}$ and $n \in \mathbb{N}^\times$? Here we can prove a general result which, in a sense, shows that such equations have few solutions.

9.6 Proposition *Any rational zero of a polynomial of the form*

$$f = X^n + a_{n-1}X^{n-1} + \cdots + a_1 X + a_0 \in \mathbb{Z}[X]$$

is an integer.

Proof Suppose $x \in \mathbb{Q} \backslash \mathbb{Z}$ is a zero of the above polynomial. Write $x = p/q$ in lowest terms. Because $x \notin \mathbb{Z}$, we have $p \in \mathbb{Z}^\times$ and $q > 1$. The statement $f(p/q) = 0$ is equivalent to

$$p^n = -q \sum_{j=0}^{n-1} a_j p^j q^{n-1-j} .$$

Because $q > 1$, there is a prime number r with $r \,|\, q$. Thus r divides p^n too and also p (see Exercise 5.7). Consequently, $p' := p/r$ and $q' := q/r$ are integers and $p'/q' = p/q = x$. Since $p' \neq 0$ and $q' < q$, this contradicts the assumption that the representation $x = p/q$ is in lowest terms. ■

9.7 Corollary *Let $n \in \mathbb{N}^\times$ and $a \in \mathbb{Z}$. If the equation $x^n = a$ has any solutions in $\mathbb{Q}$, then all such solutions are integers.*

Square Roots

We consider now the special case of the **quadratic equation** $x^2 = a$, not just in $\mathbb{Q}$, but in an arbitrary ordered field K.

9.8 Lemma *Let K be an ordered field and $a \in K^\times$. If the equation $x^2 = a$ has a solution, then $a > 0$. If $b \in K$ is a solution, then the equation has exactly two solutions, namely b and $-b$.*

Proof The first claim is clear since any solution b is nonzero and hence $a = b^2 > 0$ by Proposition 8.9(vii). Because $(-b)^2 = b^2$, if b is a solution, then so is $-b$. By Proposition 8.9(iv), $b = -b$ would imply $b = a = 0$, and so b and $-b$ are two distinct solutions. By Corollary 8.18, no further solutions can exist. ■

Let K be an ordered field and $a \in K$ with $a > 0$. If the equation $x^2 = a$ has a solution in K, then, by Lemma 9.8, it has exactly one *positive* solution. This is called the **square root** of a and is written $\sqrt{a}$. In this case we say, 'The square root of a exists in K'. In addition we set $\sqrt{0} := 0$.

9.9 Remarks (a) If $\sqrt{a}$ and $\sqrt{b}$ exist for some $a, b \geq 0$, then $\sqrt{ab}$ also exists and $\sqrt{ab} = \sqrt{a}\sqrt{b}$.

Proof From $x^2 = a$ and $y^2 = b$ it follows that $(xy)^2 = x^2y^2 = ab$. This shows the existence of $\sqrt{ab}$ as well as the equation $\sqrt{ab} = \sqrt{a}\sqrt{b}$. ∎

(b) For all $x \in K$, we have $|x| = \sqrt{x^2}$.

Proof If $x \geq 0$, then $\sqrt{x^2} = x$. Otherwise, if $x < 0$, then $\sqrt{x^2} = -x$. ∎

(c) For $a \in \mathbb{Z}$, $\sqrt{a}$ exists in $\mathbb{Q}$ if and only if a is the square of a natural number. ∎

Exercises

1 Let K be a field and $a \in K^\times$. For $m \in \mathbb{N}$, define $a^{-m} := (a^{-1})^m$.

(a) Prove that $a^{-m} = 1/a^m$ and $a^{m-n} = a^m/a^n$ for all $m, n \in \mathbb{N}$.

(b) By (a), a^k is defined for all $k \in \mathbb{Z}$. Verify the following rules:

$$a^k a^\ell = a^{k+\ell}\ , \quad a^k b^k = (ab)^k\ , \quad (a^k)^\ell = a^{k\ell}$$

for $a, b \in K^\times$ and $k, \ell \in \mathbb{Z}$.

2 For $n \in \mathbb{Z}$, $n\mathbb{Z}$ is an ideal of $\mathbb{Z}$, and so the quotient ring $\mathbb{Z}_n := \mathbb{Z}/n\mathbb{Z}$, **$\mathbb{Z}$ modulo n**, is well defined (see Exercise 8.6). Show the following:

(a) For $n \in \mathbb{N}^\times$, $\mathbb{Z}_n$ has exactly n elements. What is $\mathbb{Z}_0$?

(b) If $n \geq 2$ and $n \in \mathbb{N}$ is not a prime number, then $\mathbb{Z}_n$ has zero divisors.

(c) If $p \in \mathbb{N}$ is a prime number, then $\mathbb{Z}_p$ is a field.

(Hint: (b) Proposition 5.6. (c) For $a \in \mathbb{N}$ with $0 < a < p$ one needs to find some $x \in \mathbb{Z}$ such that $ax \in 1 + p\mathbb{Z}$. By repeated use of the division algorithm (Proposition 5.4) find positive numbers $r_0, \ldots, r_k$ and $q, q_0, \ldots, q_k$ such that $a > r_0 > r_1 > \cdots > r_k$ and

$$p = qa + r_0\ , \qquad a = q_0r_0 + r_1\ , \quad r_0 = q_1r_1 + r_2, \ldots, r_{k-2} = q_{k-1}r_{k-1} + r_k\ , \quad r_{k-1} = q_kr_k\ .$$

It follows that $r_j = m_ja + n_jp$ for $j = 0, \ldots, k$ with $m_j, n_j \in \mathbb{Z}$. Show that, since p is a prime number, $r_k = 1$.)

Remark Instead of $a \equiv b \pmod{n\mathbb{Z}}$ (see Exercise 8.6) we usually write $a \equiv b \pmod{n}$ for $n \in \mathbb{Z}$. Thus $a \equiv b \pmod{n}$ means that $a - b \in n\mathbb{Z}$.

3 Let X be an n element set. Show the following:

(a) $\mathrm{Num}(\mathcal{P}(X)) = 2^n$.

(b) $\mathrm{Num}(\mathcal{P}_{\mathrm{even}}(X)) = \mathrm{Num}(\mathcal{P}_{\mathrm{odd}}(X))$ for $n > 0$. Here $\mathcal{P}_{\mathrm{even}}(X)$ and $\mathcal{P}_{\mathrm{odd}}(X)$ are defined by

$$\mathcal{P}_{\mathrm{even}}(X) := \{\, A \subseteq X\ ;\ \mathrm{Num}(A) \equiv 0 \pmod{2} \,\}\ ,$$
$$\mathcal{P}_{\mathrm{odd}}(X) := \{\, A \subseteq X\ ;\ \mathrm{Num}(A) \equiv 1 \pmod{2} \,\}\ .$$

(Hint: Exercise 6.3 and Theorem 8.4.)

4 An ordered field K is called **Archimedean** if, for all $a, b \in K$ such that $a > 0$, there is some $n \in \mathbb{N}$ such that $b < na$. Verify that $\mathbb{Q} := (\mathbb{Q}, \leq)$ is Archimedean.

5 Show that any rational zero of a polynomial $p = \sum_{k=0}^{n} a_k X^k \in \mathbb{Z}[X]$ of degree $n \geq 1$ is in $a_n^{-1}\mathbb{Z}$. (Hint: Consider $a_n^{n-1}p$.)

6 On the symmetric group S_n define the **sign function** by

$$\operatorname{sign}\sigma := \prod_{1 \leq j < k \leq n} \frac{\sigma(j) - \sigma(k)}{j - k}, \qquad \sigma \in \mathsf{S}_n .$$

Show the following:

(a) $\operatorname{sign}(\mathsf{S}_n) \subseteq \{\pm 1\}$.

(b) $\operatorname{sign}(\sigma \circ \tau) = (\operatorname{sign}\sigma)(\operatorname{sign}\tau)$ for $\sigma, \tau \in \mathsf{S}_n$. That is, sign is a homomorphism from S_n to the multiplicative group $(\{\pm 1\}, \cdot)$. The kernel of this homomorphism is called the **alternating group** A_n, that is, $\mathsf{A}_n := \{\, \sigma \in \mathsf{S}_n \ ; \ \operatorname{sign}\sigma = 1 \,\}$. The permutations in A_n are called **even**, those in $\mathsf{S}_n \backslash \mathsf{A}_n$ are called **odd**.

(c) A_n has order $n!/2$ for $n \geq 2$, and 1 for $n = 1$.

(d) sign is surjective for $n \geq 2$.

(e) A **transposition** is a permutation which interchanges two numbers and leaves the others fixed. For $n \geq 2$, any permutation $\sigma \in \mathsf{S}_n$ can be represented as a composition of transpositions: $\sigma = \sigma_1 \circ \sigma_2 \circ \cdots \circ \sigma_N$, and then $\operatorname{sign}\sigma = (-1)^N$, independent of this representation. Thus the number of transpositions in the representation is even for even permutations and odd for odd permutations.

7 Give a complete proof of Theorem 9.5.

8 For $k \in \mathbb{N}$ and $q_0, \ldots, q_k \in \mathbb{N}^\times$, the rational number

$$q_0 + \cfrac{1}{q_1 + \cfrac{1}{q_2 + \cfrac{1}{q_3 + \cfrac{1}{\ddots \, \cfrac{}{q_{k-1} + \cfrac{1}{q_k}}}}}}$$

is called a **continued fraction**. Show that each $x \in \mathbb{Q}$ with $x \geq 0$ can be represented as a continued fraction, and that this representation is unique if $q_k \neq 1$. (Hint: Let $x = r/r_0$ in lowest terms. By the division algorithm there are unique $q_0 \in \mathbb{N}$ and $r_1 \in \mathbb{N}$ such that $r_1 < r_0$ and $r = q_0 r_0 + r_1$. If necessary, use the division algorithm on the pair (r_0, r_1). Repeating as needed, construct $q_0, \ldots, q_k$.)

9 Construct an order $\prec$ on $\mathbb{Q}$ such that $(\mathbb{Q}, \prec)$ is well ordered. (Hint: Consider Proposition 9.4 and (6.3).)

10 The Real Numbers

We have seen that the equation $x^2 = a$ for positive a is, in general, not solvable in $\mathbb{Q}$. Since x^2 is the area of a square with side x, this means, for example, that there is no square of area 2 — so long as we stay within the field of rational numbers. As is known from high school, in order to remedy this unsatisfactory situation, we must allow squares with sides whose lengths are 'irrational numbers'. This means that our field $\mathbb{Q}$ is too small, and we need a larger field which contains $\mathbb{Q}$ as a subfield, and in which the equation $x^2 = a$ for $a > 0$ always has a solution. In other words, we seek an ordered **extension field** of $\mathbb{Q}$ in which the equation $x^2 = a$ is solvable for each $a > 0$.

Order Completeness

The desired extension field is characterized by its completeness property. We say a totally ordered set X is **order complete** (or X satisfies the **completeness axiom**), if every nonempty subset of X which is bounded above has a supremum.

10.1 Proposition *Let X be a totally ordered set. Then the following are equivalent:*

(i) *X is order complete.*

(ii) *Every nonempty subset of X which is bounded below has an infimum.*

(iii) *For all nonempty subsets A, B of X such that $a \leq b$ for all $(a, b) \in A \times B$, there is some $c \in X$ such that $a \leq c \leq b$ for all $(a, b) \in A \times B$.*

Proof '(i)⇒(ii)' Let A be a nonempty subset of X which is bounded below. Then $B := \{ x \in X \ ; \ x \leq a \text{ for all } a \in A \}$ is nonempty and bounded above by any $a \in A$. By assumption, $m := \sup(B)$ exists. Since any element of A is an upper bound of B, and m is the least upper bound of B, we have $m \leq a$ for all $a \in A$. Thus m is in B, and, by Remark 4.5(c), $m = \max(B)$. By definition, this means that $m = \inf(A)$.

'(ii)⇒(iii)' Let A and B be nonempty subsets of X such that $a \leq b$ for $(a, b) \in A \times B$. Each $a \in A$ is a lower bound of B, so, by assumption, $c := \inf(B)$ exists. Since c is the greatest lower bound, we have $c \geq a$ for $a \in A$. Of course, c being a lower bound of B means $c \leq b$ for all $b \in B$.

'(iii)⇒(i)' Let A be a nonempty subset of X which is bounded above. Set $B := \{ b \in X \ ; \ b \geq a \text{ for all } a \in A \}$. Then is B nonempty and $a \leq b$ for all $a \in A$ and all $b \in B$. By hypothesis, there is some $c \in X$ such that $a \leq c \leq b$ for all $a \in A$ and $b \in B$. It follows that $c = \min(B)$, that is, $c = \sup(A)$. ■

Item (iii) of Proposition 10.1 is called the **Dedekind cut property**.

10.2 Corollary *A totally ordered set is order complete if and only if every nonempty bounded subset has a supremum and an infimum.*

The following example shows that ordered fields are not necessarily order complete.

10.3 Example $\mathbb{Q}$ is not order complete.

Proof We consider the sets

$$A := \{ x \in \mathbb{Q} \; ; \; x > 0 \text{ and } x^2 < 2 \} \; , \quad B := \{ x \in \mathbb{Q} \; ; \; x > 0 \text{ and } x^2 > 2 \} \; .$$

Clearly $1 \in A$ and $2 \in B$. From $b - a = (b^2 - a^2)/(b + a) > 0$ for $(a, b) \in A \times B$ it follows that $a < b$ for $(a, b) \in A \times B$. Now suppose that there is some $c \in \mathbb{Q}$ such that

$$a \leq c \leq b \; , \qquad (a, b) \in A \times B \; . \tag{10.1}$$

Then for $\xi := (2c + 2)/(c + 2)$ we have

$$\xi > 0 \; , \quad \xi = c - \frac{c^2 - 2}{c + 2} \; , \quad \xi^2 - 2 = \frac{2(c^2 - 2)}{(c + 2)^2} \; . \tag{10.2}$$

By Corollary 9.7 and Remark 4.3(b), either $c^2 < 2$ or $c^2 > 2$ is true. In the first case it follows from (10.2) that $\xi > c$ and $\xi^2 < 2$, that is, $\xi > c$ and $\xi \in A$, which contradicts (10.1). In the second case (10.2) implies the inequalities $\xi < c$ and $\xi^2 > 2$, that is, $\xi < c$ and $\xi \in B$, which once again contradicts (10.1). Thus there is no $c \in \mathbb{Q}$ which satisfies (10.1), and the claim follows from Proposition 10.1. ■

Dedekind's Construction of the Real Numbers

The following theorem, which shows that there is only one order complete extension field of $\mathbb{Q}$, is the most fundamental result of analysis and the starting point for all research into the 'limiting processes' which are at the center of all analytic investigation.

10.4 Theorem *There is, up to isomorphism, a unique order complete extension field $\mathbb{R}$ of $\mathbb{Q}$. This extension is called the* **field of real numbers**.

Proof For this fundamental theorem there are several proofs. The proof we present here uses **Dedekind cuts**, a concept originally due to R. Dedekind. Once again, we sketch only the essential ideas. For the (boring) technical details, see [Lan30]. Another proof, due to G. Cantor, will be given in Section II.6.

Motivated by Proposition 10.1(iii), the idea is to 'fill in' the missing number c between two subsets A and B of $\mathbb{Q}$ by simply identifying c with the ordered pair (A, B). That is, the new numbers we construct are ordered pairs (A, B) of subsets of $\mathbb{Q}$. It is then necessary to give the set of such pairs the structure of an ordered field and show that this field is order complete and contains an isomorphic copy of $\mathbb{Q}$.

It suffices to consider pairs (A, B) with $a \leq b$ for $(a, b) \in A \times B$ and such that $A \cup B = \mathbb{Q}$. Such a pair is determined by either of the sets A or B. Choosing B leads to the following formal definition. Let $\mathbb{R} \subseteq \mathcal{P}(\mathbb{Q})$ be the set of all $R \subseteq \mathbb{Q}$ with the following properties:

(D_1) $R \neq \emptyset$, $R^c = \mathbb{Q} \backslash R \neq \emptyset$.

(D_2) $R^c = \{x \in \mathbb{Q} \; ; \; x < r \text{ for all } r \in R\}$.

(D_3) R has no minimum.

The function

$$\mathbb{Q} \to \mathbb{R} \, , \quad r \mapsto \{x \in \mathbb{Q} \; ; \; x > r\} \tag{10.3}$$

is injective, so we identify $\mathbb{Q}$ with its image in $\mathbb{R}$, that is, we consider $\mathbb{Q}$ to be a subset of $\mathbb{R}$.

For $R, R' \in \mathbb{R}$, we define

$$R \leq R' :\Longleftrightarrow R \supseteq R' \; . \tag{10.4}$$

Examples 4.4(a) and (b) show that $\leq$ is a partial order on $\mathbb{R}$. If R and R' are distinct, then there is some $r \in R$ with $r \in (R')^c$ or some $r' \in R'$ with $r' \in R^c$. In the first case, $r < r'$ for each $r' \in R'$ and so $r' \in R$ for $r' \in R'$. Consequently $R' \subseteq R$, that is, $R' \geq R$. In the second case, we have similarly $R' \leq R$. Therefore $(\mathbb{R}, \leq)$ is a totally ordered set.

Let $\mathcal{R}$ be a nonempty subset of $\mathbb{R}$ which is bounded below, that is, there is some $A \in \mathbb{R}$ such that $R \subseteq A$ for all $R \in \mathcal{R}$. Set $S := \bigcup \mathcal{R}$. Then S is not empty, and, since $S \subseteq A$, we have $\emptyset \neq A^c \subseteq S^c$ which implies that S^c is nonempty. Thus S satisfies (D_1). It is clear that S also satisfies (D_2) and (D_3) and so S is in $\mathbb{R}$. Since S is itself a lower bound of $\mathcal{R}$, indeed the greatest lower bound, we have, as in Example 4.6(a), $S = \inf(\mathcal{R})$. From Proposition 10.1 we conclude that $\mathbb{R}$ is order complete.

Define addition on $\mathbb{R}$ by

$$\mathbb{R} \times \mathbb{R} \to \mathbb{R} \, , \quad (R, S) \mapsto R + S = \{r + s \; ; \; r \in R, \; s \in S\} \; .$$

It is easy to verify that this operation is well defined, associative and commutative, and has the identity element $O := \{x \in \mathbb{Q} \; ; \; x > 0\}$. Further, the additive inverse of $R \in \mathbb{R}$ is $-R := \{x \in \mathbb{Q} \; ; \; x + r > 0 \text{ for all } r \in R\}$. Thus $(\mathbb{R}, +)$ is an Abelian group and $R > O \Longleftrightarrow -R < O$.

Define multiplication on $\mathbb{R}$ by

$$R \cdot R' := \{rr' \in \mathbb{Q} \; ; \; r \in R, \; r' \in R'\} \qquad \text{for } R, R' \geq O$$

and

$$R \cdot R' := \begin{cases} -((-R) \cdot R') \, , & R < O \, , \quad R' \geq O \, , \\ -(R \cdot (-R')) \, , & R \geq O \, , \quad R' < O \, , \\ (-R) \cdot (-R') \, , & R < O \, , \quad R' < O \, . \end{cases}$$

Then one can show that $\mathbb{R} := (\mathbb{R}, +, \cdot, \leq)$ is an ordered field which contains $\mathbb{Q}$ as a subfield and that the order on $\mathbb{R}$ restricts to the usual order on $\mathbb{Q}$.

Now let $\mathbb{S}$ be some order complete extension field of $\mathbb{Q}$. Define a function by

$$\mathbb{S} \to \mathbb{R} \, , \quad r \mapsto \{x \in \mathbb{Q} \; ; \; x > r\} \; .$$

One can prove that this is an increasing isomorphism. Consequently, $\mathbb{R}$ is unique up to isomorphism. ■

The proof of Example 10.3 shows that the set

$$R := \{\, x \in \mathbb{Q} \; ; \; x > 0 \text{ and } x^2 > 2 \,\}$$

is in $\mathbb{R}$, but not in $\mathbb{Q}$. In fact, one can show that $R = \sqrt{2}$ in $\mathbb{R}$.

The Natural Order on $\mathbb{R}$

The elements of $\mathbb{R}$ are called the **real numbers** and the order on $\mathbb{R}$ is the **natural order** on the real numbers. The restriction of this order to the subsets

$$\mathbb{N} \subset \mathbb{Z} \subset \mathbb{Q} \subset \mathbb{R}$$

is, of course, the 'usual order' on each subset. A real number x is called **positive** (or **negative**) if $x > 0$ (or $x < 0$). Thus

$$\mathbb{R}^+ := \{\, x \in \mathbb{R} \; ; \; x \geq 0 \,\}$$

is the set of **nonnegative real numbers**.

Since $\mathbb{R}$ is totally ordered, we can think of the real numbers as 'points' on the **number line**[1]. Here we agree that x is 'to the left of y' when $x < y$, and that the integers $\mathbb{Z}$ are 'equally spaced'. The arrow gives the 'orientation' of the number line, that is, the direction in which 'the numbers increase'.

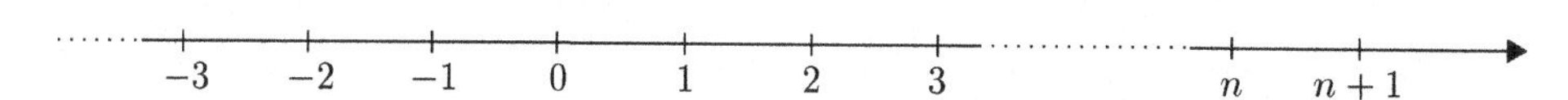

This picture of $\mathbb{R}$ is based on the intuitive ideas that the real numbers are 'unbounded in both directions' and that they form a continuum, that is, the number line has 'no holes'. The first claim will be justified in Proposition 10.6. The second is exactly the Dedekind continuity property.

The Extended Number Line

To extend our use of the symbols $\pm\infty$ to the real numbers, we set $\bar{\mathbb{R}} := \mathbb{R} \cup \{\pm\infty\}$, the **extended number line**, and make the convention that

$$-\infty < x < \infty \; , \qquad x \in \mathbb{R} \; ,$$

so that $\bar{\mathbb{R}}$ is a totally ordered set. We insist again that $\pm\infty$ are *not* real numbers.

---

[1]We use here, of course, the usual intuitive ideas of point and line. For a purely axiomatic development of these concepts, the very readable book of P. Gabriel [Gab96] is recommended (especially for the interesting historical comments).

As well as the order structure, we (partially) extend the operations $\cdot$ and $+$ to $\bar{\mathbb{R}}$ as follows: For $x \in \bar{\mathbb{R}}$, we define

$$x + \infty := \infty \quad \text{for } x > -\infty , \qquad x - \infty := -\infty \quad \text{for } x < \infty ,$$

and

$$x \cdot \infty := \begin{cases} \infty , & x > 0 , \\ -\infty , & x < 0 , \end{cases} \qquad x \cdot (-\infty) := \begin{cases} -\infty , & x > 0 , \\ \infty , & x < 0 , \end{cases}$$

and, for $x \in \mathbb{R}$, define

$$\frac{x}{\infty} := \frac{x}{-\infty} := 0 , \quad \frac{x}{0} := \begin{cases} \infty , & x > 0 , \\ -\infty , & x < 0 . \end{cases}$$

Of course, we assume also that these operations are commutative.[2] In particular, the following hold:

$$\infty + \infty = \infty , \quad -\infty - \infty = -\infty , \quad \infty \cdot \infty = \infty ,$$
$$(-\infty) \cdot \infty = \infty \cdot (-\infty) = -\infty , \quad (-\infty) \cdot (-\infty) = \infty .$$

Note that

$$\infty - \infty , \quad 0 \cdot (\pm\infty) , \quad \frac{\pm\infty}{+\infty} , \quad \frac{\pm\infty}{-\infty} , \quad \frac{0}{0} , \quad \frac{\pm\infty}{0}$$

are *not* defined, and that $\bar{\mathbb{R}}$ is *not* a field. (Why not?)

A Characterization of Supremum and Infimum

Using the extended number line, we can define a supremum and an infimum for sets of real numbers which otherwise do not have these: If M is a nonempty subset of $\mathbb{R}$ which is not bounded above (in $\mathbb{R}$), then ∞ is the least upper bound of M in $\bar{\mathbb{R}}$ and so we set $\sup(M) := \infty$. Similarly, if M is a nonempty subset of $\mathbb{R}$ which is not bounded below, then $\inf(M) := -\infty$. We define also $\sup(\emptyset) = -\infty$ and $\inf(\emptyset) = \infty$. The use of these conventions is justified by the following characterization of the supremum and infimum of sets of real numbers.

10.5 Proposition

(i) *If $A \subseteq \mathbb{R}$ and $x \in \mathbb{R}$, then*

(α) $x < \sup(A) \Longleftrightarrow \exists\, a \in A$ *such that* $x < a$.

(β) $x > \inf(A) \Longleftrightarrow \exists\, a \in A$ *such that* $x > a$.

(ii) *Every subset A of $\mathbb{R}$ has a supremum and an infimum in $\bar{\mathbb{R}}$.*

[2] Compare the footnote on page 46.

Proof (i) If $A = \emptyset$, then the claim follows directly from our convention. Suppose then that $A \neq \emptyset$. We prove only (α) since (β) is proved similarly.

'$\Rightarrow$' If, to the contrary, $x < \sup(A)$ is such that $a \leq x$ for all $a \in A$, then x is an upper bound of A, which, by the definition of $\sup(A)$, is not possible.

'$\Leftarrow$' Let $a \in A$ be such that $x < a$. Then clearly $x < a \leq \sup(A)$.

(ii) If A is a nonempty subset of $\mathbb{R}$ which is bounded above, then Theorem 10.4 guarantees the existence of $\sup(A)$ in $\mathbb{R}$, and hence also in $\bar{\mathbb{R}}$. On the other hand, if $A = \emptyset$ or A is not bounded above, then $\sup(A) = -\infty$ or $\sup(A) = \infty$ respectively. The claim about the infimum follows similarly. ■

The Archimedean Property

10.6 Proposition (Archimedes) *$\mathbb{N}$ is not bounded above in $\mathbb{R}$, that is, for each $x \in \mathbb{R}$ there is some $n \in \mathbb{N}$ such that $n > x$.*

Proof Let $x \in \mathbb{R}$. For $x < 0$, the claim is obviously true. Suppose that $x \geq 0$ and hence the set $A := \{ n \in \mathbb{N} \ ; \ n \leq x \}$ is nonempty and bounded above by x. Then $s := \sup(A)$ exists in $\mathbb{R}$. By Proposition 10.5, there is some $a \in A$ such that $s - 1/2 < a$. Now set $n := a + 1$ so that $n > s$. Then n is not in A and so $n > x$. ■

10.7 Corollary

(i) *Let $a \in \mathbb{R}$. If $0 \leq a \leq 1/n$ for all $n \in \mathbb{N}^\times$, then $a = 0$.*

(ii) *For each $a \in \mathbb{R}$ with $a > 0$ there is some $n \in \mathbb{N}^\times$ such that $1/n < a$.*

Proof If $0 < a \leq 1/n$ for all $n \in \mathbb{N}^\times$, then it follows that $n \leq 1/a$ for all $n \in \mathbb{N}^\times$. Thus $\mathbb{N}$ would be bounded above in $\mathbb{R}$, contradicting Proposition 10.6.

(ii) is an equivalent reformulation of (i). ■

The Density of the Rational Numbers in $\mathbb{R}$

The next proposition shows that $\mathbb{Q}$ is 'dense' in $\mathbb{R}$, that is, real numbers can be 'approximated' by rational numbers. We will consider this idea in much greater generality in the next chapter. In particular, we will see that the real numbers are uniquely characterized by this approximation property.

10.8 Proposition *For all $a, b \in \mathbb{R}$ such that $a < b$, there is some $r \in \mathbb{Q}$ such that $a < r < b$.*

Proof (a) By assumption we have $b - a > 0$. Thus, by Proposition 10.6, there is some $n \in \mathbb{N}$ such that $n > 1/(b - a) > 0$. This implies $nb > na + 1$.

(b) By Proposition 10.6 again, there are $m_1, m_2 \in \mathbb{N}$ such that $m_1 > na$ and $m_2 > -na$, that is, $-m_2 < na < m_1$. Consequently there is some $m \in \mathbb{Z}$ such that $m - 1 \leq na < m$ (proof?). Together with (a), this implies

$$na < m \leq 1 + na < nb \ .$$

The claim then follows by setting $r := m/n \in \mathbb{Q}$. ■

n^{th} Roots

At the beginning of this section we motivated the construction of $\mathbb{R}$ by the desire to take the square root of arbitrary positive rational numbers. The following proposition shows that we have attained this goal and considerably more.

10.9 Proposition *For all $a \in \mathbb{R}^+$ and $n \in \mathbb{N}^\times$, there is a unique $x \in \mathbb{R}^+$ such that $x^n = a$.*

Proof (a) We prove first the uniqueness claim. It suffices to show that $x^n < y^n$ if $0 < x < y$ and $n \geq 2$. This follows from

$$y^n - x^n = (y - x) \sum_{j=0}^{n-1} y^j x^{n-j} > 0 \tag{10.5}$$

(see Exercise 8.1).

(b) To prove the existence of a solution, we can, without loss of generality, assume that $n \geq 2$ and $a \notin \{0, 1\}$.

We begin with the case $a > 1$. Then, from Proposition 8.9(iii), we have

$$x^n > a^n > a > 0 \qquad \text{for all } x > a \ . \tag{10.6}$$

Now set $A := \{\, x \in \mathbb{R}^+ \ ; \ x^n \leq a \,\}$. Then $0 \in A$ and, by (10.6), $x \leq a$ for all $x \in A$. Thus $s := \sup(A)$ is a well defined real number such that $s \geq 0$. We will prove that $s^n = a$ holds by showing that $s^n \neq a$ leads to a contradiction.

Suppose first that $s^n < a$ so that $a - s^n > 0$. By Corollary 10.7 and Proposition 10.8, the inequality

$$b := \sum_{k=0}^{n-1} \binom{n}{k} s^k > 0$$

implies that there is some $\varepsilon \in \mathbb{R}$ such that $0 < \varepsilon < (a - s^n)/b$. By making ε smaller if needed, we can further suppose that $\varepsilon \leq 1$. Then $\varepsilon^k \leq \varepsilon$ for all $k \in \mathbb{N}^\times$, and, using the binomial theorem, we have

$$(s + \varepsilon)^n = s^n + \sum_{k=0}^{n-1} \binom{n}{k} s^k \varepsilon^{n-k} \leq s^n + \Bigl(\sum_{k=0}^{n-1} \binom{n}{k} s^k \Bigr) \varepsilon < a \ .$$

This shows that $s+\varepsilon \in A$, a contradiction of $\sup(A) = s < s+\varepsilon$. Therefore $s^n < a$ cannot be true.

Now suppose that $s^n > a$. Then, in particular, $s > 0$ and

$$b := \sum\nolimits^* \binom{n}{2j-1} s^{2j-1} > 0 \ ,$$

where the symbol $\sum^*$ means that we sum over all indices $j \in \mathbb{N}^\times$ such that $2j \leq n$. Proposition 10.8 implies that there is some $\varepsilon \in \mathbb{R}$ such that $0 < \varepsilon < (s^n - a)/b$ and $\varepsilon \leq 1 \wedge s$. Thus we have

$$\begin{aligned} (s-\varepsilon)^n &= s^n + \sum_{k=0}^{n-1} (-1)^{n-k} \binom{n}{k} s^k \varepsilon^{n-k} \\ &\geq s^n - \sum\nolimits^* \binom{n}{2j-1} s^{2j-1} \varepsilon^{n-2j+1} \geq s^n - \varepsilon \sum\nolimits^* \binom{n}{2j-1} s^{2j-1} \\ &> a \ . \end{aligned} \tag{10.7}$$

Now let $x \in \mathbb{R}^+$ with $x \geq s - \varepsilon$. Then it follows from (10.7) that $x^n \geq (s-\varepsilon)^n > a$, that is, $x \notin A$. This shows that $s - \varepsilon$ is an upper bound of A, which is not possible, because $s - \varepsilon < s$ and $s = \sup(A)$. Thus the assumption $s^n > a$ cannot be true.

Since $\mathbb{R}$ is totally ordered, the only remaining possibility is that $s^n = a$.

Finally we consider the case $a \in \mathbb{R}^+$ with $0 < a < 1$. Set $b := 1/a > 1$. Then, from the above, there is a unique $y > 0$ with $y^n = b$, and so $x := 1/y$ is the unique solution of $x^n = a$. ■

10.10 Remarks **(a)** If $n \in \mathbb{N}^\times$ is odd, then the equation $x^n = a$ has a unique solution $x \in \mathbb{R}$ for each $a \in \mathbb{R}$.

Proof If $a \geq 0$, then the claim follows from Proposition 10.9 and the fact that $y < 0$ implies $y^n < 0$ for $n \in 2\mathbb{N}+1$. If $a < 0$, then the claim follows from what we have just shown, and the fact that $x \mapsto -x$ is a bijection between the solution set of $x^n = a$ and the solution set of $x^n = -a$: If $x^n = a$, then, since n is odd,

$$(-x)^n = (-1)^n x^n = (-1)a = -a \ .$$

Similarly, $x^n = -a$ implies $(-x)^n = a$. ■

(b) Suppose that either $n \in \mathbb{N}$ is odd and $a \in \mathbb{R}$, or $n \in \mathbb{N}$ is even and $a \in \mathbb{R}^+$. Denote by $\sqrt[n]{a}$ the unique solution (in $\mathbb{R}$ if n is odd, or in $\mathbb{R}^+$ if n is even) of the equation $x^n = a$. We call $\sqrt[n]{a}$ the nth **root** of a.

If n is even and $a > 0$, then the equation $x^n = a$ has, besides $\sqrt[n]{a}$, the solution $-\sqrt[n]{a}$ in $\mathbb{R}$, the 'negative n^{th} root of a'.

Proof Since n is even, we have $(-1)^n = 1$, and so

$$\left(-\sqrt[n]{a}\right)^n = (-1)^n \left(\sqrt[n]{a}\right)^n = a \ ,$$

which proves the claim. ■

(c) The functions

$$\mathbb{R}^+ \to \mathbb{R}^+ , \quad x \mapsto \sqrt[n]{x} , \qquad n \in 2\mathbb{N} ,$$

and

$$\mathbb{R} \to \mathbb{R} , \quad x \mapsto \sqrt[n]{x} , \qquad n \in 2\mathbb{N}+1 ,$$

are strictly increasing.

Proof It follows from (10.5) that $0 < \sqrt[n]{x} < \sqrt[n]{y}$ for all $0 < x < y$. If $x < y < 0$ and $n \in 2\mathbb{N}+1$, then from the definition in (b) and what we have just proved, it follows that $\sqrt[n]{x} < \sqrt[n]{y}$. The remaining cases are trivial. ∎

(d) Let $a \in \mathbb{R}^+$ and $r = p/q \in \mathbb{Q}$ in lowest terms. Define the rth **power** of a by

$$a^r := \left(\sqrt[q]{a}\right)^p .$$

Note that, because of the uniqueness of the representation of r in lowest terms, a^r is well defined.

(e) Corollary 9.7 and Proposition 10.9 show, in particular, that $\sqrt{2} \in \mathbb{R}\backslash\mathbb{Q}$, that is, $\sqrt{2}$ is a real number which is not rational. The elements of $\mathbb{R}\backslash\mathbb{Q}$ are called **irrational numbers.** ∎

The Density of the Irrational Numbers in $\mathbb{R}$

In Proposition 10.8 we saw that the rational numbers $\mathbb{Q}$ are dense in $\mathbb{R}$. The next proposition shows that the irrational numbers $\mathbb{R}\backslash\mathbb{Q}$ have this same property.

10.11 Proposition *For any $a, b \in \mathbb{R}$ such that $a < b$, there is some $\xi \in \mathbb{R}\backslash\mathbb{Q}$ such that $a < \xi < b$.*

Proof Suppose $a, b \in \mathbb{R}$ satisfy $a < b$. By Proposition 10.8 there are rational numbers $r_1, r_2 \in \mathbb{Q}$ such that $a < r_1 < b$ and $r_1 < r_2 < b$. Setting $\xi := r_1 + (r_2 - r_1)/\sqrt{2}$ we have $r_1 < \xi$ and

$$r_2 - \xi = (r_2 - r_1)\left(1 - 1/\sqrt{2}\right) > 0 ,$$

and hence $\xi < r_2$. Thus $r_1 < \xi < r_2$ and also $a < \xi < b$. Finally ξ cannot be a rational number since otherwise $\sqrt{2} = (r_2 - r_1)/(\xi - r_1)$ would also be rational. ∎

By Corollary 9.7, the square root of any natural number which is not the square of a natural number, is irrational. In particular, there are 'many' irrational numbers. In Section II.7 we will show that $\mathbb{R}$ is uncountable. Since $\mathbb{Q}$ is countable, Proposition 6.8 implies that there are, in fact, uncountably many irrational numbers.

Intervals

An **interval** is a subset J of $\mathbb{R}$ such that

$$(x, y \in J,\ x < y) \Rightarrow (z \in J \text{ for } x < z < y)\ .$$

Clearly $\emptyset$, $\mathbb{R}$, $\mathbb{R}^+$, $-\mathbb{R}^+$ are intervals, but $\mathbb{R}^\times$ is not. If J is a nonempty interval, then $\inf(J) \in \bar{\mathbb{R}}$ is the **left endpoint** and $\sup(J) \in \bar{\mathbb{R}}$ is the **right endpoint** of J. It is an easy exercise to show that a nonempty interval J is determined by its endpoints and whether or not these endpoints are in J. Thus J is **closed on the left** if $a := \inf(J)$ is in J, and otherwise it is **open on the left**. Similarly, J is **closed on the right** if $b := \sup(J)$ is in J, and otherwise it is **open on the right**. The interval J is called **open** if it is empty or is open on the left and right. In this case we write (a, b) for J, that is,

$$(a, b) = \{\, x \in \mathbb{R}\ ;\ a < x < b \,\}\ , \qquad -\infty \le a \le b \le \infty\ ,$$

with the convention that $(a, a) := \emptyset$. If J is closed on the left and right or is empty, then J is called a **closed** interval which we write as

$$[a, b] = \{\, x \in \mathbb{R}\ ;\ a \le x \le b \,\}\ , \qquad -\infty < a \le b < \infty\ .$$

Further, we write $(a, b]$ (or $[a, b)$) if J is open on the left and closed on the right (or closed on the left and open on the right). Each one element subset $\{a\}$ of $\mathbb{R}$ is a closed interval. An interval is **perfect** if is contains at least two points. It is **bounded** if both endpoints are in $\mathbb{R}$, and is **unbounded** otherwise. Each unbounded interval of $\mathbb{R}$, other than $\mathbb{R}$ itself, has the form $[a, \infty)$, (a, ∞), $(-\infty, a]$ or $(-\infty, a)$ with $a \in \mathbb{R}$. If J is a bounded interval, then the nonnegative number $|J| := \sup(J) - \inf(J)$ is called the **length** of J.

Exercises

1 Determine the following subsets of $\mathbb{R}^2$:

$$\begin{aligned} A &:= \{\, (x, y) \in \mathbb{R}^2\ ;\ |x - 1| + |y + 1| \le 1 \,\}\ ,\\ B &:= \{\, (x, y) \in \mathbb{R}^2\ ;\ 2x^2 + y^2 > 1,\ |x| \le |y| \,\}\ ,\\ C &:= \{\, (x, y) \in \mathbb{R}^2\ ;\ x^2 - y^2 > 1,\ x - 2y < 1,\ y - 2x < 1 \,\}\ . \end{aligned}$$

2 (a) Show that

$$\mathbb{Q}(\sqrt{2}) := \{a + b\sqrt{2}\ ;\ a, b \in \mathbb{Q}\}$$

is a subfield of $\mathbb{R}$ which contains $\mathbb{Q}$ but which is not order complete. Is $\sqrt{3}$ in $\mathbb{Q}(\sqrt{2})$?

(b) Prove that $\mathbb{Q}$ is the smallest subfield of $\mathbb{R}$.

3 For $a, b \in \mathbb{R}^+$ and $r, s \in \mathbb{Q}$, show the following:
(a) $a^{r+s} = a^r a^s$, (b) $(a^r)^s = a^{rs}$, (c) $a^r b^r = (ab)^r$.

4 For $m, n \in \mathbb{N}^\times$ and $a, b \in \mathbb{R}^+$, show the following:

(a) $a^{1/m} < a^{1/n}$, if $m < n$ and $0 < a < 1$.

(b) $a^{1/m} > a^{1/n}$, if $m < n$ and $a > 1$.

5 Let $f : \mathbb{R} \to \mathbb{R}$ be an increasing function. Suppose that $a, b \in \mathbb{R}$ satisfy $a < b$, $f(a) > a$ and $f(b) < b$. Prove that f has at least one **fixed point**, that is, there is some $x \in \mathbb{R}$ such that $f(x) = x$. (Hint: Consider $z := \sup\{\, y \in \mathbb{R} \; ; \; a \le y \le b, \; y \le f(y) \,\}$ and $f(z)$.)

6 Prove *Bernoulli's inequality*: If $x \in \mathbb{R}$ with $x > -1$ and $n \in \mathbb{N}$, then

$$(1+x)^n \ge 1 + nx \; .$$

7 Let $M \subseteq \mathbb{R}$ be nonempty with $\inf(M) > 0$. Show that the set $M' := \{\, 1/x \; ; \; x \in M \,\}$ is bounded above and that $\sup(M') = 1/\inf(M)$.

8 For nonempty subsets A and B of $\mathbb{R}$, prove the following:

$$\sup(A+B) = \sup(A) + \sup(B) \; , \quad \inf(A+B) = \inf(A) + \inf(B) \; .$$

9 (a) For nonempty subsets A and B of $(0, \infty)$, prove the following:

$$\sup(A \cdot B) = \sup(A) \cdot \sup(B) \; , \quad \inf(A \cdot B) = \inf(A) \cdot \inf(B) \; .$$

(b) Find nonempty subsets A and B of $\mathbb{R}$ such that

$$\sup(A) \cdot \sup(B) < \sup(A \cdot B) \quad \text{and} \quad \inf(A) \cdot \inf(B) > \inf(A \cdot B) \; .$$

10 Let $n \in \mathbb{N}^\times$ and $x = (x_1, \ldots, x_n) \in [\mathbb{R}^+]^n$. Then the **geometric mean** and **arithmetic mean** of $x_1, \ldots, x_n$ are defined by $g(x) := \sqrt[n]{\prod_{j=1}^n x_j}$ and $a(x) := (1/n) \sum_{j=1}^n x_j$ respectively. Prove that $g(x) \le a(x)$ (*inequality of the geometric and arithmetic means*).

11 For $x = (x_1, \ldots, x_n)$ and $y = (y_1, \ldots, y_n)$ in $\mathbb{R}^n$, define $x \bullet y := \sum_{j=1}^n x_j y_j$. Prove that

$$\sqrt[|\alpha|]{x^\alpha} \le (x \bullet \alpha)/|\alpha| \; .$$

for all $x \in [\mathbb{R}^+]^n$ and $\alpha \in \mathbb{N}^n$ (*inequality of the* **weighted** *geometric and* **weighted** *arithmetic means*).

12 Verify that $\mathbb{R}$ is an Archimedean ordered field. See Exercise 9.4.

13 Let $(K, \le)$ be an ordered extension field of $(\mathbb{Q}, \le)$ with the property that, for each $a \in K$ such that $a > 0$, there is some $r \in \mathbb{Q}$ such that $0 < r < a$. Show that K is an Archimedean ordered field. See Exercise 9.4.

14 Prove that an ordered field K is Archimedean if and only if $\{\, n \cdot 1 \; ; \; n \in \mathbb{N} \,\}$ is not bounded above in K. See Exercise 9.4.

15 Let K be the field of rational functions with coefficients in $\mathbb{R}$ (see Remark 9.3(c)). Then for each $f \in K$ there are unique polynomials $p = \sum_{k=0}^n p_k X^k$ and $q = \sum_{k=0}^m q_k X^k$ such that $q_m = 1$ and $f = p/q$ is in lowest terms (that is, p and q have no nonconstant factors in common). With this notation let

$$\mathcal{P} := \{\, f \in K \; ; \; p_n \ge 0 \,\} \; .$$

Finally set

$$f \prec g :\Longleftrightarrow g - f \in \mathcal{P} \; .$$

Show that $(K, \prec)$ is an ordered field, but not an Archimedean ordered field.

16 For each $n \in \mathbb{N}$, let I_n be a nonempty closed interval in $\mathbb{R}$. The family $\{ I_n \ ; \ n \in \mathbb{N} \}$ is called a **nest of intervals** if the following conditions hold:

(i) $I_{n+1} \subseteq I_n$ for all $n \in \mathbb{N}$.

(ii) For each $\varepsilon > 0$, there is some $n \in \mathbb{N}$ such that $|I_n| < \varepsilon$.

(a) Show that, for each nest of intervals $\{ I_n \ ; \ n \in \mathbb{N} \}$, there is a unique $x \in \mathbb{R}$ such that $x \in \bigcap_n I_n$.

(b) For each $x \in \mathbb{R}$, show that there is a nest of intervals $\{ I_n \ ; \ n \in \mathbb{N} \}$ with rational endpoints such that $\{x\} = \bigcap_n I_n$.

11 The Complex Numbers

In Section 9 we saw that, in an ordered field K, all squares are nonnegative, that is, $x^2 \geq 0$ for $x \in K$. As a consequence, the equation $x^2 = -1$ is *not* solvable in the field of real numbers or in any other ordered field. In this section we construct an extension field of $\mathbb{R}$, the field of complex numbers $\mathbb{C}$, in which all quadratic equations (indeed, as we later see, all algebraic equations) have at least one solution. Surprisingly, in contrast to the extension of $\mathbb{Q}$ to $\mathbb{R}$ using Dedekind cuts, the extension of $\mathbb{R}$ to $\mathbb{C}$ is simple.

Constructing the Complex Numbers

Following the pattern established for the extensions of $\mathbb{N}$ to $\mathbb{Z}$ and $\mathbb{Z}$ to $\mathbb{Q}$, we suppose first that there is an extension field K of $\mathbb{R}$ and some $i \in K$ such that $i^2 = -1$. Of course, $i \notin \mathbb{R}$. From this supposition we derive properties of K which lead to an explicit construction of K.

Since K is a field, if $x, y \in \mathbb{R}$, then $z := x + iy$ is an element of K. Moreover, the representation $z = x + iy$ in K is unique, that is, if in addition, $z = a + ib$ for some $a, b \in \mathbb{R}$, then $x = a$ and $y = b$. To prove this, suppose that $y \neq b$ and $x + iy = a + ib$. Then it follows that $i = (x - a)/(b - y) \in \mathbb{R}$, which is not possible.

Motivated by these observations we set $C := \{x + iy \in K \ ; \ x, y \in \mathbb{R}\}$. For $z = x + iy$ and $w = a + ib$ in C we have (in K)

$$\begin{aligned} z + w &= x + a + i(y + b) \in C , \\ -z &= -x + i(-y) \in C , \\ zw &= xa + ixb + iya + i^2 yb = xa - yb + i(xb + ya) \in C , \end{aligned} \tag{11.1}$$

where we used $i^2 = -1$. Finally, if $z = x + iy \neq 0$, thus $x \in \mathbb{R}^\times$ or $y \in \mathbb{R}^\times$. Then we have (in K)

$$\frac{1}{z} = \frac{1}{x + iy} = \frac{x - iy}{(x + iy)(x - iy)} = \frac{x}{x^2 + y^2} + i\frac{-y}{x^2 + y^2} \in C . \tag{11.2}$$

Consequently, C is a subfield of K and an extension field of $\mathbb{R}$.

This discussion shows that C is the smallest extension field of $\mathbb{R}$ in which the equation $x^2 = -1$ is solvable, if such an extension field exists. The remaining existence question we answer by a construction.

11.1 Theorem *There is a smallest extension field $\mathbb{C}$ of $\mathbb{R}$, the* **field of complex numbers**, *in which the equation $z^2 = -1$ is solvable. It is unique up to isomorphism.*

Proof As with the constructions of $\mathbb{Z}$ from $\mathbb{N}$ and $\mathbb{Q}$ from $\mathbb{Z}$, the above discussion suggests that we consider pairs of numbers $(x,y) \in \mathbb{R}^2$ and define operations on $\mathbb{R}^2$ following (11.1) and (11.2). Specifically, we define addition and multiplication on $\mathbb{R}^2$ by

$$\mathbb{R}^2 \times \mathbb{R}^2 \to \mathbb{R}^2 , \quad \bigl((x,y),(a,b)\bigr) \mapsto (x+a, y+b)$$

and

$$\mathbb{R}^2 \times \mathbb{R}^2 \to \mathbb{R}^2 , \quad \bigl((x,y),(a,b)\bigr) \mapsto (xa-yb, xb+ya)$$

and set $\mathbb{C} := (\mathbb{R}^2, +, \cdot)$. One can easily check that $\mathbb{C}$ is a field with additive identity $(0,0)$, unity $(1,0)$, additive inverse $-(x,y) = (-x,-y)$, and multiplicative inverse $(x,y)^{-1} = \bigl(x/(x^2+y^2), -y/(x^2+y^2)\bigr)$ if $(x,y) \neq (0,0)$.

It is easy to verify that

$$\mathbb{R} \to \mathbb{C} , \quad x \mapsto (x,0) \tag{11.3}$$

is an injective homomorphism. Consequently we can identify $\mathbb{R}$ with its image in $\mathbb{C}$ and so consider $\mathbb{R}$ to be a subfield of $\mathbb{C}$.

The equation $(0,1)^2 = (0,1)(0,1) = (-1,0) = -(1,0)$ implies that $(0,1) \in \mathbb{C}$ is a solution of $z^2 = -1_{\mathbb{C}}$.

The previous discussion shows that $\mathbb{C}$ is, up to isomorphism, the smallest extension field of $\mathbb{R}$ in which the equation $z^2 = -1$ is solvable. ■

Elementary Properties

The elements of $\mathbb{C}$ are called the **complex numbers**. Since $(0,1)(y,0) = (0,y)$ for all $y \in \mathbb{R}$, we have

$$(x,y) = (x,0) + (0,1)(y,0) , \qquad (x,y) \in \mathbb{R}^2 .$$

Setting $i := (0,1) \in \mathbb{C}$ and using the identification (11.3), each $z = (x,y) \in \mathbb{C}$ has a unique representation in the form

$$z = x + iy , \qquad x,y \in \mathbb{R} , \tag{11.4}$$

where, of course, $i^2 = -1$. Then $x =: \operatorname{Re} z$ is the **real part** and $y =: \operatorname{Im} z$ is the **imaginary part** of z. The **complex conjugate** of z is defined by

$$\overline{z} := x - iy = \operatorname{Re} z - i \operatorname{Im} z .$$

Any $z \in \mathbb{C}^\times$ with $\operatorname{Re} z = 0$ is called **(pure) imaginary**.

For arbitrary $x, y \in \mathbb{C}$ we have, of course, $z = x + iy \in \mathbb{C}$. When we want to make clear that the expression $z = x + iy$ is the **decomposition of z into its real and imaginary parts**, that is, $x = \operatorname{Re} z$ and $y = \operatorname{Im} z$, we write

$$z = x + iy \in \mathbb{R} + i\mathbb{R} .$$

11.2 Remarks **(a)** If $z = x + iy \in \mathbb{R} + i\mathbb{R}$ and $w = a + ib \in \mathbb{R} + i\mathbb{R}$, then $z + w$, $-z$ and zw are given by the formulas in (11.1), and if $z \neq 0$, then $z^{-1} = 1/z$ is given by (11.2).

(b) The functions $\mathbb{C} \to \mathbb{R}$, $z \mapsto \operatorname{Re} z$ and $\mathbb{C} \to \mathbb{R}$, $z \mapsto \operatorname{Im} z$ are well defined, surjective, and $z = \operatorname{Re} z + i \operatorname{Im} z$.

(c) Let X be a set and $f \colon X \to \mathbb{C}$ a 'complex valued function'. Then

$$\operatorname{Re} f \colon X \to \mathbb{R}\,, \quad x \mapsto \operatorname{Re}\bigl(f(x)\bigr)$$

and

$$\operatorname{Im} f \colon X \to \mathbb{R}\,, \quad x \mapsto \operatorname{Im}\bigl(f(x)\bigr)$$

define two 'real valued functions', the **real part** and the **imaginary part** of f. Clearly

$$f = \operatorname{Re} f + i \operatorname{Im} f\ .$$

(d) By construction $(\mathbb{C}, +)$ is the additive group $(\mathbb{R}^2, +)$ (see Example 7.2(d)). Thus we can identify $\mathbb{C}$ with $(\mathbb{R}^2, +)$ so long as we consider only the additive structure of $\mathbb{C}$. This means that we can represent complex numbers as vectors in the coordinate plane.[1] The addition of complex numbers is then the same as vector addition and can be done geometrically using the 'parallelogram rule'. As usual, we identify a vector z with the tip of its arrow and so consider z to be a 'point' of the set $\mathbb{R}^2$ whenever we use this graphic representation.

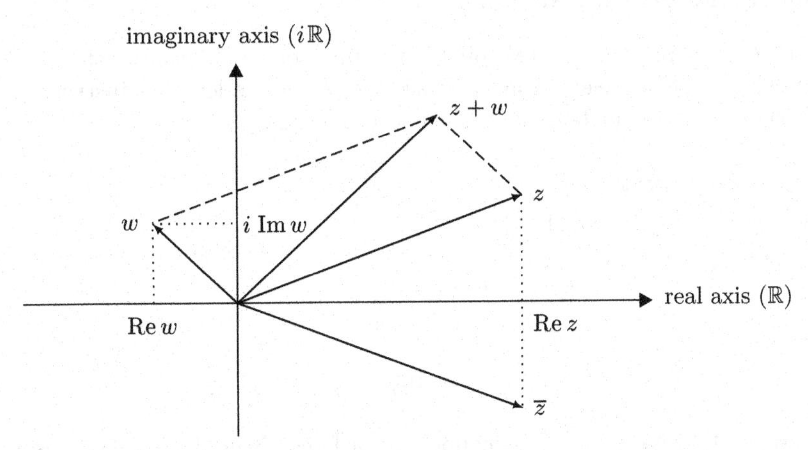

In Section III.6 we will see that multiplication in $\mathbb{C}$ also has a simple interpretation in the coordinate plane.

(e) Because $(-i)^2 = (-1)^2 i^2 = -1$, the equation $z^2 = -1$ has the two solutions $z = \pm i$. By Corollary 8.18, there are no other solutions.

[1] We refer the reader again to [Gab96] for an axiomatic treatment of these concepts (see also Section 12).

(f) For $d \in \mathbb{R}^+$ we have

$$X^2 - d = (X + \sqrt{d})(X - \sqrt{d}) \quad \text{and} \quad X^2 + d = (X + i\sqrt{d})(X - i\sqrt{d}) \ .$$

By 'completing the square' we can write $aX^2 + bX + c \in \mathbb{R}[X]$ with $a \neq 0$ in the form

$$aX^2 + bX + c = a\Big[\Big(X + \frac{b}{2a}\Big)^2 - \frac{D}{4a^2}\Big]$$

in $\mathbb{C}[X]$ where

$$D := b^2 - 4ac$$

is the **discriminant**. This implies that the **quadratic equation** $az^2 + bz + c = 0$ has the solutions

$$z_1, z_2 = \begin{cases} \dfrac{-b \pm \sqrt{D}}{2a} \in \mathbb{R} \ , & D \geq 0 \ , \\ \dfrac{-b \pm i\sqrt{-D}}{2a} \in \mathbb{C} \backslash \mathbb{R} \ , & D < 0 \ . \end{cases}$$

Moreover

$$z_1 + z_2 = -b/a \ , \quad z_1 z_2 = c/a$$

(see Exercise 8.7(g)). If $D < 0$, then $z_2 = \overline{z}_1$.

(g) Because $i^2 = -1 < 0$, the field $\mathbb{C}$ cannot be ordered. ■

Computation with Complex Numbers

In this section we present several important rules for calculating with complex numbers. The proofs are elementary and are left to the reader. It is instructive to interpret these rules geometrically in the coordinate plane.

11.3 Proposition *For all $z, w \in \mathbb{C}$,*

(i) $\operatorname{Re}(z) = (z + \overline{z})/2$, $\operatorname{Im}(z) = (z - \overline{z})/(2i)$

(ii) $z \in \mathbb{R} \Longleftrightarrow z = \overline{z}$

(iii) $\overline{\overline{z}} = z$

(iv) $\overline{z + w} = \overline{z} + \overline{w}$, $\overline{zw} = \overline{z}\,\overline{w}$

(v) $z\overline{z} = x^2 + y^2$ *where* $x := \operatorname{Re} z$, $y := \operatorname{Im} z$.

As we have already noted, $\mathbb{C}$ cannot be ordered. Nonetheless, the absolute value function on $\mathbb{R}$ which is induced from its order can be extended to a nonnegative function $|\cdot|$ on $\mathbb{C}$, also called the absolute value function,[2] by defining

$$|\cdot| : \mathbb{C} \to \mathbb{R}^+ \ , \quad z \mapsto |z| := \sqrt{z\,\overline{z}} \ .$$

[2]This fact justifies the use of the same symbol $|\cdot|$ for both absolute values. When distinct symbols are needed, we write $|\cdot|_{\mathbb{C}}$ for the absolute value in $\mathbb{C}$ and $|\cdot|_{\mathbb{R}}$ for the absolute value in $\mathbb{R}$ (see, for example, Proposition 11.4(ii)).

Hence, for $z = x + iy \in \mathbb{R} + i\mathbb{R}$, we have $|z| = \sqrt{x^2 + y^2}$, and so $|z|$ is the length of the vector z in the coordinate plane.

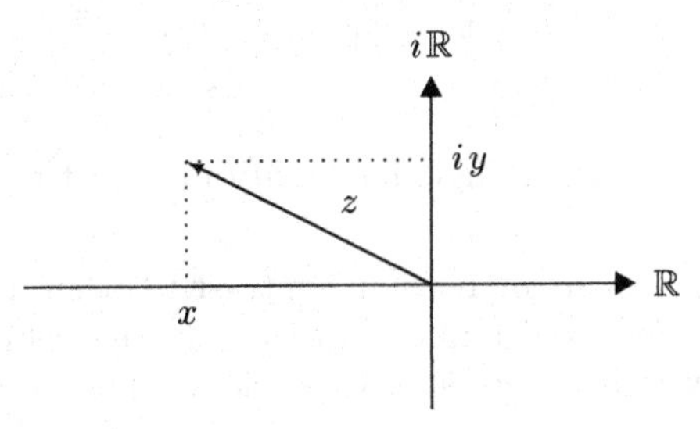

We collect in the next proposition some rules for the absolute value function.

11.4 Proposition *Let $z, w \in \mathbb{C}$.*

(i) $|zw| = |z|\,|w|$.

(ii) $|z|_{\mathbb{C}} = |z|_{\mathbb{R}}$ *for all* $z \in \mathbb{R}$.

(iii) $|\operatorname{Re}(z)| \le |z|$, $|\operatorname{Im}(z)| \le |z|$, $|z| = |\overline{z}|$.

(iv) $|z| = 0 \Leftrightarrow z = 0$.

(v) $|z + w| \le |z| + |w|$ (triangle inequality).

(vi) $z^{-1} = 1/z = \overline{z}/|z|^2$ *for all* $z \in \mathbb{C}^\times$.

Proof Let $z, w \in \mathbb{C}$ with $z = x + iy \in \mathbb{R} + i\mathbb{R}$.

(i) From Proposition 11.3(iv) and Remark 9.9(a), we have

$$|zw| = \sqrt{zw \cdot \overline{zw}} = \sqrt{z\overline{z} \cdot w\overline{w}} = \sqrt{z\overline{z}} \cdot \sqrt{w\overline{w}} = |z|\,|w| \ .$$

(ii) For $z \in \mathbb{R}$, we have $\overline{z} = z$, and so, from Remark 9.9(b),

$$|z|_{\mathbb{C}} = \sqrt{z\overline{z}} = \sqrt{z^2} = |z|_{\mathbb{R}} \ .$$

(iii) From Remark 10.10(c) we have $|\operatorname{Re}(z)| = |x| = \sqrt{x^2} \le \sqrt{x^2 + y^2} = |z|$. Similarly $|\operatorname{Im}(z)| \le |z|$. From the equation $\overline{\overline{z}} = z$ we get $|z| = \sqrt{z\overline{z}} = \sqrt{\overline{z}\,\overline{\overline{z}}} = |\overline{z}|$.

(iv) From Proposition 8.10 we have

$$|z| = 0 \Leftrightarrow |z|^2 = |x|^2 + |y|^2 = 0 \Leftrightarrow |x| = |y| = 0 \Leftrightarrow x = y = 0 \ .$$

(v) We have

$$\begin{aligned}
|z + w|^2 &= (z + w)\overline{(z + w)} = (z + w)(\overline{z} + \overline{w}) \\
&= z\overline{z} + z\overline{w} + w\overline{z} + w\overline{w} = |z|^2 + z\overline{w} + \overline{z\overline{w}} + |w|^2 \\
&= |z|^2 + 2\operatorname{Re}(z\overline{w}) + |w|^2 \le |z|^2 + 2\,|z\overline{w}| + |w|^2 \\
&= |z|^2 + 2\,|z|\,|w| + |w|^2 = (|z| + |w|)^2 \ ,
\end{aligned}$$

where we have used (iii) and Proposition 11.3.

(vi) If $z \in \mathbb{C}^\times$, then $1/z = \overline{z}/(z\overline{z}) = \overline{z}/|z|^2$. ∎

11.5 Corollary (reversed triangle inequality) *For all* $z, w \in \mathbb{C}$,

$$|z - w| \geq \big|\, |z| - |w| \,\big| \; .$$

Proof This follows from the triangle inequality in $\mathbb{C}$ just as in Corollary 8.11. ∎

Experience shows that in analysis, in contrast to other areas of mathematics, the only fields that matter are $\mathbb{R}$ and $\mathbb{C}$. Moreover, many definitions and theorems can be applied equally well to either of these fields. Thus we make the following convention:

Convention $\mathbb{K}$ *denotes either of fields* $\mathbb{R}$ *and* $\mathbb{C}$.

Balls in $\mathbb{K}$

For $a \in \mathbb{K}$ and $r > 0$ we call

$$\mathbb{B}(a,r) := \mathbb{B}_{\mathbb{K}}(a,r) := \{\, x \in \mathbb{K} \; ; \; |x - a| < r \,\}$$

the **open ball** in $\mathbb{K}$ with center a and radius r. If $\mathbb{K} = \mathbb{C}$, then $\mathbb{B}_{\mathbb{C}}(a,r)$ is the 'open disk' in the coordinate plane with center a and radius r. If $\mathbb{K}$ is the field $\mathbb{R}$, then $\mathbb{B}_{\mathbb{R}}(a,r)$ is the open interval $(a - r, a + r)$ of length $2r$ centered at a in $\mathbb{R}$.

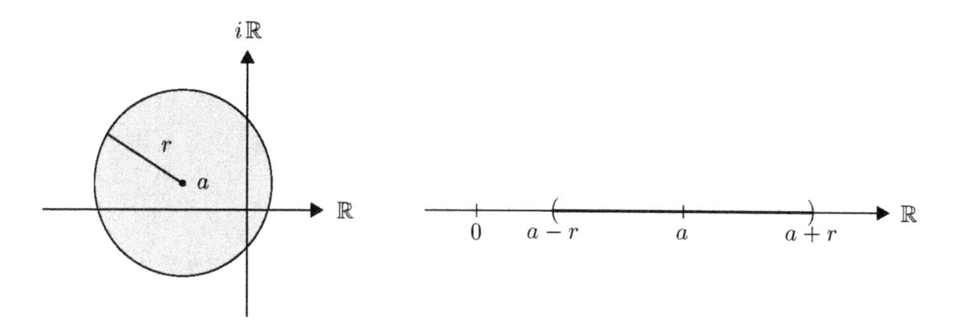

The **closed ball** in $\mathbb{K}$ with center a and radius r is defined by

$$\bar{\mathbb{B}}(a,r) := \bar{\mathbb{B}}_{\mathbb{K}}(a,r) := \{\, x \in \mathbb{K} \; ; \; |x - a| \leq r \,\} \; .$$

Thus $\bar{\mathbb{B}}_{\mathbb{R}}(a,r)$ is the closed interval $[a - r, a + r]$. Instead of $\mathbb{B}_{\mathbb{C}}(a,r)$ and $\bar{\mathbb{B}}_{\mathbb{C}}(a,r)$, we often write $\mathbb{D}(a,r)$ and $\bar{\mathbb{D}}(a,r)$ respectively. The open and closed **unit disks** in $\mathbb{C}$ are $\mathbb{D} := \mathbb{D}(0,1)$ and $\bar{\mathbb{D}} := \bar{\mathbb{D}}(0,1)$.

Exercises

1 (a) Show that for each $z \in \mathbb{C}\backslash(-\infty, 0]$ there is a unique $w \in \mathbb{C}$ such that $w^2 = z$ and $\operatorname{Re}(w) > 0$. The element w is called the **principal square root** of z and is written $\sqrt{z}$.

(b) Show that, for all $z \in \mathbb{C}\backslash(-\infty, 0]$,

$$\sqrt{z} = \sqrt{(|z| + \operatorname{Re} z)/2} + i \operatorname{sign}(\operatorname{Im} z)\sqrt{(|z| - \operatorname{Re} z)/2}\ .$$

(c) Calculate $\sqrt{i}$.

What are the other solutions of the equation $w^2 = i$?

2 Calculate $\overline{z}$, $|z|$, $\operatorname{Re} z$, $\operatorname{Im} z$, $\operatorname{Re}(1/z)$ and $\operatorname{Im}(1/z)$ for $z \in \left\{\dfrac{12+5i}{2+3i}, \sqrt{i}\right\}$.

3 Sketch the following sets in the coordinate plane:

$$\begin{aligned} A &:= \{\, z \in \mathbb{C}\ ;\ |z-1| \le |z+1| \,\} \\ B &:= \{\, z \in \mathbb{C}\ ;\ |z+1| \le |z-i| \le |z-1| \,\} \\ C &:= \{\, z \in \mathbb{C}\ ;\ 3z\overline{z} - 6z - 6\overline{z} + 9 = 0 \,\} \end{aligned}$$

4 Determine all solutions of the equations $z^4 = 1$ and $z^3 = 1$ in $\mathbb{C}$.

5 Give a proof of Proposition 11.3.

6 Let $m \in \mathbb{N}^\times$ and $U_j \subseteq \mathbb{C}$ for $0 \le j \le m$. Suppose that $a \in \mathbb{C}$ has the property that, for each j, there is some $r_j > 0$ such that $\mathbb{B}(a, r_j) \subseteq U_j$. Show that $\mathbb{B}(a, r) \subseteq \bigcap_{j=0}^m U_j$ for some $r > 0$.

7 For $a \in \mathbb{K}$ and $r > 0$, describe the set $\bar{\mathbb{B}}_{\mathbb{K}}(a, r)\backslash\mathbb{B}_{\mathbb{K}}(a, r)$.

8 Show that the identity function and $z \mapsto \overline{z}$ are the only field automorphisms of $\mathbb{C}$ which leave the elements of $\mathbb{R}$ fixed. (Hint: For an automorphism φ, consider $\varphi(i)$.)

9 Show that $S^1 := \{\, z \in \mathbb{C}\ ;\ |z| = 1 \,\}$ is a subgroup of the multiplicative group $(\mathbb{C}^\times, \cdot)$, the **circle group**.

10 Let $\mathbb{R}^{2\times 2}$ be the noncommutative ring of real 2×2 matrices. Show that the set C of matrices of the form

$$\begin{bmatrix} a & -b \\ b & a \end{bmatrix}$$

is a subfield of $\mathbb{R}^{2\times 2}$, and that the function

$$\mathbb{R} + i\mathbb{R} \to \mathbb{R}^{2\times 2}\ , \quad a + ib \mapsto \begin{bmatrix} a & -b \\ b & a \end{bmatrix}$$

is an isomorphism from $\mathbb{C}$ to C. (The necessary properties of matrices can be found in any book on linear algebra.)

11 For $p = X^n + a_{n-1}X^{n-1} + \cdots + a_1 X + a_0 \in \mathbb{C}[X]$, define $R := 1 + \sum_{k=0}^{n-1} |a_k|$. Show that $|p(z)| > R$ for all $z \in \mathbb{C}$ such that $|z| > R$.

12 Prove the **Parallelogram Identity** in $\mathbb{C}$:

$$|z+w|^2 + |z-w|^2 = 2(|z|^2 + |w|^2)\ , \qquad z, w \in \mathbb{C}\ .$$

13 Describe the function $\mathbb{C}^\times \to \mathbb{C}^\times,\ z \mapsto 1/z$ geometrically.

14 Determine all zeros of the polynomial $X^4 - 2X^3 - X^2 + 2X + 1 \in \mathbb{C}[X]$.
(Hint: Multiply the polynomial by $1/X^2$ and substitute $Y = X - 1/X$.)

15 **Cubic Equations** Let k be a cubic polynomial in $\mathbb{C}$ with leading coefficient 1, that is,

$$k = X^3 + aX^2 + bX + c\ .$$

To find the zeros of k, we first substitute $Y = X + a/3$ to get

$$Y^3 + pY + q \in \mathbb{C}[X]\ .$$

Determine the coefficients p and q in terms of a, b and c. Suppose that there exist[3] $d, u, v \in \mathbb{C}$ such that

$$d^2 = \Bigl(\frac{q}{2}\Bigr)^2 + \Bigl(\frac{p}{3}\Bigr)^3\ , \quad u^3 = -\frac{q}{2} + d\ , \quad v^3 = -\frac{q}{2} - d\ . \tag{11.5}$$

Show that $-3uv/p$ is a third root of unity, that is, $(-3uv/p)^3 = 1$, and so we can choose u and v such that $3uv = -p$. Now let $\xi \neq 1$ satisfy $\xi^3 = 1$ (see Exercise 4). Show that

$$y_1 := u + v\ , \quad y_2 := \xi u + \xi^2 v\ , \quad y_3 := \xi^2 u + \xi v$$

are the solutions of the equation $y^3 + py + q = 0$.

[3] In Section III.6 we prove that these complex numbers exist.

12 Vector Spaces, Affine Spaces and Algebras

Linear algebra is without doubt one of the most fertile of all mathematical research areas and serves as a foundation for many far-reaching theories in all parts of mathematics. In particular, linear algebra is one of the main tools of analysis, and so, in this section, we introduce the basic concepts and illustrate these with examples. Once again, the goal is to be able to recognize simple algebraic structures which appear frequently, in different forms, in the following chapters. For a deeper investigation, we direct the reader to the extensive literature of linear algebra, for example, [Art91], [Gab96], [Koe83], [Wal82] and [Wal85].

In the following, K is an arbitrary field.

Vector Spaces

A **vector space over the field** K (or simply, a K**-vector space**) is a triple $(V,+,\cdot)$ consisting of a nonempty set V, an 'inner' operation $+$ on V called **addition**, and an 'outer' operation

$$K \times V \to V\ , \quad (\lambda, v) \mapsto \lambda \cdot v\ ,$$

called **scalar multiplication** which satisfy the following axioms:

(VS_1) $(V,+)$ is an Abelian group.

(VS_2) The distributive law holds:

$$\lambda\cdot(v+w) = \lambda\cdot v+\lambda\cdot w\ , \quad (\lambda+\mu)\cdot v = \lambda\cdot v+\mu\cdot v\ , \qquad \lambda,\mu \in K\ , \quad v,w \in V\ .$$

(VS_3) $\lambda \cdot (\mu v) = (\lambda\mu)\cdot v\ , \quad 1\cdot v = v\ , \qquad \lambda,\mu\in K\ , \quad v \in V\ .$

A vector space is called **real** if $K = \mathbb{R}$ and **complex** if $K = \mathbb{C}$. We write V instead of $(V,+,\cdot)$ when the operations are clear from context.

12.1 Remarks **(a)** The elements of V are called **vectors** and the elements of K are called **scalars**. The word 'vector' is simply an abbreviation for 'element of a vector space'. Possible geometrical interpretations we leave until later.

Just as for rings, we make the convention that multiplication takes precedence over addition, and we write simply λv for $\lambda \cdot v$.

(b) The identity element of $(V,+)$ is called the **zero vector** and is denoted by 0, as is also the zero of K. For the additive inverse of $v \in V$ we write $-v$ and $v - w := v + (-w)$. This, as well as the use of the same symbols '$+$' and '$\cdot$' for the operations in K and in V do not lead to misunderstanding, since, in addition to (VS_1) and (VS_3), we have

$$0v = 0\ , \quad (-\lambda)v = \lambda(-v) = -(\lambda v) =: -\lambda v\ , \qquad \lambda \in K\ , \quad v \in V\ ,$$

and also the rule

$$\lambda v = 0 \Rightarrow (\lambda = 0 \text{ or } v = 0)\ .$$

In this implication, the first and last zeros denote zero vectors and the remaining zero stands for the zero of K.

Proof From the distributive law and the rules of arithmetic in K it follows that

$$0 \cdot v = (0+0) \cdot v = 0 \cdot v + 0 \cdot v \ .$$

Since the zero vector is the identity element of $(V,+)$, we also have $0 \cdot v = 0 \cdot v + 0$, and so Remark 7.1(c) implies $0 \cdot v = 0$. The proofs of the remaining claims are left as exercises. ■

(c) Axiom (VS_3) says that the multiplicative group $K^\times$ acts on V (from the left) (see Exercise 7.6). Indeed (VS_2) and (VS_3) can be used to define the concept of a field acting on an Abelian group. It is sometimes convenient to think of K acting on V from the right by defining $v\lambda := \lambda v$ for $(\lambda, v) \in K \times V$. ■

Linear Functions

Let V and W be vector spaces over K. Then a function $T\colon V \to W$ is (K-)**linear** if

$$T(\lambda v + \mu w) = \lambda T(v) + \mu T(w) \ , \qquad \lambda, \mu \in K \ , \quad v, w \in V \ .$$

Thus a linear function is simply a function which is compatible with the vector space operations, in other words, it is a (**vector space**) **homomorphism**. The set of all linear functions from V to W is denoted by $\mathrm{Hom}(V,W)$ or $\mathrm{Hom}_K(V,W)$, and $\mathrm{End}(V) := \mathrm{Hom}(V,V)$ is the set of all (**vector space**) **endomorphisms**. A bijective homomorphism $T \in \mathrm{Hom}(V,W)$ is a (**vector space**) **isomorphism**. A bijective endomorphism $T \in \mathrm{End}(V)$ is a (**vector space**) **automorphism**. If there is an isomorphism from V to W, then V and W are **isomorphic**, and we write $V \cong W$. Clearly $\cong$ is an equivalence relation on any set of K-vector spaces.

> Convention The statement 'V and W are vector spaces and $T\colon V \to W$ is a linear function' always implies that V and W are vector spaces over the same field.

12.2 Remarks **(a)** For a linear function $T\colon V \to W$, it is usual to write Tv instead of $T(v)$ when $v \in V$, so long as this does not lead to misunderstanding.

(b) A vector space homomorphism $T\colon V \to W$ is, in particular, a group homomorphism $T\colon (V,+) \to (W,+)$. Thus we have $T0 = 0$ and $T(-v) = -Tv$ for all $v \in V$. The **kernel** (or **null space**) of T is the kernel of this group homomorphism:

$$\ker(T) = \{\, v \in V \ ; \ Tv = 0 \,\} = T^{-1}0 \ .$$

Thus T is injective if and only if its kernel is trivial, that is, if $\ker(T) = \{0\}$ (see Remarks 7.6(a) and (d)).

(c) Let U, V and W be vector spaces over K. Then $T \circ S \in \mathrm{Hom}(U,W)$ for all $S \in \mathrm{Hom}(U,V)$ and $T \in \mathrm{Hom}(V,W)$.

(d) The set $\mathrm{Aut}(V)$ of automorphisms of V, that is, the set of bijective linear functions from V to itself, is a subgroup of the permutation group of V. It is called the **automorphism group** of V. ■

12.3 Examples Let V and W be vector spaces over K.

(a) A **zero** or **trivial (vector) space** consists of a single vector 0, and is often denoted simply by 0. Any other vector space is **nontrivial**.

(b) A nonempty subset U of V is called a **subspace** if the following holds:

(SS_1) U is a subgroup of $(V,+)$.

(SS_2) U is closed under scalar multiplication: $K \cdot U \subseteq U$.

One can easily verify that U is a subspace of V if and only if U is closed under both operations of V, that is, if

$$U + U \subseteq U\ , \quad K \cdot U \subseteq U\ .$$

(c) The kernel and image of a linear function $T\colon V \to W$ are subspaces of V and W respectively. If T is injective then $T^{-1} \in \mathrm{Hom}\bigl(\mathrm{im}(T), V\bigr)$.

(d) K is a vector space over itself when the field operations are interpreted as vector space operations.

(e) Let X be a set. Then V^X is a K-vector space with the operations (see Example 4.12)

$$(f{+}g)(x) := f(x){+}g(x)\ , \quad (\lambda f)(x) := \lambda f(x)\ , \qquad x \in X\ , \quad \lambda \in K\ , \quad f,g \in V^X\ .$$

In particular, for $m \in \mathbb{N}^\times$, K^m is a K-vector space with the operations

$$x + y = (x_1 + y_1, \ldots, x_m + y_m)\ , \quad \lambda x = (\lambda x_1, \ldots, \lambda x_m)$$

for $\lambda \in K$, and $x = (x_1, \ldots, x_m)$ and $y = (y_1, \ldots, y_m)$ in K^m. Clearly, K^1 and K are identical (as K-vector spaces).

(f) The above construction suggests the following generalization. Let $V_1, \ldots, V_m$ be vector spaces over K. Then $V := V_1 \times \cdots \times V_m$ is a vector space, the **product vector space** of $V_1, \ldots, V_m$ with operations defined by

$$v + w := (v_1 + w_1, \ldots, v_m + w_m)\ , \quad \lambda v := (\lambda v_1, \ldots, \lambda v_m)$$

for $v = (v_1, \ldots, v_m) \in V$, $w = (w_1, \ldots, w_m) \in V$ and $\lambda \in K$.

(g) On the ring of formal power series $K[\![X_1, \ldots, X_m]\!]$ in $m \in \mathbb{N}^\times$ indeterminates over K, we define a function

$$K \times K[\![X_1, \ldots, X_m]\!] \to K[\![X_1, \ldots, X_m]\!]\ , \quad (\lambda, p) \mapsto \lambda p$$

by

$$\lambda\Bigl(\sum_\alpha p_\alpha X^\alpha\Bigr) := \sum_\alpha (\lambda p_\alpha) X^\alpha .$$

With this operation as scalar multiplication and the already defined addition, $K[\![X_1,\ldots,X_m]\!]$ is a K-vector space, the **vector space of formal power series** in m indeterminates. Clearly, $K[X_1,\ldots,X_m]$ is a subspace of $K[\![X_1,\ldots,X_m]\!]$, the **vector space of polynomials** in m indeterminates.

If K is infinite, the identification of polynomials in $K[X_1,\ldots,X_m]$ with polynomial functions in $K^{(K^m)}$ (see Remark 8.20(c)) means that $K[X_1,\ldots,X_m]$ is also a subspace of $K^{(K^m)}$.

(h) $\mathrm{Hom}(V,W)$ is a subspace of W^V.

(i) Let U be a subspace of V. Then, by Proposition 7.4 and Remark 7.5(b), $(V,+)/U$ is an Abelian group. It is easy to check that

$$K\times (V,+)/U \to (V,+)/U\ ,\quad (\lambda, x+U)\mapsto \lambda x+U$$

is a well defined function which satisfies axioms (VS_2) and (VS_3). Thus $(V,+)/U$ is a K-vector space, which we denote by V/U and call the **quotient space of V modulo U**. Finally, the quotient homomorphism

$$\pi : V \to V/U\ ,\quad x\mapsto [x] := x+U$$

is a linear function.

(j) For $T\in \mathrm{Hom}(V,W)$ there is a unique linear function $\widehat{T}\colon V/\ker(T)\to W$ such that the diagram below is commutative.

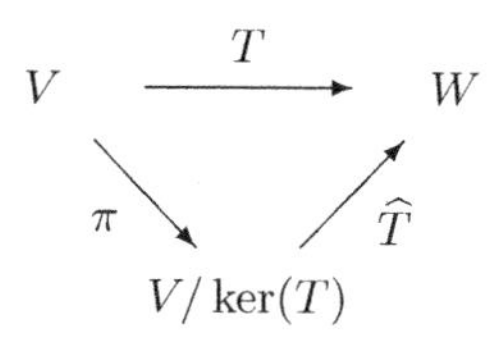

Moreover, $\widehat{T}$ is injective and $\mathrm{im}(\widehat{T}) = \mathrm{im}(T)$.

Proof This follows directly from (c), (i) and Example 4.2(c). ■

(k) Let $\{\,U_\alpha\ ;\ \alpha\in \mathsf{A}\,\}$ be a set of subspaces of V. Then $\bigcap_{\alpha\in\mathsf{A}} U_\alpha$ is a subspace of V. If M is a subset of V, then

$$\mathrm{span}(M) := \bigcap \{\,U\ ;\ U \text{ is a subspace of } V \text{ and } U\supseteq M\,\}$$

is the smallest subspace of V which contains M and is called the **span** of M.

(l) If U_1 and U_2 are subspaces of V, then the image of $U_1\times U_2$ under addition in V is a subspace of V called the **sum**, U_1+U_2, of U_1 and U_2. The sum is **direct** if $U_1\cap U_2=\{0\}$, and, in this case, it is written $U_1\oplus U_2$.

(m) If U is a subspace of V and $T \in \mathrm{Hom}(V,W)$, then $T|U$ is a linear function from U to W. In the case that $V = W$, U is said to be **invariant** under T if $T(U) \subseteq U$. So long as no confusion arises, we write T for $T|U$. ■

Vector Space Bases

Let V be a nontrivial K-vector space. An expression of the form $\sum_{j=1}^{m} \lambda_j v_j$ with $\lambda_1, \ldots, \lambda_m \in K$ and $v_1, \ldots, v_m \in V$ is called a (finite) **linear combination** of the vectors $v_1, \ldots, v_m$ (over K). The vectors $v_1, \ldots, v_m$ are **linearly dependent** if there are $\lambda_1, \ldots, \lambda_m \in K$, not all zero, such that $\lambda_1 v_1 + \cdots + \lambda_m v_m = 0$. If no such scalars exist, that is, if

$$\lambda_1 v_1 + \cdots + \lambda_m v_m = 0 \Rightarrow \lambda_1 = \cdots = \lambda_m = 0 ,$$

then the vectors $v_1, \ldots, v_m$ are **linearly independent**. A subset A of V is **linearly independent** if each finite subset of A is linearly independent. The empty set is, by convention, linearly independent. A linearly independent subset B of V such that $\mathrm{span}(B) = V$ is called a **basis** of V. A fundamental result from linear algebra is that, if V has a finite basis with m vectors, then every basis of V has exactly m vectors. In this circumstance, m is called the **dimension**, $\dim(V)$, of the vector space V and we say that V is m **dimensional**. If V has no finite basis, then V is **infinite dimensional**, $\dim(V) = \infty$. Finally, we define $\dim(0) = 0$. A very natural and useful fact about the dimension is that if W is a subspace of V, then $\dim(W) \leq \dim(V)$. For the proof of these claims about vector space dimension, the reader is referred to the linear algebra literature.

12.4 Examples **(a)** Let $m \in \mathbb{N}^\times$. For $j = 1, \ldots, m$, define

$$e_j := (0, \ldots, 0, \underset{(j)}{1}, 0, \ldots, 0) \in K^m ,$$

that is, e_j is the vector in K^m whose j^{th} component is 1 and whose other components are 0. Then $\{e_1, \ldots, e_m\}$ is a basis of K^m called the **standard basis**. Hence K^m is an m dimensional vector space called the **standard m dimensional vector space over K**.

(b) Let X be a finite set. For $x \in X$, define $e_x \in K^X$ by

$$e_x(y) := \begin{cases} 1 , & y = x , \\ 0 , & y \neq x . \end{cases} \tag{12.1}$$

Then the set $\{e_x \,;\, x \in X\}$ is a basis (the standard basis) of K^X, and hence $\dim(K^X) = \mathrm{Num}(X)$.

(c) For $n \in \mathbb{N}$ and $m \in \mathbb{N}^\times$, set

$$K_n[X_1,\ldots,X_m] := \{\, p \in K[X_1,\ldots,X_m] \; ; \; \deg(p) \leq n \,\} .$$

Then $K_n[X_1,\ldots,X_m]$ is a subspace of $K[X_1,\ldots,X_m]$ and the set of monomials $\{\, X^\alpha \; ; \; |\alpha| \leq n \,\}$ form a basis. Consequently

$$\dim\bigl(K_n[X_1,\ldots,X_m]\bigr) = \binom{m+n}{n} ,$$

and $K[X_1,\ldots,X_m]$ is an infinite dimensional space.

Proof Since the elements of $K_n[X_1,\ldots,X_m]$ are functions into K from a *finite* subset of $\mathbb{N}^m$, it follows from (b) that the monomials $\{\, X^\alpha \; ; \; |\alpha| \leq n \,\}$ are a basis. Exercise 8.8(d) shows that number of such monomials is $\binom{m+n}{n}$. If $k := \dim(K[X_1,\ldots,X_m])$ is finite, then any subspace would have dimension less than or equal to k. But this is contradicted by the subspaces $K_n[X_1,\ldots,X_m]$ which can have arbitrarily large dimension. ■

(d) For $m, n \in \mathbb{N}^\times$,

$$K_{n,\mathrm{hom}}[X_1,\ldots,X_m] := \Bigl\{ \sum_{|\alpha|=n} a_\alpha X^\alpha \; ; \; a_\alpha \in K,\ \alpha \in \mathbb{N}^m,\ |\alpha| = n \Bigr\}$$

is a subspace of $K_n[X_1,\ldots,X_m]$ called the **vector space of homogeneous polynomials of degree n in m indeterminates**. It has the dimension $\binom{m+n-1}{n}$.

Proof As in the preceding proof, the set of monomials of degree n form a basis. The claim then follows from Exercise 8.8(c). ■

12.5 Remark Let V be an m dimensional K-vector space for some $m \in \mathbb{N}^\times$ and $\{b_1,\ldots,b_m\}$ a basis of V. Then, for each $v \in V$, there is a unique m-tuple $(x_1,\ldots,x_m) \in K^m$ such that

$$v = \sum_{j=1}^{m} x_j b_j . \tag{12.2}$$

Conversely, such an m-tuple defines by (12.2) a unique vector v in V. Consequently, the function

$$K^m \to V , \quad (x_1,\ldots,x_m) \mapsto \textstyle\sum_{j=1}^m x_j b_j$$

is bijective. Since the function is clearly linear, it is an isomorphism from K^m to V. Therefore we have shown that every m dimensional K-vector space is isomorphic to the standard m dimensional space K^m. This explains, of course, the name 'standard space'. ■

Affine Spaces

The abstract concept of a vector space, which plays such a fundamental role in current mathematics, and, in particular, in modern analysis, developed from the intuitive 'vector calculus' of directed arrows in our familiar three dimensional universe. A geometrical interpretation of vector space concepts is still very useful, even in abstract situations, as we have already seen in the identification of $\mathbb{C}$ with the coordinate plane. In such an interpretation we often consider vectors to be 'points', and certain sets of vectors to be 'lines' and 'planes', etc. To give these concepts a solid foundation and to avoid confusion, we provide a short introduction to affine spaces. This will allow us to use, without further comment, the language which is most convenient for the given situation.

Let V be a K-vector space and E a nonempty set whose elements we call **points**. Then E is called an **affine space** over V if there is a function

$$V \times E \to E , \quad (v, P) \mapsto P + v$$

with following properties:

(AS_1) $P + 0 = P, \ \ P \in E.$

(AS_2) $P + (v + w) = (P + v) + w, \ \ P \in E, \ \ v, w \in V.$

(AS_3) For each $P, Q \in E$ there is a unique $v \in V$ such that $Q = P + v$.

The unique vector v provided by (AS_3) is denoted $\overrightarrow{PQ}$. It satisfies

$$Q = P + \overrightarrow{PQ} .$$

From (AS_1) we have $\overrightarrow{PP} = 0$, and from (AS_2) it follows that the function

$$E \times E \to V , \quad (P, Q) \mapsto \overrightarrow{PQ}$$

satisfies the equation

$$\overrightarrow{PQ} + \overrightarrow{QR} = \overrightarrow{PR} , \qquad P, Q, R \in E .$$

Since $\overrightarrow{PP} = 0$, this implies, in particular, that

$$\overrightarrow{PQ} = -\overrightarrow{QP} , \qquad P, Q \in E .$$

Moreover, by (AS_3), for each $P \in E$ and $v \in V$ there is a unique $Q \in E$ such that $\overrightarrow{PQ} = v$, namely $Q := P + v$. Hence V is also called **direction space** of the affine space E.

12.6 Remarks **(a)** The axioms (AS_1) and (AS_2) say that the additive group $(V,+)$ acts (from the right) on the set E (see Exercise 7.6). From Axiom (AS_3) it follows that this action has only one orbit, that is, the group **acts transitively** on E.

(b) For each $v \in V$,

$$\tau_v : E \to E\ , \quad P \mapsto P + v$$

is the **translation** of E by the vector v. It follows from (AS_1) and (AS_2) that the set of translations is a subgroup of the permutation group of E. ■

Let E be an affine space over V. Choose a fixed point O of E, the **origin**. Then the function $V \to E,\ v \mapsto O + v$ is bijective with inverse $E \to V,\ P \mapsto \overrightarrow{OP}$. The vector $\overrightarrow{OP}$ is called the **position vector** of P (with respect to O).

If $\{b_1, \ldots, b_m\}$ is a basis of V, there is a unique m-tuple $(x_1, \ldots, x_m) \in K^m$ such that

$$\overrightarrow{OP} = \sum_{j=1}^{m} x_j b_j\ .$$

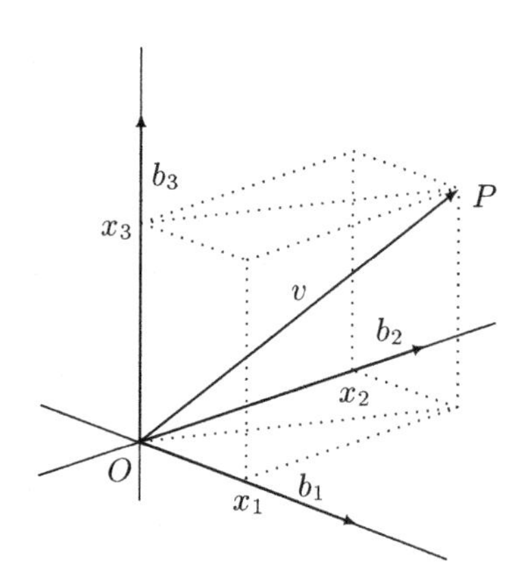

In this situation the numbers $x_1, \ldots, x_m$ are called the (affine) **coordinates** of the point P with respect to the (affine) **coordinate system** $(O; b_1, \ldots, b_m)$. The bijective function

$$E \to K^m\ , \quad P \mapsto (x_1, \ldots, x_m)\ , \quad (12.3)$$

which takes each point $P \in E$ to its coordinates, is called the **coordinate function** of E with respect to $(O; b_1, \ldots, b_m)$.

The **dimension** of an affine space is, by definition, the dimension of its direction space. A zero dimensional space contains only one point, a one dimensional space is an (affine) **line**, and a two dimensional affine space is an (affine) **plane**. An **affine subspace** of E is a set of the form $P + W = \{\, P + w\ ;\ w \in W \,\}$ where $P \in E$ and W is a subspace of the direction space V.

12.7 Example Any K-vector space V can be considered to be an affine space over itself. The operation of $(V,+)$ on V is simply addition in V. In this case, $\overrightarrow{vw} = w - v$ for $v, w \in V$. (Here v is interpreted as a point, and the w on the left and right of the equal sign are interpreted as a point and a vector respectively!) We choose, of course, the zero vector to be the origin.

If $\dim(V) = m \in \mathbb{N}^\times$ and $\{b_1, \ldots, b_m\}$ is a basis of V, then we can identify V, using the coordinate function (12.3), with the **standard space** K^m. Via (12.3), the basis $(b_1, \ldots, b_m)$ is mapped to the standard basis $e_1, \ldots, e_m$ of K^m. The operations in K^m lead then (at least in the case $m = 2$ and $K = \mathbb{R}$) to the familiar 'vector calculus' in which, for example, vector addition can be done using the 'parallelogram rule':

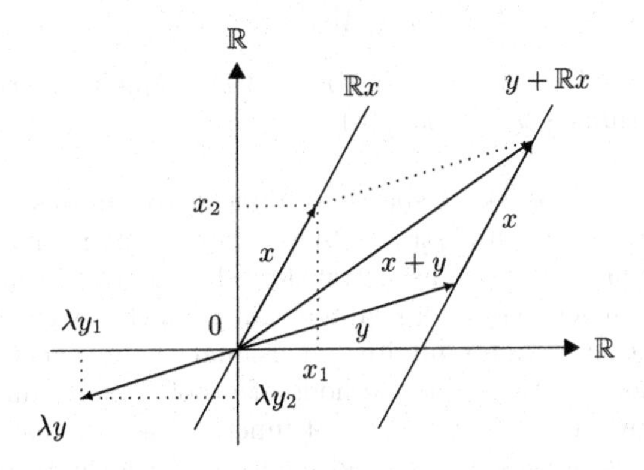

In the geometrical viewpoint, a vector is an arrow with head and tail at some points, P and Q, say, of E. That is, an arrow is an ordered pair (P, Q) of points. Two such arrows, (P, Q) and (P', Q'), are equal if $\overrightarrow{PQ} = \overrightarrow{P'Q'}$, that is, if there is some $v \in V$ such that $P = Q + v$ and $P' = Q' + v$, or, more geometrically, if (P', Q') can be obtained from (P, Q) by some translation.

Convention Unless otherwise stated, we consider any K-vector space V to be an affine space over itself with the zero vector as origin. Moreover we consider K to be a vector space over itself whenever appropriate.

Because of this convention, the elements of a vector space can be called both 'vectors' or 'points' as appropriate, and the geometrical concepts 'line', 'plane' and 'affine subspace' make sense in any vector space.

Affine Functions

Let V and W be vector spaces over K. A **function** $\alpha\colon V \to W$ is called **affine** if there is a linear function $A\colon V \to W$ such that

$$\alpha(v_1) - \alpha(v_2) = A(v_1 - v_2)\ , \qquad v_1, v_2 \in V\ . \tag{12.4}$$

When such an A exists, it is uniquely determined by α. Indeed, setting $v_1 := v$ and $v_2 := 0$ we get $A(v) = \alpha(v) - \alpha(0)$ for all $v \in V$. Conversely, α is uniquely determined by $A \in \operatorname{Hom}(V, W)$ once $\alpha(v_0)$ is known for some $v_0 \in V$. Indeed, for $v_1 := v$ and $v_2 := v_0$, it follows from (12.4) that

$$\alpha(v) = \alpha(v_0) + A(v - v_0) = \alpha(v_0) - Av_0 + Av\ , \qquad v \in V\ . \tag{12.5}$$

Therefore we have proved the following proposition.

12.8 Proposition *Let V and W be vector spaces over K. Then $\alpha\colon V \to W$ is affine if and only if it has the form*

$$\alpha(v) = w + Av \ , \qquad v \in V \ , \tag{12.6}$$

with $w \in W$ and $A \in \mathrm{Hom}(V, W)$. Moreover, A is uniquely determined by α, and α is uniquely determined by A and $\alpha(0)$.

The interpretation of vector spaces as affine spaces makes it possible to give geometric meaning to certain abstract objects. For the moment, this is not much more than a language that we have transferred from our intuitions about the three dimensional universe. In later chapters, the geometric viewpoint will become increasingly important, even for infinite dimensional vector spaces, since it suggests useful interpretations and possible methods of proof. Infinite dimensional vector spaces will frequently occur in the form of function spaces, that is, as subspaces of K^X. A deep study of these spaces, indispensable for a thorough understanding of analysis, is not within the scope of this book. This is the goal of 'higher' analysis, in particular, of functional analysis.

The interpretation of *finite dimensional* vector spaces as affine spaces has also an extremely important computational aspect. The introduction of coordinate systems leads to concrete descriptions of geometric objects in terms of equations and inequalities for the coordinates. A coordinate system is determined by the choice of an origin and basis, and it is essential to make these choices so that the calculations are as simple as possible. The right choice of the coordinate system can be decisive for a successful solution of a given problem.

Polynomial Interpolation

To illustrate the above ideas, we show how interpolation questions for polynomials can be solved easily using a clever choice of basis in $K_m[X]$. The **polynomial interpolation problem** is the following:

Given $m \in \mathbb{N}^\times$, distinct $x_0, \dots, x_m$ in $\mathbb{K}$ and a function $f : \{x_0, \dots, x_m\} \to \mathbb{K}$, find a polynomial $p \in \mathbb{K}_m[X]$ such that

$$p(x_j) = f(x_j) \ , \qquad 0 \le j \le m \ . \tag{12.7}$$

The following proposition shows that this problem has a unique solution.

12.9 Proposition *There is a unique solution $p := p_m[f; x_0, \dots, x_m] \in \mathbb{K}_m[X]$ of the polynomial interpolation problem.*

Proof The **Lagrange polynomials** $\ell_j[x_0, \dots, x_m] \in \mathbb{K}_m[X]$ are defined by

$$\ell_j[x_0, \dots, x_m] := \prod_{\substack{k=0 \\ k \ne j}}^{m} \frac{X - x_k}{x_j - x_k} \ , \qquad 0 \le j \le m \ .$$

Clearly
$$\ell_j[x_0,\dots,x_m](x_k)=\delta_{jk}\ , \qquad 0\le j,k\le m\ ,$$
where
$$\delta_{jk}:=\begin{cases}1\ , & j=k\ ,\\ 0\ , & j\ne k\ ,\end{cases} \qquad j,k\in\mathbb{Z}\ ,$$
is the **Kronecker symbol**. Then the **Lagrange interpolation polynomial**
$$L_m[f;x_0,\dots,x_m]:=\sum_{j=0}^{m} f(x_j)\ell_j[x_0,\dots,x_m]\in\mathbb{K}_m[X] \tag{12.8}$$
is a solution of the problem. If $p\in\mathbb{K}_m[X]$ is a second polynomial which satisfies (12.7), then the polynomial
$$p-L_m[f;x_0,\dots,x_m]\in\mathbb{K}_m[X]$$
has the $m+1$ distinct zeros, $x_0,\dots,x_m$, and so, by Corollary 8.18, $p=L_m[f;x_0,\dots,x_m]$. This proves the uniqueness claim. ■

12.10 Remarks (a) The above easy and explicit solution of the polynomial interpolation problem is due to our choice of the Lagrange polynomials of degree m as a basis of $\mathbb{K}_m[X]$. If we had chosen the 'canonical' basis $\{\,X^j\ ;\ 0\le j\le m\,\}$, then we would have to solve the system
$$\sum_{k=0}^{m} p_k x_j^k = f(x_j)\ , \qquad 0\le j\le m\ , \tag{12.9}$$
of $m+1$ linear equations in $m+1$ unknowns $p_0,\dots,p_m$, the coefficients of the desired polynomial. From linear algebra we know that the system (12.9) is solvable, for any choice of the right hand side, if and only if the determinant of the coefficient matrix
$$\begin{bmatrix} 1 & x_0 & x_0^2 & \cdots & x_0^m\\ 1 & x_1 & x_1^2 & \cdots & x_1^m\\ \vdots & \vdots & \vdots & \vdots & \vdots\\ 1 & x_m & x_m^2 & \cdots & x_m^m \end{bmatrix} \tag{12.10}$$
is nonzero. (12.10) is a **Vandermonde matrix** whose determinant has the value
$$\prod_{0\le j<k\le m}(x_k-x_j)$$
(see, for example, [Gab96]). Since this determinant is not zero, we get the existence and uniqueness claims of Proposition 12.9. While the proof of Proposition 12.9 gives an explicit form for $p:=p_m[f;x_0,\dots,x_m]$, solving (12.9) using standard methods of linear algebra (for example, **Gauss-Jordan elimination**) yields, in general, no such simple expression for p.

(b) If one increases the number of points and function values by one, then all of the Lagrange polynomials must be recalculated. For this reason it is often more practical to write $p_m[f;x_0,\dots,x_m]$ in the form
$$p_m[f;x_0,\dots,x_m]=\sum_{j=0}^{m} a_j\omega_j[x_0,\dots,x_{j-1}]$$

using $\omega_0 := 1$ and the **Newton polynomials**,

$$\omega_j[x_0,\dots,x_{j-1}] = (X-x_0)(X-x_1)\cdots(X-x_{j-1}) \in \mathbb{K}_j[X] , \qquad 1 \le j \le m .$$

Then (12.7) leads to a triangular system of linear equations,

$$\begin{aligned} &a_0 = f(x_0) \\ &a_0 + a_1\omega_1[x_0](x_1) = f(x_1) \\ &\dots\dots\dots\dots\dots\dots\dots\dots\dots\dots\dots\dots\dots \\ &a_0 + a_1\omega_1[x_0](x_m) + \cdots + a_m\omega_m[x_0,\dots,x_{m-1}](x_m) = f(x_m) , \end{aligned}$$

which is easy to solve using 'back substitution' (successive substitution starting from the top). In this form, $p_m[f;x_0,\dots,x_m]$ is known as the **Newton interpolation polynomial.** Thus in this case too, choosing the basis $\{\, \omega_j[x_0,\dots,x_{j-1}] \ ; \ 0 \le j \le m \,\}$ of $\mathbb{K}_m[X]$ leads to a simple solution. ■

Algebras

Let X be a nonempty set. Then K^X, the set of all functions from X to K, has, by Example 8.2(b), a ring structure and, by Example 12.3(e), a vector space structure. Moreover, ring multiplication and scalar multiplication are compatible in the sense that

$$(\lambda f)\cdot(\mu g) = (\lambda\mu) fg , \qquad \lambda,\mu \in K , \quad f,g \in K^X .$$

This situation occurs frequently enough that it has its own name.

A K-vector space $\mathcal{A}$ together with an operation

$$\mathcal{A}\times\mathcal{A}\to\mathcal{A} , \quad (a,b)\mapsto a\odot b$$

is called an **algebra** over K if the following hold:

(A_1) $(\mathcal{A},+,\odot)$ is a ring.

(A_2) The distributive law holds:

$$\begin{aligned} (\lambda a + \mu b)\odot c &= \lambda(a\odot c) + \mu(b\odot c) \\ a\odot(\lambda b + \mu c) &= \lambda(a\odot b) + \mu(a\odot c) \end{aligned}$$

for all $a,b,c\in\mathcal{A}$ and $\lambda,\mu\in K$.

For the **ring multiplication** $\odot$ **in** $\mathcal{A}$, we again write ab instead of $a\odot b$. This leads to no misunderstanding since it is always clear from context which multiplication is intended. This notation is also justified by the distributive laws which hold in $\mathcal{A}$.

In general, the algebra (that is, the ring $(\mathcal{A},\odot)$) is neither commutative nor contains a unity element.

12.11 Examples (a) Let X be a nonempty set. Then K^X is a commutative K-algebra with unity with respect to the operations of Example 8.2(b) and Example 12.3(e).

(b) For $m \in \mathbb{N}^\times$, $K[\![X_1,\ldots,X_m]\!]$ is a commutative K-algebra with unity and $K[X_1,\ldots,X_m]$ is a subalgebra with unity.

(c) Let V be a K-vector space. Then $\mathrm{End}(V)$, with composition as ring multiplication, is a K-algebra. Thus

$$ABx = A(Bx) , \qquad x \in V , \quad A, B \in \mathrm{End}(V) ,$$

and $I := \mathrm{id}_V$ is the unity element of $\mathrm{End}(V)$. In general, $\mathrm{End}(V)$, the **endomorphism algebra** of V is not commutative. ■

12.12 Remark Let V be a K-vector space. Define a function

$$K[X] \times \mathrm{End}(V) \to \mathrm{End}(V) , \quad (p, A) \mapsto p(A)$$

by

$$p(A) := \sum_k p_k A^k , \qquad p = \sum_k p_k X^k . \tag{12.11}$$

One can easily show that, for $A \in \mathrm{End}(V)$, the function

$$K[X] \to \mathrm{End}(V) , \quad p \mapsto p(A)$$

is an **algebra homomorphism**, that is, it is compatible with all algebra operations. ■

Difference Operators and Summation Formulas

We close this section with some applications illustrating the algebraic concepts introduced above.

Let E be a vector space over $\mathbb{K}$. On $E^{\mathbb{N}}$ define the **difference operator** $\triangle$ by

$$\triangle f_n := f_{n+1} - f_n , \qquad n \in \mathbb{N} , \quad f := (n \mapsto f_n) \in E^{\mathbb{N}} .$$

Obviously $\triangle \in \mathrm{End}(E^{\mathbb{N}})$. If I denotes the unity element of $\mathrm{End}(V)$, then

$$(I + \triangle) f_n = f_{n+1} , \qquad n \in \mathbb{N} , \quad f \in E^{\mathbb{N}} ,$$

that is, $I + \triangle$ is the **left shift operator**. If we write f as a 'sequence', $f = (f_0, f_1, f_2, \ldots)$, then we have $(I + \triangle)f = (f_1, f_2, f_3, \ldots)$ and, by induction,

$$(I + \triangle)^k f_n = f_{n+k} , \qquad n \in \mathbb{N} , \quad f \in E^{\mathbb{N}} , \tag{12.12}$$

hence $(I + \triangle)^k f = (f_k, f_{k+1}, f_{k+2}, \ldots)$.

Applying the binomial theorem (Proposition 8.4) to the ring $\mathrm{End}(E^{\mathbb{N}})$ we get

$$\triangle^k = \bigl(-I + (I+\triangle)\bigr)^k = \sum_{j=0}^{k} (-1)^{k-j} \binom{k}{j} (I+\triangle)^j \ , \qquad k \in \mathbb{N} \ ,$$

and so, by (12.12),

$$\triangle^k f_n = \sum_{j=0}^{k} (-1)^{k-j} \binom{k}{j} f_{n+j} \ , \qquad k, n \in \mathbb{N} \ , \quad f \in E^{\mathbb{N}} \ .$$

The binomial theorem also implies

$$(I+\triangle)^k = \sum_{j=0}^{k} \binom{k}{j} \triangle^j \ ,$$

and so

$$f_{n+k} = \sum_{j=0}^{k} \binom{k}{j} \triangle^j f_n \ , \qquad n \in \mathbb{N} \ , \quad f \in E^{\mathbb{N}} \ . \tag{12.13}$$

From the last formula we get finally

$$\sum_{k=0}^{m} f_k = \sum_{k=0}^{m} \sum_{j=0}^{k} \binom{k}{j} \triangle^j f_0 = \sum_{j=0}^{m} \sum_{k=j}^{m} \binom{k}{j} \triangle^j f_0 = \sum_{j=0}^{m} \binom{m+1}{j+1} \triangle^j f_0$$

for $m \in \mathbb{N}$. Here we have changed the order of summation as in the proof of Proposition 8.12, and, in the last step, used Exercise 5.5. Changing the indexing slightly yields the **general summation formula**

$$\sum_{k=0}^{m-1} f_k = \sum_{j=1}^{m} \binom{m}{j} \triangle^{j-1} f_0 \ , \qquad m \in \mathbb{N}^\times \ , \quad f \in E^{\mathbb{N}} \ . \tag{12.14}$$

Newton Interpolation Polynomials

Let $h \in \mathbb{K}^\times$ and $x_0 \in \mathbb{K}$. For each $m \in \mathbb{N}^\times$ and $f \in \mathbb{K}^{\mathbb{K}}$ there is, by Proposition 12.9, a unique **interpolation polynomial** $p := N_m[f; x_0; h]$ of degree $\leq m$ which satisfies

$$p(x_0 + jh) = f(x_0 + jh) \ , \qquad j = 0, \ldots, m \ ,$$

that is, f and p have equal values at the equally spaced points $x_0, x_0 + h, \ldots, x_0 + mh$. Thus

$$N_m[f; x_0; h] = N_m[f; x_0, x_0 + h, \ldots, x_0 + mh] \ .$$

By Remark 12.10(b), we can write $N_m[f; x_0; h]$ in the Newton form[1]

$$N_m[f; x_0; h] = \sum_{j=0}^{m} a_j \prod_{k=0}^{j-1} (X - x_k) \ .$$

[1]We make the convention that the 'empty product' $\prod_{k=0}^{-1}$ has the value 1.

The following proposition shows that, in this case, the coefficients a_j can be expressed easily using the difference operators $\triangle^j$. To do so we define the **divided difference operator $\triangle_h$ of length h** by

$$\triangle_h f(x) := \frac{f(x+h)-f(x)}{h} , \qquad x \in \mathbb{K} , \quad f \in \mathbb{K}^{\mathbb{K}} .$$

Obviously, $\triangle_h \in \mathrm{End}(\mathbb{K}^{\mathbb{K}})$. We set $\triangle_h^j := (\triangle_h)^j$ for $j \in \mathbb{N}$.

12.13 Proposition *The* **Newton interpolation polynomial** *for a function f and equally spaced points $x_j := x_0 + jh$, $0 \le j \le m$ has the form*

$$N_m[f;x_0;h] = \sum_{j=0}^{m} \frac{\triangle_h^j f(x_0)}{j!} \prod_{k=0}^{j-1} (X - x_k) . \tag{12.15}$$

Proof Using the notation of Remark 12.10, we need to show that $j!\,a_j = \triangle_h^j f(x_0)$ for $j = 0,\ldots,m$. Since

$$\omega_j[x_0,\ldots,x_{j-1}](x_k) = \prod_{\ell=0}^{j-1}(x_k - x_\ell) = k(k-1)\cdots(k-j+1)h^j = j!\binom{k}{j}h^j$$

for $0 \le j < k \le m$, the system of equations from Remark 12.10(b) has the form

$$\begin{aligned} &a_0 = f(x_0) \\ &a_0 + ha_1 = f(x_1) \\ &\cdots\cdots\cdots\cdots\cdots\cdots\cdots\cdots \\ &a_0 + 1!\binom{m}{1}ha_1 + \cdots + m!\binom{m}{m}h^m a_m = f(x_m) . \end{aligned} \tag{12.16}$$

Now we prove the claim by induction. For $m = 0$ the claim is clear. Suppose that $a_j = \triangle_h^j f(x_0)/j!$ for $0 \le j \le n$. For $m = n+1$ it follows from (12.16) that

$$\begin{aligned} f(x_{n+1}) &= \sum_{j=0}^{n} j!\binom{n+1}{j}h^j \frac{\triangle_h^j f(x_0)}{j!} + (n+1)!\,h^{n+1}a_{n+1} \\ &= \sum_{j=0}^{n}\binom{n+1}{j}\triangle^j f_0 + (n+1)!\,h^{n+1}a_{n+1} , \end{aligned} \tag{12.17}$$

where, for $n \in \mathbb{N}$, we define $f_0 \in \mathbb{K}^{\mathbb{N}}$ by $f_0(n) := f(x_0 + nh)$. By (12.13) we have

$$\sum_{j=0}^{n}\binom{n+1}{j}\triangle^j f_0 = \sum_{j=0}^{n+1}\binom{n+1}{j}\triangle^j f_0 - \triangle^{n+1} f_0 = f_{n+1} - \triangle^{n+1} f_0 .$$

Since $f(x_{n+1}) = f_{n+1}$, we get from (12.17) that

$$\triangle^{n+1} f_0 = (n+1)!\,h^{n+1}a_{n+1} ,$$

and hence $(n+1)!\,a_{n+1} = \triangle_h^{n+1} f(x_0)$. Thus the claim is true for each $m \in \mathbb{N}$. ■

12.14 Remarks (a) For $f \in \mathbb{K}^{\mathbb{K}}$, let

$$r_m[f; x_0; h] := f - N_m[f; x_0; h]$$

be the 'error function'. By construction, $f : \mathbb{K} \to \mathbb{K}$ is 'approximated' by the interpolation polynomial $N_m[f; x_0; h]$ so that the error is zero at the points $x_0 + jh$, $0 \le j \le m$. In Section IV.3 we will see how the error can be controlled for certain large classes of functions. In addition, we will show in Section V.4 that quite general functions can be approximated 'arbitrarily closely' (in a suitable sense) by polynomials.

(b) Obviously (12.15) also makes sense for arbitrary $f \in E^{\mathbb{K}}$, and

$$N_m[f; x_0; h](x_j) = f(x_j) , \qquad 0 \le j \le m .$$

(c) A function $f \in E^{\mathbb{N}}$ is called an **arithmetic sequence of order** $k \in \mathbb{N}^\times$ if $\triangle^k f$ is constant, that is, if $\triangle^{k+1} f = 0$. From (12.15) and Remark 8.19(c), it follows that, for each polynomial $p \in \mathbb{K}_k[X]$, each $h \in \mathbb{K}^\times$ and each $x_0 \in \mathbb{K}$, the function $\mathbb{N} \to \mathbb{K}$, $n \mapsto p(x_0 + hn)$ is an arithmetic sequence of order k. In particular, for each $k \in \mathbb{N}$, the 'power sequence' $\mathbb{N} \to \mathbb{N}$, $n \mapsto n^k$ is an arithmetic sequence of order k.

For arithmetic sequences of order k, the summation formula (12.14) has a simple form:

$$\sum_{j=0}^{n} f_j = \sum_{i=0}^{k} \binom{n+1}{i+1} \triangle^i f_0 , \qquad n \in \mathbb{N} .$$

In particular, for the '**power summations**' we have

$$\sum_{j=0}^{n} j = \binom{n+1}{2} = \frac{n(n+1)}{2} ,$$

$$\sum_{j=0}^{n} j^2 = \binom{n+1}{2} + 2\binom{n+1}{3} = \frac{n(n+1)(2n+1)}{6} ,$$

$$\sum_{j=0}^{n} j^3 = \binom{n+1}{2} + 6\binom{n+1}{3} + 6\binom{n+1}{4} = \frac{n^2(n+1)^2}{4} = \Bigl(\sum_{j=0}^{n} j\Bigr)^2 ,$$

which the reader can easily confirm. ■

Exercises

In the following K is a field and E, E_j, F, F_j, $1 \le j \le m$ are vector spaces over K.

1 (a) Determine all subspaces of K.

(b) What is the dimension of $\mathbb{C}$ over $\mathbb{R}$?

2 (a) Show that the projections $\mathrm{pr}_k : E_1 \times \cdots \times E_m \to E_k$ and the **canonical injections**

$$i_k : E_k \to E_1 \times \cdots \times E_m , \quad x \mapsto (0, \ldots, 0, \underset{(k)}{x}, 0, \ldots, 0)$$

are linear and determine the corresponding kernels and images.

(b) Show that $E_k \cong \mathrm{im}(i_k)$, $1 \le k \le m$.

3 Show that $T\colon \mathbb{R}^2 \to \mathbb{R}^2$, $(x,y) \mapsto (x-y, y-x)$ is linear. Determine the subspaces $\ker(T)$ and $\operatorname{im}(T)$.

4 Suppose that the diagram below is commutative.

$$\begin{array}{ccc} E & \xrightarrow{\ T\ } & F \\ {\scriptstyle P}\downarrow & & \downarrow{\scriptstyle Q} \\ E_1 & \xrightarrow{\ S\ } & F_1 \end{array}$$

If T, P and Q are linear and P is surjective, is S also linear?

5 Let X be a nonempty set and $x_0 \in X$. Show that the function $\delta_{x_0} : E^X \to E$ defined by

$$\delta_{x_0}(f) := f(x_0) , \qquad f \in E^X ,$$

is linear.

6 Let E and F be finite dimensional. Show that $\dim(E \times F) = \dim(E)\dim(F)$.

7 For $T \in \operatorname{Hom}(E,F)$, prove that $E/\ker(T) \cong \operatorname{im}(T)$.

8 Let $x_0,\ldots,x_m \in \mathbb{K}$ be distinct. Show that the following **Cauchy equations** hold for the Lagrange polynomials $\ell_j := \ell_j[x_0,\ldots,x_m] \in \mathbb{K}_m[X]$:

(a) $\sum_{j=0}^m \ell_j = 1 \ (= X^0)$.

(b) $(X-y)^k = \sum_{j=0}^m (x_j - y)^k \ell_j,\ \ y \in \mathbb{K},\ \ 1 \le k \le m.$

9 Show that, for distinct $x_0,\ldots,x_m \in \mathbb{K}$, the Lagrange polynomials $\ell_j,\ 0 \le j \le m$, and the Newton polynomials $\omega_j,\ 0 \le j \le m$, form bases of $\mathbb{K}_m[X]$.

10 Let $x_0,\ldots,x_m \in \mathbb{K}$ be distinct and $f \in \mathbb{K}^{\mathbb{K}}$. Prove the following:

(a) The coefficients a_j of the Newton polynomials in Remark 12.10(b) are given by

$$a_n = \sum_{j=0}^n \frac{f(x_j)}{\prod_{\substack{k=0\\k\ne j}}^n (x_j - x_k)} =: f[x_0,\ldots,x_n] , \qquad 0 \le n \le m .$$

(b) The coefficients $f[x_0,\ldots,x_n]$ are symmetric in their arguments. That is, if $0 \le n \le m$ and σ is a permutation of $\{0,1,\ldots,n\}$, then $f[x_0,\ldots,x_n] = f[x_{\sigma(0)},\ldots,x_{\sigma(n)}]$.

(c) $f[x_0,\ldots,x_n] = \dfrac{f[x_0,\ldots,x_{n-1}] - f[x_1,\ldots,x_n]}{x_0 - x_n},\ \ 1 \le n \le m.$

Remark Because of (c), the numbers $f[x_0,\ldots,x_n]$ are easy to calculate recursively.

(Hint: (a) $p_n[f,x_0,\ldots,x_n] = L_n[f,x_0,\ldots,x_n],\ \ 0 \le n \le m.$
(c) $p_n[f,x_0,\ldots,x_n] = b_0 + b_1(X-x_n) + \cdots + b_n(X-x_n)(X-x_{n-1})\cdots(X-x_1)$ with $a_n = b_n$ for $1 \le n \le m$. From this one can show that $b_n = f[x_n,x_{n-1},\ldots,x_1]$ and $(x_n - x_0)a_n + a_{n-1} - b_{n-1} = 0$.)

11 For $f \in E^{\mathbb{N}}$, show that $f_n = \sum_{j=0}^k (-1)^j \binom{k}{j} \triangle^j f_{n+k-j},\ \ n \in \mathbb{N}.$

(Hint: $I = (I + \triangle) - \triangle$.)

12 For $h \in \mathbb{K}^\times$ and $k, m \in \mathbb{N}$, show that $\triangle_h^k \in \mathrm{Hom}\bigl(\mathbb{K}_m[X], \mathbb{K}_{m-k}[X]\bigr)$ where we set $\mathbb{K}_j[X] := 0$ if j is negative.

What are the leading coefficients of $\triangle_h^k X^m$?

13 Verify the identity $\sum_{j=0}^n j^4 = n(n+1)(2n+1)(3n^2+3n-1)/30$ for $n \in \mathbb{N}$.

14 Show that $\mathbb{Q}(\sqrt{2}) := \{\, a + b\sqrt{2}\ ;\ a, b \in \mathbb{Q} \,\}$ (see Exercise 10.2) is a vector space over $\mathbb{Q}$. What is its dimension?

15 $\mathbb{R}$ can be considered as a vector space over the field $\mathbb{Q}(\sqrt{2})$. Are 1 and $\sqrt{3}$ linearly independent over $\mathbb{Q}(\sqrt{2})$?

16 For $m \in \mathbb{N}$ and an $m+1$ element subset $\{x_0, \ldots, x_m\}$ of $\mathbb{K}$ consider the function

$$e : \mathbb{K}_m[X] \to \mathbb{K}^{m+1}, \quad p \mapsto \bigl(p(x_0), \ldots, p(x_m)\bigr) .$$

Show that e is an isomorphism from $\mathbb{K}_m[X]$ to $\mathbb{K}^{m+1}$. What is e^{-1}?

17 Let $T : K \to E$ be linear. Prove that there is a unique $m \in E$ such that $T(x) = xm$ for all $x \in K$.

Chapter II

Convergence

With this chapter we enter at last the realm of analysis. This branch of mathematics is largely build upon the concept of convergence which allows us, in a certain sense, to add together infinite sets of numbers (or vectors). This ability to consider infinite operations is the essential difference between analysis and algebra.

The attempt to axiomatize naive ideas about the convergence of sequences of numbers leads naturally to the concepts of distance, the neighborhood of a point, and metric spaces — the subject of Section 1. In the special case of a sequence of numbers we can exploit the vector space structure of $\mathbb{K}$. An analysis of the proofs in this situation shows that most can be applied to sequences of vectors in a vector space, so long as some analog of absolute value is available. Thus we are naturally led to define normed vector spaces, a particularly important class of metric spaces.

Among normed vector spaces, inner product spaces are distinguished by the richness of their structure, as well as by the fact that their geometry is much like the familiar Euclidean geometry of the plane. Indeed, for elementary analysis, the most important classes of inner product spaces are the m-dimensional Euclidean spaces $\mathbb{R}^m$ and $\mathbb{C}^m$.

In Sections 4 and 5 we return to the simplest situation, namely convergence in $\mathbb{R}$. Using the order structure, and in particular, the order completeness of $\mathbb{R}$, we derive our first concrete convergence criteria. These allow us to calculate the limits of a number of important sequences. In addition, from the order completeness of $\mathbb{R}$, we derive a fundamental existence principle, the Bolzano-Weierstrass theorem.

Section 6 is devoted to the concept of completeness in metric spaces. Specialization to normed vector spaces leads to the definition of a Banach space. The basic example of such a space is $\mathbb{K}^m$, but we also show that sets of bounded functions are Banach spaces.

Banach spaces are ubiquitous in analysis and so play a central role in our presentation. Even so, their structure is simple enough that a beginner can go with little difficulty from understanding real numbers to understanding Banach spaces. Moreover, the early introduction of these spaces makes possible short and elegant proofs in later chapters.

For completeness and for the general (mathematical) education of the reader, we present in Section 6 Cantor's proof of the existence of an order complete ordered field using a 'completion' of $\mathbb{Q}$.

In the remaining sections of this chapter, we discuss the convergence of series. In Section 7 we learn the basic properties of series and discuss the most important examples. We are then able to investigate the decimal and other representations of real numbers, which enables us to prove that the real numbers form an uncountable set.

Among convergent series, those which converge absolutely play a particularly important role. Absolute convergence is often easy to recognize and such series are relatively easy to manipulate. Moreover, many series which are important in practice converge absolutely. This is particularly true about power series which we introduce and study in the last section of this chapter. The most important of these is the exponential series, whose significance will become clear in following chapters.

1 Convergence of Sequences

In this section we consider functions which are defined on the natural numbers and hence take on only a countable number of values. For such a function $\varphi\colon \mathbb{N} \to X$, we are particularly interested in the behavior of the values $\varphi(n)$ 'as n goes to infinity'. Because we can evaluate φ only finitely many times, that is, we can never 'reach infinity', we must develop methods which allow us to prove statements about infinitely many function values 'near infinity'. Such methods form the theory of convergent sequences, which we present in this section.

Sequences

Let X be a set. A **sequence** (in X) is simply a function from $\mathbb{N}$ to X. If $\varphi\colon \mathbb{N} \to X$ is a sequence, we write also

$$(x_n),\ (x_n)_{n\in\mathbb{N}} \quad \text{or} \quad (x_0, x_1, x_2, \ldots)$$

for φ, where $x_n := \varphi(n)$ is the nth **term** of the sequence $\varphi = (x_0, x_1, x_2, \ldots)$.

Sequences in $\mathbb{K}$ are called **number sequences**, and the **$\mathbb{K}$-vector space $\mathbb{K}^{\mathbb{N}}$ of all number sequences** is denoted by s or $s(\mathbb{K})$ (see Example I.12.3(e)). More precisely, one says (x_n) is a **real** (or **complex**) **sequence** if $\mathbb{K} = \mathbb{R}$ (or $\mathbb{K} = \mathbb{C}$).

1.1 Remarks (a) It is vital to distinguish a sequence (x_n) from its image $\{x_n \ ;\ n \in \mathbb{N}\}$. For example, if $x_n = x \in X$ for all n, that is, (x_n) is a constant sequence, then $(x_n) = (x, x, x, \ldots) \in X^{\mathbb{N}}$ whereas $\{\, x_n \ ;\ n \in \mathbb{N}\,\}$ is the one element set $\{x\}$.

(b) Let (x_n) be a sequence in X and E a property. Then we say that E holds for **almost all** terms of (x_n) if there is some $m \in \mathbb{N}$ such that $E(x_n)$ is true for all $n \geq m$, that is, if E holds for all but finitely many of the x_n. Of course, $E(x_n)$ could also be true for several (or all) $n < m$. If there is a subset $N \subseteq \mathbb{N}$ with $\operatorname{Num}(N) = \infty$ and $E(x_n)$ is true for each $n \in N$ then E is true for **infinitely many** terms. For example, the real sequence

$$\Bigl(-5, 4, -3, 2, -1, 0, -\frac{1}{2}, \frac{1}{3}, -\frac{1}{4}, \frac{1}{5}, \ldots, -\frac{1}{2n}, \frac{1}{2n+1}, \ldots\Bigr)$$

has infinitely many positive terms, infinitely many negative terms, and has absolute value less than 1 for almost all terms.

(c) For $m \in \mathbb{N}^{\times}$, a function $\psi\colon m + \mathbb{N} \to X$ is also called a sequence in X. That is, $(x_j)_{j\geq m} = (x_m, x_{m+1}, x_{m+2}, \ldots)$ is a sequence in X even though the indexing does not start with 0. This convention is justified, since after 're-indexing' using the function $\mathbb{N} \to m + \mathbb{N},\ n \mapsto m + n$, the 'shifted sequence' $(x_j)_{j\geq m}$ can be identified with the (usual) sequence $(x_{m+k})_{k\in\mathbb{N}} \in X^{\mathbb{N}}$. ■

If one graphs the first few terms of the complex sequence $(z_n)_{n\geq 1}$ defined by $z_n := (1 - 1/n)(1 + i)$, one observes that, as n increases, the points z_n get 'arbitrarily close' to $z := 1 + i$. In other words, the distance from z_n to z becomes 'arbitrarily small' with increasing n. The goal of this section is to axiomatize our intuitive and geometrical ideas about the convergence of such number sequences so that they can be applied to sequences in vector spaces and in other more abstract sets.

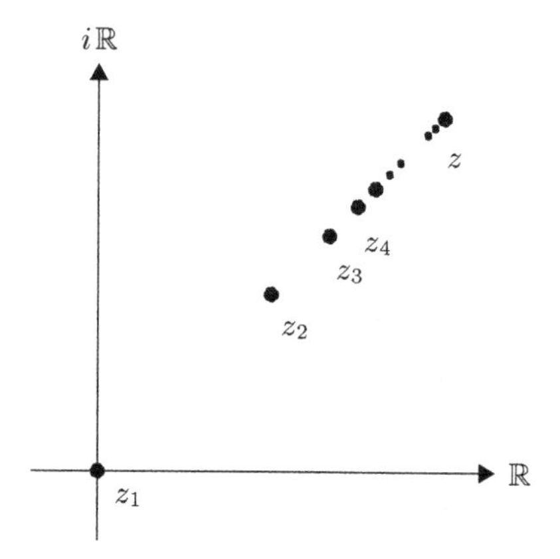

First we recognize that the concept of distance is of central importance. In $\mathbb{K}$ we can, with the help of the absolute value function, determine the distance between two points. To investigate the convergence of sequences in an arbitrary set X, we first need to endow X with a structure which permits the 'distance' between two elements in X to be determined.

Metric Spaces

Let X be a set. A function $d\colon X \times X \to \mathbb{R}^+$ is called a **metric** on X if the following hold:

(M$_1$) $d(x, y) = 0 \Leftrightarrow x = y$.

(M$_2$) $d(x, y) = d(y, x)$, $x, y \in X$ (symmetry).

(M$_3$) $d(x, y) \leq d(x, z) + d(z, y)$, $x, y, z \in X$ (triangle inequality).

If d is a metric on X, then (X, d) is called a **metric space**. When the metric is clear from context, we write simply X for (X, d). Finally we call $d(x, y)$ the **distance** between the **points** x and y in the metric space X.

The axioms (M$_1$)–(M$_3$) are clearly quite natural properties for a distance function. For example, (M$_3$) can be seen as an axiomatic formulation of the rule that 'the direct path from x to y is shorter than the path which goes from x to z and then to y'.

In the metric space (X, d), for $a \in X$ and $r > 0$, the set

$$\mathbb{B}(a, r) := \mathbb{B}_X(a, r) := \{\, x \in X \ ;\ d(a, x) < r \,\}$$

is called the **open ball** with center at a and radius r, while

$$\bar{\mathbb{B}}(a, r) := \bar{\mathbb{B}}_X(a, r) := \{\, x \in X \ ;\ d(a, x) \leq r \,\}$$

is called the **closed ball** with center at a and radius r.

1.2 Examples (a) $\mathbb{K}$ is a metric space with the **natural metric**

$$\mathbb{K} \times \mathbb{K} \to \mathbb{R}^+ , \quad (x, y) \mapsto |x - y| .$$

Unless otherwise stated, we consider $\mathbb{K}$ to be a metric space with the natural metric.[1]

Proof The validity of (M_1)–(M_3) follows directly from Proposition I.11.4. ■

(b) Let (X, d) be a metric space and Y a nonempty subset of X. Then the restriction of d to $Y \times Y$, $d_Y := d|Y \times Y$, is a metric on Y, the **induced metric**, and (Y, d_Y) is a metric space, a **metric subspace** of X. When no misunderstanding is possible, we write d instead of d_Y.

(c) Any nonempty subset of $\mathbb{C}$ is a metric space with the metric induced from the natural metric on $\mathbb{C}$. The metric on $\mathbb{R}$ induced in this way is the natural metric as defined in (a).

(d) Let X be a nonempty set. Then the function $d(x, y) := 1$ for $x \neq y$ and $d(x, x) := 0$ is a metric, called the **discrete metric**, on X.

(e) Let (X_j, d_j), $1 \leq j \leq m$, be metric spaces and $X := X_1 \times \cdots \times X_m$. Then the function

$$d(x, y) := \max_{1 \leq j \leq m} d_j(x_j, y_j)$$

for $x := (x_1, \ldots, x_m) \in X$ and $y := (y_1, \ldots, y_m) \in X$ is a metric on X called the **product metric**. The metric space $X := (X, d)$ is called the **product of the metric spaces** (X_j, d_j). One can check that

$$\mathbb{B}_X(a, r) = \prod_{j=1}^{m} \mathbb{B}_{X_j}(a_j, r) , \quad \bar{\mathbb{B}}_X(a, r) = \prod_{j=1}^{m} \bar{\mathbb{B}}_{X_j}(a_j, r)$$

for all $a := (a_1, \ldots, a_m) \in X$ and $r > 0$. ■

An important consequence of the metric space axioms is the reversed triangle inequality (see Corollary I.11.5).

1.3 Proposition *Let (X, d) be a metric space. Then for all $x, y, z \in X$ we have*

$$d(x, y) \geq |d(x, z) - d(z, y)| .$$

Proof From (M_3) we get the inequality $d(x, y) \geq d(x, z) - d(y, z)$. Interchanging x and y yields

$$d(x, y) = d(y, x) \geq d(y, z) - d(x, z) = -\big(d(x, z) - d(y, z)\big) ,$$

from which the claim follows. ■

[1] Note that, with this convention, the definitions of open and closed balls given above (as applied to $\mathbb{K}$) coincide with those of Section I.11.

A subset U of a metric space X is called a **neighborhood** of $a \in X$ if there is some $r > 0$ such that $\mathbb{B}(a,r) \subseteq U$. The **set of all neighborhoods of the point** a is denoted by $\mathcal{U}(a)$, that is,

$$\mathcal{U}(a) := \mathcal{U}_X(a) := \{U \subseteq X \ ; \ U \text{ is a neighborhood of } a\} \subseteq \mathcal{P}(X) .$$

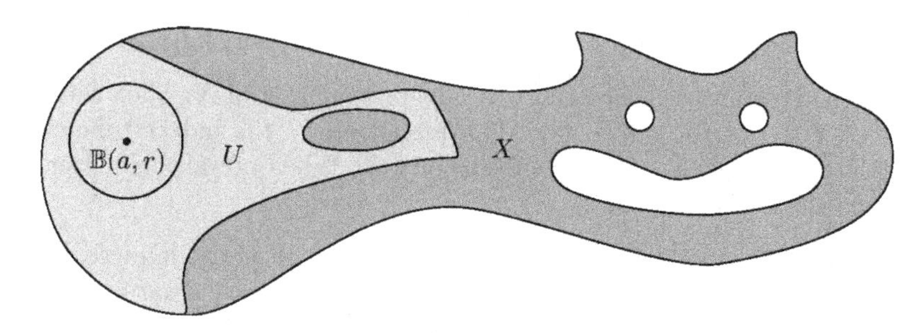

1.4 Examples Let X be a metric space and $a \in X$.

(a) For each $\varepsilon > 0$, $\mathbb{B}(a,\varepsilon)$ and $\bar{\mathbb{B}}(a,\varepsilon)$ are neighborhoods of a called the **open** and the **closed ε-neighborhoods** of a.

(b) Obviously X is in $\mathcal{U}(a)$. If $U_1, U_2 \in \mathcal{U}(a)$, then $U_1 \cap U_2$ and $U_1 \cup U_2$ are also in $\mathcal{U}(a)$. Any $U \subseteq X$ which contains a neighborhood of $a \in X$ is also in $\mathcal{U}(a)$.

Proof By supposition there are $r_j > 0$ with $\mathbb{B}(a,r_j) \subseteq U_j$ for $j = 1,2$. Define $r > 0$ by $r := \min\{r_1, r_2\}$, then $\mathbb{B}(a,r) \subseteq U_1 \cap U_2 \subseteq U_1 \cup U_2$. The other claims are clear. ■

(c) For $X := [0,1]$ with metric induced from $\mathbb{R}$, $[1/2, 1]$ is a neighborhood of 1, but not of $1/2$. ■

For the remainder of this section, $X := (X,d)$ is a metric space and (x_n) is a sequence in X.

Cluster Points

We call $a \in X$ a **cluster point** of (x_n) if every neighborhood of a contains infinitely many terms of the sequence.

Before we consider some examples, it is useful to have the following characterization of cluster points:

1.5 Proposition *The following are equivalent:*

(i) *a is a cluster point of (x_n).*

(ii) *For each $U \in \mathcal{U}(a)$ and $m \in \mathbb{N}$, there is some $n \geq m$ such that $x_n \in U$.*

(iii) *For each $\varepsilon > 0$ and $m \in \mathbb{N}$, there is some $n \geq m$ such that $x_n \in \mathbb{B}(a,\varepsilon)$.*

Proof This follows directly from the definitions. ■

1.6 Examples **(a)** The real sequence $\big((-1)^n\big)_{n\in\mathbb{N}}$ has two cluster points, namely, 1 and -1.

(b) The complex sequence $(i^n)_{n\in\mathbb{N}}$ has four cluster points, namely, ± 1 and $\pm i$.

(c) The constant sequence $(x, x, x, \ldots)$ has the unique cluster point x.

(d) The sequence of the natural numbers $(n)_{n\in\mathbb{N}}$ has no cluster points.

(e) Let φ be a bijection from $\mathbb{N}$ to $\mathbb{Q}$ (such functions exist by Proposition I.9.4). Define a sequence (x_n) by $x_n := \varphi(n)$ for all $n \in \mathbb{N}$. Then all real numbers are cluster points of (x_n).

Proof Suppose that there is some $a \in \mathbb{R}$ which is not a cluster point of (x_n). Then, by Proposition 1.5, there are $\varepsilon > 0$ and $m \in \mathbb{N}$ such that

$$x_n \notin \mathbb{B}(a, \varepsilon) = (a - \varepsilon, a + \varepsilon) \ , \qquad n \geq m \ .$$

That is, the interval $(a - \varepsilon, a + \varepsilon)$ contains only finitely many rational numbers. But this is not possible because of Proposition I.10.8. ∎

Convergence

A sequence (x_n) **converges** (or is **convergent**) with **limit** a if each neighborhood of a contains almost all terms of the sequence. In this case we write[2]

$$\lim_{n\to\infty} x_n = a \qquad \text{or} \qquad x_n \to a \ \ (n \to \infty) \ ,$$

and we say that (x_n) **converges to** a **as** n **goes to** ∞. A sequence (x_n) which is not convergent is called **divergent** and we say that (x_n) **diverges**.

The essential part of the definition is the requirement that *each* neighborhood of the limit contains almost all terms of the sequence. This requirement corresponds, in the case that $X = \mathbb{K}$, to the geometric intuition that the distance from x_n to a 'becomes arbitrarily small'. If a is a cluster point of (x_n) and U is a neighborhood of a, then, of course, U contains infinitely many terms of the sequence, but it is also possible that infinitely many terms of the sequence are not in U.

The next proposition is again simply a reformulation of the corresponding definitions.

1.7 Proposition *The following statements are equivalent:*

(i) $\lim x_n = a$.

(ii) *For each* $U \in \mathcal{U}(a)$, *there is some*[3] $N := N(U)$ *such that* $x_n \in U$ *for all* $n \geq N$.

(iii) *For each* $\varepsilon > 0$, *there is some*[3] $N := N(\varepsilon)$ *such that* $x_n \in \mathbb{B}(a, \varepsilon)$ *for all* $n \geq N$.

---

[2] When no misunderstanding is possible, we write also $\lim_n x_n = a$, $\lim x_n = a$ or $x_n \to a$.

[3] We use this notation to indicate that the number N, in general, depends on U (or ε).

The following examples are rather simple. For more complicated examples we need the methods to be developed starting in Section 4.

1.8 Examples **(a)** For the real sequence $(1/n)_{n\in\mathbb{N}^\times}$, we have $\lim(1/n) = 0$.

Proof Let $\varepsilon > 0$. By Corollary I.10.7, there is some $N \in \mathbb{N}^\times$ such that $1/N < \varepsilon$. Then $1/n \leq 1/N < \varepsilon$ for all $n \geq N$, that is, $1/n \in (0,\varepsilon) \subseteq \mathbb{B}(0,\varepsilon)$ for all $n \geq N$. ■

(b) For the complex sequence (z_n) defined by

$$z_n := \frac{n+2}{n+1} + i\,\frac{2n}{n+2}\;,$$

we have $\lim z_n = 1 + 2i$.

Proof Let $\varepsilon > 0$. By Corollary I.10.7, there is some $N \in \mathbb{N}$ such that $1/N < \varepsilon/8$. Then, for all $n \geq N$, we have

$$\frac{n+2}{n+1} - 1 = \frac{1}{n+1} < \frac{1}{N} < \frac{\varepsilon}{8} < \frac{\varepsilon}{2}$$

and

$$2 - \frac{2n}{n+2} = \frac{4}{n+2} < \frac{4}{N} < \frac{\varepsilon}{2}\;.$$

Consequently

$$|z_n - (1+2i)|^2 = \left|\frac{n+2}{n+1} - 1\right|^2 + \left|\frac{2n}{n+2} - 2\right|^2 < \frac{\varepsilon^2}{4} + \frac{\varepsilon^2}{4} < \varepsilon^2\;, \qquad n \geq N\;.$$

This shows that $z_n \in \mathbb{B}_{\mathbb{C}}\big((1+2i),\varepsilon\big)$ for all $n \geq N$. ■

(c) The constant sequence $(a,a,a,\ldots)$ converges to a.

(d) The real sequence $\big((-1)^n\big)_{n\in\mathbb{N}}$ is divergent.

(e) Let X be the product of the metric spaces (X_j,d_j), $1 \leq j \leq m$. Then the sequence[4] $(x_n) = \big((x_n^1,\ldots,x_n^m)\big)_{n\in\mathbb{N}}$ converges in X to the point $a := (a^1,\ldots,a^m)$ if and only if, for each $j \in \{1,\ldots,m\}$, the sequence $(x_n^j)_{n\in\mathbb{N}}$ converges in X_j to $a^j \in X_j$.

Proof For each given $\varepsilon > 0$, almost all x_n are in $\mathbb{B}_X(a,\varepsilon) = \prod_{j=1}^m \mathbb{B}_{X_j}(a^j,\varepsilon)$ if and only if for each $j = 1,\ldots,m$, almost all x_n^j are in $\mathbb{B}_{X_j}(a^j,\varepsilon)$ (see Example 1.2(e)). ■

[4] In the following we often write $x^j := \mathrm{pr}_j(x)$ for $x \in X$ and $1 \leq j \leq m$. Even in the case $X_j = \mathbb{K}$ it will be clear from context whether x^j is the component of a point in a product space or a power of x.

Bounded Sets

A subset $Y \subseteq X$ is called **d-bounded** or **bounded in X** (with respect to the metric d) if there is some $M > 0$ such that $d(x,y) \le M$ for all $x, y \in Y$. In this circumstance the **diameter** of Y, defined by

$$\operatorname{diam}(Y) := \sup_{x,y \in Y} d(x,y) ,$$

is finite. A sequence (x_n) is **bounded** if its image $\{ x_n \ ; \ n \in \mathbb{N} \}$ is bounded.

1.9 Examples **(a)** For all $a \in X$ and $r > 0$, $\mathbb{B}(a,r)$ and $\bar{\mathbb{B}}(a,r)$ are bounded in X.

(b) Each subset of a bounded set is bounded. Finite unions of bounded sets are bounded.

(c) A subset Y of X is bounded in X if and only if there are some $x_0 \in X$ and $r > 0$ such that $Y \subseteq \mathbb{B}_X(x_0, r)$. If $Y \neq \emptyset$ then there is some $x_0 \in Y$ with this property.

(d) Bounded intervals are bounded.

(e) A subset Y of $\mathbb{K}$ is bounded if and only if there is some $M > 0$ such that $|y| \le M$ for all $y \in Y$. ∎

1.10 Proposition *Any convergent sequence is bounded.*

Proof Suppose that $x_n \to a$. Then there is some N such that $x_n \in \mathbb{B}(a,1)$ for all $n \ge N$. It follows from the triangle inequality that

$$d(x_n, x_m) \le d(x_n, a) + d(a, x_m) \le 2 , \qquad m, n \ge N .$$

Since there is also some $M \ge 0$ such that $d(x_j, x_k) \le M$ for all $j, k \le N$, we have $d(x_n, x_m) \le M + 2$ for all $m, n \in \mathbb{N}$. ∎

Uniqueness of the Limit

1.11 Proposition *Let (x_n) be convergent with limit a. Then a is the unique cluster point of (x_n).*

Proof It is clear that a is a cluster point of (x_n). To show uniqueness, suppose that $b \neq a$ is some point of X. Then, by (M_1), $\varepsilon := d(b,a)/2$ is positive. Since $a = \lim x_n$, there is some N such that $d(a, x_n) < \varepsilon$ for all $n \ge N$. Proposition 1.3 then implies that

$$d(b, x_n) \ge |d(b,a) - d(a,x_n)| \ge d(b,a) - d(a,x_n) > 2\varepsilon - \varepsilon = \varepsilon , \qquad n \ge N .$$

That is, almost all terms of (x_n) are outside of $\mathbb{B}(b, \varepsilon)$. Thus b is not a cluster point of (x_n). ∎

1.12 Remark The converse of Proposition 1.11 is false, that is, there are divergent sequences with exactly one cluster point, for example, $\left(\frac{1}{2}, 2, \frac{1}{3}, 3, \frac{1}{4}, 4, \ldots\right)$. ∎

As a direct consequence of Proposition 1.11 we have the following:

1.13 Corollary *The limit of a convergent sequence is unique.*

Subsequences

Let $\varphi = (x_n)$ be a sequence in X and $\psi \colon \mathbb{N} \to \mathbb{N}$ a strictly increasing function. Then $\varphi \circ \psi \in X^{\mathbb{N}}$ is called a **subsequence** of φ. Extending the notation $(x_n)_{n\in\mathbb{N}}$ introduced above for the sequence φ, we write $(x_{n_k})_{k\in\mathbb{N}}$ for the subsequence $\varphi \circ \psi$ where $n_k := \psi(k)$. Since ψ is strictly increasing we have $n_0 < n_1 < n_2 < \cdots$.

1.14 Example The sequence $\left((-1)^n\right)_{n\in\mathbb{N}}$ has the two constant subsequences, $\left((-1)^{2k}\right)_{k\in\mathbb{N}} = (1, 1, 1, \ldots)$ and $\left((-1)^{2k+1}\right)_{k\in\mathbb{N}} = (-1, -1, -1, \ldots)$. ∎

1.15 Proposition *If (x_n) is a convergent sequence with limit a, then each subsequence $(x_{n_k})_{k\in\mathbb{N}}$ of (x_n) is convergent with $\lim_{k\to\infty} x_{n_k} = a$.*

Proof Let $(x_{n_k})_{k\in\mathbb{N}}$ be a subsequence of (x_n) and U a neighborhood of a. Because $a = \lim x_n$, there is some N such that $x_n \in U$ for all $n \geq N$. From the definition of a subsequence, $n_k \geq k$ for all $k \in \mathbb{N}$, and so, in particular, $n_k \geq N$ for all $k \geq N$. Thus $x_{n_k} \in U$ for all $k \geq N$. This means that (x_{n_k}) converges to a. ∎

1.16 Example For $m \geq 2$,

$$\frac{1}{k^m} \to 0 \quad (k \to \infty) \qquad \text{and} \qquad \frac{1}{m^k} \to 0 \quad (k \to \infty)\ .$$

Proof Set $\psi_1(k) := k^m$ and $\psi_2(k) := m^k$ for all $k \in \mathbb{N}^\times$. Since $\psi_i : \mathbb{N}^\times \to \mathbb{N}^\times$, $i = 1, 2$, are strictly increasing, $(k^{-m})_{k\in\mathbb{N}^\times}$ and $(m^{-k})_{k\in\mathbb{N}^\times}$ are subsequences of $(1/n)_{n\in\mathbb{N}^\times}$. The claim then follows from Proposition 1.15 and Example 1.8(a). ∎

The next proposition provides a further characterization of the cluster points of a sequence.

1.17 Proposition *A point a is a cluster point of a sequence (x_n) if and only if there is some subsequence $(x_{n_k})_{k\in\mathbb{N}}$ of (x_n) which converges to a.*

Proof Let a be a cluster point of (x_n). We define recursively a sequence of natural numbers $(n_k)_{k\in\mathbb{N}}$ by

$$n_0 := 0\ , \quad n_k := \min\{\, m \in \mathbb{N}\ ;\ m > n_{k-1},\ x_m \in \mathbb{B}(a, 1/k)\,\}\ , \qquad k \in \mathbb{N}^\times\ .$$

Since a is a cluster point of (x_n), the sets

$$\{\, m \in \mathbb{N} \; ; \; m > n_{k-1},\ x_m \in \mathbb{B}(a, 1/k) \,\} \ , \qquad k \in \mathbb{N}^\times \ ,$$

are nonempty. By the well ordering principle, n_k is well defined for each $k \in \mathbb{N}^\times$. Thus $\psi : \mathbb{N} \to \mathbb{N},\ k \mapsto n_k$ is well defined and strictly increasing.

We next show that the subsequence $(x_{n_k})_{k\in\mathbb{N}}$ converges to a. Let $\varepsilon > 0$. By Corollary I.10.7 there is some $K := K(\varepsilon) \in \mathbb{N}^\times$ such that $1/k < \varepsilon$ for all $k \geq K$. By the construction of n_k we have

$$x_{n_k} \in \mathbb{B}(a, 1/k) \subseteq \mathbb{B}(a, \varepsilon) \ , \qquad k \geq K \ .$$

Thus $a = \lim_{k\to\infty} x_{n_k}$.

Conversely, let $(x_{n_k})_{k\in\mathbb{N}}$ be a subsequence of (x_n) such that $a = \lim_{k\to\infty} x_{n_k}$. Then, by Proposition 1.11, a is a cluster point of $(x_{n_k})_{k\in\mathbb{N}}$ and hence also of (x_n). ■

Exercises

1 Let d be the discrete metric on $\mathbb{K}$ and $X := (\mathbb{K}, d)$.

(a) Give explicit descriptions of $\mathbb{B}_X(a,r)$ and $\bar{\mathbb{B}}_X(a,r)$ for $a \in X$ and $r > 0$.

(b) Describe the cluster points of an arbitrary sequence in X.

(c) For $a \in X$, describe all sequences (x_n) in X such that $x_n \to a$.

2 Prove the claims of Example 1.2(e).

3 Prove that the sequence $(z_n)_{n\geq 1}$ where $z_n := (1 - 1/n)(1 + i)$ converges to $1 + i$ (as suggested by the graph following Remarks 1.1).

4 Prove the claims of Examples 1.9.

5 Determine *all* cluster points of the complex sequence (z_n) in the following cases:

(a) $z_n := \big((1+i)/\sqrt{2}\big)^n$.

(b) $z_n := \big(1 + (-1)^n\big)(n+1)n^{-1} + (-1)^n$.

(c) $z_n := (-1)^n n/(n+1)$.

6 For $n \in \mathbb{N}$, define

$$a_n := n + \frac{1}{k} - \frac{k^2 + k - 2}{2} \ ,$$

where $k \in \mathbb{N}^\times$ satisfies

$$k^2 + k - 2 \leq 2n \leq k^2 + 3k - 2 \ .$$

Show that (a_n) is well defined and determine all cluster points of (a_n). (Hint: Calculate the first few terms of the sequence explicitly to understand the complete sequence.)

7 For $m, n \in \mathbb{N}^\times$, define

$$d(m,n) := \begin{cases} (m+n)/mn \ , & m \neq n \ , \\ 0 \ , & m = n \ . \end{cases}$$

Show that $(\mathbb{N}^\times, d)$ is a metric space and describe $A_n := \bar{\mathbb{B}}(n, 1 + 1/n)$ for $n \in \mathbb{N}^\times$.

8 Let $X := \{\, z \in \mathbb{C} \,;\, |z| \le 3 \,\}$ with the natural metric. Describe $\bar{\mathbb{B}}_X(0,3)$ and $\bar{\mathbb{B}}_X(2,4)$. Show that $\bar{\mathbb{B}}_X(2,4) \subset \bar{\mathbb{B}}_X(0,3)$.

9 Two metrics d_1 and d_2 on a set X are called **equivalent** if, for each $x \in X$ and $\varepsilon > 0$, there are positive numbers r_1 and r_2 such that

$$\mathbb{B}_1(x, r_1) \subseteq \mathbb{B}_2(x, \varepsilon) \ , \quad \mathbb{B}_2(x, r_2) \subseteq \mathbb{B}_1(x, \varepsilon) \ .$$

Here $\mathbb{B}_j$ denotes the ball in (X, d_j), $j = 1, 2$. Now let (X, d) be a metric space and

$$\delta(x,y) := \frac{d(x,y)}{1 + d(x,y)} \ , \qquad x, y \in X \ .$$

Prove that d and δ are equivalent metrics on X. (Hint: The function $t \mapsto t/(1+t)$ is increasing.)

10 For $X := (0,1)$, prove the following:

(a) $d(x,y) := |(1/x) - (1/y)|$ is a metric on X.

(b) The natural metric and d are equivalent.

(c) There is no metric on $\mathbb{R}$ which is equivalent to the natural metric and which induces the metric d on X.

11 Let (X_j, d_j), $j = 1, \ldots, n$, be metric spaces, $X := X_1 \times \cdots \times X_n$ and d the product metric on X. Show that

$$\delta(x,y) := \sum_{j=1}^{n} d_j(x_j, y_j) \ , \qquad x := (x_1, \ldots, x_n) \in X \ , \quad y := (y_1, \ldots, y_n) \in X \ ,$$

is a metric on X which is equivalent to d.

12 For $z, w \in \mathbb{C}$, set

$$\delta(z,w) := \begin{cases} |z - w| \ , & \text{if } z = \lambda w \text{ for some } \lambda > 0 \ , \\ |z| + |w| & \text{otherwise} \ . \end{cases}$$

Show that δ defines a metric on $\mathbb{C}$, the **SNCF-metric.**[5]

13 Let (x_n) be a sequence in $\mathbb{C}$ with $\mathrm{Re}\, x_n = 0$ for all $n \in \mathbb{N}$. Show that, if (x_n) converges to x, then $\mathrm{Re}\, x = 0$.

[5]Users of the French railway system (the SNCF) will have noticed that the fastest connection between two cities (for example, Bordeaux and Lyon) often goes through Paris.

2 Real and Complex Sequences

In this section we derive the most important rules for calculating with convergent sequences of numbers. If we interpret these sequences as vectors in the vector space $s = s(\mathbb{K}) = \mathbb{K}^{\mathbb{N}}$, these rules show that the convergent sequences form a subspace of s. In the case of *real* sequences, we use the order structure of $\mathbb{R}$ to derive the comparison test which is the main tool for investigating convergence in $s(\mathbb{R})$.

Null Sequences

A sequence (x_n) in $\mathbb{K}$ is called a **null sequence** if it converges to zero, that is, if, for each $\varepsilon > 0$, there is some $N \in \mathbb{N}$ such that $|x_n| < \varepsilon$ for all $n \geq N$. The set of all null sequences in $\mathbb{K}$ we denote by c_0, that is,

$$c_0 := c_0(\mathbb{K}) := \{\, (x_n) \in s \; ; \; (x_n) \text{ converges with } \lim x_n = 0 \,\} .$$

2.1 Remarks Let (x_n) be a sequence in $\mathbb{K}$ and $a \in \mathbb{K}$.

(a) (x_n) is a null sequence if and only if $(|x_n|)$, the sequence of absolute values, is a null sequence in $\mathbb{R}$.

Proof This comes directly from the definition. ■

(b) (x_n) converges to a if and only if the 'shifted sequence' $(x_n - a)$ is a null sequence.

Proof From Proposition 1.7 we know that (x_n) converges to a if and only if, for each $\varepsilon > 0$, there is some N such that $|x_n - a| < \varepsilon$ for all $n \geq N$. Hence the claim follows from (a). ■

(c) If there is a real null sequence (r_n) such that $|x_n| \leq r_n$ for almost all $n \in \mathbb{N}$ then (x_n) is a null sequence.

Proof Let $\varepsilon > 0$. By assumption there are $M, N \in \mathbb{N}$ such that $|x_n| \leq r_n$ for all $n \geq M$ and $r_n < \varepsilon$ for all $n \geq N$. Consequently $|x_n| < \varepsilon$ for all $n \geq \max\{M, N\}$. ■

Elementary Rules

2.2 Proposition *Let (x_n) and (y_n) be convergent sequences in $\mathbb{K}$ with $\lim x_n = a$ and $\lim y_n = b$. Let $\alpha \in \mathbb{K}$.*

(i) *The sequence $(x_n + y_n)$ converges with $\lim(x_n + y_n) = a + b$.*

(ii) *The sequence (αx_n) converges with $\lim(\alpha x_n) = \alpha a$.*

Proof Let $\varepsilon > 0$.

(i) Because $x_n \to a$ and $y_n \to b$, there are $M, N \in \mathbb{N}$ such that $|x_n - a| < \varepsilon/2$

for all $n \geq M$, and $|y_n - b| < \varepsilon/2$ for all $n \geq N$. Hence

$$|x_n + y_n - (a+b)| \leq |x_n - a| + |y_n - b| < \frac{\varepsilon}{2} + \frac{\varepsilon}{2} = \varepsilon \;, \qquad n \geq \max\{M, N\} \;.$$

This shows that $(x_n + y_n)$ converges to $a + b$.

(ii) Since the case $\alpha = 0$ is obvious, we suppose that $\alpha \neq 0$. By assumption (x_n) converges with limit a. Thus there is some N such that $|x_n - a| < \varepsilon/|\alpha|$ for all $n \geq N$. It follows that

$$|\alpha x_n - \alpha a| = |\alpha|\,|x_n - a| \leq |\alpha| \frac{\varepsilon}{|\alpha|} = \varepsilon \;, \qquad n \geq N \;,$$

which proves the claim. ■

2.3 Remark Denote the set of all convergent sequences in $\mathbb{K}$ by

$$c := c(\mathbb{K}) := \big\{\, (x_n) \in s \;;\; (x_n) \text{ converges} \,\big\} \;.$$

Then Proposition 2.2 has the following interpretation:

c is a subspace of s, and the function

$$\lim : c \to \mathbb{K} \;, \quad (x_n) \mapsto \lim x_n$$

is linear.

Clearly $\ker(\lim) = c_0$, and so, by Example I.12.3(c), c_0 is a subspace of c. ■

The next proposition shows, in particular, that convergent sequences can be multiplied 'termwise'.

2.4 Proposition *Let (x_n) and (y_n) be sequences in $\mathbb{K}$.*

(i) *If (x_n) is a null sequence and (y_n) is a bounded sequence, then $(x_n y_n)$ is a null sequence.*

(ii) *If $\lim x_n = a$ and $\lim y_n = b$, then $\lim(x_n y_n) = ab$.*

Proof (i) Since (y_n) is bounded, there is some $M > 0$ such that $|y_n| \leq M$ for all $n \in \mathbb{N}$. Since (x_n) is a null sequence, for each $\varepsilon > 0$, there is some $N \in \mathbb{N}$ such that $|x_n| < \varepsilon/M$ for all $n \geq N$. It now follows that

$$|x_n y_n| = |x_n|\,|y_n| < \frac{\varepsilon}{M} M = \varepsilon \;, \qquad n \geq N \;.$$

Thus $(x_n y_n)$ is a null sequence.

(ii) Since $x_n \to a$, $(x_n - a)$ is a null sequence. By Proposition 1.10, (y_n) is bounded. From (i), $\big((x_n - a)y_n\big)_{n\in\mathbb{N}}$ is a null sequence. Since $\big(a(y_n - b)\big)_{n\in\mathbb{N}}$ is also a null sequence, Proposition 2.2 implies that

$$x_n y_n - ab = (x_n - a)y_n + a(y_n - b) \to 0 \quad (n \to \infty) \;.$$

Therefore the sequence $(x_n y_n)$ converges to ab. ■

2.5 Remarks **(a)** The hypothesis in Proposition 2.4(i), that the sequence (y_n) is bounded, cannot be removed.

Proof Let $x_n := 1/n$ and $y_n := n^2$ for all $n \in \mathbb{N}^\times$. Then (x_n) is a null sequence but the sequence $(x_n y_n) = (n)_{n\in\mathbb{N}}$ is divergent. ∎

(b) From Example I.12.11(a) we know that $s = s(\mathbb{K}) = \mathbb{K}^{\mathbb{N}}$ is an algebra (over $\mathbb{K}$). So, with Remark 2.3, Proposition 2.4(ii) can be reformulated as follows:

c is a subalgebra of s and the function

$\lim : c \to \mathbb{K}$ is an algebra homomorphism .

Finally, it follows from Proposition 1.10 and Proposition 2.4(i) that c_0 is also an ideal of c. ∎

The next proposition and Remark 2.5(b) show that the limit of a sequence of quotients is the limit of the numerators divided by the limit of the denominators, if these limits exist.

2.6 Proposition *Let (x_n) be a convergent sequence in $\mathbb{K}$ with limit $a \in \mathbb{K}^\times$. Then almost all terms of (x_n) are nonzero and $1/x_n \to 1/a \quad (n \to \infty)$.*

Proof Since $|a| > 0$, there is some $N \in \mathbb{N}$ such that $|x_n - a| < |a|/2$ for all $n \geq N$. Hence, by the reversed triangle inequality,

$$|a| - |x_n| \leq |x_n - a| \leq \frac{|a|}{2} , \qquad n \geq N ,$$

that is, $|x_n| \geq |a|/2 > 0$ for almost all n. This proves the first claim. It also follows from $|x_n| \geq |a|/2$ that

$$\left| \frac{1}{x_n} - \frac{1}{a} \right| = \frac{|x_n - a|}{|x_n|\,|a|} \leq \frac{2}{|a|^2} |x_n - a| , \qquad n \geq N . \tag{2.1}$$

By hypothesis and Remark 2.1(b), $(|x_n - a|)$ is a null sequence, and so, by Proposition 2.2, $\bigl(2\,|x_n - a|/|a|^2\bigr)$ is also a null sequence. The claim then follows from (2.1) and Remarks 2.1(b) and (c). ∎

The Comparison Test

We investigate next the relationship between convergent real sequences and the order structure of $\mathbb{R}$. In particular, in Proposition 2.9 we derive the comparison test, a simple, but very useful, method of determining the limits of real sequences.

2.7 Proposition *Let (x_n) and (y_n) be convergent sequences in $\mathbb{R}$ such that $x_n \leq y_n$ for infinitely many $n \in \mathbb{N}$. Then*

$$\lim x_n \leq \lim y_n .$$

Proof Set $a := \lim x_n$ and $b := \lim y_n$ and suppose, contrary to our claim, that $a > b$. Then $\varepsilon := a - b$ is positive and so there is some $n \in \mathbb{N}$ such that

$$a - \varepsilon/4 < x_n \leq y_n < b + \varepsilon/4 ,$$

that is, $\varepsilon = a - b < \varepsilon/2$, which is not possible. ∎

2.8 Remark Proposition 2.7 does not hold for strict inequalities, that is, $x_n < y_n$ for infinitely many $n \in \mathbb{N}$ does not imply that $\lim x_n < \lim y_n$.

Proof Let $x_n := -1/n$ and $y_n := 1/n$ for all $n \in \mathbb{N}^\times$. Then $x_n < y_n$ for all $n \in \mathbb{N}^\times$, but $\lim x_n = \lim y_n = 0$. ∎

2.9 Proposition *Suppose that (x_n), (y_n) and (z_n) are real sequences with the property that $x_n \leq y_n \leq z_n$ for almost all $n \in \mathbb{N}$. If $\lim x_n = \lim z_n =: a$, then (y_n) also converges to a.*

Proof Let m_0 be such that $x_n \leq y_n \leq z_n$ for all $n \geq m_0$. Given $\varepsilon > 0$, let m_1 and m_2 be such that

$$x_n > a - \varepsilon , \quad n \geq m_1 \qquad \text{and} \qquad z_n < a + \varepsilon , \quad n \geq m_2 .$$

Set $N := \max\{m_0, m_1, m_2\}$. Then

$$a - \varepsilon < x_n \leq y_n \leq z_n < a + \varepsilon , \qquad n \geq N ,$$

that is, almost all terms of (y_n) are in the ε-neighborhood $\mathbb{B}(a, \varepsilon)$ of a. ∎

Complex Sequences

If (x_n) is a convergent sequence in $\mathbb{R}$ with $\lim x_n = a$, then $\lim |x_n| = |a|$. Indeed, if (x_n) is a null sequence, then this is Example 2.1(a). If $a > 0$, then almost all terms of (x_n) are positive (see Exercise 3), and so $\lim |x_n| = \lim x_n = a = |a|$. Finally, if $a < 0$, then almost all terms of the sequence (x_n) are negative, and we have

$$\lim |x_n| = \lim(-x_n) = -\lim x_n = -a = |a| .$$

The next proposition shows that the same is true of complex sequences.

2.10 Proposition *Let (x_n) be a convergent sequence in $\mathbb{K}$ such that $\lim x_n = a$. Then $(|x_n|)$ converges and $\lim |x_n| = |a|$.*

Proof Let $\varepsilon > 0$. Then there is some N such that $|x_n - a| < \varepsilon$ for all $n \geq N$. From the reversed triangle inequality we have

$$\big| |x_n| - |a| \big| \leq |x_n - a| < \varepsilon , \qquad n \geq N .$$

Thus $|x_n| \in \mathbb{B}_{\mathbb{R}}(|a|, \varepsilon)$ for all $n \geq N$. This implies that $(|x_n|)$ converges to $|a|$. ∎

Convergent sequences in $\mathbb{C}$ can be characterized by the convergence of the corresponding real and imaginary parts.

2.11 Proposition *For a sequence (x_n) in $\mathbb{C}$ the following are equivalent:*

(i) *(x_n) converges.*

(ii) *$(\mathrm{Re}(x_n))$ and $(\mathrm{Im}(x_n))$ converge.*

In this circumstance,

$$\lim x_n = \lim \mathrm{Re}(x_n) + i \lim \mathrm{Im}(x_n) .$$

Proof '(i)⇒(ii)' Suppose that (x_n) converges with $x = \lim x_n$. Then, by Remark 2.1(b), $(|x_n - x|)$ is a null sequence. From Proposition I.11.4 we have

$$|\mathrm{Re}(x_n) - \mathrm{Re}(x)| \le |x_n - x| .$$

By Remark 2.1(c), $(\mathrm{Re}(x_n) - \mathrm{Re}(x))$ is also a null sequence, that is, $(\mathrm{Re}(x_n))$ converges to $\mathrm{Re}(x)$. Similarly $(\mathrm{Im}(x_n))$ converges to $\mathrm{Im}(x)$.

'(ii)⇒(i)' Suppose that $(\mathrm{Re}(x_n))$ and $(\mathrm{Im}(x_n))$ converge with $a := \lim \mathrm{Re}(x_n)$ and $b := \lim \mathrm{Im}(x_n)$. Set $x := a + ib$. Then

$$|x_n - x| = \sqrt{|\mathrm{Re}(x_n) - a|^2 + |\mathrm{Im}(x_n) - b|^2} \le |\mathrm{Re}(x_n) - a| + |\mathrm{Im}(x_n) - b| .$$

It follows easily from this inequality that (x_n) converges to x in $\mathbb{C}$. ∎

We close this section with some examples which illustrate the above propositions.

2.12 Examples (a) $\displaystyle\lim_{n\to\infty} \frac{n+1}{n+2} = 1.$

Proof Write $(n+1)/(n+2)$ in the form $(1+1/n)/(1+2/n)$. Since

$$\lim(1+1/n) = \lim(1+2/n) = 1$$

(why?), the claim follows from Propositions 2.4 and 2.6. ∎

(b) $\displaystyle\lim_{n\to\infty} \left(\frac{3n}{(2n+1)^2} + i\,\frac{2n^2}{n^2+1} \right) = 2i.$

Proof Let

$$x_n := \frac{3n}{(2n+1)^2} + i\,\frac{2n^2}{n^2+1} , \qquad n \in \mathbb{N} .$$

Write the real part of x_n in the form

$$\frac{3/n}{(2+1/n)^2} .$$

Since $\lim(2+1/n)=2$, it follows from Proposition 2.4 that $\lim(2+1/n)^2=4$. Since $(3/n)$ is a null sequence, we have from Propositions 2.4 and 2.6 that

$$\operatorname{Re}(x_n)=\frac{3n}{(2n+1)^2}\to 0\quad (n\to\infty)\ .$$

By Example 1.8(a) and Proposition 2.6, the sequence of the imaginary parts of x_n satisfies

$$\frac{2n^2}{n^2+1}=\frac{2}{1+1/n^2}\to 2\quad (n\to\infty)\ .$$

The claim now follows from Proposition 2.11. ■

(c) $\left(\dfrac{i^n}{1+in}\right)$ is a null sequence in $\mathbb{C}$.

Proof We write

$$\frac{i^n}{1+in}=\frac{1}{n}\,\frac{i^n}{i+1/n}\ ,\qquad n\in\mathbb{N}^\times\ .$$

Then, by Proposition 2.4, it suffices to show that the sequence $\bigl(i^n/(i+1/n)\bigr)_{n\in\mathbb{N}^\times}$ is bounded. Since

$$\left|i+\frac{1}{n}\right|=\sqrt{1+\frac{1}{n^2}}\ge 1\ ,\qquad n\in\mathbb{N}^\times\ ,$$

we get the inequality

$$\left|\frac{i^n}{i+1/n}\right|=\frac{|i^n|}{|i+1/n|}=\frac{1}{|i+1/n|}\le 1\ ,\qquad n\in\mathbb{N}^\times\ ,$$

which shows the claimed boundedness. ■

Exercises

1 Determine whether the following sequences (x_n) in $\mathbb{R}$ converge. Calculate the limit in the case of convergence.

(a) $x_n:=\sqrt{n+1}-\sqrt{n}$.

(b) $x_n:=(-1)^n\sqrt{n}\bigl(\sqrt{n+1}-\sqrt{n}\bigr)$.

(c) $x_n:=\dfrac{1+2+3+\cdots+n}{n+2}-\dfrac{n}{2}$.

(d) $x_n:=\dfrac{(2-1/\sqrt{n})^{10}-(1+1/n^2)^{10}}{1-1/n^2-1/\sqrt{n}}$.

(e) $x_n:=(100+1/n)^2$.

2 Using the binomial expansion of $(1+1)^n$, prove that $\bigl(n^3/2^n\bigr)$ is a null sequence.

3 Let (x_n) be a convergent real sequence with positive limit. Show that almost all terms of the sequence are positive.

4 Let (x_j) be a convergent sequence in $\mathbb{K}$ with limit a. Prove that

$$\lim_{n\to\infty} \frac{1}{n} \sum_{j=1}^{n} x_j = a .$$

5 For $m \in \mathbb{N}^\times$, consider $\mathbb{K}^m$ to be a metric space with the product metric (see Example 1.2(e)). Let

$$s(\mathbb{K}^m) := \operatorname{Funct}(\mathbb{N}, \mathbb{K}^m) = (\mathbb{K}^m)^{\mathbb{N}}$$

and

$$c(\mathbb{K}^m) := \{\, (x_n) \in s(\mathbb{K}^m) \ ; \ (x_n) \text{ converges} \,\} .$$

Show the following:

(a) $c(\mathbb{K}^m)$ is a subspace of $s(\mathbb{K}^m)$.

(b) The function

$$\lim : c(\mathbb{K}^m) \to \mathbb{K}^m , \quad (x_n) \mapsto \lim_{n\to\infty} (x_n)$$

is linear.

(c) Let $(\lambda_n) \in c(\mathbb{K})$ and $(x_n) \in c(\mathbb{K}^m)$ be such that $\lambda_n \to \alpha$ and $x_n \to a$. Then $\lambda_n x_n \to \alpha a$ in $\mathbb{K}^m$ (Hint: Example 1.8(e)).

6 Let (x_n) be a convergent sequence in $\mathbb{K}$ with limit a. Let $p, q \in \mathbb{K}[X]$ be such that $q(a) \neq 0$. Prove that, for the rational function $r := p/q$, we have

$$r(x_n) \to r(a) \quad (n \to \infty) .$$

In particular, for each polynomial p, the sequence $\big(p(x_n)\big)_{n\in\mathbb{N}}$ converges to $p(a)$.

7 Let (x_n) be a convergent sequence in $(0,\infty)$ with limit $x \in (0,\infty)$. For $r \in \mathbb{Q}$ prove that

$$(x_n)^r \to x^r \quad (n \to \infty) .$$

(Hint: For $r = 1/q$, let $y_n := (x_n)^r$ and $y := x^r$. Then

$$x_n - x = (y_n - y) \sum_{k=0}^{q-1} y_n^k y^{q-1-k}$$

by Exercise I.8.1.)

8 Let (x_n) be a sequence in $(0,\infty)$. Show that $(1/x_n)$ is a null sequence if and only if, for each $K > 0$, there is some N such that $x_n > K$ for all $n \geq N$.

9 Let (a_n) be a sequence in $(0,\infty)$ and

$$x_n := \sum_{k=0}^{n} (a_k + 1/a_k) , \qquad n \in \mathbb{N} .$$

Show that $(1/x_n)$ is a null sequence. (Hint: For $a > 0$, show that $a + 1/a \geq 2$ (see Exercise I.8.10). Now use Exercise 8.)

3 Normed Vector Spaces

In this section we consider metrics on vector spaces. We want, of course, that such metrics be compatible with the vector space structure, and so we begin by investigating the vector space $\mathbb{R}^2$ for which we already have a concept of distance. Specifically, if we denote the length of a vector x in $\mathbb{R}^2$ by $\|x\|$, then the distance between two points $x, y \in \mathbb{R}^2$ is $\|x-y\|$. We will see later that this defines a metric on $\mathbb{R}^2$ (see Remark 3.1(a)). As well as this relationship to the metric, the function $x \mapsto \|x\|$ has certain properties with respect to the vector space structure:

First we note that the length of a vector in $\mathbb{R}^2$ is nonnegative, that is, $\|x\| \geq 0$ for all $x \in \mathbb{R}^2$, and that the only vector of zero length is the zero vector.

For $x \in \mathbb{R}^2$ and $\alpha > 0$, we can view αx as the vector x stretched (or shrunk) by the factor α. If $\alpha < 0$, then αx is x stretched (or shrunk) by the factor $-\alpha$ and then reversed in direction.

$5x/2$ $\quad$ x $\quad$ 0 $\quad$ $-2x$

In either case, the length of the vector αx is $\|\alpha x\| = |\alpha|\,\|x\|$.

Finally, for all vectors x and y in $\mathbb{R}^2$, we have the *triangle inequality*, $\|x+y\| \leq \|x\| + \|y\|$.

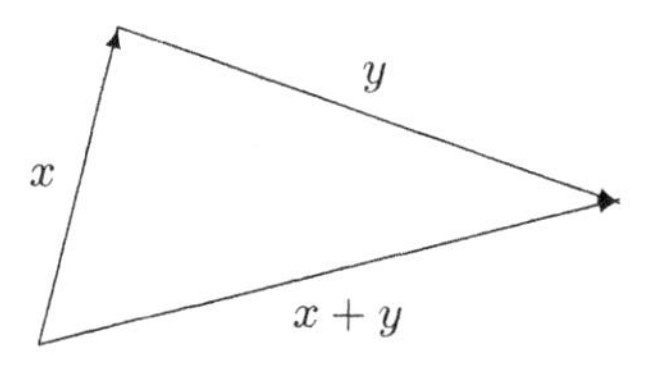

These three properties suffice for $\|x-y\|$ to be a metric on $\mathbb{R}^2$. Since they also generalize easily to arbitrary vector spaces, we are led naturally to the following definition of a normed vector space.

Norms

Let E be a vector space over $\mathbb{K}$. A function $\|\cdot\| : E \to \mathbb{R}^+$ is called a **norm** if the following hold:

(N_1) $\|x\| = 0 \Longleftrightarrow x = 0$.

(N_2) $\|\lambda x\| = |\lambda|\,\|x\|$, $x \in E$, $\lambda \in \mathbb{K}$ (positive homogeneity).

(N_3) $\|x+y\| \leq \|x\| + \|y\|$, $x, y \in E$ (triangle inequality).

A pair $(E, \|\cdot\|)$ consisting of a vector space E and a norm $\|\cdot\|$ is called a **normed vector space**.[1] If the norm is clear from context, we write E instead of $(E, \|\cdot\|)$.

3.1 Remarks Let $E := (E, \|\cdot\|)$ be a normed vector space.

(a) The function

$$d : E \times E \to \mathbb{R}^+ , \qquad (x, y) \mapsto \|x - y\|$$

is a metric on E, the **metric induced from the norm**. *Hence any normed vector space is also a metric space.*

[1] Unless otherwise stated, a vector space is henceforth assumed to be a $\mathbb{K}$-vector space.

Proof The axioms (M_1) and (M_2) follow immediately from (N_1) and (N_2). The axiom (M_3) follows from (N_3) since

$$d(x,y) = \|x-y\| = \|(x-z)+(z-y)\| \leq \|x-z\| + \|z-y\| = d(x,z) + d(z,y)$$

for all $x, y, z \in E$. ■

(b) The **reversed triangle inequality** holds for the norm:

$$\|x-y\| \geq \big|\|x\| - \|y\|\big| , \qquad x, y \in E .$$

Proof Proposition 1.3 implies the reversed triangle inequality for the induced metric. Hence

$$\|x-y\| = d(x,y) \geq |d(x,0) - d(0,y)| = \big|\|x\| - \|y\|\big|$$

for all $x, y \in E$. ■

(c) Because of (a), all statements from Section 1 about metric spaces hold also for E. In particular, the concepts 'neighborhood', 'cluster point' and 'convergence' are well defined in E.

For example, the convergence of a sequence (x_n) in E with limit x has the meaning

$$x_n \to x \text{ in } E \Leftrightarrow \forall \varepsilon > 0 \; \exists N \in \mathbb{N} \colon \|x_n - x\| < \varepsilon \; \forall n \geq N .$$

Further, a review of Section 2 shows that any statement whose proof does not use the field structure or order structure of $\mathbb{K}$, holds also for sequences in E.

In particular, Remarks 2.1 *and Propositions* 2.2 *and* 2.10 *hold in any normed vector space.* ■

Balls

For $a \in E$ and $r > 0$, we define the **open** and **closed balls** with center at a and radius r by

$$\mathbb{B}_E(a,r) := \mathbb{B}(a,r) := \{ x \in E \, ; \, \|x-a\| < r \}$$

and

$$\bar{\mathbb{B}}_E(a,r) := \bar{\mathbb{B}}(a,r) := \{ x \in E \, ; \, \|x-a\| \leq r \} .$$

Note that these definitions agree with those for the metric space (E,d) when d is induced from the norm. We write also

$$\mathbb{B} := \mathbb{B}(0,1) = \{ x \in E \, ; \, \|x\| < 1 \} \quad \text{and} \quad \bar{\mathbb{B}} := \bar{\mathbb{B}}(0,1) = \{ x \in E \, ; \, \|x\| \leq 1 \}$$

for the **open** and **closed unit balls** in E. Using the notation of (I.4.1) we have

$$r\mathbb{B} = \mathbb{B}(0,r) , \quad r\bar{\mathbb{B}} = \bar{\mathbb{B}}(0,r) , \quad a + r\mathbb{B} = \mathbb{B}(a,r) , \quad a + r\bar{\mathbb{B}} = \bar{\mathbb{B}}(a,r) .$$

Bounded Sets

A subset X of E is called **bounded in** E (or **norm bounded**) if it is bounded in the induced metric space.

3.2 Remarks Let $E := (E, \|\cdot\|)$ be a normed vector space.

(a) $X \subseteq E$ is bounded if and only if there is some $r > 0$ such that $X \subseteq r\mathbb{B}$, that is, $\|x\| < r$ for all $x \in X$.

(b) If X and Y are nonempty bounded subsets of E, then so are $X \cup Y$, $X + Y$, and λX with $\lambda \in \mathbb{K}$.

(c) Example 1.2(d) shows that, on each vector space V, there is a metric with respect to which V is bounded. But, if V is nonzero, then (N_2) implies that there is no norm on V with this property. ■

Examples

We now define suitable norms for the vector spaces introduced in Section I.12.

3.3 Examples **(a)** The absolute value $|\cdot|$ is a norm on the vector space $\mathbb{K}$.

> Convention Unless otherwise stated, we will henceforth consider $\mathbb{K}$ to be a normed vector space with norm as above.

(b) Let F be a subspace of a normed vector space $E := (E, \|\cdot\|)$. Then the restriction $\|\cdot\|_F := \|\cdot\| \,|\, F$ of $\|\cdot\|$ to F is a norm on F. Thus $F := (F, \|\cdot\|_F)$ is a normed vector space with this **induced norm**. When no confusion is possible, we use the symbol $\|\cdot\|$ for the induced norm on F.

(c) Let $(E_j, \|\cdot\|_j)$, $1 \leq j \leq m$, be normed vector spaces over $\mathbb{K}$. Then

$$\|x\|_\infty := \max_{1\leq j\leq m} \|x_j\|_j \ , \qquad x = (x_1, \ldots, x_m) \in E := E_1 \times \cdots \times E_m \ , \tag{3.1}$$

defines a norm, called the **product norm**, on the product vector space E. The metric on E induced from this norm coincides with the product metric from Example 1.2(e) when d_j is the metric induced on E_j from $\|\cdot\|_j$.

Proof It is clear that (N_1) is satisfied. From the positive homogeneity of $\|\cdot\|_j$ for each $\lambda \in \mathbb{K}$ and $x \in E$ we get

$$\|\lambda x\|_\infty = \max_{1\leq j\leq m} \|\lambda x_j\|_j = \max_{1\leq j\leq m} |\lambda| \, \|x_j\|_j = |\lambda| \max_{1\leq j\leq m} \|x_j\|_j = |\lambda| \, \|x\|_\infty \ ,$$

and hence (N_2) holds. Finally, it follows from $x + y = (x_1 + y_1, \ldots, x_m + y_m)$ and the triangle inequality for the norms $\|\cdot\|_j$ that

$$\|x+y\|_\infty = \max_{1\le j\le m} \|x_j+y_j\|_j \le \max_{1\le j\le m} (\|x_j\|_j + \|y_j\|_j) \le \|x\|_\infty + \|y\|_\infty$$

for all $x,y \in E$, that is, (N_3) holds. Consequently (3.1) defines a norm on the product vector space E. The last claim is clear. ■

(d) For $m \in \mathbb{N}^\times$, $\mathbb{K}^m$ is a normed vector space with the **maximum norm**

$$|x|_\infty := \max_{1\le j\le m} |x_j| , \qquad x = (x_1,\ldots,x_m) \in \mathbb{K}^m .$$

In the case $m=1$, $(\mathbb{K}^1, |\cdot|_\infty) = (\mathbb{K}, |\cdot|) = \mathbb{K}$.

Proof This is a special case of (c). ■

The Space of Bounded Functions

Let X be a nonempty set and $(E, \|\cdot\|)$ a normed vector space. A function $u \in E^X$ is called **bounded** if the image of u in E is bounded. For $u \in E^X$, define

$$\|u\|_\infty := \|u\|_{\infty,X} := \sup_{x\in X} \|u(x)\| \in \mathbb{R}^+ \cup \{\infty\} . \tag{3.2}$$

3.4 Remarks **(a)** For $u \in E^X$, the following are equivalent:

(i) u is bounded.

(ii) $u(X)$ is bounded in E.

(iii) There is some $r > 0$ such that $\|u(x)\| \le r$ for all $x \in X$.

(iv) $\|u\|_\infty < \infty$.

(b) Clearly $\mathrm{id} \in \mathbb{K}^\mathbb{K}$ is *not* bounded, that is, $\|\mathrm{id}\|_\infty = \infty$. ■

Remark 3.4(b) shows that $\|\cdot\|_\infty$ may not be a norm on the vector space E^X when E is not trivial. We therefore set

$$B(X,E) := \{\, u \in E^X \ ; \ u \text{ is bounded} \,\} ,$$

and call $B(X,E)$ the **space of bounded functions** from X to E.

3.5 Proposition *$B(X,E)$ is a subspace of E^X and $\|\cdot\|_\infty$ is a norm, called the* **supremum norm**, *on $B(X,E)$.*

Proof The first statement follows from Remark 3.2(b). By Remark 3.4(a), the function $\|\cdot\|_\infty : B(X,E) \to \mathbb{R}^+$ is well defined. Axiom (N_1) for $\|\cdot\|_\infty$ follows from

$$\|u\|_\infty = 0 \Leftrightarrow \bigl(\|u(x)\| = 0 \ , \ x \in X\bigr) \Leftrightarrow \bigl(u(x) = 0 \ , \ x \in X\bigr) \Leftrightarrow \bigl(u = 0 \text{ in } E^X\bigr) .$$

Here we have, of course, used the fact that $\|\cdot\|$ is a norm on E. For $u \in B(X,E)$ and $\alpha \in \mathbb{K}$, we have

$$\|\alpha u\|_\infty = \sup\{\, \|\alpha u(x)\| \ ;\ x \in X \,\} = \sup\{\, |\alpha|\, \|u(x)\| \ ;\ x \in X \,\} = |\alpha|\, \|u\|_\infty \ .$$

Thus $\|\cdot\|_\infty$ satisfies also (N_2).

Finally, for all $u, v \in B(X,E)$ and $x \in X$, we have $\|u(x)\| \le \|u\|_\infty$ and also $\|v(x)\| \le \|v\|_\infty$. Thus

$$\begin{aligned} \|u+v\|_\infty &= \sup\{\, \|u(x)+v(x)\| \ ;\ x \in X \,\} \\ &\le \sup\{\, \|u(x)\| + \|v(x)\| \ ;\ x \in X \,\} \le \|u\|_\infty + \|v\|_\infty \ , \end{aligned}$$

and so $\|\cdot\|_\infty$ satisfies the axiom (N_3). ■

Convention Henceforth, $B(X,E)$ denotes the space of bounded functions from X to E together with the supremum norm $\|\cdot\|_\infty$, that is,

$$B(X,E) := \big(B(X,E), \|\cdot\|_\infty\big) \ . \tag{3.3}$$

3.6 Remarks (a) If $X := \mathbb{N}$, then $B(X,E)$ is the normed vector space of bounded sequences in E. In the special case $E := \mathbb{K}$, $B(\mathbb{N},\mathbb{K})$ is denoted by ℓ_∞, that is,

$$\ell_\infty := \ell_\infty(\mathbb{K}) := B(\mathbb{N},\mathbb{K})$$

is the normed **vector space of bounded sequences** with the supremum norm

$$\|(x_n)\|_\infty = \sup_{n\in\mathbb{N}} |x_n| \ , \qquad (x_n) \in \ell_\infty \ .$$

(b) Since, by Proposition 1.10, any convergent sequence is bounded, it follows from Remark 2.3 that c_0 and c are subspaces of ℓ_∞. Thus *c_0 and c are normed vector spaces with respect to the supremum norm and $c_0 \subseteq c \subseteq \ell_\infty$ as subspaces.*

(c) If $X = \{1,\ldots,m\}$ for some $m \in \mathbb{N}^\times$, then

$$B(X,E) = (E^m, \|\cdot\|_\infty) \ ,$$

where $\|\cdot\|_\infty$ is the product norm of Example 3.3(c) (with the obvious identifications). Thus the notation here and in Example 3.3(c) are consistent. ■

Inner Product Spaces

We consider now the normed vector space $E := (\mathbb{R}^2, |\cdot|_\infty)$. In view of the above notation, the unit ball of E is

$$\mathbb{B}_E = \{\, x \in \mathbb{R}^2 \ ;\ |x|_\infty \le 1 \,\} = \{\, (x_1, x_2) \in \mathbb{R}^2 \ ;\ -1 \le x_1, x_2 \le 1 \,\} .$$

Thus $\mathbb{B}_E$ is a square in the plane with sides of length 2 and center 0. In any normed vector space $(F, \|\cdot\|)$, the set $\{\, x \in F \ ;\ \|x\| = 1 \,\}$, that is, the 'boundary' of the unit ball, is called the **unit sphere** in $(F, \|\cdot\|)$. For our space E, this is the boundary of the square in the diagram. Every point on this unit sphere is 1 unit from the origin. This distance is, of course, measured in the induced metric $|\cdot|_\infty$ and so the geometric appearance of the 'ball' and 'sphere' may be contrary to our previous experience. In school we learn that we get 'round' circles if the distance between a point and the origin is defined, following Pythagoras, to be the square root of the sum of the squares of its components (see also Section I.11 for $\mathbb{B}_{\mathbb{C}}$). We want to extend this idea of distance to $\mathbb{K}^m$ by defining a new norm on $\mathbb{K}^m$, the Euclidean norm, which is important for both historical and practical reasons. To do so, we need a certain amount of preparation.

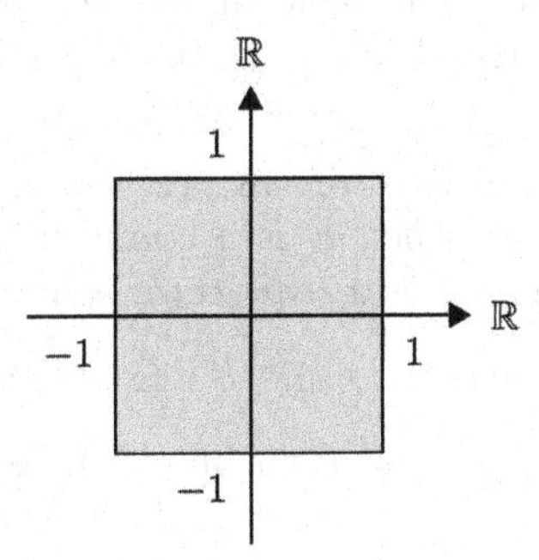

Let E be a vector space over the field $\mathbb{K}$. A function

$$(\cdot\,|\,\cdot) : E \times E \to \mathbb{K} , \quad (x, y) \mapsto (x\,|\,y) \tag{3.4}$$

is called a **scalar product** or **inner product** on E if the following hold:[2]

(SP_1) $(x\,|\,y) = \overline{(y\,|\,x)}$, $x, y \in E$.

(SP_2) $(\lambda x + \mu y\,|\,z) = \lambda(x\,|\,z) + \mu(y\,|\,z)$, $x, y, z \in E$, $\lambda, \mu \in \mathbb{K}$.

(SP_3) $(x\,|\,x) \ge 0$, $x \in E$, and $(x\,|\,x) = 0 \Longleftrightarrow x = 0$.

A vector space E with a scalar product $(\cdot\,|\,\cdot)$ is called an **inner product space** and is written $\big(E, (\cdot\,|\,\cdot)\big)$. Once again, when no confusion is possible, we write E for $\big(E, (\cdot\,|\,\cdot)\big)$.

3.7 Remarks **(a)** In the real case $\mathbb{K} = \mathbb{R}$, (SP_1) can be written as

$$(x\,|\,y) = (y\,|\,x) , \qquad x, y \in E .$$

In other words, the function (3.4) is **symmetric** when E is a real vector space. In the case $\mathbb{K} = \mathbb{C}$, the function (3.4) is said to be **Hermitian** when (SP_1) holds.

[2] If $\mathbb{K} = \mathbb{R}$, then $\overline{\alpha} := \alpha$ and $\operatorname{Re}\alpha := \alpha$ for all $\alpha \in \mathbb{R}$ by Proposition I.11.3. Thus we can ignore the complex conjugation symbol and the symbol Re in the following definition.

(b) From (SP_1) and (SP_2) it follows that

$$(x\,|\,\lambda y + \mu z) = \overline{\lambda}(x\,|\,y) + \overline{\mu}(x\,|\,z)\ , \qquad x, y, z \in E\ , \quad \lambda, \mu \in \mathbb{K}\ , \tag{3.5}$$

that is, for each fixed $x \in E$, the function $(x\,|\,\cdot) : E \to \mathbb{K}$ is **conjugate linear**. Since (SP_1) means that $(\cdot\,|\,x) : E \to \mathbb{K}$ is linear for each fixed $x \in E$, one says that (3.4) is a **sesquilinear form**. In the real case $\mathbb{K} = \mathbb{R}$, (3.5) means simply that $(x\,|\,\cdot) : E \to \mathbb{R}$ is linear for $x \in E$. In this case, (3.4) is called a **bilinear form** on E.

Finally, (SP_3) means that the form (3.4) is **positive** (**definite**). With these definitions we can say: *A scalar product is a positive Hermitian sesquilinear form on E when E is a complex vector space, or a positive symmetric bilinear form when E is a real vector space.*

(c) For all $x, y \in E$, $(x \pm y\,|\,x \pm y) = (x\,|\,x) \pm 2\,\mathrm{Re}(x\,|\,y) + (y\,|\,y)$.[3]

(d) $(x\,|\,0) = 0$ for all $x \in E$. ■

Let $m \in \mathbb{N}^\times$. For $x = (x_1, \ldots, x_m)$ and $y = (y_1, \ldots, y_m)$ in $\mathbb{K}^m$, define

$$(x\,|\,y) := \sum_{j=1}^{m} x_j \overline{y}_j\ .$$

It is easy to check that this defines a scalar product on $\mathbb{K}^m$. This is called the **Euclidean inner product** on $\mathbb{K}^m$.

The Cauchy-Schwarz Inequality

After these preliminaries, we can now prove one of the most useful theorems about inner product spaces.

3.8 Theorem (Cauchy-Schwarz inequality) *Let $\big(E, (\cdot\,|\,\cdot)\big)$ be an inner product space. Then*

$$|(x\,|\,y)|^2 \le (x\,|\,x)(y\,|\,y)\ , \qquad x, y \in E\ , \tag{3.6}$$

and equality occurs in (3.6) *if and only if x and y are linearly dependent.*

Proof (a) For $y = 0$, the claim follows from Remark 3.7(d). Suppose then that $y \ne 0$. For any $\alpha \in \mathbb{K}$, we have

$$\begin{aligned} 0 \le (x - \alpha y\,|\,x - \alpha y) &= (x\,|\,x) - 2\,\mathrm{Re}(x\,|\,\alpha y) + (\alpha y\,|\,\alpha y) \\ &= (x\,|\,x) - 2\,\mathrm{Re}\big(\overline{\alpha}(x\,|\,y)\big) + |\alpha|^2\,(y\,|\,y)\ . \end{aligned} \tag{3.7}$$

[3] As already mentioned, using the symbols $\pm$ and $\mp$ one can write two equations as if they were one. For one of these equations, the upper symbol ($+$ or $-$) is used throughout, and for the other, the lower symbol is used throughout.

Setting $\alpha := (x\,|\,y)/(y\,|\,y)$ yields

$$0 \le (x\,|\,x) - 2\,\mathrm{Re}\Big(\frac{\overline{(x\,|\,y)}}{(y\,|\,y)}(x\,|\,y)\Big) + \frac{|(x\,|\,y)|^2}{(y\,|\,y)^2}(y\,|\,y) = (x\,|\,x) - \frac{|(x\,|\,y)|^2}{(y\,|\,y)}\ ,$$

and so (3.6) holds. If $x \ne \alpha y$, then, from (3.7), we see that (3.6) is a strict inequality.

(b) Finally, let x and y be linear dependent vectors in E. Then there is some $(\alpha,\beta) \in \mathbb{K}^2 \setminus \{(0,0)\}$ such that $\alpha x + \beta y = 0$. If $\alpha \ne 0$, then $x = -(\beta/\alpha)y$ and we have

$$|(x\,|\,y)|^2 = \left|\frac{\beta}{\alpha}\right|^2 |(y\,|\,y)|^2 = \Big(-\frac{\beta}{\alpha}y\,\Big|\,-\frac{\beta}{\alpha}y\Big)(y\,|\,y) = (x\,|\,x)(y\,|\,y)\ .$$

If $\beta \ne 0$, then $y = -(\alpha/\beta)x$ and a similar calculation gives $|(x\,|\,y)|^2 = (x\,|\,x)(y\,|\,y)$. ∎

3.9 Corollary (classical Cauchy-Schwarz inequality) *Let $\xi_1,\ldots,\xi_m$ and $\eta_1,\ldots,\eta_m$ be elements of $\mathbb{K}$. Then*

$$\Big|\sum_{j=1}^m \xi_j\overline{\eta}_j\Big|^2 \le \Big(\sum_{j=1}^m |\xi_j|^2\Big)\Big(\sum_{j=1}^m |\eta_j|^2\Big) \tag{3.8}$$

with equality if and only if there are numbers $\alpha,\beta \in \mathbb{K}$ such that $(\alpha,\beta) \ne (0,0)$ and $\alpha\xi_j + \beta\eta_j = 0$ for all $j = 1,\ldots,m$.

Proof This follows by applying Theorem 3.8 to $\mathbb{K}^m$ with the Euclidean inner product. ∎

Let $\big(E,(\cdot\,|\,\cdot)\big)$ be an arbitrary inner product space. Then it follows from $(x\,|\,x) \ge 0$ that $\|x\| := \sqrt{(x\,|\,x)} \ge 0$ is well defined for all $x \in E$ and

$$\|x\| = 0 \Longleftrightarrow \|x\|^2 = 0 \Longleftrightarrow (x\,|\,x) = 0 \Longleftrightarrow x = 0\ .$$

Thus $\|\cdot\|$ satisfies the norm axiom (N_1). The proof of (N_2) for $\|\cdot\|$ is also easy since, for $\alpha \in \mathbb{K}$ and $x \in E$,

$$\|\alpha x\| = \sqrt{(\alpha x\,|\,\alpha x)} = \sqrt{|\alpha|^2\,(x\,|\,x)} = |\alpha|\,\|x\|\ .$$

The next proposition shows that (N_3), the triangle inequality, follows from the Cauchy-Schwarz inequality and hence that $\|\cdot\| : E \to \mathbb{R}^+$ is a norm on E.

3.10 Theorem *Let $\big(E,(\cdot\,|\,\cdot)\big)$ be an inner product space and*

$$\|x\| := \sqrt{(x\,|\,x)}\ , \qquad x \in E\ .$$

Then $\|\cdot\|$ is a norm on E, the **norm induced from the scalar product** $(\cdot\,|\,\cdot)$.

Proof In view of the above discussion, it suffices to prove the triangle inequality for $\|\cdot\|$. From the Cauchy-Schwarz inequality we have

$$|(x\,|\,y)| \le \sqrt{(x\,|\,x)(y\,|\,y)} = \sqrt{\|x\|^2\,\|y\|^2} = \|x\|\,\|y\| \ .$$

Hence

$$\begin{aligned}\|x+y\|^2 &= (x+y\,|\,x+y) = (x\,|\,x) + 2\,\mathrm{Re}(x\,|\,y) + (y\,|\,y)\\ &\le \|x\|^2 + 2\,|(x\,|\,y)| + \|y\|^2 \le \|x\|^2 + 2\,\|x\|\,\|y\| + \|y\|^2\\ &= (\|x\| + \|y\|)^2\ ,\end{aligned}$$

that is, we have shown that $\|x+y\| \le \|x\| + \|y\|$. ■

Because of Theorem 3.10 we make the following convention:

Convention Any inner product space $\bigl(E, (\cdot\,|\,\cdot)\bigr)$ is considered to be a normed vector space with the norm induced from $(\cdot\,|\,\cdot)$ as above.

A norm which is induced from a scalar product is also called a **Hilbert norm**.

Using the norm we can reformulate the Cauchy-Schwarz inequality of Theorem 3.8 as follows:

3.11 Corollary *Let $\bigl(E, (\cdot\,|\,\cdot)\bigr)$ be an inner product space. Then*

$$|(x\,|\,y)| \le \|x\|\,\|y\|\ , \qquad x, y \in E\ .$$

Euclidean Spaces

A particularly important example is the Euclidean inner product on $\mathbb{K}^m$. Since we so frequently work with this inner product, it is convenient to make the following convention:

Convention Unless otherwise stated, we consider $\mathbb{K}^m$ to be endowed with the Euclidean inner product $(\cdot\,|\,\cdot)$ and the induced norm[4]

$$|x| := \sqrt{(x\,|\,x)} = \sqrt{\textstyle\sum_{j=1}^m |x_j|^2}\ , \qquad x = (x_1,\ldots,x_m) \in \mathbb{K}^m\ ,$$

the **Euclidean norm**. In the real case, we write also $x \bullet y$ for $(x\,|\,y)$.

[4]In the case $m=1$, this notation is consistent with the notation $|\cdot|$ for the absolute value in $\mathbb{K}$ because $(x\,|\,y) = x\overline{y}$ for all $x, y \in \mathbb{K}^1 = \mathbb{K}$. It is *not* consistent with the notation $|\alpha|$ for the length of a multi-index $\alpha \in \mathbb{N}^m$. It should be clear from context which meaning is intended.

We now have two norms on the vector space $\mathbb{K}^m$, namely the maximum norm

$$|x|_\infty = \max_{1\le j\le m} |x_j| , \qquad x = (x_1,\dots,x_m) \in \mathbb{K}^m ,$$

and the Euclidean norm $|\cdot|$. We define a further norm by

$$|x|_1 := \sum_{j=1}^m |x_j| , \qquad x = (x_1,\dots,x_m) \in \mathbb{K}^m .$$

Checking that this is, in fact, a norm is easy and left to the reader. The next proposition shows, using a further application of the Cauchy-Schwarz inequality, that the Euclidean norm is 'comparable' with the norms $|\cdot|_1$ and $|\cdot|_\infty$.

3.12 Proposition *Let $m \in \mathbb{N}^\times$. Then*

$$|x|_\infty \le |x| \le \sqrt{m}\,|x|_\infty , \quad \frac{1}{\sqrt{m}}\,|x|_1 \le |x| \le |x|_1 , \qquad x \in \mathbb{K}^m .$$

Proof From the inequality $|x_k|^2 \le \sum_{j=1}^m |x_j|^2$ for $k = 1,\dots,m$ it follows immediately that $|x|_\infty \le |x|$. The inequalities

$$\sum_{j=1}^m |x_j|^2 \le \Bigl(\sum_{j=1}^m |x_j|\Bigr)^2 \quad \text{and} \quad \sum_{j=1}^m |x_j|^2 \le m \max_{1\le j\le m} |x_j|^2 = m\bigl(\max_{1\le j\le m} |x_j|\bigr)^2$$

are trivially true, and so have we shown that $|x| \le |x|_1$ and $|x| \le \sqrt{m}\,|x|_\infty$. From Corollary 3.9 it follows that

$$|x|_1 = \sum_{j=1}^m 1\cdot|x_j| \le \Bigl(\sum_{j=1}^m 1^2\Bigr)^{1/2} \bigl(\sum_{j=1}^m |x_j|^2\bigr)^{1/2} = \sqrt{m}\,|x| ,$$

which finishes the proof. ■

Equivalent Norms

Let E be a vector space. Two norms $\|\cdot\|_1$ and $\|\cdot\|_2$ on E are **equivalent** if there is some $K \ge 1$ such that

$$\frac{1}{K}\,\|x\|_1 \le \|x\|_2 \le K\,\|x\|_1 , \qquad x \in E . \tag{3.9}$$

In this case we write $\|\cdot\|_1 \sim \|\cdot\|_2$.

3.13 Remarks **(a)** It is not difficult to prove that $\sim$ is an equivalence relation on the set of all norms of a fixed vector space.

(b) The qualitative claim of Proposition 3.12 can now be expressed in the form

$$|\cdot|_1 \sim |\cdot| \sim |\cdot|_\infty \quad \text{on } \mathbb{K}^m .$$

(c) To make the quantitative claim of Proposition 3.12 clearer, we write $\mathbb{B}^m$ for the **real open Euclidean unit ball**, that is,

$$\mathbb{B}^m := \mathbb{B}_{\mathbb{R}^m} ,$$

and $\mathbb{B}_1^m$ and $\mathbb{B}_\infty^m$ for the unit balls in $(\mathbb{R}^m, |\cdot|_1)$ and in $(\mathbb{R}^m, |\cdot|_\infty)$ respectively. Then Proposition 3.12 says

$$\mathbb{B}^m \subseteq \mathbb{B}_\infty^m \subseteq \sqrt{m}\mathbb{B}^m , \quad \mathbb{B}_1^m \subseteq \mathbb{B}^m \subseteq \sqrt{m}\mathbb{B}_1^m .$$

In the case $m = 2$, these inclusions are shown in the following diagram:

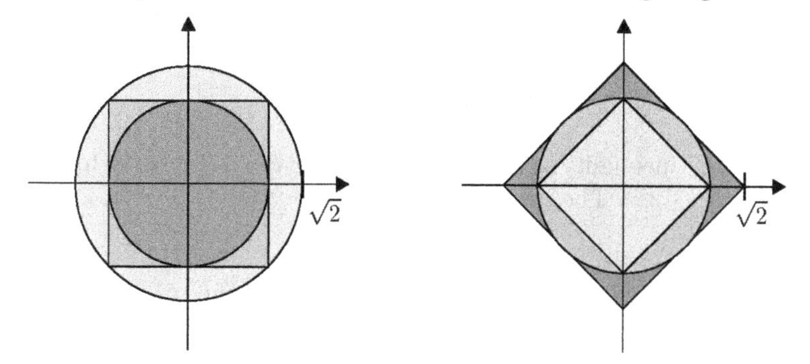

Note that

$$\mathbb{B}_\infty^m = \underbrace{\mathbb{B}_\infty^1 \times \cdots \times \mathbb{B}_\infty^1}_{m} = (-1,1)^m , \tag{3.10}$$

but, for $\mathbb{B}^m$ and $\mathbb{B}_1^m$, there is no analogous representation.

(d) Let $E = (E, \|\cdot\|)$ be a normed vector space and $\|\cdot\|_1$ a norm on E which is equivalent to $\|\cdot\|$. Set $E_1 := (E, \|\cdot\|_1)$. Then

$$\mathcal{U}_E(a) = \mathcal{U}_{E_1}(a) , \qquad a \in E ,$$

that is, *the set of neighborhoods of a depends only on the equivalence class of the norm. Equivalent norms produce the same set of neighborhoods.*

Proof (i) By Remark 3.1(a), the sets $\mathcal{U}_E(a)$ and $\mathcal{U}_{E_1}(a)$ are well defined for each $a \in E$.

(ii) From (3.9) it follows that $K^{-1}\mathbb{B}_{E_1} \subseteq \mathbb{B}_E \subseteq K\mathbb{B}_{E_1}$, and so, for $a \in E$ and $r > 0$, we have

$$\mathbb{B}_{E_1}(a, K^{-1}r) \subseteq \mathbb{B}_E(a,r) \subseteq \mathbb{B}_{E_1}(a, Kr) . \tag{3.11}$$

(iii) For each $U \in \mathcal{U}_E(a)$, there exists a $r > 0$ such that $\mathbb{B}_E(a,r) \subseteq U$. From (3.11) we get $\mathbb{B}_{E_1}(a, K^{-1}r) \subseteq U$, that is, $U \in \mathcal{U}_{E_1}(a)$. This shows $\mathcal{U}_E(a) \subseteq \mathcal{U}_{E_1}(a)$.

Conversely, if $U \in \mathcal{U}_{E_1}(a)$, then there is some $\delta > 0$ such that $\mathbb{B}_{E_1}(a,\delta) \subseteq U$. Set $r := \delta/K > 0$. Then, from (3.11), we have $\mathbb{B}_E(a,r) \subseteq U$, and hence $U \in \mathcal{U}_E(a)$. Thus we have shown that $\mathcal{U}_{E_1}(a) \subseteq \mathcal{U}_E(a)$. ■

(e) Using the bijection

$$\mathbb{C} \ni z = x + iy \longleftrightarrow (x,y) \in \mathbb{R}^2 ,$$

the complex numbers $\mathbb{C} := \mathbb{R} + i\mathbb{R}$ can be identified with the set $\mathbb{R}^2$ (or even with the Abelian group $(\mathbb{R}^2, +)$, as Remark I.11.2(c) shows). More generally, for $m \in \mathbb{N}^\times$, the sets $\mathbb{C}^m$ and $\mathbb{R}^{2m}$ can be identified using the bijection[5]

$$\mathbb{C}^m \ni (z_1,\ldots,z_m) = (x_1 + iy_1,\ldots,x_m + iy_m) \longleftrightarrow (x_1,y_1,\ldots,x_m,y_m) \in \mathbb{R}^{2m} .$$

With respect to this **canonical identification**,

$$\mathbb{B}_{\mathbb{C}^m} = \mathbb{B}^{2m} = \mathbb{B}_{\mathbb{R}^{2m}}$$

and hence

$$\mathcal{U}_{\mathbb{C}^m} = \mathcal{U}_{\mathbb{R}^{2m}} .$$

Thus for **topological questions**, that is, statements about neighborhoods of points, the sets $\mathbb{C}^m$ and $\mathbb{R}^{2m}$ can be identified.

(f) The notions 'cluster point' and 'convergence' are **topological concepts**, that is, they are defined in terms of neighborhoods. Thus they are invariant under changes to equivalent norms. ■

Convergence in Product Spaces

As a consequence of the above and earlier discussions we now have a simple, but very useful, description of convergent sequences in $\mathbb{K}^m$.

3.14 Proposition *Let $m \in \mathbb{N}^\times$ and $x_n = (x_n^1,\ldots,x_n^m) \in \mathbb{K}^m$ for $n \in \mathbb{N}$. Then the following are equivalent:*

(i) *The sequence $(x_n)_{n\in\mathbb{N}}$ converges to $x = (x^1,\ldots,x^m)$ in $\mathbb{K}^m$.*

(ii) *For each $k \in \{1,\ldots,m\}$, the sequence $(x_n^k)_{n\in\mathbb{N}}$ converges to x^k in $\mathbb{K}$.*

Proof This follows directly from Example 1.8(e) and Remarks 3.13(c) and (d). ■

[5] We emphasize that the complex vector space $\mathbb{C}^m$ *cannot* be identified with the real vector space $\mathbb{R}^{2m}$ (Why not?)!

Claim (ii) of Proposition 3.14 is often called *componentwise convergence* of the sequence (x_n), so Proposition 3.14 can be formulated, somewhat imprecisely, as: *A sequence in* $\mathbb{K}^m$ *converges if and only if it converges componentwise.* Thus it suffices, in principle, to study the convergence of sequences in $\mathbb{K}$ — indeed, because of Remark 3.13(e), it suffices to study convergence in $\mathbb{R}$. For many reasons, which the reader will find for him/herself in further study, there is little to be gained by making such a 'simplification' in our presentation.

Exercises

1 Let $\|\cdot\|$ be a norm on a $\mathbb{K}$-vector space E. Show that, for each $T \in \mathrm{Aut}(E)$, the function $\|x\|_T := \|Tx\|$, $x \in E$, defines a norm $\|\cdot\|_T$ on E. In particular, for each $\alpha \in \mathbb{K}^\times$, the function $E \to \mathbb{R}^+$, $x \mapsto \|\alpha x\|$ is a norm on E.

2 Suppose that a sequence (x_n) in a normed vector space $E = (E, \|\cdot\|)$ converges to x. Prove that the sequence $(\|x_n\|)$ in $[0, \infty)$ converges to $\|x\|$.

3 Verify the claims of Remark 3.4(a).

4 Prove that the **parallelogram identity**,

$$2(\|x\|^2 + \|y\|^2) = \|x+y\|^2 + \|x-y\|^2 , \qquad x, y \in E ,$$

holds in any inner product space $\big(E, (\cdot|\cdot)\big)$.

5 For which $\lambda := (\lambda_1, \ldots, \lambda_m) \in \mathbb{K}^m$ is

$$(\cdot|\cdot)_\lambda : \mathbb{K}^m \times \mathbb{K}^m \to \mathbb{K} , \quad (x,y) \mapsto \sum_{k=1}^m \lambda_k x_k \overline{y}_k$$

a scalar product on $\mathbb{K}^m$?

6 Let $\big(E, (\cdot|\cdot)\big)$ be a real inner product space. Prove the inequality

$$(\|x\| + \|y\|) \frac{(x|y)}{\|x\|\,\|y\|} \le \|x+y\| \le \|x\| + \|y\| , \qquad x, y \in E \backslash \{0\} .$$

When do we get equality? (Hint: Square the first inequality.)

7 Let X be a metric space. A subset Y of X is called **closed** if every sequence (y_n) in Y which converges in X, converges in Y, that is, $\lim y_n \in Y$.

Show that c_0 is a closed subspace of ℓ_∞.

8 Let $\|\cdot\|_1$ and $\|\cdot\|_2$ be equivalent norms on a vector space E. Define

$$d_j(x,y) := \|x-y\|_j , \qquad x, y \in E , \quad j = 1, 2 .$$

Show that d_1 and d_2 are equivalent metrics on E.

9 Let (X_j, d_j), $1 \le j \le n$, be metric spaces. Show that the function defined by

$$(x,y) \mapsto \Bigl(\sum_{j=1}^{n} d_j(x_j,y_j)^2\Bigr)^{1/2}$$

for all $x = (x_1,\ldots,x_n)$, $y = (y_1,\ldots,y_n)$ and $x,y \in X := X_1 \times \cdots \times X_n$ is a metric which is equivalent to the product metric on X.

10 Let $\bigl(E, (\cdot\,|\,\cdot)\bigr)$ be an inner product space. Two elements $x, y \in E$ are called **orthogonal** if $(x\,|\,y) = 0$. In this case we write $x \perp y$. A subset $M \subseteq E$ is called an **orthogonal system** if $x \perp y$ for all $x, y \in M$ with $x \ne y$. Finally M is called an **orthonormal system** if M is an orthogonal system such that $\|x\| = 1$ for all $x \in M$.

Let $\{x_0,\ldots,x_m\} \subseteq E$ be an orthogonal system with $x_j \ne 0$ for $0 \le j \le m$. Show the following:

(a) $\{x_0,\ldots,x_m\}$ is linearly independent.

(b) $\bigl\|\sum_{k=0}^m x_k\bigr\|^2 = \sum_{k=0}^m \|x_k\|^2$ (Pythagoras' theorem).

11 Let F be a subspace of an inner product space E. Prove that the **orthogonal complement** of F, that is,

$$F^\perp := \{\, x \in E \; ; \; x \perp y = 0, \ y \in F \,\} \ ,$$

is a closed subspace of E (see Exercise 7).

12 Let $B = \{u_0,\ldots,u_m\}$ be an orthonormal system in an inner product space $\bigl(E, (\cdot\,|\,\cdot)\bigr)$ and $F := \mathrm{span}(B)$. Define

$$p_F : E \to F \ , \quad x \mapsto \sum_{k=0}^{m} (x\,|\,u_k) u_k \ .$$

Prove the following:

(a) $x - p_F(x) \in F^\perp$, $x \in E$.

(b) $\|x - p_F(x)\| = \inf_{y \in F} \|x - y\|$, $x \in E$.

(c) $\|x - p_F(x)\|^2 = \|x\|^2 - \sum_{k=0}^m |(x\,|\,u_k)|^2$, $x \in E$.

(d) $p_F \in \mathrm{Hom}(E,F)$ with $p_F^2 = p_F$.

(e) $\mathrm{im}(p_F) = F$, $\ker(p_F) = F^\perp$ and $E = F \oplus F^\perp$.

(Hint: (b) For $y \in F$, we have $\|x-y\|^2 = \|x - p_F(x)\|^2 + \|p_F(x) - y\|^2$ which follows from Exercise 10 and (a).)

13 With B and F as in Exercise 12, prove the following:

(a) For all $x \in E$, $\sum_{k=0}^m |(x\,|\,u_k)|^2 \le \|x\|^2$.

(b) For all $x \in F$,

$$x = \sum_{k=0}^{m} (x\,|\,u_k) u_k \quad \text{and} \quad \|x\|^2 = \sum_{k=0}^{m} |(x\,|\,u_k)|^2 \ .$$

(Hint: To prove (a), use the Cauchy-Schwarz inequality.)

14 For $m, n \in \mathbb{N}^\times$, let $\mathbb{K}^{m\times n}$ be the set of all $m \times n$ matrices with entries in $\mathbb{K}$. We can consider $\mathbb{K}^{m\times n}$ to be the set of all functions from $\{1,\ldots,m\} \times \{1,\ldots,n\}$ to $\mathbb{K}$, and so, by Example I.12.3(e), $\mathbb{K}^{m\times n}$ is a vector space. Here αA and $A+B$ for $\alpha \in \mathbb{K}$ and $A, B \in \mathbb{K}^{m\times n}$ are the usual scalar multiplication and matrix addition. Show the following:

(a) The function on $\mathbb{K}^{m\times n}$ defined by

$$|A| := \Bigl(\sum_{j=1}^{m}\sum_{k=1}^{n} |a_{jk}|^2\Bigr)^{1/2}, \qquad A = [a_{jk}] \in \mathbb{K}^{m\times n},$$

is a norm.

(b) The following functions define equivalent norms:

(α) $[a_{jk}] \mapsto \sum_{j=1}^{m}\sum_{k=1}^{n} |a_{jk}|$

(β) $[a_{jk}] \mapsto \max_{1\le j\le m} \sum_{k=1}^{n} |a_{jk}|$

(γ) $[a_{jk}] \mapsto \max_{1\le k\le n} \sum_{j=1}^{m} |a_{jk}|$

(δ) $[a_{jk}] \mapsto \max_{\substack{1\le j\le m\\ 1\le k\le n}} |a_{jk}|$

15 Let E and F be normed vector spaces. Show that $\mathcal{B}(E,F) \cap \mathrm{Hom}(E,F) = \{0\}$.

4 Monotone Sequences

Any sequence in $\mathbb{R}$, that is, any element of $s(\mathbb{R}) = \mathbb{R}^{\mathbb{N}}$, is a function between ordered sets. In this section we again consider the relationship between this order and the convergence of sequences. Specifically, we investigate the convergence of **monotone sequences** as defined preceding Example I.4.7. Thus a sequence (x_n) is **increasing**[1] if $x_n \le x_{n+1}$ for all $n \in \mathbb{N}$, and (x_n) is **decreasing** if $x_n \ge x_{n+1}$ for all $n \in \mathbb{N}$.

Bounded Monotone Sequences

It follows from the completeness property of $\mathbb{R}$ that every bounded monotone sequence converges.

4.1 Theorem *Every increasing (or decreasing) bounded sequence (x_n) in $\mathbb{R}$ converges, and*

$$x_n \uparrow \sup\{\, x_n \ ;\ n \in \mathbb{N}\,\} \quad (\text{or } x_n \downarrow \inf\{\, x_n \ ;\ n \in \mathbb{N}\,\}) \ .$$

Proof (i) Let (x_n) be an increasing bounded sequence. Then $X := \{\, x_n \ ;\ n \in \mathbb{N}\,\}$ is bounded and nonempty. Since $\mathbb{R}$ is order complete, $x := \sup(X)$ is well defined.

(ii) Let $\varepsilon > 0$. By Proposition I.10.5 there is some N such that $x_N > x - \varepsilon$. Since (x_n) is increasing, we have $x_n \ge x_N > x - \varepsilon$ for all $n \ge N$. Together with $x_n \le x$, this implies that

$$x_n \in (x - \varepsilon, x + \varepsilon) = \mathbb{B}_{\mathbb{R}}(x, \varepsilon) \ , \qquad n \ge N \ .$$

Thus x_n converges to x in $\mathbb{R}$.

(iii) If (x_n) is a decreasing bounded sequence, we set $x := \inf\{\, x_n \ ;\ n \in \mathbb{N}\,\}$. Then $(y_n) := (-x_n)$ is increasing and bounded, and $-x = \sup\{\, y_n \ ;\ n \in \mathbb{N}\,\}$. It follows from (ii) that $-x_n = y_n \to -x \quad (n \to \infty)$, and so, by Proposition 2.2, $x_n = -y_n \to x$. ■

In Proposition 1.10 we saw that boundedness is a necessary condition for the convergence of a sequence. Theorem 4.1 shows that boundedness suffices for the convergence of a *monotone* sequence. Of course, a convergent sequence does not have to be monotone, as the null sequence $(-1)^n/n$ shows.

---

[1] For an increasing (or decreasing) sequence (x_n) we often use the symbol $(x_n)\uparrow$ (or $(x_n)\downarrow$). If, in addition, (x_n) converges with limit x, we write $x_n \uparrow x$ (or $x_n \downarrow x$) instead of $x_n \to x$.

Some Important Limits

4.2 Examples (a) Let $a \in \mathbb{C}$. Then

$$\begin{aligned} &a^n \to 0\,, && \text{if } |a| < 1\,,\\ &a^n \to 1\,, && \text{if } a = 1\,,\\ &(a^n)_{n\in\mathbb{N}} \text{ diverges}\,, && \text{if } |a| \ge 1\,,\ \ a \ne 1\,. \end{aligned}$$

Proof (i) Suppose first that the sequence $(a^n)_{n\in\mathbb{N}}$ converges. From Proposition 2.2 we have

$$\lim_{n\to\infty} a^n = \lim_{n\to\infty} a^{n+1} = a \lim_{n\to\infty} a^n\,,$$

and so either $\lim_{n\to\infty} a^n = 0$ or $a = 1$.

(ii) Consider the case $|a| < 1$. Then the sequence $\bigl(|a|^n\bigr) = \bigl(|a^n|\bigr)$ is decreasing and bounded. By Theorem 4.1 and (i), $\bigl(|a^n|\bigr)$ is a null sequence, that is, $a^n \to 0\ \ (n \to \infty)$.

(iii) If $a = 1$, then $a^n = 1$ for $n \in \mathbb{N}$, and, of course, $a^n \to 1$.

(iv) Now let $|a| \ge 1$ and $a \ne 1$. If (a^n) converges, then, by (i), $\bigl(|a^n|\bigr)_{n\in\mathbb{N}}$ is a null sequence. But this is not possible because $|a^n| = |a|^n \ge 1$ for all n. ∎

(b) Let $k \in \mathbb{N}$ and $a \in \mathbb{C}$ be such that $|a| > 1$. Then

$$\lim_{n\to\infty} \frac{n^k}{a^n} = 0\,,$$

that is, *for $|a| > 1$ the function $n \mapsto a^n$ increases faster than any power function $n \mapsto n^k$.*

Proof For $\alpha := 1/|a| \in (0,1)$ and $x_n := n^k\alpha^n$, we have

$$\frac{x_{n+1}}{x_n} = \Bigl(\frac{n+1}{n}\Bigr)^k \alpha = \Bigl(1 + \frac{1}{n}\Bigr)^k \alpha\,, \qquad n \in \mathbb{N}^\times\,,$$

and so $x_{n+1}/x_n \downarrow \alpha$ as $n \to \infty$. Fix some $\beta \in (\alpha, 1)$. Then there is some N such that $x_{n+1}/x_n < \beta$ for all $n \ge N$. Consequently,

$$x_{N+1} < \beta x_N\,, \quad x_{N+2} < \beta x_{N+1} < \beta^2 x_N\,, \quad \ldots$$

A simple induction argument yields $x_n < \beta^{n-N} x_N$ for all $n \ge N$ and so

$$\Bigl|\frac{n^k}{a^n}\Bigr| = x_n < \beta^{n-N} x_N = \frac{x_N}{\beta^N}\beta^n\,, \qquad n \ge N\,.$$

The claim then follows from Remark 2.1(c) since, by (a), $(\beta^n)_{n\in\mathbb{N}}$ is a null sequence. ∎

(c) For all $a \in \mathbb{C}$,

$$\lim_{n\to\infty} \frac{a^n}{n!} = 0\,.$$

The factorial function $n \mapsto n!$ increases faster than the function[2] $n \mapsto a^n$.

[2] In Section 8 we provide a very short proof of this fact.

Proof For $n > N > |a|$ we have

$$\left|\frac{a^n}{n!}\right| = \frac{|a|^N}{N!} \prod_{k=N+1}^{n} \frac{|a|}{k} \le \frac{|a|^N}{N!}\Big(\frac{|a|}{N+1}\Big)^{n-N} < \frac{N^N}{N!}\Big(\frac{|a|}{N}\Big)^n .$$

The claim then follows from (a) and Remark 2.1(c). ■

(d) $\lim_{n\to\infty} \sqrt[n]{n} = 1$.

Proof Let $\varepsilon > 0$. Then, by (b), the sequence $\big(n(1+\varepsilon)^{-n}\big)$ is a null sequence. Thus there is some N such that

$$\frac{n}{(1+\varepsilon)^n} < 1 , \qquad n \ge N ,$$

that is,

$$1 \le n \le (1+\varepsilon)^n , \qquad n \ge N .$$

By Remark I.10.10(c), the nth root function is increasing, and so

$$1 \le \sqrt[n]{n} \le 1+\varepsilon , \qquad n \ge N .$$

This proves the claim. ■

(e) For all $a > 0$, $\lim_{n\to\infty} \sqrt[n]{a} = 1$.

Proof By the Archimedean property of $\mathbb{R}$, there is some N such that $1/n < a < n$ for all $n \ge N$. Thus

$$\frac{1}{\sqrt[n]{n}} = \sqrt[n]{\frac{1}{n}} \le \sqrt[n]{a} \le \sqrt[n]{n} , \qquad n \ge N .$$

Set $x_n := 1/\sqrt[n]{n}$ and $y_n := \sqrt[n]{n}$ for all $n \in \mathbb{N}^\times$. Then, from (d) and Proposition 2.6, we have $\lim x_n = \lim y_n = 1$. The claim now follows from Theorem 2.9. ■

(f) The sequence $\big((1+1/n)^n\big)$ converges and its limit,

$$e := \lim_{n\to\infty}\Big(1+\frac{1}{n}\Big)^n ,$$

the **Euler number**, satisfies $2 < e \le 3$.

Proof For all $n \in \mathbb{N}^\times$, set $e_n := (1+1/n)^n$.

(i) In the first step we prove that the sequence (e_n) is increasing. Consider

$$\begin{aligned}\frac{e_{n+1}}{e_n} &= \Big(\frac{n+2}{n+1}\Big)^{n+1}\cdot\Big(\frac{n}{n+1}\Big)^n \\ &= \Big(\frac{n^2+2n}{(n+1)^2}\Big)^{n+1}\cdot\frac{n+1}{n} = \Big(1-\frac{1}{(n+1)^2}\Big)^{n+1}\cdot\frac{n+1}{n} .\end{aligned} \tag{4.1}$$

The first factor after the last equal sign in (4.1) can be approximated using the Bernoulli inequality (see Exercise I.10.6):

$$\Big(1-\frac{1}{(n+1)^2}\Big)^{n+1} \ge 1-\frac{1}{n+1} = \frac{n}{n+1} .$$

Thus (4.1) implies $e_n \le e_{n+1}$ as claimed.

(ii) We show that $2 < e_n < 3$. From the binomial theorem we have

$$e_n = \Bigl(1+\frac{1}{n}\Bigr)^n = \sum_{k=0}^{n} \binom{n}{k} \frac{1}{n^k} = 1 + \sum_{k=1}^{n} \binom{n}{k} \frac{1}{n^k} . \tag{4.2}$$

Further, for $1 \le k \le n$, we have

$$\binom{n}{k} \frac{1}{n^k} = \frac{1}{k!} \frac{n \cdot (n-1) \cdot \cdots \cdot (n-k+1)}{n \cdot n \cdot \cdots \cdot n} \le \frac{1}{k!} \le \frac{1}{2^{k-1}} .$$

It then follows from (4.2) (see Exercise I.8.1) that

$$e_n \le 1 + \sum_{k=1}^{n} \Bigl(\frac{1}{2}\Bigr)^{k-1} = 1 + \frac{1-(\frac{1}{2})^n}{1-\frac{1}{2}} < 1 + \frac{1}{\frac{1}{2}} = 3 .$$

Finally $2 = e_1 < e_n$ for $n \ge 2$. The claim now follows from (i), (ii) and Theorem 4.1. ■

The number e plays an important role in analysis. Its value can theoretically be determined from the sequence (e_n). Unfortunately, the sequence (e_n) does not converge very quickly. The numerical value of e is approximately[3]

$$2.71828\ 18284\ 59045\ 23536\ \ldots$$

Comparing this value with several terms of (e_n),

$$e_1 = 2 , \quad e_{10} = 2.59374\ldots , \quad e_{100} = 2.70481\ldots , \quad e_{1000} = 2.71692\ldots ,$$

we see that, even for $n = 1000$, the error $e - e_n$ is $0.0014\ldots$ (see also the next example).

(g) We can represent e as the limit of a sequence which converges much faster:

$$e = \lim_{n\to\infty} \sum_{k=0}^{n} \frac{1}{k!} .$$

Proof (i) Set $x_n := \sum_{k=0}^{n} 1/k!$. Obviously, the sequence (x_n) is increasing. The proof of (f) shows that $e_n \le x_n < 3$ for all $n \in \mathbb{N}^\times$ and so, by Theorem 4.1, the sequence (x_n) converges and its limit e' satisfies $e \le e' \le 3$.

(ii) To complete the proof we need to show $e' \le e$. Fix some $m \in \mathbb{N}^\times$. Then for all $n \ge m$, we have

$$\begin{aligned} e_n = \Bigl(1+\frac{1}{n}\Bigr)^n &= \sum_{k=0}^{n} \binom{n}{k} \frac{1}{n^k} \ge \sum_{k=0}^{m} \binom{n}{k} \frac{1}{n^k} \\ &= 1 + \sum_{k=1}^{m} \frac{1}{k!} \frac{n \cdot (n-1) \cdot \cdots \cdot (n-k+1)}{n \cdot n \cdot \cdots \cdot n} \\ &= 1 + \sum_{k=1}^{m} \frac{1}{k!} 1 \cdot \Bigl(1-\frac{1}{n}\Bigr) \cdot \cdots \cdot \Bigl(1-\frac{k-1}{n}\Bigr) . \end{aligned}$$

[3] We are somewhat premature in using such 'decimal representations'. A complete discussion of this way of representing numbers is in Section 7.

If we set

$$x'_{m,n} := 1 + \sum_{k=1}^{m} \frac{1}{k!}\Big(1-\frac{1}{n}\Big)\cdot\cdots\cdot\Big(1-\frac{k-1}{n}\Big)\,, \qquad n \geq m\,,$$

then $x'_{m,n} \uparrow x_m$ as $n \to \infty$ and $x'_{m,n} \leq e_n$ for all $n \geq m$. Because $e_n \uparrow e$, it follows from Proposition 2.7 that $x_m \leq e$. Since this holds for any $m \in \mathbb{N}^\times$, e is an upper bound for $X := \{\, x_m \,;\, m \in \mathbb{N} \,\}$. Since $e' = \lim x_n = \sup X$, Theorem 4.1 implies that $e' \leq e$. ∎

As we have already mentioned, the sequence (x_n) converges much faster to e than the sequence (e_n). In fact, one can prove the following error estimate (see Exercise 7):

$$0 < e - x_n \leq \frac{1}{nn!}\,, \qquad n \in \mathbb{N}^\times\,.$$

For $n = 6$ we have $1/(6!\,6) = 0.00023\ldots$ which is already a smaller error than in Example 4.2(f) for $n = 1000$.

Exercises

1 Let $a_1, \ldots, a_k \in \mathbb{R}^+$. Prove that

$$\lim_{n\to\infty} \sqrt[n]{a_1^n + \cdots + a_k^n} = \max\{a_1, \ldots, a_k\}\,.$$

2 Prove that

$$(1 - 1/n)^n \to 1/e \quad (n \to \infty)\,.$$

(Hint: Consider $\lim(1 - 1/n^2)^n = 1$ and Proposition 2.6.)

3 Show that, for all $r \in \mathbb{Q}$,

$$(1 + r/n)^n \to e^r \quad (n \to \infty)\,.$$

(Hint: Consider the cases $r > 0$ and $r < 0$ separately. Use Exercises 2 and 2.7.)

4 For $a \in (0, \infty)$, define a real sequence (x_n) recursively by $x_0 \geq a$ and

$$x_{n+1} := (x_n + a/x_n)/2\,, \qquad n \in \mathbb{N}\,.$$

Prove that (x_n) is decreasing and converges to $\sqrt{a}$.[4]

5 Let $a, x_0 \in (0, \infty)$ and

$$x_{n+1} := a/(1 + x_n)\,, \qquad n \in \mathbb{N}\,.$$

Show that the sequence (x_n) converges and determine its limit.

---

[4]This procedure to determine $\sqrt{a}$ is called the **Babylonian algorithm** or **Heron's method**. The sequence (x_n) converges very quickly to $\sqrt{a}$. For example, if $x_0 = a = 4$, then

$$x_1 = 2.5\,, \qquad x_2 = 2.05\,, \qquad x_3 = 2.006\,09\ldots\,, \qquad x_4 = 2.000\,000\,093\ldots$$

Note also that all the x_n are rational if a and the 'initial value' x_0 are rational. In Section IV.4 we give a geometric interpretation of Heron's method and estimate the rate of convergence.

6 Prove the convergence of the sequence

$$x_0 > 0 , \quad x_1 > 0 , \quad x_{n+2} := \sqrt{x_{n+1} x_n} , \qquad n \in \mathbb{N} .$$

7 (a) For $n \in \mathbb{N}^\times$, prove the following error estimate:

$$0 < e - \sum_{k=0}^{n} \frac{1}{k!} < \frac{1}{n n!} .$$

(b) Use the inequality in (a) to show that e is irrational.
(Hint: (a) For $n \in \mathbb{N}^\times$, let $y_m := \sum_{k=n+1}^{n+m} 1/k!$. Show that $y_m \to e - \sum_{k=0}^{n} 1/k!$ and $(m+n)!\, y_m < \sum_{k=1}^{m} (n+1)^{1-k}$. (b) Prove by contradiction.)

8 Let (x_n) be defined recursively by

$$x_0 := 1 , \quad x_{n+1} := 1 + 1/x_n , \qquad n \in \mathbb{N} .$$

Show that the sequence (x_n) converges and determine its limit.

9 The **Fibonacci numbers** f_n are defined recursively by

$$f_0 := 0 , \quad f_1 := 1 , \quad f_{n+1} := f_n + f_{n-1} , \qquad n \in \mathbb{N}^\times .$$

Prove that $\lim(f_{n+1}/f_n) = g$, where g is the limit of Exercise 8.

10 Let

$$x_0 := 5 , \quad x_1 := 1 , \quad x_{n+1} := \frac{2}{3} x_n + \frac{1}{3} x_{n-1} , \qquad n \in \mathbb{N}^\times .$$

Verify that (x_n) converges and determine $\lim x_n$.
(Hint: Derive an expression for $x_n - x_{n+1}$.)

5 Infinite Limits

Certain sequences in $\mathbb{R}$ can usefully considered to converge to $+\infty$ or $-\infty$ in the extended number line $\bar{\mathbb{R}}$. For this extension of the limit concept, it suffices to define appropriate neighborhoods of the elements $\pm\infty$ in $\bar{\mathbb{R}}$.

Convergence to $\pm\infty$

Because there is no suitable metric[1] on $\bar{\mathbb{R}}$, we extend the set of all neighborhoods in $\mathbb{R}$ using the following ad hoc definition: A subset $U \subseteq \bar{\mathbb{R}}$ is called a **neighborhood of** ∞ (or **of** $-\infty$) if there is some $K > 0$ such that $(K, \infty) \subseteq U$ (or such that $(-\infty, -K) \subseteq U$). The set of neighborhoods of $\pm\infty$ is denoted by $\mathcal{U}(\pm\infty)$, that is,

$$\mathcal{U}(\pm\infty) := \{ U \subseteq \bar{\mathbb{R}} \; ; \; U \text{ is neighborhood of } \pm\infty \} .$$

Now let (x_n) be a sequence in $\mathbb{R}$. Then $\pm\infty$ is called a **cluster point** (or **limit**) of (x_n), if each neighborhood U of $\pm\infty$ contains infinitely many (or almost all) terms of (x_n). If $\pm\infty$ is the limit of (x_n), we usually write

$$\lim_{n\to\infty} x_n = \pm\infty \qquad \text{or} \qquad x_n \to \pm\infty \ (n \to \infty) .$$

The sequence (x_n) **converges** in $\bar{\mathbb{R}}$ if there is some $x \in \bar{\mathbb{R}}$ such that $\lim_{n\to\infty} x_n = x$. The sequence (x_n) **diverges** in $\bar{\mathbb{R}}$, if it does not converge in $\bar{\mathbb{R}}$. With this definition, any sequence which converges in $\mathbb{R}$, also converges in $\bar{\mathbb{R}}$, and any sequence which diverges in $\bar{\mathbb{R}}$, also diverges in $\mathbb{R}$. On the other hand there are divergent sequences in $\mathbb{R}$ which converge in $\bar{\mathbb{R}}$ (to $\pm\infty$). In this case the sequence is said to converge **improperly**. This distinction is significant because our understanding about convergence in metric spaces does not apply to improper convergence, and hence such convergence needs a separate study.

5.1 Examples **(a)** Let (x_n) be a sequence in $\mathbb{R}$. Then $x_n \to \infty$ if and only if, for each $K > 0$, there is some $N_K \in \mathbb{N}$ such that $x_n > K$ for all $n \geq N_K$.

(b) $\lim_{n\to\infty}(n^2 - n) = \infty$ and $\lim_{n\to\infty}(-2^n) = -\infty$.

(c) The sequence $\left((-n)^n\right)_{n\in\mathbb{N}}$ has the cluster points ∞ and $-\infty$, and so diverges in $\bar{\mathbb{R}}$. ∎

5.2 Proposition *Let (x_n) be a sequence in $\mathbb{R}^\times$.*

(i) $1/x_n \to 0$, if $x_n \to \infty$ or $x_n \to -\infty$.

(ii) $1/x_n \to \infty$, if $x_n \to 0$ and $x_n > 0$ for almost all $n \in \mathbb{N}$.

(iii) $1/x_n \to -\infty$, if $x_n \to 0$ and $x_n < 0$ for almost all $n \in \mathbb{N}$.

[1] Of course, various metrics could be defined on $\bar{\mathbb{R}}$, but none of the metrics we have defined so far are suitable for our purposes (see Exercise 5).

Proof (i) Let $\varepsilon > 0$. Then there is some N such that $|x_n| > 1/\varepsilon$ for all $n \geq N$, and we have the inequality

$$|1/x_n| = 1/|x_n| < \varepsilon , \qquad n \geq N .$$

Hence $(1/x_n)$ converges to 0.

(ii) Let $K > 0$. Then there is some N such that $0 < x_n < 1/K$ for all $n \geq N$. Hence

$$1/x_n > K , \qquad n \geq N ,$$

and so the claim follows from Example 5.1(a).

Claim (iii) can be proved similarly. ■

5.3 Proposition *Every monotone sequence (x_n) in $\mathbb{R}$ converges in $\bar{\mathbb{R}}$, and*

$$\lim x_n = \begin{cases} \sup\{ x_n \ ; \ n \in \mathbb{N} \} , & \textit{if } (x_n) \textit{ is increasing} , \\ \inf\{ x_n \ ; \ n \in \mathbb{N} \} , & \textit{if } (x_n) \textit{ is decreasing} . \end{cases}$$

Proof We consider an increasing sequence (x_n). If $\{ x_n \ ; \ n \in \mathbb{N} \}$ is bounded above, then, by Theorem 4.1, (x_n) converges in $\mathbb{R}$ to $\sup\{ x_n \ ; \ n \in \mathbb{N} \}$. Otherwise, if $\{ x_n \ ; \ n \in \mathbb{N} \}$ is not bounded above, then for each $K > 0$ there is some m such that $x_m > K$. Since (x_n) is increasing we have also $x_n > K$ for all $n \geq m$, that is, (x_n) converges to ∞. The case of a decreasing sequence is proved similarly. ■

The Limit Superior and Limit Inferior

5.4 Definition Let (x_n) be a sequence in $\mathbb{R}$. We can define two new sequences (y_n) and (z_n) by

$$y_n := \sup_{k \geq n} x_k := \sup\{ x_k \ ; \ k \geq n \} ,$$

$$z_n := \inf_{k \geq n} x_k := \inf\{ x_k \ ; \ k \geq n \} .$$

Clearly (y_n) is a *decreasing* sequence and (z_n) is an *increasing* sequence in $\bar{\mathbb{R}}$. By Proposition 5.3, these sequences converge in $\bar{\mathbb{R}}$:

$$\limsup_{n \to \infty} x_n := \varlimsup_{n \to \infty} x_n := \lim_{n \to \infty} \bigl(\sup_{k \geq n} x_k \bigr) ,$$

the **limit superior**, and

$$\liminf_{n \to \infty} x_n := \varliminf_{n \to \infty} x_n := \lim_{n \to \infty} \bigl(\inf_{k \geq n} x_k \bigr) ,$$

the **limit inferior** of the sequence (x_n). We also have

$$\limsup x_n = \inf_{n \in \mathbb{N}} \bigl(\sup_{k \geq n} x_k \bigr) \quad \text{and} \quad \liminf x_n = \sup_{n \in \mathbb{N}} \bigl(\inf_{k \geq n} x_k \bigr) ,$$

which follow again from Proposition 5.3. ■

We see next that the limit superior and limit inferior of a sequence are, in particular, cluster points.

5.5 Theorem *Any sequence (x_n) in $\mathbb{R}$ has a smallest cluster point x_* and a greatest cluster point x^* in $\bar{\mathbb{R}}$ and these satisfy*

$$\liminf x_n = x_* \quad \textit{and} \quad \limsup x_n = x^* .$$

Proof Set $x^* := \limsup x_n$ and $y_n := \sup_{k \geq n} x_k$ for $n \in \mathbb{N}$. Then (y_n) is a decreasing sequence such that

$$x^* = \inf_{n \in \mathbb{N}} y_n . \tag{5.1}$$

We consider three cases:

(i) Suppose that $x^* = -\infty$. Then for each $K > 0$, there is some n such that

$$-K > y_n = \sup_{k \geq n} x_k ,$$

since otherwise we would have $x^* \geq -K_0$ for some $K_0 \geq 0$. Hence $x_k \in (-\infty, -K)$ for all $k \geq n$, that is, $x^* = -\infty$ is the only cluster point of (x_n).

(ii) Suppose that $x^* \in \mathbb{R}$. By Proposition I.10.5 and (5.1), we have for each $\xi > x^*$ some n such that $\xi > y_n \geq x_k$ for all $k \geq n$. Consequently, no cluster point of (x_n) is larger than x^*. It remains to show only that x^* is itself a cluster point of (x_n). Let $\varepsilon > 0$. Since

$$\sup_{k \geq n} x_k = y_n \geq x^* , \qquad n \in \mathbb{N} ,$$

we have, once again from Proposition I.10.5, for each n, some $k \geq n$ such that $x_k > x^* - \varepsilon$. Since we already know that no cluster point of (x_n) is larger than x^*, the interval $(x^* - \varepsilon, x^* + \varepsilon)$ must contain infinitely many terms of the sequence of (x_n), that is, x^* is a cluster point of (x_n).

(iii) Finally we consider the case $x^* = \infty$. Because of (5.1) we have $y_n = \infty$ for all $n \in \mathbb{N}$. Hence for each $K > 0$ and n, there is some $k \geq n$ such that $x_k > K$. This means that $x^* = \infty$ is a cluster point of (x_n), and clearly the largest such point.

Showing that $x_* := \liminf x_n$ is the smallest cluster point of (x_n) is similar. ∎

5.6 Examples

(a) $\overline{\lim} \dfrac{(-1)^n n}{n+1} = 1$ and $\underline{\lim} \dfrac{(-1)^n n}{n+1} = -1$.

(b) $\overline{\lim}\, n^{(-1)^n} = \infty$ and $\underline{\lim}\, n^{(-1)^n} = 0$. ∎

5.7 Theorem *Let (x_n) be a sequence in $\mathbb{R}$. Then*

$$(x_n) \text{ converges in } \bar{\mathbb{R}} \iff \overline{\lim}\, x_n \leq \underline{\lim}\, x_n \ .$$

When the sequence converges, the limit x satisfies

$$x = \lim x_n = \underline{\lim}\, x_n = \overline{\lim}\, x_n \ .$$

Proof '$\Rightarrow$' If (x_n) converges to x in $\bar{\mathbb{R}}$, then x is the unique cluster point of (x_n) and so the claim follows from Theorem 5.5.

'$\Leftarrow$' Suppose that $\overline{\lim}\, x_n \leq \underline{\lim}\, x_n$. Then, from Theorem 5.5 again, the only cluster point of (x_n) is $x := \overline{\lim}\, x_n = \underline{\lim}\, x_n$. If $x = -\infty$ (or $x = \infty$), then, for each $K > 0$, there is some k such that $x_n < -K$ (or $x_n > K$) for all $n \geq k$. Thus $\lim x_n = -\infty$ (or $\lim x_n = \infty$).

If x is in $\mathbb{R}$, then, from Theorem 5.5 and Proposition I.10.5, for a given $\varepsilon > 0$, there are at most finitely many $j \in \mathbb{N}$ and finitely many $k \in \mathbb{N}$ such that $x_j < x - \varepsilon$ and $x_k > x + \varepsilon$. Thus each neighborhood U of x contains almost all terms of the sequence (x_n), that is, $\lim x_n = x$. ■

The Bolzano-Weierstrass Theorem

For a bounded sequence in $\mathbb{R}$, Theorem 5.5 is called the *Bolzano-Weierstrass theorem*. We actually prove this theorem in somewhat more generality.

5.8 Theorem (Bolzano-Weierstrass) *Every bounded sequence in $\mathbb{K}^m$ has a convergent subsequence, that is, a cluster point.*

Proof We consider first the case $\mathbb{K} = \mathbb{R}$ and prove the claim by induction on m. The case $m = 1$ follows from Theorem 5.5 and Proposition 1.17. For the induction step $m \to m+1$, suppose that (z_n) is a bounded sequence in $\mathbb{R}^{m+1}$. Then the bound $M := \sup\{\,|z_n| \ ; \ n \in \mathbb{N}\,\}$ exists in $[0, \infty)$. Since $\mathbb{R}^{m+1} = \mathbb{R}^m \times \mathbb{R}$, we can write each $z \in \mathbb{R}^{m+1}$ in the form $z = (x, y)$ with $x \in \mathbb{R}^m$ and $y \in \mathbb{R}$. Thus, from $z_n = (x_n, y_n)$, we get a sequence (x_n) in $\mathbb{R}^m$ and a sequence (y_n) in $\mathbb{R}$. Since

$$\max\{|x_n|, |y_n|\} \leq |z_n| = \sqrt{|x_n|^2 + |y_n|^2} \leq M \ , \qquad n \in \mathbb{N} \ ,$$

(x_n) and (y_n) are bounded. We now use our induction hypothesis to find a subsequence (x_{n_k}) of (x_n) and some $x \in \mathbb{R}^m$ such that $x_{n_k} \to x$ as $k \to \infty$. Since the subsequence (y_{n_k}) is also bounded, it follows from the induction hypothesis, that there is a subsequence $(y_{n_{k_j}})$ of (y_{n_k}) and some $y \in \mathbb{R}$ such that $y_{n_{k_j}} \to y$ as $j \to \infty$. Finally, by Propositions 1.15 and 3.14, the subsequence $(x_{n_{k_j}}, y_{n_{k_j}})$ of (z_n) converges to $z := (x, y) \in \mathbb{R}^{m+1}$ as $j \to \infty$. This completes the proof of the case $\mathbb{K} = \mathbb{R}$.

The case $\mathbb{K} = \mathbb{C}$ follows from what we have just proved using the identification of $\mathbb{C}^m$ with $\mathbb{R}^{2m}$. ■

Exercises

1 Let (x_n) be a sequence in $\mathbb{R}$, $x_* := \underline{\lim}\, x_n$ and $x^* := \overline{\lim}\, x_n$. Suppose that x_* and x^* are in $\mathbb{R}$. Prove that, for each $\varepsilon > 0$, there is some N such that

$$x_* - \varepsilon < x_n < x^* + \varepsilon \,, \qquad n \ge N \,.$$

How must the claim be modified in the cases $x_* = -\infty$ and $x^* = \infty$?

2 Let (x_n) and (y_n) be sequences in $\mathbb{R}$ and

$$x_* := \underline{\lim}\, x_n \,, \quad x^* := \overline{\lim}\, x_n \,, \quad y_* := \underline{\lim}\, y_n \,, \quad y^* := \overline{\lim}\, y_n \,.$$

Show the following:

(a) $\overline{\lim}(-x_n) = -x_*$.

(b) If (x^*, y^*) and (x_*, y_*) are not equal to $(\infty, -\infty)$ or $(-\infty, \infty)$, then

$$\overline{\lim}(x_n + y_n) \le x^* + y^* \text{ and } \underline{\lim}(x_n + y_n) \ge x_* + y_* \,.$$

(c) If $x_n \ge 0$ and $y_n \ge 0$ for all $n \in \mathbb{N}$, and $(x_*, y_*) \notin \{(0,\infty), (\infty, 0)\}$, $(x_*, y^*) \ne (\infty, 0)$ and $(x^*, y^*) \ne (0, \infty)$, then

$$0 \le x_* y_* \le \underline{\lim}(x_n y_n) \le x_* y^* \le \overline{\lim}(x_n y_n) \le x^* y^* \,.$$

(d) If (y_n) converges to $y \in \mathbb{R}$, then

$$\overline{\lim}(x_n + y_n) = x^* + y \,, \quad \underline{\lim}(x_n + y_n) = x_* + y \,,$$

and

$$\begin{aligned} \overline{\lim}(x_n y_n) &= y x^* \,, \qquad & y > 0 \,, \\ \overline{\lim}(x_n y_n) &= y x_* \,, \qquad & y < 0 \,. \end{aligned}$$

(e) If $x_n \le y_n$ for all $n \in \mathbb{N}$, then $\underline{\lim}\, x_n \le \underline{\lim}\, y_n$ and $\overline{\lim}\, x_n \le \overline{\lim}\, y_n$.

3 For $n \in \mathbb{N}$, let $x_n := 2^n\big(1 + (-1)^n\big) + 1$. Determine the following:

$$\underline{\lim}\, x_n, \ \ \overline{\lim}\, x_n, \ \ \underline{\lim}(x_{n+1}/x_n), \ \ \overline{\lim}(x_{n+1}/x_n), \ \ \underline{\lim}\, \sqrt[n]{x_n}, \ \ \overline{\lim}\, \sqrt[n]{x_n} \,.$$

4 Let (x_n) be a sequence in $\mathbb{R}$ such that $x_n > 0$ for all $n \in \mathbb{N}$. Prove

$$\underline{\lim}\, \frac{x_{n+1}}{x_n} \le \underline{\lim}\, \sqrt[n]{x_n} \le \overline{\lim}\, \sqrt[n]{x_n} \le \overline{\lim}\, \frac{x_{n+1}}{x_n} \,.$$

(Hint: If $q < \underline{\lim}(x_{n+1}/x_n)$, then $x_{n+1}/x_n \ge q$ for all $n \ge n(q)$.)

5 (a) Show that the function

$$\varphi \colon \bar{\mathbb{R}} \to [-1, 1] \,, \quad \varphi(x) := \begin{cases} -1 \,, & x = -\infty \,, \\ x/(1+|x|) \,, & x \in \mathbb{R} \,, \\ 1 \,, & x = \infty \,, \end{cases}$$

is strictly increasing and bijective.

(b) Show that the function

$$d \colon \bar{\mathbb{R}} \times \bar{\mathbb{R}} \to \mathbb{R}^+ \,, \quad (x, y) \mapsto |\varphi(x) - \varphi(y)|$$

is a metric on $\bar{\mathbb{R}}$.

6 For the sequences

$$(x_n) := (0,1,2,1,0,1,2,1,0,1,2,1,\ldots) \quad \text{and} \quad (y_n) := (2,1,1,0,2,1,1,0,2,1,1,0,\ldots)$$

determine the following:

$$\begin{gathered}
\overline{\lim}\, x_n + \overline{\lim}\, y_n\ , \quad \overline{\lim}(x_n + y_n)\ , \\
\underline{\lim}\, x_n + \underline{\lim}\, y_n\ , \quad \underline{\lim}(x_n + y_n)\ , \quad \underline{\lim}\, x_n + \overline{\lim}\, y_n\ .
\end{gathered}$$

6 Completeness

In Section 1 we defined convergence using the concept of neighborhoods. In this definition, the limit of the sequence appears explicitly, and so, in principal, to show that a sequence converges it is necessary to know what its limit is. In this section we show that in certain 'complete' spaces it is possible to recognize the convergence of a sequence without knowing the limit. Sequences in such metric spaces are convergent if and only if they are Cauchy sequences. These sequences are an important tool in the theoretical investigation of convergence. In addition, they are used in Cantor's construction the real numbers which we mentioned in Section I.10 and carry out in this section.

Cauchy Sequences

In the following $X = (X, d)$ is a metric space.

A sequence (x_n) in X is called a **Cauchy sequence** if, for each $\varepsilon > 0$, there is some $N \in \mathbb{N}$ such that $d(x_n, x_m) < \varepsilon$ for all $m, n \geq N$.

If (x_n) is a sequence in a normed vector space $E = (E, \|\cdot\|)$, then (x_n) is a Cauchy sequence if and only if for each $\varepsilon > 0$ there is some N such that $\|x_n - x_m\| < \varepsilon$ for all $m, n \geq N$. In particular, we notice that Cauchy sequences in E are 'translation invariant', that is, if (x_n) is a Cauchy sequence and a is an arbitrary vector in E, then the 'translated' sequence $(x_n + a)$ is also a Cauchy sequence. This shows, in particular, that Cauchy sequences *cannot* be defined using neighborhoods.

6.1 Proposition *Every convergent sequence is a Cauchy sequence.*

Proof Let (x_n) be a convergent sequence in X with limit x. Then, for each $\varepsilon > 0$, there is some N such that $d(x_n, x) < \varepsilon/2$ for all $n \geq N$. From the triangle inequality it follows that

$$d(x_n, x_m) \leq d(x_n, x) + d(x, x_m) < \frac{\varepsilon}{2} + \frac{\varepsilon}{2} = \varepsilon \ , \qquad m, n \geq N \ .$$

Hence (x_n) is a Cauchy sequence. ■

The converse of Proposition 6.1 is not true, that is, there are metric spaces in which not every Cauchy sequence converges.

6.2 Example Define (x_n) recursively by $x_0 := 2$ and $x_{n+1} := \frac{1}{2}(x_n + 2/x_n)$ for all $n \in \mathbb{N}$. Then (x_n) is a Cauchy sequence in $\mathbb{Q}$ which does not converge in $\mathbb{Q}$.

Proof Clearly $x_n \in \mathbb{Q}$ for all $n \in \mathbb{N}$. Moreover, from Exercise 4.4, we know that (x_n) converges to $\sqrt{2}$ in $\mathbb{R}$. Thus, by Proposition 6.1, (x_n) is a Cauchy sequence in $\mathbb{R}$, and hence in $\mathbb{Q}$ too.

On the other hand, (x_n) cannot converge in $\mathbb{Q}$. Indeed, if $x_n \to a$ for some $a \in \mathbb{Q}$, then $x_n \to a$ in $\mathbb{R}$ also. But then the uniqueness of the limit implies $a = \sqrt{2} \in \mathbb{R}\backslash\mathbb{Q}$, a contradiction. ■

6.3 Proposition *Every Cauchy sequence is bounded.*

Proof Let (x_n) be a Cauchy sequence. Then there is some $N \in \mathbb{N}$ such that $d(x_n, x_m) < 1$ for all $m, n \geq N$. In particular, $d(x_n, x_N) \leq 1$ for all $n \geq N$. Set $M = \max_{n<N}\{d(x_n, x_N)\}$. Then for all n we have $d(x_n, x_N) \leq 1 + M$, and so, by Example 1.9(c), (x_n) is bounded in X. ■

6.4 Proposition *If a Cauchy sequence has a convergent subsequence, then it is itself convergent.*

Proof Let (x_n) be a Cauchy sequence and $(x_{n_k})_{k\in\mathbb{N}}$ a convergent subsequence with limit x. Suppose that $\varepsilon > 0$. Then there is some N such that $d(x_n, x_m) < \varepsilon/2$ for all $m, n \geq N$. There is also some K such that $d(x_{n_k}, x) < \varepsilon/2$ for all $k \geq K$. Set $M := \max\{K, N\}$. Then

$$d(x_n, x) \leq d(x_n, x_{n_M}) + d(x_{n_M}, x) < \frac{\varepsilon}{2} + \frac{\varepsilon}{2} = \varepsilon \ , \qquad n \geq M \ ,$$

that is, (x_n) converges to x. ■

Banach Spaces

A metric space X is called **complete** if every Cauchy sequence in X converges. A complete normed vector space is called a **Banach space**.

Using the Bolzano-Weierstrass theorem we can show that complete metric spaces exist.

6.5 Theorem $\mathbb{K}^m$ *is a Banach space.*

Proof We know already from Section 3 that $\mathbb{K}^m$ is a normed vector space, so it remains to show completeness. Let (x_n) be a Cauchy sequence in $\mathbb{K}^m$. By Proposition 6.3, (x_n) is bounded. The Bolzano-Weierstrass theorem implies the existence of a convergent subsequence, and then Proposition 6.4 implies that (x_n) itself converges. ■

6.6 Theorem *Let X be a nonempty set and $E = (E, \|\cdot\|)$ a Banach space. Then $B(X, E)$ is also a Banach space.*

Proof Let (u_n) be a Cauchy sequence in the normed vector space $B(X, E)$ (see Proposition 3.5). Suppose that $\varepsilon > 0$. Then there is some $N := N(\varepsilon)$ such that

$\|u_n - u_m\|_\infty \le \varepsilon$ for all $m, n \ge N$. In particular,

$$\|u_n(x) - u_m(x)\|_E \le \|u_n - u_m\|_\infty \le \varepsilon\ , \qquad m, n \ge N\ , \quad x \in X\ . \tag{6.1}$$

This shows that, for each $x \in X$, the sequence $\big(u_n(x)\big)$ is a Cauchy sequence in E. The completeness of E implies that, for each $x \in X$, there is some vector $a_x \in E$ such that $u_n(x) \to a_x$ as $n \to \infty$. By Corollary 1.13, a_x is unique and we define $u \in E^X$ by $u(x) := a_x$ for $x \in X$.

We will prove that the Cauchy sequence (u_n) converges to u in $B(X, E)$. We show first that $u \in E^X$ is bounded. Indeed, taking the limit $m \to \infty$ in (6.1) yields

$$\|u_n(x) - u(x)\|_E \le \varepsilon\ , \qquad n \ge N\ , \quad x \in X \tag{6.2}$$

(see Proposition 2.10 and Remark 3.1(c)), and so we have

$$\|u(x)\|_E \le \varepsilon + \|u_N(x)\|_E \le \varepsilon + \|u_N\|_\infty\ , \qquad x \in X\ .$$

This shows that the function $u\colon X \to E$ is bounded, that is, it is in $B(X, E)$. Finally, taking the supremum over all $x \in X$ in (6.2) we get $\|u_n - u\|_\infty \le \varepsilon$ for all $n \ge N$, that is, (u_n) converges to u in $B(X, E)$. ■

As a direct consequence of the previous two theorems we have the following: *For every nonempty set X, $B(X, \mathbb{R})$, $B(X, \mathbb{C})$ and $B(X, \mathbb{K}^m)$ are Banach spaces.*

6.7 Remarks **(a)** The completeness of a normed vector space E is invariant under changes to equivalent norms, that is, if $\|\cdot\|_1$ and $\|\cdot\|_2$ are equivalent norms on E, then $(E, \|\cdot\|_1)$ is complete if and only if $(E, \|\cdot\|_2)$ is complete.

(b) The vector space $\mathbb{K}^m$ with either of the norms $|\cdot|_1$ or $|\cdot|_\infty$ is complete. (We will prove in Section III.3 that *all* norms on $\mathbb{K}^m$ are equivalent.)

(c) A complete inner product space (see Theorem 3.10) is called a **Hilbert space**. In particular, Theorem 6.5 shows that $\mathbb{K}^m$ is a Hilbert space. ■

Cantor's Construction of the Real Numbers

We close this section with a second construction of the real numbers $\mathbb{R}$. Since we make no further use of this construction in the following, this discussion can be omitted on a first reading of this book.

First we note that all statements in this chapter about sequences remain true if we replace 'for each $\varepsilon > 0$' by 'for each $\varepsilon = 1/N$ with $N \in \mathbb{N}^\times$' in Proposition 1.7(iii), in the definitions of null sequences and Cauchy sequences, and in the corresponding proofs. This is a consequence of Corollary I.10.7.

This puts us back in the situation where only the rational numbers have been constructed. By Theorem I.9.5, $\mathbb{Q} = (\mathbb{Q}, \le)$ is an ordered field, and so Proposition I.8.10

implies that $\mathbb{Q}$, with the metric induced from the absolute value $|\cdot|$, is a metric space. Because of the above discussion,

$$\mathcal{R} := \{\, r \in \mathbb{Q}^{\mathbb{N}} \ ;\ r \text{ is a Cauchy sequence} \,\}$$

and

$$\mathbf{c}_0 := \{\, r \in \mathbb{Q}^{\mathbb{N}} \ ;\ r \text{ is a null sequence} \,\}$$

are well defined sets. From Proposition 6.1 we have $\mathbf{c}_0 \subseteq \mathcal{R}$.

From Example I.8.2(b) we know that $\mathbb{Q}^{\mathbb{N}}$ is a commutative ring with unity. We denote by $\overline{a}$ the constant sequence $(a, a, \ldots)$ in $\mathbb{Q}^{\mathbb{N}}$. Then $\overline{1}$ is the unity element of the ring $\mathbb{Q}^{\mathbb{N}}$. By Example I.4.4(c), $\mathbb{Q}^{\mathbb{N}}$ is also a partially ordered set. Since this partial order is not a total order, $\mathbb{Q}^{\mathbb{N}}$ is not an ordered ring.

6.8 Lemma *$\mathcal{R}$ is a subring of $\mathbb{Q}^{\mathbb{N}}$ containing $\overline{1}$ and $\mathbf{c}_0$ is a nontrivial proper ideal of $\mathcal{R}$.*

Proof Let $r = (r_n)$ and $s = (s_n)$ be elements of $\mathcal{R}$, and $N \in \mathbb{N}^{\times}$. Since every Cauchy sequence is bounded, there is some $B \in \mathbb{N}^{\times}$ such that

$$|r_n| \le B\ , \qquad |s_n| \le B\ , \qquad n \in \mathbb{N}\ .$$

Set $M := 2BN \in \mathbb{N}^{\times}$. Then there is some $n_0 \in \mathbb{N}$ such that

$$|r_n - r_m| < 1/M\ , \quad |s_n - s_m| < 1/M\ , \qquad m, n \ge n_0\ .$$

Thus we have the inequalities

$$|r_n + s_n - (r_m + s_m)| \le |r_n - r_m| + |s_n - s_m| < 2/M \le 1/N$$

and

$$|r_n s_n - r_m s_m| \le |r_n|\,|s_n - s_m| + |r_n - r_m|\,|s_m| < 2B/M = 1/N$$

for all $m, n \ge n_0$. Consequently $r + s$ and $r \cdot s$ are in $\mathcal{R}$, that is, $\mathcal{R}$ is a subring of $\mathbb{Q}^{\mathbb{N}}$. It is clear that $\mathcal{R}$ contains the unity element $\overline{1}$. From Propositions 2.2 and 2.4 (with $\mathbb{K}$ replaced by $\mathbb{Q}$) and from Proposition 6.3, it follows that $\mathbf{c}_0$ is an ideal of $\mathcal{R}$. Since

$$\Big(\frac{1}{n+1}\Big)_{n \in \mathbb{N}} \in \mathbf{c}_0 \backslash \{0\}\ , \quad \overline{1} \in \mathcal{R} \backslash \mathbf{c}_0\ ,$$

$\mathbf{c}_0$ is a nontrivial proper ideal. ■

From Exercise I.8.6, we know that $\mathcal{R}$ cannot be a field. Let R be the quotient ring of $\mathcal{R}$ by the ideal $\mathbf{c}_0$, that is, $R = \mathcal{R}/\mathbf{c}_0$ (see Exercise I.8.6). It is clear that the function

$$\mathbb{Q} \to R\ , \quad a \mapsto [\overline{a}] = \overline{a} + \mathbf{c}_0\ , \tag{6.3}$$

which maps each rational number a to the coset $[\overline{a}]$ of the constant sequence $\overline{a}$ in $\mathcal{R}$, is an injective ring homomorphism. Thus we will consider $\mathbb{Q}$ to be a subring of R by identifying $\mathbb{Q}$ with its image under the function (6.3).

We next define an order on R. We say $r = (r_n) \in \mathcal{R}$ is strictly positive if there is some $N \in \mathbb{N}^{\times}$ and $n_0 \in \mathbb{N}$ such that $r_n > 1/N$ for all $n \ge n_0$. Let $\mathcal{P}$ be the set of strictly positive Cauchy sequences, that is, $\mathcal{P} := \{\, r \in \mathcal{R} \ ;\ r \text{ is strictly positive} \,\}$. Define a relation $\le$ on R by

$$[r] \le [s] :\Longleftrightarrow s - r \in \mathcal{P} \cup \mathbf{c}_0\ . \tag{6.4}$$

6.9 Lemma *$(R, \leq)$ is an ordered ring which induces the natural order on $\mathbb{Q}$.*

Proof It is easy to see that (6.4) defines a relation on R, that is, the definition is independent of the choice of representative. It is also clear that the relation $\leq$ is reflexive, and one can readily show transitivity. To prove antisymmetry, let $[r] \leq [s]$ and $[s] \leq [r]$. Then $r - s$ must belong to $\mathbf{c}_0$, since otherwise both $r - s$ and $s - r$ would be strictly positive, which is not possible. Hence $[r]$ and $[s]$ coincide, and we have shown that $\leq$ is a partial order on R.

Let $r, s \in \mathcal{R}$, and suppose that neither $r - s$ nor $s - r$ is strictly positive. Then for each $N \in \mathbb{N}^\times$, there is some $n \geq N$ such that $|r_n - s_n| < 1/N$. Hence $r - s$ has a subsequence which converges to 0 in $\mathbb{Q}$. By Proposition 6.4, $r - s$ is itself a null sequence, that is, $r - s \in \mathbf{c}_0$. This implies that R is totally ordered by $\leq$.

We leave to the reader the simple proof that $\leq$ is compatible with the ring structure of R.

Finally, let $p, q \in \mathbb{Q}$ be such that $[\overline{p}] \leq [\overline{q}]$. Then either $p < q$ or $\overline{q - p}$ is a null sequence, which implies $p = q$. Thus the order in R induces the natural order on $\mathbb{Q}$. ∎

6.10 Proposition *R is a field.*

Proof Let $[r] \in R^\times$. We need to show that $[r]$ is invertible. We can suppose (why?) that r is in $\mathcal{P}$. Hence there are $n_0 \in \mathbb{N}$ and $M \in \mathbb{N}^\times$ such that $r_n \geq 1/M$ for all $n \geq n_0$. Thus $s := (s_n)$, defined by

$$s_n := \begin{cases} 0 , & n < n_0 , \\ 1/r_n , & n \geq n_0 , \end{cases}$$

is an element of $\mathbb{Q}^{\mathbb{N}}$. Since r is a Cauchy sequence, for $N \in \mathbb{N}^\times$, there is some $n_1 \geq n_0$ such that $|r_n - r_m| < 1/(NM^2)$ for all $m, n \geq n_1$. This implies

$$|s_n - s_m| = \left|\frac{r_n - r_m}{r_n r_m}\right| \leq M^2 \, |r_n - r_m| < 1/N , \qquad m, n \geq n_1 .$$

Thus s is in $\mathcal{R}$. Since $[r]\,[s] = [rs] = \overline{1}$, $[r]$ is invertible with $[r]^{-1} = [s]$. ∎

We now want to show that R is order complete. To do so, we need first the following two lemmas:

6.11 Lemma *Every increasing sequence in $\mathbb{Q}$ which is bounded above is a Cauchy sequence. Similarly, every decreasing sequence in $\mathbb{Q}$ which is bounded below is a Cauchy sequence.*

Proof Let $r = (r_n)$ be an increasing sequence in $\mathbb{Q}$ with an upper bound $M \in \mathbb{N}^\times$, that is, $r_n < M$ for all $n \in \mathbb{N}$. We can suppose that $r_0 = 0$ (why?).

Let $N \in \mathbb{N}^\times$. Then not all of the sets

$$I_k := \{\, n \in \mathbb{N} \ ; \ (k-1)/N \leq r_n < k/N \,\} , \qquad k = 1, \ldots, MN ,$$

are empty. Hence

$$K := \max\{\, k \in \{1, \ldots, MN\} \ ; \ I_k \neq \emptyset \,\}$$

is well defined and the following hold:

$$r_n < K/N \ , \quad n \in \mathbb{N} \ , \qquad \exists\, n_0 \in \mathbb{N} : r_{n_0} \geq (K-1)/N \ .$$

From the monotonicity of the sequence (r_n) we get the inequalities

$$0 \leq r_n - r_m < \frac{K}{N} - \frac{K-1}{N} = \frac{1}{N} \ , \qquad n > m \geq n_0 \ ,$$

which proves that r is in $\mathcal{R}$. The proof for decreasing sequences is similar. ■

6.12 Lemma *Every increasing sequence* (ρ_k) *in* R *which is bounded above has a supremum* $\sup\{\,\rho_k \ ; \ k \in \mathbb{N}\,\}$. *Similarly, every decreasing sequence* (ρ_k) *in* R *which is bounded below has an infimum* $\inf\{\,\rho_k \ ; \ k \in \mathbb{N}\,\}$.

Proof It suffices to consider the case of increasing sequences. If there is some $m \in \mathbb{N}$ such that $\rho_k = \rho_m$ for all $k \geq m$, then $\sup\{\,\rho_k \ ; \ k \in \mathbb{N}\,\} = \rho_m$. Otherwise we can construct recursively a subsequence $(\rho_{k_j})_{j\in\mathbb{N}}$ of (ρ_k) such that $\rho_{k_j} < \rho_{k_{j+1}}$ for all $j \in \mathbb{N}$. Because of the monotonicity of the sequence (ρ_k), it suffices to prove the existence of $\sup\{\,\rho_{k_j} \ ; \ j \in \mathbb{N}\,\}$. Thus we suppose that $\rho_k < \rho_{k+1}$ for all $k \in \mathbb{N}$.

Each ρ_k has the form $[r^k]$ with $r^k = (r_n^k)_{n\in\mathbb{N}} \in \mathcal{R}$. For $k \in \mathbb{N}$ we have $\rho_{k+1} - \rho_k \in \mathcal{P}$ and so there are $n_k \in \mathbb{N}$ and $N_k \in \mathbb{N}^\times$ such that $r_n^{k+1} - r_n^k \geq 1/N_k$ for all $n \geq n_k$. Without loss of generality we can suppose that the sequence $(n_k)_{k\in\mathbb{N}}$ is increasing. Since r^k and r^{k+1} are Cauchy sequences, there are $m_k \geq n_k$ such that

$$r_n^k - r_{m_k}^k < \frac{1}{4N_k} \ , \quad r_{m_k}^{k+1} - r_n^{k+1} < \frac{1}{4N_k} \ , \qquad n \geq m_k \ .$$

Hence for $s_k := r_{m_k}^k + 1/(2N_k)$, we have

$$r_n^{k+1} - s_k > \frac{1}{4N_k} \ , \quad s_k - r_n^k > \frac{1}{4N_k} \ , \qquad n \geq m_k \ .$$

Consequently

$$\rho_k = [r^k] < [\overline{s_k}] = s_k[\overline{1}] < [r^{k+1}] = \rho_{k+1} \ , \qquad k \in \mathbb{N} \ . \tag{6.5}$$

Set $s := (s_k)$. By construction, s is an increasing sequence in $\mathbb{Q}$. Since the sequence (ρ_k) is bounded above, by (6.5), so is s. It follows from Lemma 6.11 that s is in $\mathcal{R}$, and then (6.5) shows that $\rho_k \leq [s]$ for all $k \in \mathbb{N}$.

Finally, let $\rho \in R$ with $\rho_k \leq \rho < [s]$ for all $k \in \mathbb{N}$. Then it follows from (6.5) that

$$s_k[\overline{1}] < \rho_{k+1} \leq \rho < [s] \ , \qquad k \in \mathbb{N} \ ,$$

which is a contradiction. Therefore we have $[s] = \sup\{\,\rho_k \ ; \ k \in \mathbb{N}\,\}$. ■

To finish Cantor's construction we can now easily prove that R is an order complete ordered extension field of $\mathbb{Q}$. Then the uniqueness statement of Theorem I.10.4 ensures that we have once again constructed the real numbers.

6.13 Theorem *R is an order complete ordered extension field of $\mathbb{Q}$.*

Proof Because of Lemma 6.9 and Proposition 6.10, we need to show only the order completeness of R.

Hence let A be a nonempty subset of R which is bounded above by $\gamma \in R$. We construct recursively an increasing sequence (α_j) and a decreasing sequence (β_j) as follows: Choose some $\alpha_0 \in A$, then set $\beta_0 := \gamma$ and $\gamma_0 := (\alpha_0 + \beta_0)/2$. If there is some $a \in A$ such that $a \geq \gamma_0$, then set $\alpha_1 := \gamma_0$ and $\beta_1 := \beta_0$, otherwise set $\alpha_1 := \alpha_0$ and $\beta_1 := \gamma_0$. In the next step we repeat the above procedure, replacing α_0 and β_0 by α_1 and β_1 to get α_2 and β_2. Iterating this process produces sequences (α_j) and (β_j) with the claimed properties, as well as

$$0 < \beta_j - \alpha_j \leq (\beta_0 - \alpha_0)/2^j \ , \qquad j \in \mathbb{N} \ . \tag{6.6}$$

Since (α_j) is bounded above by γ and (β_j) is bounded below by α_0, Lemma 6.12 implies that $\alpha := \sup\{\, \alpha_j \ ; \ j \in \mathbb{N} \,\}$ and $\beta := \inf\{\, \beta_j \ ; \ j \in \mathbb{N} \,\}$ exist. Moreover, taking the infimum of (6.6) yields

$$0 \leq \beta - \alpha \leq \inf\{\, (\beta_0 - \alpha_0)/2^j \ ; \ j \in \mathbb{N} \,\} = 0 \ .$$

Hence $\alpha = \beta$.

Finally, by construction, we have $a \leq \beta_j$ for all $a \in A$ and $j \in \mathbb{N}$. Hence

$$a \leq \inf\{\, \beta_j \ ; \ j \in \mathbb{N} \,\} = \beta = \alpha = \sup\{\, \alpha_j \ ; \ j \in \mathbb{N} \,\} \leq \gamma \ , \qquad a \in A \ .$$

Since this holds for every upper bound γ of A, it follows that $\alpha = \sup(A)$. ■

Exercises

1 Let $(\alpha, \beta) \in \mathbb{R}^2$. For $k \in \mathbb{N}$, set

$$x_k := \begin{cases} (\alpha, \beta) \ , & k \text{ even} \ , \\ (\beta, \alpha) \ , & k \text{ odd} \ , \end{cases}$$

and $s_n := \sum_{k=1}^{n} k^{-2} x_k$ for all $n \in \mathbb{N}^\times$. Show that (s_n) converges.

2 Let $X := (X, d)$ be a complete metric space and (x_n) a sequence in X. Suppose that

$$d(x_{n+1}, x_n) \leq \alpha d(x_n, x_{n-1}) \ , \qquad n \in \mathbb{N}^\times \ ,$$

for some $\alpha \in (0,1)$. Prove that (x_n) converges.

3 Show that every sequence in $\mathbb{R}$ has a monotone subsequence.

4 Prove the following (see Exercise 3.7):

(a) Every closed subset of a complete metric space is a complete metric space (with the induced metric).

(b) Every closed subspace of a Banach space is itself a Banach space (with the induced norm).

(c) ℓ_∞, c and c_0 are Banach spaces.

(d) Let M be a complete metric space and $D \subseteq M$ a subset which is complete (with respect to the induced metric). Then D is closed in M.

5 Verify that the order $\leq$ on $R = \mathcal{R}/\mathbf{c}_0$ is transitive and compatible with the ring structure of R.

6 For all $n \in \mathbb{N}^\times$, set $x_n := \sum_{k=1}^n k^{-1}$. Prove the following:

(a) The sequence (x_n) is not a Cauchy sequence in $\mathbb{R}$.

(b) For each $m \in \mathbb{N}^\times$, $\lim_n(x_{n+m} - x_n) = 0$.

(Hint: (a) Show that (x_n) is not bounded.)

7 Let $x_n := \sum_{k=1}^n k^{-2}$ for all $n \in \mathbb{N}^\times$. Prove or disprove that (x_n) is a Cauchy sequence in $\mathbb{Q}$.

7 Series

So far we have two ways to prove the convergence of a sequence (x_n) in a Banach space[1] $(E, |\cdot|)$. Either we make some guess about the limit x and then show directly that $|x - x_n|$ converges to zero, or we prove that (x_n) is a Cauchy sequence and then use the completeness of E.

We will use both of these techniques in the following two sections for the investigation of special sequences called *series*. We will see that the simple recursive structure of series leads to very convenient convergence criteria. In particular, we will discuss the root and ratio tests in arbitrary Banach spaces, and the Leibniz test for alternating real series.

Convergence of Series

Let (x_k) be a sequence in E. Then we define a new sequence (s_n) in E by

$$s_n := \sum_{k=0}^{n} x_k \ , \qquad n \in \mathbb{N} \ .$$

The sequence (s_n) is called a **series** in E and it is written $\sum x_k$ or $\sum_k x_k$. The element s_n is called the nth **partial sum** and x_k is called the kth **summand** of the series $\sum x_k$. Thus a series is simply a sequence whose terms are defined recursively by

$$s_0 := x_0 \ , \qquad s_{n+1} = s_n + x_{n+1} \ , \qquad n \in \mathbb{N} \ .$$

A series is the sequence of its partial sums.

The series $\sum x_k$ **converges** (or is **convergent**) if the sequence (s_n) of its partial sums converges. Then the limit of (s_n) is called the **value of the series** $\sum x_k$ and is written $\sum_{k=0}^{\infty} x_k$.[2] Finally, the series $\sum x_k$ **diverges** (or is **divergent**) if the sequence (s_n) of its partial sums diverges in E.

7.1 Examples **(a)** The series $\sum 1/k!$ converges in $\mathbb{R}$. By Example 4.2(g), it has the value e, that is, $e = \sum_{k=0}^{\infty} 1/k!$.

(b) The series $\sum 1/k^2$ converges in $\mathbb{R}$.

Proof Clearly the sequence (s_n) of partial sums is increasing. Since for each $n \in \mathbb{N}^\times$,

$$s_n = \sum_{k=1}^{n} \frac{1}{k^2} \leq 1 + \sum_{k=2}^{n} \frac{1}{k(k-1)} = 1 + \sum_{k=2}^{n} \Bigl(\frac{1}{(k-1)} - \frac{1}{k}\Bigr) = 1 + 1 - \frac{1}{n} < 2 \ ,$$

the sequence (s_n) is bounded. Thus the claim follows from Theorem 4.1. ■

---

[1] In the following, we often denote the norm of a Banach space E by $|\cdot|$ instead of $\|\cdot\|$. The attentive reader should have no trouble avoiding confusion with the Euclidean norm.

[2] It is occasionally useful to delete the first m terms of the series $\sum_k x_k$. This new series is written $\sum_{k \geq m} x_k$ or $(s_n)_{n \geq m}$. It often happens that x_0 is not defined (for example, if $x_k = 1/k$). In this case, $\sum x_k$ means $\sum_{k \geq 1} x_k$.

It is intuitively clear that a series can converge only if the summands form a null sequence. We prove this *necessary* criterion in the following proposition.

7.2 Proposition *If the series $\sum x_k$ converges, then (x_k) is a null sequence.*

Proof Let $\sum x_k$ be a convergent series. By Proposition 6.1, the sequence (s_n) of partial sums is a Cauchy sequence. Thus, for each $\varepsilon > 0$, there is some $N \in \mathbb{N}$ such that $|s_n - s_m| < \varepsilon$ for all $m, n \geq N$. In particular,

$$|s_{n+1} - s_n| = \Bigl|\sum_{k=0}^{n+1} x_k - \sum_{k=0}^{n} x_k\Bigr| = |x_{n+1}| < \varepsilon , \qquad n \geq N ,$$

that is, (x_n) is a null sequence. ■

Harmonic and Geometric Series

The following example shows that the converse of Proposition 7.2 is false.

7.3 Example The **harmonic series** $\sum 1/k$ diverges in $\mathbb{R}$.

Proof From the inequality

$$|s_{2n} - s_n| = \sum_{k=n+1}^{2n} \frac{1}{k} \geq \frac{n}{2n} = \frac{1}{2} , \qquad n \in \mathbb{N}^\times ,$$

it follows that (s_n) is not a Cauchy sequence. Thus, by Proposition 6.1, the sequence (s_n) diverges, meaning that the harmonic series diverges. ■

As a simple application of Proposition 7.2 we provide a complete description of the convergence behavior of the **geometric series** $\sum a^k$, $a \in \mathbb{K}$.

7.4 Example Let $a \in \mathbb{K}$. Then

$$\sum_{k=0}^{\infty} a^k = \frac{1}{1-a} , \qquad |a| < 1 .$$

For $|a| \geq 1$, the geometric series diverges.

Proof From Exercise I.8.1 we have

$$s_n = \sum_{k=0}^{n} a^k = \frac{1-a^{n+1}}{1-a} , \qquad n \in \mathbb{N} .$$

If $|a| < 1$, then it follows from Example 4.2(a) that (s_n) converges to $1/(1-a)$ as $n \to \infty$.

Otherwise, if $|a| \geq 1$, then $|a^k| = |a|^k \geq 1$, and the series $\sum_k a^k$ diverges by Proposition 7.2. ■

Calculating with Series

Series are special sequences and so all the rules that we have derived for convergent sequences apply also to series. In particular, the linearity of the limit function holds for series (see Section 2 and Remark 3.1(c)).

7.5 Proposition *Let $\sum a_k$ and $\sum b_k$ be convergent series in a normed vector space E and $\alpha \in \mathbb{K}$.*

(i) *The series $\sum(a_k + b_k)$ converges and*

$$\sum_{k=0}^{\infty}(a_k + b_k) = \sum_{k=0}^{\infty} a_k + \sum_{k=0}^{\infty} b_k \ .$$

(ii) *The series $\sum(\alpha a_k)$ converges and*

$$\sum_{k=0}^{\infty}(\alpha a_k) = \alpha \sum_{k=0}^{\infty} a_k \ .$$

Proof Set $s_n := \sum_{k=0}^{n} a_k$ and $t_n := \sum_{k=0}^{n} b_k$ for $n \in \mathbb{N}$. By assumption, there is some $s, t \in E$ such that $s_n \to s$ and $t_n \to t$. In view of the identities

$$s_n + t_n = \sum_{k=0}^{n}(a_k + b_k) \ , \quad \alpha s_n = \sum_{k=0}^{n}(\alpha a_k) \ ,$$

both claims follow from Proposition 2.2 and Remark 3.1(c). ∎

Convergence Tests

The fact that a sequence in a Banach space converges if and only if it is a Cauchy sequence takes the following form for series.

7.6 Theorem (Cauchy criterion) *For a series $\sum x_k$ in a Banach space $(E, |\cdot|)$, the following are equivalent:*

(i) *$\sum x_k$ converges.*

(ii) *For each $\varepsilon > 0$ there is some $N \in \mathbb{N}$ such that*

$$\Big|\sum_{k=n+1}^{m} x_k\Big| < \varepsilon \ , \qquad m > n \geq N \ .$$

Proof Clearly $s_m - s_n = \sum_{k=n+1}^{m} x_k$ for all $m > n$. Thus (s_n) is a Cauchy sequence in E if and only if (ii) is true. The claim then follows from the completeness of E. ∎

For real series with nonnegative summands we have the following simple convergence test:

7.7 Theorem *Let $\sum x_k$ be a series in $\mathbb{R}$ such that $x_k \geq 0$ for all $k \in \mathbb{N}$. Then $\sum x_k$ converges if and only if (s_n) is bounded. In this case, the series has the value $\sup_{n\in\mathbb{N}} s_n$.*

Proof Since the summands are nonnegative, the sequence (s_n) of partial sums is increasing. By Theorem 4.1, (s_n) converges if and only if (s_n) is bounded. The final claim comes from the same theorem. ■

If $\sum x_k$ is a series in $\mathbb{R}$ with nonnegative summands, we write $\sum x_k < \infty$ if the sequence of partial sums is bounded. With this notation, the first claim of Theorem 7.7 can be expressed as

$$\sum x_k < \infty \iff \sum x_k \text{ converges} .$$

Alternating Series

A series $\sum y_k$ in $\mathbb{R}$ is called **alternating** if y_k and y_{k+1} have opposite signs for all k. An alternating series can always be written in the form $\pm\sum(-1)^k x_k$ with $x_k \geq 0$.

7.8 Theorem (Leibniz criterion) *Let (x_k) be a decreasing null sequence with nonnegative terms. Then the alternating series $\sum(-1)^k x_k$ converges in $\mathbb{R}$.*

Proof Because of the inequality

$$s_{2n+2} - s_{2n} = -x_{2n+1} + x_{2n+2} \leq 0 , \qquad n \in \mathbb{N} ,$$

the sequence of partial sums with even indices $(s_{2n})_{n\in\mathbb{N}}$ is decreasing. Similarly,

$$s_{2n+3} - s_{2n+1} = x_{2n+2} - x_{2n+3} \geq 0 , \qquad n \in \mathbb{N} ,$$

and so $(s_{2n+1})_{n\in\mathbb{N}}$ is increasing. Moreover, $s_{2n+1} \leq s_{2n}$, and so

$$s_{2n+1} \leq s_0 \quad \text{and} \quad s_{2n} \geq 0 , \qquad n \in \mathbb{N} .$$

By Theorem 4.1, there are real numbers s and t such that $s_{2n} \to s$ and $s_{2n+1} \to t$ as $n \to \infty$. Our goal is to show that the sequence (s_n) of partial sums converges. We note first that

$$t - s = \lim_{n\to\infty} (s_{2n+1} - s_{2n}) = \lim_{n\to\infty} x_{2n+1} = 0 .$$

Hence, for each $\varepsilon > 0$, there are $N_1, N_2 \in \mathbb{N}$ such that

$$|s_{2n} - s| < \varepsilon , \quad 2n \geq N_1 , \qquad \text{and} \qquad |s_{2n+1} - s| < \varepsilon , \quad 2n+1 \geq N_2 .$$

Thus $|s_n - s| < \varepsilon$ for all $n \geq \max\{N_1, N_2\}$, which proves the claim. ■

7.9 Corollary *With the notation of Theorem 7.8 we have* $|s - s_n| \leq x_{n+1}$, $n \in \mathbb{N}$.

Proof In the proof of Theorem 7.8 we showed that

$$\inf_{n\in\mathbb{N}} s_{2n} = s = \sup_{n\in\mathbb{N}} s_{2n+1} \ .$$

This implies the inequalities

$$0 \leq s_{2n} - s \leq s_{2n} - s_{2n+1} = x_{2n+1} \ , \qquad n \in \mathbb{N} \ , \tag{7.1}$$

and

$$0 \leq s - s_{2n-1} \leq s_{2n} - s_{2n-1} = x_{2n} \ , \qquad n \in \mathbb{N} \ . \tag{7.2}$$

Combining (7.1) and (7.2) yields $|s - s_n| \leq x_{n+1}$. ■

Corollary 7.9 shows that the error made when the value of an alternating series is replaced by its nth partial sum, is at most the absolute value of the 'first omitted summand'. Moreover, (7.1) and (7.2) show that the nth partial sum is alternately less than and greater than the value of the series.

7.10 Examples By the Leibniz criterion, the alternating series

(a) $\displaystyle\sum_{k=1}^{\infty} \frac{(-1)^{k+1}}{k} = 1 - \frac{1}{2} + \frac{1}{3} - \frac{1}{4} + - \cdots$ **(alternating harmonic series)**

(b) $\displaystyle\sum_{k=0}^{\infty} \frac{(-1)^{k}}{2k+1} = 1 - \frac{1}{3} + \frac{1}{5} - \frac{1}{7} + - \cdots$

converge. Their values are $\log 2$ and $\pi/4$ respectively (see Application IV.3.9(d) and Exercise V.3.11). ■

Decimal, Binary and Other Representations of Real Numbers

What we have proved about series can be used to justify the representation of real numbers by decimal expansions. For example, the rational number

$$24 + \frac{1}{10^1} + \frac{3}{10^2} + \frac{0}{10^3} + \frac{7}{10^4} + \frac{1}{10^5}$$

has a unique decimal representation:

$$24.13071 := 2 \cdot 10^1 + 4 \cdot 10^0 + \frac{1}{10^1} + \frac{3}{10^2} + \frac{0}{10^3} + \frac{7}{10^4} + \frac{1}{10^5} \ .$$

We also want to make sense of 'infinite decimal expansions' such as

$$7.52341043\ldots$$

when an algorithm is specified which determines all further digits of the expansion. The following example shows that such representations need to be viewed with caution:

$$3.999\ldots = 3 + \sum_{k=1}^{\infty} \frac{9}{10^k} = 3 + \frac{9}{10} \sum_{k=0}^{\infty} 10^{-k} = 3 + \frac{9}{10} \cdot \frac{1}{1 - \frac{1}{10}} = 4 \ .$$

The choice of the number 10 as the 'basis' of the above representation may have some historical, cultural or practical justification, but it does not follow from any mathematical consideration. We can also consider, for example, *binary representations*, such as

$$101.10010\ldots = 1 \cdot 2^2 + 0 \cdot 2^1 + 1 \cdot 2^0 + 1 \cdot 2^{-1} + 0 \cdot 2^{-2} + 0 \cdot 2^{-3} + 1 \cdot 2^{-4} + 0 \cdot 2^{-5} + \cdots$$

In the following we make this preliminary discussion more precise. For a real number $x \in \mathbb{R}$, let $\lfloor x \rfloor := \max\{\, k \in \mathbb{Z} \ ;\ k \le x \,\}$ denote the largest integer less than or equal to x. It is a simple consequence of the well ordering principle I.5.5 that the **floor function**,

$$\lfloor \cdot \rfloor : \mathbb{R} \to \mathbb{Z} \ , \quad x \mapsto \lfloor x \rfloor \ ,$$

is well defined.

Fix some $g \in \mathbb{N}$ with $g \ge 2$. We call the g elements of the set $\{0, 1, \ldots, g-1\}$, the **base g digits**. Thus $\{0, 1\}$ is the set of binary (base 2) digits, $\{0, 1, 2\}$ is the set of ternary (base 3) digits, and $\{0, 1, 2, 3, 4, 5, 6, 7, 8, 9\}$ is the set of decimal (base 10) digits. For any sequence $(x_k)_{k \in \mathbb{N}^\times}$ of base g digits, that is, for $x_k \in \{0, 1, \ldots, g-1\}$, $k \in \mathbb{N}^\times$, we have the inequality

$$0 \le \sum_{k=1}^{n} x_k g^{-k} \le (g-1) \sum_{k=1}^{\infty} g^{-k} = 1 \ , \qquad n \in \mathbb{N}^\times \ .$$

By Theorem 7.7, the series $\sum x_k g^{-k}$ converges and its value x satisfies $0 \le x \le 1$. This series is called the **base g expansion** of the real number $x \in [0, 1]$. In the special cases $g = 2$, $g = 3$ and $g = 10$, this series is called the **binary expansion**, the **ternary expansion** and the **decimal expansion** of x respectively.

It is usual to write the base g expansion of the number $x \in [0, 1]$ in the form

$$0.x_1 x_2 x_3 x_4 \ldots := \sum_{k=1}^{\infty} x_k g^{-k},$$

assuming that the choice of g is clear. It is easy to see that any $m \in \mathbb{N}$ has a unique representation in the form[3]

$$m = \sum_{j=0}^{\ell} y_j g^j \ , \qquad y_k \in \{0, 1, \ldots, g-1\} \ , \quad 0 \le k \le \ell \ . \tag{7.3}$$

[3] See Exercise I.5.11. To get uniqueness we have to ignore leading zeros. For example, we consider $0\ 3^3 + 0\ 3^2 + 1\ 3^1 + 2\ 3^0$ and $1\ 3^1 + 2\ 3^0$ to be identical ternary representations of 5.

Then

$$x = m + \sum_{k=1}^{\infty} x_k g^{-k} = \sum_{j=0}^{\ell} y_j g^j + \sum_{k=1}^{\infty} x_k g^{-k}$$

is a nonnegative real number. The right hand side of this equation is called the base g expansion of x and is written

$$y_\ell y_{\ell-1} \cdots y_0 . x_1 x_2 x_3 \cdots$$

(if g is clear). Similarly,

$$-y_\ell y_{\ell-1} \cdots y_0 . x_1 x_2 x_3 \cdots$$

is called the base g expansion of $-x$. Finally, a base g expansion is called **periodic** if there are $\ell \in \mathbb{N}$ and $p \in \mathbb{N}^\times$ such that $x_{k+p} = x_k$ for all $k \geq \ell$.

7.11 Theorem *Suppose that $g \geq 2$. Then every real number x has a base g expansion. This expansion is unique if expansions satisfying $x_k = g - 1$ for almost all $k \in \mathbb{N}$ are excluded. Moreover, x is a rational number if and only if its base g expansion is periodic.*

Proof (a) It suffices to consider only the case $x \geq 0$. Then there is some $r \in [0, 1)$ such that $x = \lfloor x \rfloor + r$. Because of the above remarks, it suffices, in fact, to consider only the case that x is in the interval $[0, 1)$.

(b) In order to prove the existence of a base g expansion of $x \in [0, 1)$, we define a sequence $x_1, x_2, \ldots$ recursively by

$$x_1 := \lfloor gx \rfloor , \quad x_k := \Bigl\lfloor g^k \Bigl(x - \sum_{j=1}^{k-1} x_j g^{-j} \Bigr) \Bigr\rfloor , \qquad k \geq 2 . \tag{7.4}$$

Of course, by construction, $x_k \in \mathbb{N}$. We show that the $x_k \in \mathbb{N}$ are, in fact, base g digits, that is,

$$x_k \in \{0, 1, \ldots, g-1\} , \qquad k \in \mathbb{N}^\times . \tag{7.5}$$

We write first

$$\begin{aligned} g^k \Bigl(x - \sum_{j=1}^{k-1} x_j g^{-j} \Bigr) &= g^k x - x_1 g^{k-1} - x_2 g^{k-2} - \cdots - x_{k-2} g^2 - x_{k-1} g \\ &= g^{k-2} \bigl(g(gx - x_1) - x_2 \bigr) - \cdots - x_{k-2} g^2 - x_{k-1} g \\ &= g \Bigl(\cdots g \bigl(g(gx - x_1) - x_2 \bigr) - \cdots - x_{k-1} \Bigr) \end{aligned} \tag{7.6}$$

(see Remark I.8.14(f)). Set $r_0 := x$ and $r_k := g r_{k-1} - x_k$ for all $k \in \mathbb{N}^\times$. Then from (7.6) we get

$$g^k \Bigl(x - \sum_{j=1}^{k-1} x_j g^{-j} \Bigr) = g r_{k-1} , \qquad k \in \mathbb{N}^\times . \tag{7.7}$$

Thus $x_k = \lfloor g r_{k-1} \rfloor$ for all $k \in \mathbb{N}^\times$. Since

$$r_k = g r_{k-1} - x_k = g r_{k-1} - \lfloor g r_{k-1} \rfloor \in [0,1) , \qquad k \in \mathbb{N} ,$$

this proves that the $x_k \in \mathbb{N}$ are base g digits.

Our next goal is to show that the value of the series $\sum x_k g^{-k}$ is x. Indeed, from $x_k = \lfloor g r_{k-1} \rfloor$ and (7.7) it follows that

$$0 \leq x_k \leq g r_{k-1} = g^k \Bigl(x - \sum_{j=1}^{k-1} x_j g^{-j} \Bigr) , \qquad k \in \mathbb{N}^\times ,$$

and hence

$$x - \sum_{j=1}^{k-1} x_j g^{-j} \geq 0 , \qquad k \geq 2 . \tag{7.8}$$

On the other hand, we have $r_k = g^k \bigl(x - \sum_{j=1}^{k-1} x_j g^{-j} \bigr) - x_k < 1$, and so

$$x - \sum_{j=1}^{k-1} x_j g^{-j} < g^{-k}(1 + x_k) , \qquad k \geq 2 . \tag{7.9}$$

Combining (7.8) and (7.9), we have

$$0 \leq x - \sum_{j=1}^{k-1} x_j g^{-j} < g^{-k+1} , \qquad k \geq 2 .$$

Since $\lim_{k\to\infty} g^{-k+1} = 0$, this implies that $x = \sum_{k=1}^{\infty} x_k g^{-k}$.[4]

(c) To show uniqueness we suppose that there are $x_k, y_k \in \{0, 1, \ldots, g-1\}$, $k \in \mathbb{N}^\times$, and some $k_0 \in \mathbb{N}^\times$ such that

$$\sum_{k=1}^{\infty} x_k g^{-k} = \sum_{k=1}^{\infty} y_k g^{-k} ,$$

with $x_{k_0} \neq y_{k_0}$ and $x_k = y_k$ for $1 \leq k \leq k_0 - 1$. This implies

$$(x_{k_0} - y_{k_0}) g^{-k_0} = \sum_{k=k_0+1}^{\infty} (y_k - x_k) g^{-k} . \tag{7.10}$$

Without loss of generality, we can suppose that $x_{k_0} > y_{k_0}$ and so $1 \leq x_{k_0} - y_{k_0}$. Moreover for all x_k and y_k we have $y_k - x_k \leq g - 1$ and, since we have excluded

[4]One should also check that no series constructed by this algorithm satisfies the condition $x_k = g - 1$ for almost all k.

the case that almost all base g digits are equal to $g-1$, there is some $k_1 > k_0$ such that $y_{k_1} - x_{k_1} < g-1$. Thus, from (7.10), we get the inequalities

$$g^{-k_0} \le (x_{k_0} - y_{k_0}) g^{-k_0} < (g-1) \sum_{k=k_0+1}^{\infty} g^{-k} = g^{-k_0} ,$$

which are clearly impossible. Therefore we have proved the uniqueness claim.

(d) Let $\sum_{k=1}^{\infty} x_k g^{-k}$ be a periodic base g expansion of $x \in [0,1)$. Then there are $\ell \in \mathbb{N}$ and $p \in \mathbb{N}^{\times}$ such that $x_{k+p} = x_k$ for all $k \ge \ell$. It suffices to show that $x' := \sum_{k=\ell}^{\infty} x_k g^{-k}$ is a rational number. Set

$$x_0 := \sum_{k=\ell}^{\ell+p-1} x_k g^{-k} \in \mathbb{Q} .$$

Since $x_{k+p} = x_k$ for all $k \ge \ell$, we have

$$\begin{aligned} g^p x' - x' &= g^p x_0 + \sum_{k=\ell+p}^{\infty} x_k g^{-k+p} - \sum_{k=\ell}^{\infty} x_k g^{-k} \\ &= g^p x_0 + \sum_{k=\ell}^{\infty} x_{k+p} g^{-k} - \sum_{k=\ell}^{\infty} x_k g^{-k} = g^p x_0 . \end{aligned}$$

Thus $x' = g^p x_0 (g^p - 1)^{-1}$ is rational.

Now let $x \in [0,1)$ be a rational number, that is, $x = p/q$ for some positive natural numbers p and q with $p < q$. Let $\sum_{k=1}^{\infty} x_k g^{-k}$ be the base g expansion of x. Set $r_0 := x$ and $r_k := g r_{k-1} - x_k$ for all $k \in \mathbb{N}^{\times}$ as in (b).

We claim that for each $k \in \mathbb{N}$,

$$\text{there is some } s_k \in \{0,1,\ldots,q-1\} \text{ such that } r_k = s_k/q . \tag{7.11}$$

For $k=0$, the claim is true with $s_0 := p$. Suppose that (7.11) is true for some $k \in \mathbb{N}$, that is, $r_k = s_k/q$ with $0 \le s_k \le q-1$. Since $x_{k+1} = \lfloor g r_k \rfloor = \lfloor g s_k / q \rfloor$, there is some $s_{k+1} \in \{0,1,\ldots,q-1\}$ such that $g s_k = q x_{k+1} + s_{k+1}$, and so

$$r_{k+1} = g r_k - x_{k+1} = \frac{g s_k}{q} - x_{k+1} = \frac{s_{k+1}}{q} .$$

Consequently (7.11) is true for $k+1$, and by induction, for all k. Since, for s_k, only the q values $0,1,\ldots,q-1$ are available, there are some $k_0 \in \{1,\ldots,q-1\}$ and $j_0 \in \{k_0, k_0+1, \ldots, k_0+q\}$ such that $s_{j_0} = s_{k_0}$. Hence $r_{j_0+1} = r_{k_0}$, which implies $r_{j_0+i} = r_{k_0+i}$ for all $1 \le i \le j_0 - k_0$. Thus, from $x_{k+1} = \lfloor g r_k \rfloor$ for $k \in \mathbb{N}^{\times}$, it follows that the base g expansion of x is periodic. ■

The Uncountability of $\mathbb{R}$

With the help of Theorem 7.11 it is now easy to prove that $\mathbb{R}$ is uncountable.

7.12 Theorem *The set of real numbers $\mathbb{R}$ is uncountable.*

Proof Suppose that $\mathbb{R}$ is countable. The subset $\{\, 1/n \ ;\ n \geq 2 \,\} \subseteq (0,1)$ is countably infinite, and so, by Example I.6.1(a) and Proposition I.6.7, the interval $(0,1)$ is also countably infinite. Hence $(0,1) = \{\, x_n \ ;\ n \in \mathbb{N} \,\}$ for some sequence $(x_n)_{n\in\mathbb{N}}$. By Theorem 7.11, each $x_n \in (0,1)$ has a unique ternary expansion of the form $x_n = 0.x_{n,1}x_{n,2}\ldots$, where, for infinitely many $k \in \mathbb{N}^\times$, $x_{n,k} \in \{0,1,2\}$ is not equal to 2. In particular, by Proposition I.6.7, the set

$$X := \{\, 0.x_{n,1}x_{n,2}\cdots \ ;\ x_{n,k} \neq 2,\ n \in \mathbb{N},\ k \in \mathbb{N}^\times \,\}$$

is countable. Since X is clearly equinumerous with $\{0,1\}^{\mathbb{N}}$, we have shown that $\{0,1\}^{\mathbb{N}}$ is countable. This contradicts Proposition I.6.11. ■

Exercises

1 Determine the values of the following series:

$$\text{(a)} \sum \frac{(-1)^k}{2^k}\ , \quad \text{(b)} \sum \frac{1}{4k^2-1}\ .$$

2 Determine whether the following series converge or diverge:

$$\text{(a)} \sum \frac{\sqrt{k+1}-\sqrt{k}}{\sqrt{k}}\ , \quad \text{(b)} \sum (-1)^k\bigl(\sqrt{k+1}-\sqrt{k}\bigr)\ , \quad \text{(c)} \sum \frac{k!}{k^k}\ , \quad \text{(d)} \sum \frac{(k+1)^{k-1}}{(-k)^k}\ .$$

3 An infinitesimally small snail crawls with a constant speed of 5cm/hour along a 1 meter long rubber band. At the end of the first and all subsequent hours, the rubber band is stretched uniformly an extra meter. If the snail starts at the left end of the rubber band, will it reach the right end in a finite amount of time?

4 Let $\sum a_k$ be a convergent series in a Banach space E. Show that the sequence (r_n) with $r_n := \sum_{k=n}^{\infty} a_k$ is a null sequence.

5 Let (x_k) be a decreasing sequence such that $\sum x_k$ converges. Prove that (kx_k) is a null sequence.

6 Let (x_k) be a sequence in $[0,\infty)$. Prove that

$$\sum x_k < \infty \Leftrightarrow \sum \frac{x_k}{1+x_k} < \infty\ .$$

7 Let (d_k) be a sequence in $[0,\infty)$ such that $\sum_{k=0}^{\infty} d_k = \infty$.
(a) What can be said about the convergence of the following series?

$$\text{(i)} \sum \frac{d_k}{1+d_k}\ , \quad \text{(ii)} \sum \frac{d_k}{1+k^2 d_k}\ .$$

Is the hypothesis on the sequence (d_k) needed in both cases?

(b) Show by example that the series

$$\text{(i)} \sum \frac{d_k}{1+kd_k} , \quad \text{(ii)} \sum \frac{d_k}{1+d_k^2}$$

can both converge and diverge.

(Hint: (a) Consider separately the cases $\overline{\lim}\, d_k < \infty$ and $\overline{\lim}\, d_k = \infty$.)

8 Let $s := \sum_{k=1}^{\infty} k^{-2}$. Show that

$$1 - \frac{1}{2^2} - \frac{1}{4^2} + \frac{1}{5^2} + \frac{1}{7^2} - \frac{1}{8^2} - \frac{1}{10^2} + + - - \cdots = \frac{4}{9} s .$$

9 For $(j,k) \in \mathbb{N} \times \mathbb{N}$, let

$$x_{jk} := \begin{cases} 1/(j^2 - k^2) , & j \neq k , \\ 0 , & j = k . \end{cases}$$

For each $j \in \mathbb{N}^\times$, determine the value of the series $\sum_{k=0}^{\infty} x_{jk}$. (Hint: Factor x_{jk} suitably.)

10 The series $\sum c_k/k!$ is called a **Cantor series** if the coefficients c_k are integers such that $0 \leq c_{k+1} \leq k$ for all $k \in \mathbb{N}^\times$.

Prove the following:

(a) Every nonnegative real number x can be represented as the value of a Cantor series, that is, there is a Cantor series with $x = \sum_{k=1}^{\infty} c_k/k!$. This representation is unique if almost all of the c_k are not equal to $k-1$.

(b) Show that

$$\sum_{k=n+1}^{\infty} \frac{k-1}{k!} = \frac{1}{n!} , \qquad n \in \mathbb{N} .$$

(c) Let $x \in [0,1)$ be represented by the Cantor series $\sum c_k/k!$. Then x is rational[5] if and only if there is some $k_0 \in \mathbb{N}^\times$ such that $c_k = 0$ for all $k \geq k_0$.

11 Prove the **Cauchy condensation theorem**: If (x_k) is a decreasing sequence in $[0,\infty)$, then $\sum x_k$ converges if and only if $\sum 2^k x_{2^k}$ converges.

12 Let $s \geq 0$ be rational. Show that the series $\sum_k k^{-s}$ converges if and only if $s > 1$. (Hint: Exercise 11 and Example 7.4.)

13 Prove the claim of (7.3).

14 Let

$$x_n := \begin{cases} n^{-1} , & n \text{ odd} , \\ -n^{-2} , & n \text{ even} . \end{cases}$$

Show that $\sum x_n$ diverges. Why does the Leibniz criterion not apply to this series?

---

[5] Compare Exercise 4.7(b).

15 Let (z_n) be a sequence in $(0, \infty)$ with $\underline{\lim}\, z_n = 0$. Show that there are null sequences (x_n) and (y_n) in $(0, \infty)$ such that

(a) $\sum x_n < \infty$ and $\overline{\lim}\, x_n/z_n = \infty$.

(b) $\sum y_n = \infty$ and $\underline{\lim}\, y_n/z_n = 0$.

In particular, for any slowly converging null sequence (z_n) there is a null sequence (x_n) which converges quickly enough so that $\sum x_n < \infty$, but, even so, has a subsequence (x_{n_k}) which converges more slowly to zero than the corresponding subsequence (z_{n_k}) of (z_n). And, for any quickly converging null sequence (z_n) there is a null sequence (y_n) which converges slowly enough so that $\sum y_n = \infty$, but, even so, has a subsequence (y_{n_k}) which converges more quickly to zero than the corresponding subsequence (z_{n_k}) of (z_n).

(Hint: Let (z_n) be a sequence in $(0, \infty)$ such that $\underline{\lim}\, z_n = 0$.
(a) For each $k \in \mathbb{N}^\times$ choose some $n_k \in \mathbb{N}$ such that $z_{n_k} < k^{-3}$. Now set $x_{n_k} = k^{-2}$ for all $k \in \mathbb{N}$, and $x_n = n^{-2}$ otherwise.
(b) Choose a subsequence (z_{n_k}) with $\lim_k z_{n_k} = 0$. Set $y_{n_k} = z_{n_k}^2$ for all $k \in \mathbb{N}$, and $y_n = 1/n$ otherwise.)

8 Absolute Convergence

Since series are a special type of sequences, the rules which we have derived for general sequences apply also to series. But because the summands of a series belong to some underlying normed vector space, we can derive other rules which make use of this fact. For example, for a given series $\sum x_n$, we can investigate the series $\sum |x_n|$. Even though the convergence of a sequence (y_n) implies the convergence of the sequence of its norms $(|y_n|)$, the convergence of a series $\sum x_n$ does not imply the convergence of $\sum |x_n|$. This is seen, for example, in the different convergence behaviors of the alternating harmonic series, $\sum (-1)^{k+1}/k$, and the harmonic series, $\sum 1/k$.

Moreover, we should not expect that the associative law holds for 'infinitely many' additions:

$$1 = 1 + (-1+1) + (-1+1) + \cdots = (1-1) + (1-1) + (1-1) + \cdots = 0 .$$

This situation is considerably improved if we restrict our attention to convergent series in $\mathbb{R}$ with positive summands, or, more generally, to series with the property that the series of the absolute values (norms) of its summands converges.

In this section $\sum x_k$ is a series in a Banach space $E := (E, |\cdot|)$.

The series $\sum x_k$ **converges absolutely** or is **absolutely convergent** if $\sum |x_k|$ converges in $\mathbb{R}$, that is, $\sum |x_k| < \infty$.

The next proposition justifies the word 'convergent' in this definition.

8.1 Proposition *Every absolutely convergent series converges.*

Proof Let $\sum x_k$ be an absolutely convergent series in E. Then $\sum |x_k|$ converges in $\mathbb{R}$. By Theorem 7.6, $\sum |x_k|$ satisfies the Cauchy criterion, that is, for all $\varepsilon > 0$ there is some N such that

$$\sum_{k=n+1}^{m} |x_k| < \varepsilon , \qquad m > n \geq N .$$

Since

$$\Bigl|\sum_{k=n+1}^{m} x_k\Bigr| \leq \sum_{k=n+1}^{m} |x_k| < \varepsilon , \qquad m > n \geq N , \tag{8.1}$$

the series $\sum x_k$ also satisfies the Cauchy criterion. It follows from Theorem 7.6 that $\sum x_k$ converges. ■

8.2 Remarks **(a)** The alternating harmonic series $\sum(-1)^{k+1}/k$ shows that the converse of Proposition 8.1 is false. This series converges (see Example 7.10(a)), whereas the corresponding series of the absolute values, that is, the harmonic series $\sum k^{-1}$, diverges (see Example 7.3).

(b) The series $\sum x_k$ is called **conditionally convergent** if $\sum x_k$ converges but $\sum |x_k|$ does not. The alternating harmonic series is a conditionally convergent series.

(c) For every absolutely convergent series $\sum x_k$ we have the 'generalized triangle inequality',

$$\Bigl|\sum_{k=0}^{\infty} x_k\Bigr| \le \sum_{k=0}^{\infty} |x_k| \ .$$

Proof The triangle inequality implies

$$\Bigl|\sum_{k=0}^{n} x_k\Bigr| \le \sum_{k=0}^{n} |x_k| \ , \qquad n \in \mathbb{N} \ .$$

The claim now follows from Propositions 2.7, 2.10 and 5.3 (see also Remark 3.1(c)).

Majorant, Root and Ratio Tests

Absolute convergence plays a particularly significant role in the study of series. Because of this, the *majorant criterion* is of key importance, since it provides an easy and flexible means to show the absolute convergence of a series.

Let $\sum x_k$ be a series in E and $\sum a_k$ a series in $\mathbb{R}^+$. Then the series $\sum a_k$ is called a **majorant** (or **minorant**[1]) for $\sum x_k$ if there is some $K \in \mathbb{N}$ such that $|x_k| \le a_k$ (or $a_k \le |x_k|$) for all $k \ge K$.

8.3 Theorem (majorant criterion) *If a series in a Banach space has a convergent majorant, then it converges absolutely.*

Proof Let $\sum x_k$ be a series in E and $\sum a_k$ a convergent majorant. Then there is some K such that $|x_k| \le a_k$ for all $k \ge K$. By Theorem 7.6, for $\varepsilon > 0$, there is some $N \ge K$ such that $\sum_{k=n+1}^{m} a_k < \varepsilon$ for all $m > n \ge N$. Since $\sum a_k$ is a majorant for $\sum x_k$, we have

$$\sum_{k=n+1}^{m} |x_k| \le \sum_{k=n+1}^{m} a_k < \varepsilon \ , \qquad m > n \ge N \ .$$

Since the series $\sum |x_k|$ satisfies the Cauchy criterion, $\sum |x_k|$ converges. This means that the series $\sum x_k$ converges absolutely. ∎

[1]Note that, by definition, a minorant has nonnegative terms.

8.4 Examples (a) For $m \geq 2$, $\sum_k k^{-m}$ converges in $\mathbb{R}$.

Proof Because $m \geq 2$ we have $k^{-m} \leq k^{-2}$ for all $k \in \mathbb{N}^\times$. Example 7.1(b) shows that $\sum k^{-2}$ is a convergent majorant for $\sum k^{-m}$. ■

(b) For any $z \in \mathbb{C}$ such that $|z| < 1$, the series $\sum z^k$ converges absolutely.

Proof We have $|z^k| = |z|^k$ for all $k \in \mathbb{N}$. Because of $|z| < 1$ and Example 7.4, the geometric series $\sum |z|^k$ is a convergent majorant for $\sum z^k$. ■

Using the majorant criterion we can derive other important tests for the convergence of series. We start with the *root test*, a sufficient condition for the absolute convergence of series in an *arbitrary* Banach space.

8.5 Theorem (root test) *Let $\sum x_k$ be a series in E and*

$$\alpha := \overline{\lim} \sqrt[k]{|x_k|} \ .$$

Then the following hold:

$\sum x_k$ converges absolutely if $\alpha < 1$.

$\sum x_k$ diverges if $\alpha > 1$.

For $\alpha = 1$, both convergence and divergence of $\sum x_k$ are possible.

Proof (a) If $\alpha < 1$, then the interval $(\alpha, 1)$ is not empty and we can choose some $q \in (\alpha, 1)$. By Theorem 5.5, α is the greatest cluster point of the sequence $\bigl(\sqrt[k]{|x_k|}\bigr)$. Hence there is some K such that $\sqrt[k]{|x_k|} < q$ for all $k \geq K$, that is, for all $k \geq K$, we have $|x_k| < q^k$. Therefore the geometric series $\sum q^k$ is a convergent majorant for $\sum x_k$, and the claim follows from Theorem 8.3.

(b) If $\alpha > 1$, then, by Theorem 5.5 again, there are infinitely many $k \in \mathbb{N}$ such that $\sqrt[k]{|x_k|} \geq 1$. Thus $|x_k| \geq 1$ for infinitely many $k \in \mathbb{N}$. In particular, (x_k) is not a null sequence and the series $\sum x_k$ diverges by Proposition 7.2.

(c) To prove the claim for the case $\alpha = 1$ it suffices to provide a conditionally convergent series in $E = \mathbb{R}$ such that $\alpha = 1$. For the alternating harmonic series, $x_k := (-1)^{k+1}/k$, we have, by Example 4.2(d),

$$\sqrt[k]{|x_k|} = \sqrt[k]{\frac{1}{k}} = \frac{1}{\sqrt[k]{k}} \to 1 \quad (k \to \infty) \ .$$

Thus $\alpha = \overline{\lim} \sqrt[k]{|x_k|} = 1$ follows from Theorem 5.7. ■

The essential idea in this proof is the use of a geometric series as a convergent majorant. This suggests a further useful convergence condition, the *ratio test*.

8.6 Theorem (ratio test) *Let $\sum x_k$ be a series in E and K_0 be such that $x_k \neq 0$ for all $k \geq K_0$. Then the following hold:*

(i) *If there are $q \in (0,1)$ and $K \geq K_0$ such that*

$$\frac{|x_{k+1}|}{|x_k|} \leq q \ , \qquad k \geq K \ ,$$

then the series $\sum x_k$ converges absolutely.

(ii) *If there is some $K \geq K_0$ such that*

$$\frac{|x_{k+1}|}{|x_k|} \geq 1 \ , \qquad k \geq K \ ,$$

then the series $\sum x_k$ diverges.

Proof (i) By hypothesis we have $|x_{k+1}| \leq q\,|x_k|$ for all $k \geq K$. A simple induction argument yields the inequality

$$|x_k| \leq q^{k-K}\,|x_K| = \frac{|x_K|}{q^K} q^k \ , \qquad k > K \ .$$

Set $c := |x_K|/q^K$. Then $c\sum q^k$ is a convergent majorant for the series $\sum x_k$, and the claim follows from Theorem 8.3.

(ii) The hypothesis implies that (x_k) is not a null sequence. By Proposition 7.2, the series $\sum x_k$ must be divergent. ■

8.7 Examples **(a)** $\sum k^2 2^{-k} < \infty$ since, from $x_k := k^2 2^{-k}$, we get

$$\frac{|x_{k+1}|}{|x_k|} = \frac{(k+1)^2}{2^{k+1}} \cdot \frac{2^k}{k^2} = \frac{1}{2}\Bigl(1+\frac{1}{k}\Bigr)^2 \to \frac{1}{2} \ \ (k \to \infty) \ .$$

Thus there is some K with $|x_{k+1}|/|x_k| \leq 3/4$ for all $k \geq K$. The claimed convergence then follows from the ratio test.

(b) Consider the series

$$\sum \Bigl(\frac{1}{2}\Bigr)^{k+(-1)^k} = \frac{1}{2} + 1 + \frac{1}{8} + \frac{1}{4} + \frac{1}{32} + \frac{1}{16} + \cdots$$

with summands $x_k := \bigl(\frac{1}{2}\bigr)^{k+(-1)^k}$ for all $k \in \mathbb{N}$. Then

$$\frac{|x_{k+1}|}{|x_k|} = \begin{cases} 2 \ , & k \text{ even} \ , \\ 1/8 \ , & k \text{ odd} \ , \end{cases}$$

and we recognize that neither hypothesis of Proposition 8.6 is satisfied.[2] Even so, the series converges since

$$\overline{\lim}\sqrt[k]{|x_k|} = \overline{\lim}\sqrt[k]{\left(\frac{1}{2}\right)^{k+(-1)^k}} = \frac{1}{2}\,\overline{\lim}\sqrt[k]{\left(\frac{1}{2}\right)^{(-1)^k}} = \frac{1}{2}\,,$$

as Example 4.2(e) shows.

(c) For each $z \in \mathbb{C}$, the series $\sum z^k/k!$ converges absolutely.[3]

Proof Let $z \in \mathbb{C}^\times$. With $x_k := z^k/k!$ for all $k \in \mathbb{N}$, we have

$$\frac{|x_{k+1}|}{|x_k|} = \frac{|z|}{k+1} \le \frac{1}{2}\,, \qquad k \ge 2\,|z|\;,$$

and so the claim follows from Theorem 8.6. ■

The Exponential Function

Because of the previous example, we can define a function, exp, by

$$\exp : \mathbb{C} \to \mathbb{C}\,, \qquad z \mapsto \sum_{k=0}^{\infty} \frac{z^k}{k!}\;.$$

This is called the **exponential function**, and the series $\sum z^k/k!$ is called the **exponential series**. The exponential function is extremely important in all of mathematics and we make a thorough study of its properties in the following. We already notice that the exponential function of a real number is a real number, that is, $\exp(\mathbb{R}) \subseteq \mathbb{R}$. For the restriction of the exponential function to $\mathbb{R}$ we use again the symbol exp.

Rearrangements of Series

Let $\sigma : \mathbb{N} \to \mathbb{N}$ be a permutation. Then the series $\sum_k x_{\sigma(k)}$ is called a **rearrangement** of $\sum x_k$. The summands of the rearrangement $\sum_k x_{\sigma(k)}$ are the same as those of the original series, but they occur in different order. If σ is a permutation of $\mathbb{N}$ with $\sigma(k) = k$ for almost all $k \in \mathbb{N}$, then $\sum x_k$ and $\sum_k x_{\sigma(k)}$ have the same convergence behavior, and their values are equal if the series converge. For a permutation $\sigma : \mathbb{N} \to \mathbb{N}$ with $\sigma(k) \ne k$ for infinitely many $k \in \mathbb{N}$, this may not be true, as the following example demonstrates:

---

[2]For practical reasons it is advisable to try the ratio test first. If this test fails, it is still possible that the root test may determine the convergence behavior of the series (see Exercise 5.4).

[3]This, together with Proposition 7.2, provides a further proof of the claim of Example 4.2(c).

8.8 Example Let $x_k := (-1)^{k+1}/k$, and let $\sigma : \mathbb{N}^\times \to \mathbb{N}^\times$ be defined by $\sigma(1) := 1$, $\sigma(2) := 2$ and

$$\sigma(k) := \begin{cases} k + k/3 \ , & \text{if } 3 \mid k \ , \\ k - (k-1)/3 \ , & \text{if } 3 \mid (k-1) \ , \\ k + (k-2)/3 \ , & \text{if } 3 \mid (k-2) \ , \end{cases}$$

for all $k \geq 3$. It is easy to check that σ is a permutation of $\mathbb{N}^\times$ and so

$$\sum x_{\sigma(k)} = 1 - \frac{1}{2} - \frac{1}{4} + \frac{1}{3} - \frac{1}{6} - \frac{1}{8} + \frac{1}{5} - \frac{1}{10} - \frac{1}{12} + - - \cdots$$

is a rearrangement of the alternating harmonic series

$$\sum x_k = 1 - \frac{1}{2} + \frac{1}{3} - \frac{1}{4} + - \cdots \ .$$

We will show that this rearrangement converges. Denote the nth partial sums of $\sum x_k$ and $\sum_k x_{\sigma(k)}$ by s_n and t_n respectively. Let $s = \lim s_n$, the value of $\sum x_k$. Since

$$\sigma(3n) = 4n \ , \quad \sigma(3n-1) = 4n-2 \ , \quad \sigma(3n-2) = 2n-1 \ , \qquad n \in \mathbb{N}^\times \ ,$$

we have

$$\begin{aligned} t_{3n} &= 1 - \frac{1}{2} - \frac{1}{4} + \frac{1}{3} - \frac{1}{6} - \frac{1}{8} + - - \cdots + \frac{1}{2n-1} - \frac{1}{4n-2} - \frac{1}{4n} \\ &= \Bigl(1 - \frac{1}{2} - \frac{1}{4}\Bigr) + \Bigl(\frac{1}{3} - \frac{1}{6} - \frac{1}{8}\Bigr) + \cdots + \Bigl(\frac{1}{2n-1} - \frac{1}{4n-2} - \frac{1}{4n}\Bigr) \\ &= \Bigl(\frac{1}{2} - \frac{1}{4}\Bigr) + \Bigl(\frac{1}{6} - \frac{1}{8}\Bigr) + \cdots + \Bigl(\frac{1}{4n-2} - \frac{1}{4n}\Bigr) \\ &= \frac{1}{2}\Bigl(1 - \frac{1}{2} + \frac{1}{3} - \frac{1}{4} + - \cdots + \frac{1}{2n-1} - \frac{1}{2n}\Bigr) \\ &= \frac{1}{2} s_n \ . \end{aligned}$$

Thus the subsequence $(t_{3n})_{n \in \mathbb{N}^\times}$ of $(t_m)_{m \in \mathbb{N}^\times}$ converges to $s/2$. Since we also have

$$\lim_{n\to\infty} |t_{3n+1} - t_{3n}| = \lim_{n\to\infty} |t_{3n+2} - t_{3n}| = 0 \ ,$$

it follows that (t_m) is a Cauchy sequence. By Propositions 6.4 and 1.15, the sequence (t_m) converges to $s/2$, that is,

$$\sum_{k=1}^{\infty} x_{\sigma(k)} = 1 - \frac{1}{2} - \frac{1}{4} + \frac{1}{3} - \frac{1}{6} - \frac{1}{8} + \frac{1}{5} - \frac{1}{10} - \frac{1}{12} + - - \cdots = \frac{s}{2} \ .$$

Note that s is not zero since $|s - 1| = |s - s_1| \leq -x_2 = \frac{1}{2}$ by Corollary 7.9. ■

This example shows that addition is not commutative when there are 'infinitely many summands', that is, a convergent series cannot be arbitrarily rearranged without changing its value.[4] In contrast, the next proposition shows that the value of an absolutely convergent series is invariant under rearrangements.

8.9 Theorem (rearrangement theorem) *Every rearrangement of an absolutely convergent series $\sum x_k$ is absolutely convergent and has the same value as $\sum x_k$.*

Proof For each $\varepsilon > 0$, there is, by Theorem 7.6, some $N \in \mathbb{N}$ such that

$$\sum_{k=N+1}^{m} |x_k| < \varepsilon , \qquad m > N .$$

Taking the limit $m \to \infty$ yields the inequality $\sum_{k=N+1}^{\infty} |x_k| \le \varepsilon$.

Now let σ be a permutation of $\mathbb{N}$. For $M := \max\{\sigma^{-1}(0), \ldots, \sigma^{-1}(N)\}$ we have $\{\sigma(0), \ldots, \sigma(M)\} \supseteq \{0, \ldots, N\}$. Thus, for each $m \ge M$,

$$\Big|\sum_{k=0}^{m} x_{\sigma(k)} - \sum_{k=0}^{N} x_k\Big| \le \sum_{k=N+1}^{\infty} |x_k| \le \varepsilon \tag{8.2}$$

and also

$$\Big|\sum_{k=0}^{m} |x_{\sigma(k)}| - \sum_{k=0}^{N} |x_k|\Big| \le \varepsilon . \tag{8.3}$$

The inequality (8.3) implies the absolute convergence of $\sum x_{\sigma(k)}$. Taking the limit $m \to \infty$ in (8.2), and then using Proposition 2.10 and Remark 3.1(c), we see that

$$\Big|\sum_{k=0}^{\infty} x_{\sigma(k)} - \sum_{k=0}^{N} x_k\Big| \le \varepsilon ,$$

and so the values of the two series agree. ■

Double Series

As an application of the rearrangement theorem we consider *double series* $\sum x_{jk}$ in a Banach space E. Thus we have a function $x \colon \mathbb{N} \times \mathbb{N} \to E$ and, just as in Section 1, we abbreviate $x(j,k)$ by x_{jk}. The function x can be represented by a doubly infinite array

$$\begin{array}{ccccc} x_{00} & x_{01} & x_{02} & x_{03} & \cdots \\ x_{10} & x_{11} & x_{12} & x_{13} & \cdots \\ x_{20} & x_{21} & x_{22} & x_{23} & \cdots \\ x_{30} & x_{31} & x_{32} & x_{33} & \cdots \\ \vdots & \vdots & \vdots & \vdots & \ddots \end{array} \tag{8.4}$$

[4] See Exercise 4.

There are many ways that the entries in this array can be summed, that is, there are many different ways of ordering the entries so as to form a series. It is not at all clear under what conditions such series converge and to what extent the value of these series are independent of the choice of ordering.

By Proposition I.6.9, the set $\mathbb{N} \times \mathbb{N}$ is countable, that is, there is a bijection $\alpha \colon \mathbb{N} \to \mathbb{N} \times \mathbb{N}$. If α is such a bijection, we call the series $\sum_n x_{\alpha(n)}$ an **ordering** of the double series $\sum x_{jk}$. If we fix $j \in \mathbb{N}$ (or $k \in \mathbb{N}$), then the series $\sum_k x_{jk}$ (or $\sum_j x_{jk}$) is called the jth **row series** (or j^{th} **column series**) of $\sum x_{jk}$. If every row series (or column series) converges, then we can consider the **series of row sums** $\sum_j \bigl(\sum_{k=0}^{\infty} x_{jk}\bigr)$ (or the **series of column sums** $\sum_k \bigl(\sum_{j=0}^{\infty} x_{jk}\bigr)$). Finally we say that the double series $\sum x_{jk}$ is **summable**[5] if

$$\sup_{n \in \mathbb{N}} \sum_{j,k=0}^{n} |x_{jk}| < \infty .$$

8.10 Theorem (double series theorem) *Let $\sum x_{jk}$ be a summable double series.*

(i) *Every ordering $\sum_n x_{\alpha(n)}$ of $\sum x_{jk}$ converges absolutely to a value $s \in E$ which is independent of α.*

(ii) *The series of row sums $\sum_j \bigl(\sum_{k=0}^{\infty} x_{jk}\bigr)$ and column sums $\sum_k \bigl(\sum_{j=0}^{\infty} x_{jk}\bigr)$ converge absolutely, and*

$$\sum_{j=0}^{\infty} \Bigl(\sum_{k=0}^{\infty} x_{jk} \Bigr) = \sum_{k=0}^{\infty} \Bigl(\sum_{j=0}^{\infty} x_{jk} \Bigr) = s .$$

Proof (i) Set $M = \sup_{n \in \mathbb{N}} \sum_{j,k=0}^{n} |x_{jk}| < \infty$. Let $\alpha \colon \mathbb{N} \to \mathbb{N} \times \mathbb{N}$ be a bijection and $N \in \mathbb{N}$. Then there is some $K \in \mathbb{N}$ such that

$$\bigl\{\alpha(0), \ldots, \alpha(N)\bigr\} \subseteq \bigl\{(0,0), (1,0), \ldots, (K,0), \ldots, (0,K), \ldots, (K,K)\bigr\} . \tag{8.5}$$

Together with the summability of $\sum x_{jk}$ this implies

$$\sum_{n=0}^{N} |x_{\alpha(n)}| \le \sum_{j,k=0}^{K} |x_{jk}| \le M .$$

Hence $\sum_n x_{\alpha(n)}$ is absolutely convergent by Theorem 7.7.

[5]We have defined only the convergence of the row (and column) series and the convergence of an arbitrary ordering of a double series. For the double series $\sum x_{jk}$ itself, we have *no* definition of convergence. Note that the convergence of each row (or column) series must be proved before one can consider the series of row (or column) sums.

Now let $\beta\colon \mathbb{N} \to \mathbb{N} \times \mathbb{N}$ be another bijection. Then $\sigma := \alpha^{-1} \circ \beta$ is a permutation of $\mathbb{N}$. Set $y_m := x_{\alpha(m)}$ for all $m \in \mathbb{N}$. Then

$$y_{\sigma(n)} = x_{\alpha(\sigma(n))} = x_{\beta(n)} \;, \qquad n \in \mathbb{N} \;,$$

that is, $\sum_n x_{\beta(n)}$ is a rearrangement of $\sum_n x_{\alpha(n)}$. Since we already know that $\sum_n x_{\alpha(n)}$ converges absolutely, the remaining claim follows from Theorem 8.9.

(ii) Note first that the row series $\sum_{k=0}^{\infty} x_{jk}$, $j \in \mathbb{N}$, and the column series $\sum_{j=0}^{\infty} x_{jk}$, $k \in \mathbb{N}$, converge absolutely. Indeed, this follows directly from the summability of $\sum x_{jk}$ and Theorem 7.7. So the series of row sums $\sum_j \bigl(\sum_{k=0}^{\infty} x_{jk}\bigr)$ and the series of column sums $\sum_k \bigl(\sum_{j=0}^{\infty} x_{jk}\bigr)$ are well defined.

We next prove that these series converge absolutely. Consider the inequalities

$$\sum_{j=0}^{\ell}\Bigl|\sum_{k=0}^{m} x_{jk}\Bigr| \le \sum_{j=0}^{\ell}\sum_{k=0}^{m} |x_{jk}| \le \sum_{j,k=0}^{m} |x_{jk}| \le M \;, \qquad \ell \le m \;.$$

In the limit $m \to \infty$ we get $\sum_{j=0}^{\ell}\bigl|\sum_{k=0}^{\infty} x_{jk}\bigr| \le M$, $\ell \in \mathbb{N}$, which proves the absolute convergence of the series of row sums $\sum_j \bigl(\sum_{k=0}^{\infty} x_{jk}\bigr)$. A similar argument shows the absolute convergence of the series of column sums.

Now let $\alpha\colon \mathbb{N} \to \mathbb{N} \times \mathbb{N}$ be a bijection and $s := \sum_{n=0}^{\infty} x_{\alpha(n)}$. For any $\varepsilon > 0$, there is some $N \in \mathbb{N}$ such that $\sum_{n=N+1}^{\infty} |x_{\alpha(n)}| < \varepsilon/2$. Also there is some $K \in \mathbb{N}$ so that (8.5) holds. Hence we have

$$\Bigl|\sum_{j=0}^{\ell}\sum_{k=0}^{m} x_{jk} - \sum_{n=0}^{N} x_{\alpha(n)}\Bigr| \le \sum_{n=N+1}^{\infty} |x_{\alpha(n)}| < \varepsilon/2 \;, \qquad \ell, m \ge K \;.$$

Taking the limits $m \to \infty$ and $\ell \to \infty$, we get

$$\Bigl|\sum_{j=0}^{\infty}\Bigl(\sum_{k=0}^{\infty} x_{jk}\Bigr) - \sum_{n=0}^{N} x_{\alpha(n)}\Bigr| \le \varepsilon/2 \;.$$

Applying the triangle inequality to

$$\Bigl|s - \sum_{n=0}^{N} x_{\alpha(n)}\Bigr| \le \sum_{n=N+1}^{\infty} |x_{\alpha(n)}| < \varepsilon/2$$

yields

$$\Bigl|\sum_{j=0}^{\infty}\Bigl(\sum_{k=0}^{\infty} x_{jk}\Bigr) - s\Bigr| \le \varepsilon \;.$$

Since this holds for each $\varepsilon > 0$, the series of row sums has the value s. A similar argument shows that the value of $\sum_k \bigl(\sum_{j=0}^{\infty} x_{jk}\bigr)$ is also s. ■

Cauchy Products

Double series appear naturally when one forms the product of two series. If $\sum x_j$ and $\sum y_k$ are series in $\mathbb{K}$, then multiplying the summands together we get the following doubly infinite array:

$$\begin{array}{ccccc} x_0y_0 & x_0y_1 & x_0y_2 & x_0y_3 & \cdots \\ x_1y_0 & x_1y_1 & x_1y_2 & x_1y_3 & \cdots \\ x_2y_0 & x_2y_1 & x_2y_2 & x_2y_3 & \cdots \\ x_3y_0 & x_3y_1 & x_3y_2 & x_3y_3 & \cdots \\ \vdots & \vdots & \vdots & \vdots & \vdots\vdots\vdots \end{array} \tag{8.6}$$

If $\sum x_j$ and $\sum y_k$ both converge, then the series of row sums is $\sum_j x_j \cdot \sum_{k=0}^{\infty} y_k$ and the series of column sums is $\sum_k y_k \cdot \sum_{j=0}^{\infty} x_j$. Set $x_{jk} := x_j y_k$ for all $(j,k) \in \mathbb{N} \times \mathbb{N}$. Let $\delta : \mathbb{N} \to \mathbb{N} \times \mathbb{N}$ be the bijection from (I.6.3), so that, with the nth diagonal sum defined by

$$z_n := \sum_{k=0}^{n} x_k y_{n-k} \ , \qquad n \in \mathbb{N} \ , \tag{8.7}$$

we have

$$\sum_j x_{\delta(j)} = \sum_n z_n = \sum_n \Bigl(\sum_{k=0}^{n} x_k y_{n-k} \Bigr) \ .$$

This particular ordering $\sum_n x_{\delta(n)}$ is called the **Cauchy product** of the series $\sum x_j$ and $\sum y_k$ (compare (8.8) in Section I.8).

In order to make use of the Cauchy product of $\sum x_j$ and $\sum y_k$ it is necessary that the double series $\sum x_j y_k$ be summable. A simple sufficient criterion for this is the absolute convergence of $\sum x_j$ and $\sum y_k$.

8.11 Theorem (Cauchy product of series) *Suppose that the series $\sum x_j$ and $\sum y_k$ in $\mathbb{K}$ converge absolutely. Then the Cauchy product $\sum_n \sum_{k=0}^{n} x_k y_{n-k}$ of $\sum x_j$ and $\sum y_k$ converges absolutely, and*

$$\Bigl(\sum_{j=0}^{\infty} x_j \Bigr) \Bigl(\sum_{k=0}^{\infty} y_k \Bigr) = \sum_{n=0}^{\infty} \sum_{k=0}^{n} x_k y_{n-k} \ .$$

Proof Setting $x_{jk} := x_j y_k$ for all $(j,k) \in \mathbb{N} \times \mathbb{N}$, we have

$$\sum_{j,k=0}^{n} |x_{jk}| = \sum_{j=0}^{n} |x_j| \cdot \sum_{k=0}^{n} |y_k| \le \sum_{j=0}^{\infty} |x_j| \cdot \sum_{k=0}^{\infty} |y_k| \ , \qquad n \in \mathbb{N} \ .$$

Hence, because of the absolute convergence of $\sum x_j$ and $\sum y_k$, the double series $\sum x_{jk}$ is summable. The claims now follow from Theorem 8.10. ■

8.12 Examples **(a)** For the exponential function we have

$$\exp(x) \cdot \exp(y) = \exp(x+y) , \qquad x, y \in \mathbb{C} . \tag{8.8}$$

Proof By Example 8.7(c), the series $\sum x^j/j!$ and $\sum y^k/k!$ are absolutely convergent, and so, Theorem 8.11 implies

$$\exp(x) \cdot \exp(y) = \Bigl(\sum_{j=0}^{\infty} \frac{x^j}{j!}\Bigr)\Bigl(\sum_{k=0}^{\infty} \frac{y^k}{k!}\Bigr) = \sum_{n=0}^{\infty} \Bigl(\sum_{k=0}^{n} \frac{x^k}{k!} \frac{y^{n-k}}{(n-k)!}\Bigr) . \tag{8.9}$$

From the binomial formula we get

$$\sum_{k=0}^{n} \frac{x^k}{k!} \frac{y^{n-k}}{(n-k)!} = \frac{1}{n!} \sum_{k=0}^{n} \frac{n!}{k!\,(n-k)!} x^k y^{n-k} = \frac{1}{n!} \sum_{k=0}^{n} \binom{n}{k} x^k y^{n-k} = \frac{1}{n!}(x+y)^n .$$

So, from (8.9), we get

$$\exp(x) \cdot \exp(y) = \sum_{n=0}^{\infty} \frac{(x+y)^n}{n!} = \exp(x+y)$$

as claimed. ■

(b) As an application of this property of the exponential function, we determine the values of the exponential function for rational arguments.[6] Namely,

$$\exp(r) = e^r , \qquad r \in \mathbb{Q} ,$$

that is, for a rational number r, $\exp(r)$ is the rth power of e.

Proof (i) From Example 7.1(a) we have $\exp(1) = \sum_{k=0}^{\infty} 1/k! = e$. Thus (8.8) implies

$$\exp(2) = \exp(1+1) = \exp(1) \cdot \exp(1) = \bigl[\exp(1)\bigr]^2 = e^2 .$$

A simple induction argument yields

$$\exp(k) = e^k , \qquad k \in \mathbb{N} .$$

(ii) For $k \in \mathbb{N}$, (8.8) implies that $\exp(-k) \cdot \exp(k) = \exp(0)$. Since $\exp(0) = 1$ we have

$$\exp(-k) = \bigl[\exp(k)\bigr]^{-1} , \qquad k \in \mathbb{N} .$$

Using (i) we then have (see Exercise I.9.1)

$$\exp(-k) = \frac{1}{\exp(k)} = \frac{1}{e^k} = (e^{-1})^k = e^{-k} , \qquad k \in \mathbb{N} ,$$

that is, $\exp(k) = e^k$ for all $k \in \mathbb{Z}$.

---

[6] In Section III.6 we prove a generalization of this statement.

(iii) For $q \in \mathbb{N}^\times$, (8.8) implies that

$$e = \exp(1) = \exp\Bigl(q \cdot \frac{1}{q}\Bigr) = \exp\Bigl(\underbrace{\frac{1}{q} + \cdots + \frac{1}{q}}_{q \text{ times}}\Bigr) = \Bigl[\exp\Bigl(\frac{1}{q}\Bigr)\Bigr]^q$$

and hence $\exp(1/q) = e^{1/q}$. Finally let $p \in \mathbb{N}$ and $q \in \mathbb{N}^\times$. Then, using Remark I.10.10(b), we get

$$\exp\Bigl(\frac{p}{q}\Bigr) = \exp\Bigl(\underbrace{\frac{1}{q} + \cdots + \frac{1}{q}}_{p \text{ times}}\Bigr) = \Bigl[\exp\Bigl(\frac{1}{q}\Bigr)\Bigr]^p = \bigl[e^{1/q}\bigr]^p = e^{p/q}$$

(see Exercise I.10.3). From (8.8) and $\exp(0) = 1$ it follows also that

$$\exp\Bigl(-\frac{p}{q}\Bigr) = \Bigl[\exp\Bigl(\frac{p}{q}\Bigr)\Bigr]^{-1} .$$

By what we have already proved and Exercise I.10.3, we obtain finally

$$\exp\Bigl(-\frac{p}{q}\Bigr) = \Bigl[\exp\Bigl(\frac{p}{q}\Bigr)\Bigr]^{-1} = \bigl[e^{p/q}\bigr]^{-1} = e^{-p/q} .$$

This completes the proof. ∎

(c) For conditionally convergent series, Theorem 8.11 is false in general.

Proof For the Cauchy product of the conditionally convergent series $\sum x_k$ and $\sum y_k$ defined by $x_k := y_k := (-1)^k/\sqrt{k+1}$ for all $k \in \mathbb{N}$ we have

$$z_n := \sum_{k=0}^{n} \frac{(-1)^k(-1)^{n-k}}{\sqrt{k+1}\,\sqrt{n-k+1}} = (-1)^n \sum_{k=0}^{n} \frac{1}{\sqrt{k(n-k)}} , \qquad n \in \mathbb{N}^\times .$$

From the inequality

$$(k+1)(n-k+1) \le (n+1)^2$$

for $0 \le k \le n$, we get

$$|z_n| = \sum_{k=0}^{n} \frac{1}{\sqrt{(k+1)(n-k+1)}} \ge \frac{n+1}{n+1} = 1 .$$

Thus, by Proposition 7.2, the series $\sum_{k=1}^\infty z_n$ cannot converge. ∎

(d) Consider the double series $\sum x_{jk}$ with

$$x_{jk} := \begin{cases} 1 , & j-k=1 , \\ -1 , & j-k=-1 , \\ 0 & \text{otherwise} , \end{cases}$$

represented by the doubly infinite array[7]

$$\begin{bmatrix} 0 & -1 & & & & & & & \\ 1 & 0 & -1 & & & & \text{\Large 0} & & \\ & 1 & 0 & -1 & & & & & \\ & & 1 & 0 & -1 & & & & \\ & & & 1 & 0 & -1 & & & \\ & & & & 1 & 0 & -1 & & \\ & & \text{\Large 0} & & & 1 & 0 & -1 & \\ & & & & & & 1 & 0 & \ddots \\ & & & & & & & 1 & \ddots \\ & & & & & & & & \ddots \end{bmatrix}$$

This double series is not summable and the values of the row and column series disagree:

$$\sum_j \Bigl(\sum_{k=0}^{\infty} x_{jk}\Bigr) = -1\,, \quad \sum_k \Bigl(\sum_{j=0}^{\infty} x_{jk}\Bigr) = 1\,.$$

The series $\sum_n x_{\delta(n)}$, where $\delta : \mathbb{N} \to \mathbb{N} \times \mathbb{N}$ denotes the bijection of (I.6.3), is divergent.

Exercises

1 Determine whether the following series converge or diverge:

$$\text{(a)} \sum \frac{k^4}{3^k}\,, \quad \text{(b)} \sum \frac{k}{(\sqrt[3]{k}+1)^k}\,, \quad \text{(c)} \sum \Bigl(1 - \frac{1}{k}\Bigr)^{k^2}\,,$$

$$\text{(d)} \sum \binom{2k}{k}^{-1}\,, \quad \text{(e)} \sum \binom{2k}{k} 2^{-k}\,, \quad \text{(f)} \sum \binom{2k}{k} 5^{-k}\,.$$

2 For what values of $a \in \mathbb{R}$ do the series

$$\sum \frac{a^{2k}}{(1+a^2)^{k-1}} \quad \text{and} \quad \sum \frac{1-a^{2k}}{1+a^{2k}}$$

converge?

3 Let $\sum x_k$ be a conditionally convergent series in $\mathbb{R}$. Show that the series[8] $\sum x_k^+$ and $\sum x_k^-$ diverge.

4 Prove **Riemann's rearrangement theorem**: If $\sum x_k$ is a conditionally convergent series in $\mathbb{R}$, then, for any $s \in \mathbb{R}$, there is a permutation σ of $\mathbb{N}$ such that $\sum_k x_{\sigma(k)} = s$. Further, there is a permutation τ of $\mathbb{N}$ such that $\sum_k x_{\tau(k)}$ diverges. (Hint: Use Exercise 3 and approximate $s \in \mathbb{R}$ above and below by suitable combinations of the partial sums of the series $\sum x_k^+$ and $-\sum x_k^-$.)

---

[7] The large zeros indicate that all entries which otherwise not specified are 0.

[8] For $x \in \mathbb{R}$, define $x^+ := \max\{x, 0\}$ and $x^- := \max\{-x, 0\}$.

5 For all $(j,k) \in \mathbb{N} \times \mathbb{N}$, let

$$x_{jk} := \begin{cases} (j^2 - k^2)^{-1} , & j \neq k , \\ 0 , & j = k . \end{cases}$$

Show that the double series $\sum x_{jk}$ is not summable. (Hint: Using Exercise 7.9, determine the values of the series of row sums and the series of column sums.)

6 Let

$$\ell_1 := \ell_1(\mathbb{K}) := \Bigl(\{ (x_k) \in s \ ; \ \textstyle\sum x_k \text{ is absolutely convergent} \}, \ \|\cdot\|_1 \Bigr)$$

where

$$\|(x_k)\|_1 := \sum_{k=0}^{\infty} |x_k| .$$

Prove the following:

(a) ℓ_1 is a Banach space.

(b) ℓ_1 is a proper subspace of ℓ_∞ with $\|\cdot\|_\infty \leq \|\cdot\|_1$.

(c) The norm induced on ℓ_1 from ℓ_∞ is not equivalent to the ℓ_1-norm. (Hint: Consider the sequence (ξ_j) with $\xi_j := (x_{j,k})_{k \in \mathbb{N}}$ where $x_{j,k} = 1$ for $k \leq j$, and $x_{j,k} = 0$ for $k > j$.)

7 Let $\sum x_n$, $\sum y_n$ and $\sum z_n$ be series in $(0, \infty)$ with $\sum y_n < \infty$ and $\sum z_n = \infty$. Prove the following:

(a) If there is some N such that

$$\frac{x_{n+1}}{x_n} \leq \frac{y_{n+1}}{y_n} , \qquad n \geq N ,$$

then $\sum x_n$ converges.

(b) If there is some N such that

$$\frac{x_{n+1}}{x_n} \geq \frac{z_{n+1}}{z_n} , \qquad n \geq N ,$$

then $\sum x_n$ diverges.

8 Determine whether the following series converge or diverge:

$$\sum \frac{(-1)^{n+1}}{3n + (-1)^n n} , \quad \sum \frac{(-1)^{n+1}}{3n + 6(-1)^n} .$$

9 Let $a, b > 0$ with $a - b = 1$. Show that the Cauchy product of the series[9]

$$a + \sum_{n \geq 1} a^n \quad \text{and} \quad -b + \sum_{n \geq 1} b^n$$

converges absolutely. In particular, the Cauchy product of

$$2 + 2 + 2^2 + 2^3 + \cdots \quad \text{and} \quad -1 + 1 + 1 + \cdots$$

converges absolutely.

[9]Note that the series $a + \sum a^n$ diverges.

10 Prove the following properties of the exponential function:

(a) $\exp(x) > 0, \quad x \in \mathbb{R}$.

(b) $\exp : \mathbb{R} \to \mathbb{R}$ is strictly increasing.

(c) For each $\varepsilon > 0$, there are $x < 0$ and $y > 0$ such that

$$\exp(x) < \varepsilon \quad \text{and} \quad \exp(y) > 1/\varepsilon \ .$$

(Hint: Consider Examples 8.12(a) and (b).)

9 Power Series

We investigate next the conditions under which formal power series can be considered to be well defined functions. As we have already seen in Remark I.8.14(e), for a power series which is *not* a polynomial, this is a question about the convergence of series.

Let

$$a := \sum a_k X^k := \sum_k a_k X^k \tag{9.1}$$

be a (formal) power series in one indeterminate with coefficients in $\mathbb{K}$. Then, for each $x \in \mathbb{K}$, $\sum a_k x^k$ is a series in $\mathbb{K}$. When this series converges we denote its value by $\underline{a}(x)$, the **value of the** (formal) **power series** (9.1) **at** x. Set

$$\mathrm{dom}(\underline{a}) := \left\{ x \in \mathbb{K} \; ; \; \textstyle\sum a_k x^k \text{ converges in } \mathbb{K} \right\} .$$

Then $\underline{a} \colon \mathrm{dom}(\underline{a}) \to \mathbb{K}$ is a well defined function:

$$\underline{a}(x) := \sum_{k=0}^{\infty} a_k x^k , \qquad x \in \mathrm{dom}(\underline{a}) . \tag{9.2}$$

Note that $0 \in \mathrm{dom}(\underline{a})$ for any $a \in \mathbb{K}[\![X]\!]$. The following examples show that each of the cases

$$\mathrm{dom}(\underline{a}) = \mathbb{K} , \quad \{0\} \subset \mathrm{dom}(\underline{a}) \subset \mathbb{K} , \quad \mathrm{dom}(\underline{a}) = \{0\}$$

is possible.

9.1 Examples (a) Let $a \in \mathbb{K}[X] \subseteq \mathbb{K}[\![X]\!]$, that is, $a_k = 0$ for almost all $k \in \mathbb{N}$. Then $\mathrm{dom}(\underline{a}) = \mathbb{K}$ and $\underline{a}$ coincides with the polynomial function introduced in Section I.8.

(b) The exponential series $\sum x^k/k!$ converges absolutely for each $x \in \mathbb{C}$. Thus, for the power series

$$a := \sum \frac{1}{k!} X^k \in \mathbb{C}[\![X]\!] ,$$

we have $\mathrm{dom}(\underline{a}) = \mathbb{C}$ and $\underline{a} = \exp$.

(c) By Example 7.4, the geometric series $\sum_k x^k$ converges absolutely to the value $1/(1-x)$ for each $x \in \mathbb{B}_{\mathbb{K}}$, and it diverges if x is not in $\mathbb{B}_{\mathbb{K}}$. Thus for the geometric series

$$a := \sum X^k \in \mathbb{K}[\![X]\!]$$

we have $\mathrm{dom}(\underline{a}) = \mathbb{B}_{\mathbb{K}}$ and $\underline{a}(x) = 1/(1-x)$ for all $x \in \mathrm{dom}(\underline{a})$.

(d) The series $\sum_k k!\,x^k$ diverges for all $x \in \mathbb{K}^\times$. Consequently, the domain of the function $\underline{a}$ represented by the power series $a := \sum k!\,X^k$ is $\{0\}$.

Proof For all $x \in \mathbb{K}^\times$ and $k \in \mathbb{N}$, let $x_k := k!\,x^k$. Then

$$\frac{|x_{k+1}|}{|x_k|} = (k+1)\,|x| \to \infty \quad (k \to \infty)\ .$$

Hence the series $\sum x_k = \sum k!\,x^k$ diverges by the ratio test. ∎

The Radius of Convergence

For power series, the convergence tests of the previous section can be put in a particularly useful form.

9.2 Theorem *For a power series $a = \sum a_k X^k$ with coefficients in $\mathbb{K}$ there is a unique $\rho := \rho_a \in [0,\infty]$ with the following properties:*

(i) *The series $\sum a_k x^k$ converges absolutely if $|x| < \rho$ and diverges if $|x| > \rho$.*

(ii) **Hadamard's formula** holds:

$$\rho_a = \frac{1}{\overline{\lim}_{k\to\infty} \sqrt[k]{|a_k|}}\ . \tag{9.3}$$

The number[1] $\rho_a \in [0,\infty]$ is called the **radius of convergence** *of a, and*

$$\rho_a \mathbb{B}_{\mathbb{K}} = \{\, x \in \mathbb{K}\ ;\ |x| < \rho_a \,\}$$

is the **disk of convergence** *of a.*

Proof Define ρ_a by (9.3). Then $\rho_a \in [0,\infty]$ and

$$\overline{\lim_{k\to\infty}} \sqrt[k]{|a_k x^k|} = |x| \overline{\lim_{k\to\infty}} \sqrt[k]{|a_k|} = |x|/\rho_a\ .$$

Then all claims follow from the root test. ∎

9.3 Corollary *For $a = \sum a_k X^k \in \mathbb{K}[\![X]\!]$, we have $\rho_a \mathbb{B}_{\mathbb{K}} \subseteq \mathrm{dom}(\underline{a}) \subseteq \rho_a \bar{\mathbb{B}}_{\mathbb{K}}$. In particular, the power series a represents the function $\underline{a}$ on its disk of convergence.*[2]

For some power series the ratio test can also be used to determine the radius of convergence.

---

[1] Of course in (9.3) we use the conventions of Section I.10 for the extended number line $\bar{\mathbb{R}}$.

[2] In Remark 9.6 we see that $\rho_a \mathbb{B}_{\mathbb{K}}$ is, in general, a proper subset of $\mathrm{dom}(\underline{a})$.

9.4 Proposition *Let $a = \sum a_k X^k$ be a power series such that $\lim|a_k/a_{k+1}|$ exists in $\bar{\mathbb{R}}$. Then the radius of convergence of a is given by the formula*

$$\rho_a = \lim_{k\to\infty}\left|\frac{a_k}{a_{k+1}}\right| .$$

Proof Since $\alpha := \lim|a_k/a_{k+1}|$ exists in $\bar{\mathbb{R}}$, we have

$$\left|\frac{a_{k+1}x^{k+1}}{a_k x^k}\right| = \left|\frac{a_{k+1}}{a_k}\right| |x| \to \frac{|x|}{\alpha} \quad (k\to\infty) . \tag{9.4}$$

Now if $x, y \in \mathbb{K}$ are such that $|x| < \alpha$ and $|y| > \alpha$, then (9.4) and the ratio test imply that the series $\sum a_k x^k$ converges absolutely and the series $\sum a_k y^k$ diverges. Hence, by Theorem 9.2, we have $\alpha = \rho_a$. ■

9.5 Examples **(a)** The radius of convergence of the exponential series $\sum (1/k!)X^k$ is ∞.

Proof Since

$$\left|\frac{a_k}{a_{k+1}}\right| = \left|\frac{1/k!}{1/(k+1)!}\right| = k+1 \to \infty \quad (k\to\infty) ,$$

the claim follows from Proposition 9.4. ■

(b) Let $m \in \mathbb{Q}$. Then[3] the radius of convergence of $\sum k^m X^k \in \mathbb{K}[\![X]\!]$ is 1.

Proof From Propositions 2.4 and 2.6 we get

$$\left|\frac{a_k}{a_{k+1}}\right| = \Big(\frac{k}{k+1}\Big)^m \to 1 \quad (k\to\infty) .$$

Thus the claim follows from Proposition 9.4. ■

(c) Let $a \in \mathbb{K}[\![X]\!]$ be defined by

$$a = \sum \frac{1}{k!} X^{k^2} = 1 + X + \frac{1}{2!}X^4 + \frac{1}{3!}X^9 + \cdots$$

Then $\rho_a = 1$.

Proof[4] The coefficients a_k of a satisfy

$$a_k = \begin{cases} 1/j! , & k = j^2 , \ j \in \mathbb{N} , \\ 0 & \text{otherwise} . \end{cases}$$

From $1 \le j! \le j^j$, Remark I.10.10(c) and Exercise I.10.3 we get the inequality

$$1 \le \sqrt[j^2]{j!} \le \sqrt[j^2]{j^j} = (j^j)^{1/j^2} = j^{1/j} = \sqrt[j]{j} .$$

Since $\lim_j \sqrt[j]{j} = 1$ (see Example 4.2(d)) we conclude that $\rho_a = \overline{\lim}_k \sqrt[k]{|a_k|} = 1$. ■

---

[3] Here (and in similar situations) we make the convention that the zeroth coefficient a_0 of the power series a has the value 0 when not otherwise stated.

[4] Note that Proposition 9.4 cannot be used here. Why not?

9.6 Remark No general statement can be made about the convergence of a power series on the 'boundary', $\{x \in \mathbb{K} ; |x| = \rho\}$, of the disk of convergence. We demonstrate this using the power series obtained by setting $m = 0, -1, -2$ in Example 9.5(b):

$$\text{(i)} \ \sum X^k , \qquad \text{(ii)} \ \sum \frac{1}{k} X^k , \qquad \text{(iii)} \ \sum \frac{1}{k^2} X^k .$$

These series have radius of convergence $\rho = 1$. On the boundary of the disk of convergence we have the following behavior:

(i) By Example 7.4, the geometric series $\sum x^k$ diverges for each $x \in \mathbb{K}$ such that $|x| = 1$. Thus, is in this case, $\mathrm{dom}(\underline{a}) = \mathbb{B}_{\mathbb{K}}$.

(ii) By the Leibniz criterion of Theorem 7.8, the series $\sum (-1)^k/k$ converges conditionally in $\mathbb{R}$. On the other hand, in Example 7.3 we saw that the harmonic series $\sum 1/k$ diverges. Thus we have $-1 \in \mathrm{dom}(\underline{a})$ and $1 \notin \mathrm{dom}(\underline{a})$.

(iii) Let $x \in \mathbb{K}$ be such that $|x| = 1$. Then the majorant criterion of Theorem 8.3 and Example 7.1(b) ensure the absolute convergence of $\sum k^{-2} x^k$. Consequently $\mathrm{dom}(a) = \bar{\mathbb{B}}_{\mathbb{K}}$. ■

Addition and Multiplication of Power Series

From Section I.8 we know that $\mathbb{K}[\![X]\!]$ is a ring when addition is defined 'termwise' and multiplication is defined by convolution. The following proposition shows that these operations are compatible with the addition and multiplication of the corresponding functions.

9.7 Proposition *Let $a = \sum a_k X^k$ and $b = \sum b_k X^k$ be power series with radii of convergence ρ_a and ρ_b respectively. Set $\rho := \min(\rho_a, \rho_b)$. Then for all $x \in \mathbb{K}$ such that $|x| < \rho$ we have*

$$\sum_{k=0}^{\infty} a_k x^k + \sum_{k=0}^{\infty} b_k x^k = \sum_{k=0}^{\infty} (a_k + b_k) x^k ,$$

$$\Bigl[\sum_{k=0}^{\infty} a_k x^k\Bigr]\Bigl[\sum_{k=0}^{\infty} b_k x^k\Bigr] = \sum_{k=0}^{\infty} \Bigl(\sum_{j=0}^{k} a_j b_{k-j}\Bigr) x^k .$$

In particular, the radii of convergence ρ_{a+b} and $\rho_{a \cdot b}$ of the power series $a + b$ and $a \cdot b$ satisfy $\rho_{a+b} \geq \rho$ and $\rho_{a \cdot b} \geq \rho$.

Proof Because of Theorem 9.2, all the claims follow directly from Proposition 7.5 and Theorem 8.11. ■

The Uniqueness of Power Series Representations

Let $p \in \mathbb{K}[X]$. In Remark I.8.19(c) we showed that if p has at least $\deg(p)+1$ zeros then p is the zero polynomial. The following theorem extends this result to power series.

9.8 Theorem *Let $\sum a_k X^k$ be a power series with positive radius of convergence ρ_a. If there is a null sequence (y_j) such that $0 < |y_j| < \rho_a$ and*

$$\underline{a}(y_j) = \sum_{k=0}^{\infty} a_k y_j^k = 0 , \qquad j \in \mathbb{N} , \tag{9.5}$$

then $a_k = 0$ for all $k \in \mathbb{N}$, that is, $a = 0 \in \mathbb{K}[\![X]\!]$.

Proof (i) For an arbitrary $n \in \mathbb{N}$, we derive an estimate of $\sum_{k \geq n} a_k x^k$. Choose $r \in (0, \rho_a)$ and $x \in r\bar{\mathbb{B}}_{\mathbb{K}}$. The absolute convergence of a on $\rho_a \mathbb{B}_{\mathbb{K}}$ implies that

$$\Big|\sum_{k=n}^{\infty} a_k x^k\Big| \leq \sum_{k=n}^{\infty} |a_k|\,|x|^k = |x|^n \sum_{k=n}^{\infty} |a_k|\,|x|^{k-n} \leq |x|^n \sum_{j=0}^{\infty} |a_{j+n}|\, r^j .$$

So, for each $r \in (0, \rho_a)$ and $n \in \mathbb{N}$, there is some

$$C := C(r,n) := \sum_{j=0}^{\infty} |a_{j+n}|\, r^j \in [0, \infty)$$

such that

$$\Big|\sum_{k=n}^{\infty} a_k x^k\Big| \leq C\,|x|^n , \qquad x \in \bar{\mathbb{B}}_{\mathbb{K}}(0,r) . \tag{9.6}$$

(ii) Since (y_j) is a null sequence, there is some $r \in (0, \rho_a)$ such that all y_j are in $r\bar{\mathbb{B}}_{\mathbb{K}}$. Suppose that there is some $n \in \mathbb{N}$ such that $a_n \neq 0$. Then, by the well ordering principle, there is a least $n_0 \in \mathbb{N}$ such that $a_{n_0} \neq 0$. From (9.6) we have the inequality

$$|\underline{a}(x) - a_{n_0} x^{n_0}| \leq C\,|x|^{n_0+1} , \qquad x \in \bar{\mathbb{B}}_{\mathbb{K}}(0,r) ,$$

and so (9.5) implies $|a_{n_0}| \leq C\,|y_j|$ for all $j \in \mathbb{N}$. But $y_j \to 0$ and so, by Corollary I.10.7, we have the contradiction $a_{n_0} = 0$. ■

9.9 Corollary (identity theorem for power series) *Let*

$$a = \sum a_k X^k \quad \text{and} \quad b = \sum b_k X^k$$

be power series with positive radii of convergence ρ_a and ρ_b respectively. If there is a null sequence (y_j) such that $0 < |y_j| < \min(\rho_a, \rho_b)$ and $\underline{a}(y_j) = \underline{b}(y_j)$ for all $j \in \mathbb{N}$, then $a = b$ in $\mathbb{K}[\![X]\!]$, that is, $a_k = b_k$ for all $k \in \mathbb{N}$.

Proof This follows directly from Proposition 9.7 and Theorem 9.8. ■

9.10 Remarks (a) If a power series $a = \sum a_k X^k$ has positive radius of convergence, then, by Corollary 9.9, the coefficients a_k of a are uniquely determined by $\underline{a}$ in the disk of convergence. In other words, if a function $f\colon \mathrm{dom}(f) \subseteq \mathbb{K} \to \mathbb{K}$ can be represented by a power series on a disk around the origin, then this power series is unique.

(b) The function $\underline{a}$ represented by $a = \sum a_k X^k$ on $\rho_a \mathbb{B}_{\mathbb{K}}$ is bounded on any closed ball $r\bar{\mathbb{B}}_{\mathbb{K}}$ with $r \in (0, \rho_a)$. More precisely,

$$\sup_{|x| \le r} |\underline{a}(x)| \le \sum_{k=0}^{\infty} |a_k|\, r^k \ .$$

Proof This follows directly from (9.6) with $n = 0$. ■

(c) In Section III.6 we will investigate nonzero power series which have infinitely many zeros. Thus the hypothesis of Theorem 9.8, that the sequence of zeros converges, cannot be omitted.

(d) Let $a = \sum a_k X^k$ be a *real* power series, that is, an element of $\mathbb{R}[\![X]\!]$. Because $\mathbb{R}[\![X]\!] \subseteq \mathbb{C}[\![X]\!]$, a can also be considered as a *complex* power series. If we denote by $a_{\mathbb{C}}$ the function represented by $a \in \mathbb{C}[\![X]\!]$, then $a_{\mathbb{C}} \supseteq \underline{a}$, that is, $a_{\mathbb{C}}$ is an extension of $\underline{a}$. In view of Theorem 9.2, the radius of convergence ρ_a is independent of whether a is thought of as a real or complex power series. Hence

$$(-\rho_a, \rho_a) = \mathrm{dom}(\underline{a}) \cap \rho_a \mathbb{B}_{\mathbb{C}} \subseteq \rho_a \mathbb{B}_{\mathbb{C}} \subseteq \mathrm{dom}(a_{\mathbb{C}}) \ .$$

Thus it suffices, in fact, to consider only complex power series. *If a convergent series has real coefficients, then the corresponding function is real valued on real arguments.* ■

Exercises

1 Determine the radius of convergence of the power series $\sum a_k X^k$ when a_k is given by each of the following:

$$\text{(a)}\ \frac{\sqrt{k2^k}}{(k+1)^6}\ , \quad \text{(b)}\ (-1)^k \frac{k!}{k^k}\ , \quad \text{(c)}\ \frac{1}{\sqrt{1+k^2}}\ , \quad \text{(d)}\ \frac{1}{\sqrt{k!}}\ , \quad \text{(e)}\ \frac{1}{k^k}\ , \quad \text{(f)}\ \Bigl(1+\frac{1}{k^2}\Bigr)^k \ .$$

2 Show that the power series $a = \sum (1+k) X^k$ has radius of convergence 1 and that $\underline{a}(z) = (1-z)^{-2}$ for all $|z| < 1$.

3 Suppose that the power series $\sum a_k X^k$ has radius of convergence $\rho > 0$. Show that the series $\sum (k+1) a_{k+1} X^k$ has the same radius of convergence ρ.

4 Suppose that $\sum a_k$ is a divergent series in $(0,\infty)$ such that $\sum a_k X^k$ has radius of convergence 1. Define

$$f_n := \sum_{k=0}^{\infty} a_k \Bigl(1-\frac{1}{n}\Bigr)^k , \qquad n \in \mathbb{N}^\times .$$

Prove that the sequence (f_n) converges to ∞. (Hint: Use the Bernoulli inequality to get an upper bound for terms of the form $1-(1-1/n)^m$.)

5 Suppose that a sequence (a_k) in $\mathbb{K}$ satisfies

$$0 < \underline{\lim}\, |a_k| \le \overline{\lim}\, |a_k| < \infty .$$

Determine the radius of convergence of $\sum a_k X^k$.

6 Show that the radius of convergence ρ of a power series $\sum a_k X^k$ such that $a_k \neq 0$ for all $k \in \mathbb{N}$ satisfies

$$\underline{\lim}\Bigl|\frac{a_k}{a_{k+1}}\Bigr| \le \rho \le \overline{\lim}\Bigl|\frac{a_k}{a_{k+1}}\Bigr| .$$

7 A subset D of a vector space is called **symmetric** with respect to 0 if $x \in D$ implies $-x \in D$ for all x. If D is symmetric and $f : D \to E$ is a function to a vector space E, then f is called **even** (or **odd**) if $f(x) = f(-x)$ (or $f(x) = -f(-x)$) for all $x \in D$. Now let $f : \mathbb{K} \to \mathbb{K}$ be a function which can be represented by power series on a suitable disk centered at 0. What conditions on the coefficients of this power series determine whether f is even or odd?

8 Let a and b be power series with radii of convergence ρ_a and ρ_b respectively. Show, by example, that $\rho_{a+b} > \max(\rho_a, \rho_b)$ and $\rho_{ab} > \max(\rho_a, \rho_b)$ are possible.

9 Let $a = \sum a_k X^k \in \mathbb{C}[\![X]\!]$ with $a_0 = 1$.

(a) Show that there is some $b = \sum b_k X^k \in \mathbb{C}[\![X]\!]$ such that $ab = 1 \in \mathbb{C}[\![X]\!]$. Provide a recursive algorithm for calculating the coefficients b_k.

(b) Show that the radius of convergence ρ_b of b is positive if the radius of convergence of a is positive.

10 Suppose that $b = \sum b_k X^k \in \mathbb{C}[\![X]\!]$ satisfies $(1 - X - X^2)b = 1 \in \mathbb{C}[\![X]\!]$.

(a) Show that the coefficients b_k satisfy

$$b_0 = 1 , \quad b_1 = 1 , \quad b_{k+1} = b_k + b_{k-1} , \qquad k \in \mathbb{N}^\times ,$$

that is, (b_k) is the Fibonacci sequence (see Exercise 4.9).

(b) What is the radius of convergence of b?

Chapter III

Continuous Functions

In this chapter we investigate the topological foundations of analysis and give some of its first applications. We limit ourselves primarily to the topology of metric spaces because the theory of metric spaces is the framework for a huge part of analysis, yet is simple and concrete enough so as to minimize difficulties for beginners. Even so, the concept of a metric space is not general enough for deeper mathematical investigations, and so, when possible, we have provided proofs which are valid in general topological spaces. The extent to which the theorems are true in general topological spaces is discussed at the end of each section. These comments, which can be neglected on the first reading of this book, provide the reader with an introduction to abstract topology.

In the first section we consider continuous functions between metric spaces. In particular, we use the results about convergent sequences from the previous chapter to investigate continuity.

Section 2 is dedicated to the concept of openness. One key result here is the characterization of continuous functions as functions with the property that the preimage of each open set is open.

In the next section we discuss compact metric spaces. In particular, we show that, for metric spaces, compactness is the same as sequential compactness. The great importance of compactness is already apparent in the applications we present in this section. For example, using the extreme value theorem for continuous real valued functions on compact metric spaces, we show that all norms on $\mathbb{K}^n$ are equivalent, and give a proof of the fundamental theorem of algebra.

In Section 4 we investigate connected and path connected spaces. In particular, we show that these concepts coincide for open subsets of normed vector spaces. As an important application of connectivity, we prove a generalized version of the intermediate value theorem.

After this excursion into abstract topology, laying the foundation for the analytic investigations in the following chapters, we turn in the two remaining sections of this chapter to the study of real functions. In the short fifth section we discuss the behavior of monotone functions of real variables and prove, in particular, the inverse function theorem for continuous monotone functions.

In contrast to the relatively abstract nature of the first five sections of this chapter, in the last, comparatively long, section we study the exponential function and its relatives: the logarithm, the power and the trigonometric functions. In this investigation we put into action practically all of the methods and theorems that are introduced in this chapter.

1 Continuity

Experience shows that, even though functions can, in general, be very complicated and hard to describe, the functions that occur in applications share some important qualitative properties. One of these is continuity. For a function $f : X \to Y$, being (or not being) continuous measures how 'small changes' in the image $f(X) \subseteq Y$ arise from corresponding 'small changes' in the domain X. For this to make sense, the sets X and Y must be endowed with some extra structure that allows a precise meaning for 'small changes'. Metric spaces are the obvious candidates for sets with this extra structure.

Elementary Properties and Examples

Let $f : X \to Y$ be a function between metric spaces[1] (X, d_X) and (Y, d_Y). Then f is **continuous at** $x_0 \in X$ if, for *each* neighborhood V of $f(x_0)$ in Y, there is a neighborhood U of x_0 in X such that $f(U) \subseteq V$.

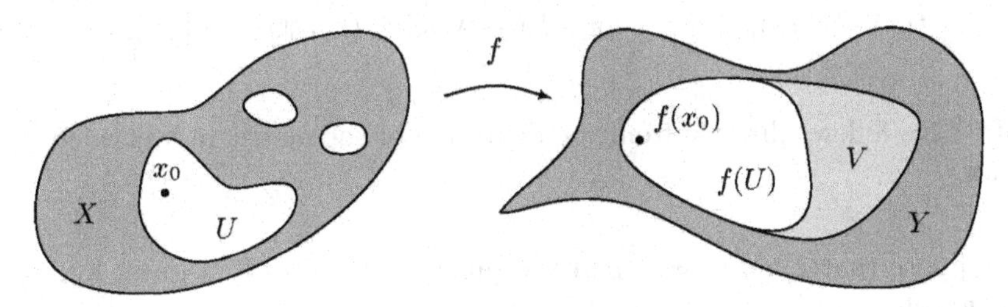

Hence to prove the continuity of f at x_0, one supposes that an *arbitrary* neighborhood V of $f(x_0)$ is given and then shows that there is a neighborhood U of x_0 such that $f(U) \subseteq V$, that is, $f(x) \in V$ for all $x \in U$.

The function $f : X \to Y$ is **continuous** if it is continuous at each point of X. We say f is **discontinuous at** x_0 if f is not continuous at x_0. Finally f is **discontinuous** if it is discontinuous at (at least) one point of X, that is, if f is not continuous. The set of all continuous functions from X to Y is denoted $C(X, Y)$. Obviously $C(X, Y)$ is a subset of Y^X.

This definition of continuity uses the concept of neighborhoods and so is quite simple. In concrete situations the following equivalent formulation is often more useful.

1.1 Proposition *A function $f : X \to Y$ is continuous at $x_0 \in X$ if and only if, for each $\varepsilon > 0$, there is some*[2] *$\delta := \delta(x_0, \varepsilon) > 0$ with the property that*

$$d(f(x_0), f(x)) < \varepsilon \quad \text{for all } x \in X \text{ such that } d(x_0, x) < \delta . \tag{1.1}$$

[1]We usually write d for both the metric d_X in X and the metric d_Y in Y.
[2]The notation $\delta := \delta(x_0, \varepsilon)$ indicates that δ depends, in general, on $x_0 \in X$ and $\varepsilon > 0$.

Proof '$\Rightarrow$' Let f be continuous at x_0 and $\varepsilon > 0$. Then, for the neighborhood $V := \mathbb{B}_Y\bigl(f(x_0), \varepsilon\bigr) \in \mathcal{U}_Y\bigl(f(x_0)\bigr)$, there is some $U \in \mathcal{U}_X(x_0)$ such that $f(U) \subseteq V$. By definition, there is some $\delta := \delta(x_0, \varepsilon) > 0$ such that $\mathbb{B}_X(x_0, \delta) \subseteq U$. Thus

$$f\bigl(\mathbb{B}_X(x_0, \delta)\bigr) \subseteq f(U) \subseteq V = \mathbb{B}_Y\bigl(f(x_0), \varepsilon\bigr) .$$

These inclusions imply (1.1).

'$\Leftarrow$' Suppose that (1.1) is true and $V \in \mathcal{U}_Y\bigl(f(x_0)\bigr)$. Then there is some $\varepsilon > 0$ such that $\mathbb{B}_Y\bigl(f(x_0), \varepsilon\bigr) \subseteq V$. Because of (1.1), there is some $\delta > 0$ such that the image of $U := \mathbb{B}_X(x_0, \delta)$ is contained in $\mathbb{B}_Y\bigl(f(x_0), \varepsilon\bigr)$, and hence also in V. Thus f is continuous at x_0. ■

1.2 Corollary *Let E and F be normed vector spaces and $X \subseteq E$. Then $f : X \to F$ is continuous at $x_0 \in X$ if and only if, for each $\varepsilon > 0$, there is some $\delta := \delta(x_0, \varepsilon) > 0$ satisfying*

$$\|f(x) - f(x_0)\|_F < \varepsilon \qquad \textit{for all } x \in X \textit{ such that } \|x - x_0\|_E < \delta .$$

Proof This follows directly from the definition of the metric in a normed vector space. ■

Suppose that $E := F := \mathbb{R}$ and the function $f : X \to \mathbb{R}$ is given by the following graph.

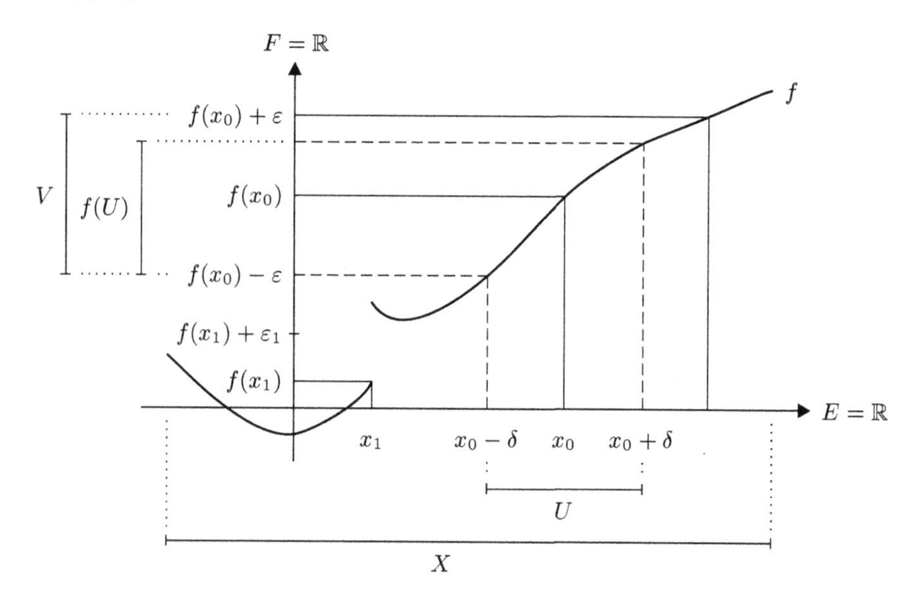

Then f is continuous at x_0 since, for each $\varepsilon > 0$, there is some $\delta > 0$ such that the image of $U := (x_0 - \delta, x_0 + \delta)$ is contained in $V := \big(f(x_0) - \varepsilon, f(x_0) + \varepsilon\big)$.

On the other hand, there is no $\delta > 0$ such that $|f(x) - f(x_1)| < \varepsilon_1$ for all $x \in (x_1, x_1 + \delta)$, and so f is discontinuous at x_1.

1.3 Examples In the following examples, X and Y are metric spaces.

(a) The square root function $\mathbb{R}^+ \to \mathbb{R}^+$, $x \mapsto \sqrt{x}$ is continuous.

Proof Let $x_0 \in \mathbb{R}^+$ and $\varepsilon > 0$. If $x_0 = 0$, we set $\delta := \varepsilon^2 > 0$. Then

$$\left|\sqrt{x} - \sqrt{x_0}\right| = \sqrt{x} < \varepsilon , \qquad x \in [0, \delta) .$$

Otherwise $x_0 > 0$, and we choose $\delta := \delta(x_0, \varepsilon) := \min\{\varepsilon\sqrt{x_0}, x_0\}$. Then

$$\left|\sqrt{x} - \sqrt{x_0}\right| = \left|\frac{x - x_0}{\sqrt{x} + \sqrt{x_0}}\right| < \frac{|x - x_0|}{\sqrt{x_0}} \le \varepsilon$$

for all $x \in (x_0 - \delta, x_0 + \delta)$. ■

(b) The floor function $\lfloor\cdot\rfloor : \mathbb{R} \to \mathbb{R}$, $x \mapsto \lfloor x \rfloor := \max\{\, k \in \mathbb{Z} \ ; \ k \le x \,\}$ is continuous at $x_0 \in \mathbb{R}\backslash\mathbb{Z}$ and discontinuous at $x_0 \in \mathbb{Z}$.

Proof If $x_0 \in \mathbb{R}\backslash\mathbb{Z}$, then there is a unique $k \in \mathbb{Z}$ such that $x_0 \in (k, k+1)$. If we choose $\delta := \min\{x_0 - k, k + 1 - x_0\} > 0$, then we clearly have

$$\big|\lfloor x \rfloor - \lfloor x_0 \rfloor\big| = 0 , \qquad x \in (x_0 - \delta, x_0 + \delta) .$$

Thus the floor function $\lfloor\cdot\rfloor$ is continuous at x_0.

Otherwise, for $x_0 \in \mathbb{Z}$, we have the inequality $\big|\lfloor x \rfloor - \lfloor x_0 \rfloor\big| = \lfloor x_0 \rfloor - \lfloor x \rfloor \ge 1$ for all $x < x_0$. So there is no neighborhood U of x_0 such that $\big|\lfloor x \rfloor - \lfloor x_0 \rfloor\big| < 1/2$ for all $x \in U$. That is, $\lfloor\cdot\rfloor$ is discontinuous at x_0. ■

(c) The **Dirichlet function** $f : \mathbb{R} \to \mathbb{R}$ defined by

$$f(x) := \begin{cases} 1 , & x \in \mathbb{Q} , \\ 0 , & x \in \mathbb{R}\backslash\mathbb{Q} , \end{cases}$$

is nowhere continuous, that is, it is discontinuous at every $x_0 \in \mathbb{R}$.

Proof Let $x_0 \in \mathbb{R}$. Since both the rational numbers $\mathbb{Q}$ and the irrational numbers $\mathbb{R}\backslash\mathbb{Q}$ are dense in $\mathbb{R}$ (see Propositions I.10.8 and I.10.11), in each neighborhood of x_0 there is some x such that $|f(x) - f(x_0)| = 1$. Thus f is discontinuous at x_0. ■

(d) Suppose that $f : X \to \mathbb{R}$ is continuous at $x_0 \in X$ and $f(x_0) > 0$. Then there is a neighborhood U of x_0 such that $f(x) > 0$ for all $x \in U$.

Proof Set $\varepsilon := f(x_0)/2 > 0$. Then there is a neighborhood U of x_0 such that

$$f(x_0) - f(x) \le |f(x) - f(x_0)| < \varepsilon = \frac{f(x_0)}{2} , \qquad x \in U .$$

Thus we have $f(x) > f(x_0)/2 > 0$ for all $x \in U$. ■

(e) A function $f : X \to Y$ is **Lipschitz continuous** with **Lipschitz constant** $\alpha > 0$ if

$$d\big(f(x), f(y)\big) \le \alpha d(x,y) , \qquad x, y \in X .$$

Every Lipschitz continuous function is continuous.[3]

Proof Given $x_0 \in X$ and $\varepsilon > 0$, set $\delta := \varepsilon/\alpha$. The continuity of f then follows from Proposition 1.1. Note that, in this case, δ is independent of $x_0 \in X$. ∎

(f) Any constant function $X \to Y,\ x \mapsto y_0$ is Lipschitz continuous.

(g) The identity function $\mathrm{id} : X \to X,\ x \mapsto x$ is Lipschitz continuous.

(h) If $E_1, \ldots, E_m$ are normed vector spaces, then $E := E_1 \times \cdots \times E_m$ is a normed vector space with respect to the product norm $\|\cdot\|_\infty$ of Example II.3.3(c). The canonical projections

$$\mathrm{pr}_k : E \to E_k , \quad x = (x_1, \ldots, x_m) \mapsto x_k , \qquad 1 \le k \le m ,$$

are Lipschitz continuous. In particular, the projections $\mathrm{pr}_k : \mathbb{K}^m \to \mathbb{K}$ are Lipschitz continuous.

Proof For $x = (x_1, \ldots, x_m)$ and $y = (y_1, \ldots, y_m)$, we have

$$\|\mathrm{pr}_k(x) - \mathrm{pr}_k(y)\|_{E_k} = \|x_k - y_k\|_{E_k} \le \|x - y\|_\infty ,$$

which implies the Lipschitz continuity of pr_k. For the remaining claim, see Proposition II.3.12. ∎

(i) Each of the functions $z \mapsto \mathrm{Re}(z),\ z \mapsto \mathrm{Im}(z)$ and $z \mapsto \overline{z}$ is Lipschitz continuous on $\mathbb{C}$.

Proof This follows from the inequality

$$\max\big\{|\mathrm{Re}(z_1) - \mathrm{Re}(z_2)|, |\mathrm{Im}(z_1) - \mathrm{Im}(z_2)|\big\} \le |z_1 - z_2| = |\overline{z}_1 - \overline{z}_2| , \qquad z_1, z_2 \in \mathbb{C} ,$$

which comes from Proposition I.11.4. ∎

(j) Let E be a normed vector space. Then the norm function

$$\|\cdot\| : E \to \mathbb{R} , \quad x \mapsto \|x\|$$

is Lipschitz continuous.

Proof The reversed triangle inequality,

$$\big|\, \|x\| - \|y\| \,\big| \le \|x - y\| , \qquad x, y \in E ,$$

implies the claim. ∎

---

[3] The converse is not true. See Exercise 18.

(k) If $A \subseteq X$ and $f : X \to Y$ is continuous at $x_0 \in A$, then $f \,|\, A : A \to Y$ is continuous at x_0. Here A has the metric induced from X.

Proof This follows directly from the continuity of f and the definition of the induced metric. ■

(l) Let $M \subseteq X$ be a nonempty subset of X. For each $x \in X$,

$$d(x, M) := \inf_{m \in M} d(x, m)$$

is called the **distance** from x to M. The **distance function**

$$d(\cdot, M) : X \to \mathbb{R} , \quad x \mapsto d(x, M)$$

is Lipschitz continuous.

Proof Let $x, y \in X$. From the triangle inequality we have $d(x, m) \leq d(x, y) + d(y, m)$ for each $m \in M$. Since $d(x, M) \leq d(x, m)$ for all $m \in M$ this implies

$$d(x, M) \leq d(x, y) + d(y, m) , \qquad m \in M .$$

Taking the infimum over all $m \in M$ yields

$$d(x, M) \leq d(x, y) + d(y, M) .$$

Combining this equation and the same equation with x and y interchanged gives

$$|d(x, M) - d(y, M)| \leq d(x, y) ,$$

which shows the Lipschitz continuity of $d(\cdot, M)$. ■

(m) For any inner product space $\big(E, (\cdot \,|\, \cdot)\big)$, the scalar product $(\cdot \,|\, \cdot) : E \times E \to \mathbb{K}$ is continuous.

Proof Let $(x, y), (x_0, y_0) \in E \times E$ and $\varepsilon \in (0, 1)$. From the triangle and Cauchy-Schwarz inequalities we get

$$\begin{aligned} \big|(x \,|\, y) - (x_0 \,|\, y_0)\big| &\leq \big|(x - x_0 \,|\, y)\big| + \big|(x_0 \,|\, y - y_0)\big| \\ &\leq \|x - x_0\| \, \|y\| + \|x_0\| \, \|y - y_0\| \\ &\leq d\big((x, y), (x_0, y_0)\big) \, (\|y\| + \|x_0\|) \\ &\leq d\big((x, y), (x_0, y_0)\big) \, (\|x_0\| + \|y_0\| + \|y - y_0\|) , \end{aligned}$$

where d is the product metric. Set $M := \max\{1, \|x_0\|, \|y_0\|\}$ and $\delta := \varepsilon/(1 + 2M)$. Then, for all $(x, y) \in \mathbb{B}_{E \times E}\big((x_0, y_0), \delta\big)$, it follows from the above inequality that

$$\big|(x \,|\, y) - (x_0 \,|\, y_0)\big| < \delta(2M + \delta) < \varepsilon ,$$

which proves the continuity of the scalar product at the point (x_0, y_0). ■

(n) Let E and F be normed vector spaces and $X \subseteq E$. Then the continuity of $f : X \to F$ at $x_0 \in X$ is independent of the choice of equivalent norms on E and on F.

Proof This follows easily from Corollary 1.2. ■

(o) A function f between metric spaces X and Y is **isometric** (or an **isometry**) if $d\big(f(x), f(x')\big) = d(x, x')$ for all $x, x' \in X$, that is, if f 'preserves distances'. Clearly, such a function is Lipschitz continuous and is a bijection from X to its image $f(X)$. If E and F are normed vector spaces and $T : E \to F$ is linear, then T is isometric if and only if $\|Tx\| = \|x\|$ for all $x \in E$. If, in addition, T is surjective then T is an **isometric isomorphism** from E to F, and T^{-1} is also isometric. ∎

Sequential Continuity

The neighborhood concept is central for both the definition of continuity and the definition of the convergence of a sequence. This suggests that the continuity of a function could be defined using sequences: A function $f : X \to Y$ between metric spaces X and Y is called **sequentially continuous** at $x \in X$, if, for *every* sequence (x_k) in X such that $\lim x_k = x$, we have $\lim f(x_k) = f(x)$.

1.4 Theorem (sequence criterion) *Let X, Y be metric spaces. Then a function $f : X \to Y$ is continuous at x if and only if it is sequentially continuous at x.*

Proof '$\Longrightarrow$' Let (x_k) be a sequence in X such that $x_k \to x$. Let V be a neighborhood of $f(x)$ in Y. By supposition there is a neighborhood U of x in X such that $f(U) \subseteq V$. Since $x_k \to x$, there is some $N \in \mathbb{N}$ such that $x_k \in U$ for all $k \geq N$. Thus $f(x_k) \in V$ for all $k \geq N$, that is, $f(x_k)$ converges to $f(x)$.

'$\Longleftarrow$' Suppose, to the contrary, that f is sequentially continuous but discontinuous at x. Then there is a neighborhood V of $f(x)$ such that no neighborhood U of x satisfies $f(U) \subseteq V$. In particular, we have

$$f\big(\mathbb{B}(x, 1/k)\big) \cap V^c \neq \emptyset\ , \qquad k \in \mathbb{N}^\times\ .$$

Hence, for each $k \in \mathbb{N}^\times$, we can choose some $x_k \in X$ such that $d(x, x_k) < 1/k$ and $f(x_k) \notin V$. By construction, (x_k) converges to x but $\big(f(x_k)\big)$ does not converge to $f(x)$. This contradicts the sequential continuity of f. ∎

Let $f : X \to Y$ be a continuous function between metric spaces. Then for any convergent sequence (x_k) in X we have

$$\lim f(x_k) = f(\lim x_k)\ .$$

Thus one says that 'continuous functions respect the taking of limits'.

Addition and Multiplication of Continuous Functions

Theorem 1.4 makes it possible to apply theorems about convergent sequences to continuous functions. To do so, it is first useful to introduce a few definitions.

Let M be an arbitrary set and F a vector space. Let f and g be functions with $\mathrm{dom}(f), \mathrm{dom}(g) \subseteq M$ and values in F. Then the **sum** of f and g is the function $f+g$ defined by

$$f+g\colon \mathrm{dom}(f+g) := \mathrm{dom}(f) \cap \mathrm{dom}(g) \to F\ , \quad x \mapsto f(x)+g(x)\ .$$

Similarly, for $\lambda \in \mathbb{K}$, we define λf by[4]

$$\lambda f\colon \mathrm{dom}(f) \to F\ , \quad x \mapsto \lambda f(x)\ .$$

Finally, in the special case $F = \mathbb{K}$, we set

$$\begin{aligned} \mathrm{dom}(f \cdot g) &:= \mathrm{dom}(f) \cap \mathrm{dom}(g)\ , \\ \mathrm{dom}(f/g) &:= \mathrm{dom}(f) \cap \{\, x \in \mathrm{dom}(g)\ ;\ g(x) \neq 0 \,\}\ , \end{aligned}$$

and define the **product** and **quotient** of f and g by

$$f \cdot g\colon \mathrm{dom}(f \cdot g) \to \mathbb{K}\ , \quad x \mapsto f(x) \cdot g(x)$$

and

$$f/g\colon \mathrm{dom}(f/g) \to \mathbb{K}\ , \quad x \mapsto f(x)/g(x)\ .$$

1.5 Proposition *Suppose that X is a metric space, F is a normed vector space, and*

$$f\colon \mathrm{dom}(f) \subseteq X \to F\ , \quad g\colon \mathrm{dom}(g) \subseteq X \to F$$

are continuous at $x_0 \in \mathrm{dom}(f) \cap \mathrm{dom}(g)$.

(i) *$f+g$ and λf are continuous at x_0.*

(ii) *If $F = \mathbb{K}$, then $f \cdot g$ is continuous at x_0.*

(iii) *If $F = \mathbb{K}$ and $g(x_0) \neq 0$, then f/g is continuous at x_0.*

Proof These claims follow from the sequence criterion of Theorem 1.4, Proposition II.2.2 and Remark II.3.1(c), together with Propositions II.2.4(ii) and II.2.6 and Example 1.3(d). ■

1.6 Corollary

(i) *Rational functions are continuous.*

(ii) *Polynomials in n variables are continuous (on $\mathbb{K}^n$).*

(iii) *$C(X,F)$ is a subspace of F^X, the* **vector space of continuous functions**[5] *from X to F.*

Proof Claims (i) and (iii) are immediate consequences of Proposition 1.5. For (ii), Example 1.3(h) is also needed. ■

[4]The definitions of $f+g$ and λf coincide with those of Example I.12.3(e) if f and g are defined on all of M.

[5]When no confusion is possible, we often write $C(X)$ instead of $C(X,\mathbb{K})$.

1.7 Proposition *Let $a = \sum a_k X^k$ be a power series with positive radius of convergence ρ_a. Then the function $\underline{a}$ represented by a is continuous on $\rho_a \mathbb{B}$.*

Proof Let $x_0 \in \rho_a \mathbb{B}_{\mathbb{C}}$, $\varepsilon > 0$, and $|x_0| < r < \rho_a$. Since, by Theorem II.9.2, the series $\sum |a_k| \, r^k$ converges, there is some $K \in \mathbb{N}$ such that

$$\sum_{k=K+1}^{\infty} |a_k| \, r^k < \varepsilon/4 \ . \tag{1.2}$$

Thus, for $|x| \le r$, we have

$$\begin{aligned} |\underline{a}(x) - \underline{a}(x_0)| &\le \Bigl| \sum_{k=0}^{K} a_k x^k - \sum_{k=0}^{K} a_k x_0^k \Bigr| + \sum_{k=K+1}^{\infty} |a_k| \, |x|^k + \sum_{k=K+1}^{\infty} |a_k| \, |x_0|^k \\ &\le |p(x) - p(x_0)| + 2 \sum_{k=K+1}^{\infty} |a_k| \, r^k \ , \end{aligned} \tag{1.3}$$

where we have set

$$p := \sum_{k=0}^{K} a_k X^k \in \mathbb{C}[X] \ .$$

By Corollary 1.6, there is some $\delta \in (0, r - |x_0|)$ such that

$$|p(x) - p(x_0)| < \varepsilon/2 \ , \qquad |x - x_0| < \delta \ .$$

Together with (1.2) and (1.3), this implies $|\underline{a}(x) - \underline{a}(x_0)| < \varepsilon$ for all $|x - x_0| < \delta$. Since $\mathbb{B}(x_0, \delta) \subseteq \rho_a \mathbb{B}_{\mathbb{C}}$, we have proved the claim. ∎

The following important theorem often provides a simple proof of the continuity of certain functions. This we illustrate in the examples following the theorem.

1.8 Theorem (continuity of compositions) *Let X, Y and Z be metric spaces. Suppose that $f : X \to Y$ is continuous at $x \in X$, and $g : Y \to Z$ is continuous at $f(x) \in Y$. Then the composition $g \circ f : X \to Z$ is continuous at x.*

Proof Let W be a neighborhood of $g \circ f(x) = g\bigl(f(x)\bigr)$ in Z. Because of the continuity of g at $f(x)$, there is a neighborhood V of $f(x)$ in Y such that $g(V) \subseteq W$. Since f is continuous at x, there is a neighborhood U of x in X such that $f(U) \subseteq V$. Thus

$$g \circ f(U) = g\bigl(f(U)\bigr) \subseteq g(V) \subseteq W \ ,$$

from which the claim follows. ∎

1.9 Examples In the following, X is a metric space and E is a normed vector space.

(a) Let $f : X \to E$ be continuous at x_0. Then the **norm of** f,

$$\|f\| : X \to \mathbb{R} , \quad x \mapsto \|f(x)\| ,$$

is continuous at x_0.

Proof By Example 1.3(j), $\|\cdot\| : E \to \mathbb{R}$ is Lipschitz continuous. Since $\|f\| = \|\cdot\| \circ f$, the claim follows from Theorem 1.8. ■

(b) Let $g : \mathbb{R} \to X$ be continuous. Then the function $\widehat{g} : E \to X,\ x \mapsto g(\|x\|)$ is continuous.

Proof It suffices to note that $\widehat{g} = g \circ \|\cdot\|$ is a composition of continuous functions. ■

(c) The converse of Theorem 1.8 is false, that is, the continuity of $g \circ f$ does *not* imply that f or g is continuous.

Proof Set $Z := [-3/2, -1/2] \cup (1/2, 3/2]$ and $I := [-1, 1]$. Define functions $f : Z \to \mathbb{R}$ and $g : I \to \mathbb{R}$ by

$$f(x) := \begin{cases} x + 1/2 , & x \in [-3/2, -1/2] , \\ x - 1/2 , & x \in (1/2, 3/2] , \end{cases}$$

and

$$g(y) := \begin{cases} y - 1/2 , & y \in [-1, 0] , \\ y + 1/2 , & y \in (0, 1] . \end{cases}$$

It is not difficult to check that $f : Z \to \mathbb{R}$ is continuous and $g : I \to \mathbb{R}$ is discontinuous at 0, whereas the compositions $f \circ g = \mathrm{id}_I$ and $g \circ f = \mathrm{id}_Z$ are both continuous. We leave the reader the task of constructing a similar example in which f is also discontinuous. ■

(d) The function $f : \mathbb{R} \to \mathbb{R},\ x \mapsto 1/\sqrt{1 + x^2}$ is continuous.

Proof Since $1/\sqrt{1+x^2} = \sqrt{1/(1+x^2)}$, the claim follows from Corollary 1.6.(i), Proposition 1.5.(iii), Theorem 1.8 and Example 1.3(a). ■

(e) The exponential function $\exp : \mathbb{C} \to \mathbb{C}$ is continuous.

Proof This follows from Proposition 1.7 and Example II.9.5(a). ■

1.10 Proposition *Let X be a metric space. Then a function $f = (f_1, \ldots, f_m)$ from X to $\mathbb{K}^m$ is continuous at x if and only if $f_k : X \to \mathbb{K}$ is continuous at x for each k. In particular, $f : X \to \mathbb{C}$ is continuous at x if and only if $\mathrm{Re}\, f$ and $\mathrm{Im}\, f$ are continuous at x.*

Proof Let (x_n) be a sequence in X such that $x_n \to x$. From Proposition II.3.14 we have

$$f(x_n) \to f(x) \iff f_k(x_n) \to f_k(x) , \quad k = 1, \ldots, m .$$

The claim now follows from the sequence criterion. ■

One-Sided Continuity

Let X be a subset of $\mathbb{R}$ and $x_0 \in X$. The order structure of $\mathbb{R}$ allows us to consider *one-sided* neighborhoods of x_0. Specifically, for $\delta > 0$, the set $X \cap (x_0 - \delta, x_0]$ (or $X \cap [x_0, x_0 + \delta)$) is called a **left** (or **right**) **δ-neighborhood** of x_0.

Now let Y be a metric space. Then $f : X \to Y$ is **left** (or **right**) **continuous** at x_0, if, for each neighborhood V of $f(x_0)$ in Y, there is some $\delta > 0$ such that $f\bigl(X \cap (x_0 - \delta, x_0]\bigr) \subseteq V$ (or $f\bigl(X \cap [x_0, x_0 + \delta)\bigr) \subseteq V$).

As in Proposition 1.1, it suffices to consider ε-neighborhoods of $f(x_0)$ in Y, that is, $f : X \to Y$ is left (or right) continuous at x_0 if and only if, for each $\varepsilon > 0$, there is some $\delta > 0$ such that $d\bigl(f(x_0), f(x)\bigr) < \varepsilon$ for all x in the left (or right) δ-neighborhood of x.

It is clear that continuous functions are left and right continuous. On the other hand, one-sided continuity does not imply continuity, as we see in the following examples.

1.11 Examples **(a)** The floor function $\lfloor\cdot\rfloor : \mathbb{R} \to \mathbb{R}$ is continuous at $x \in \mathbb{R}\backslash\mathbb{Z}$ and right, but not left, continuous at $x \in \mathbb{Z}$.

(b) The function

$$\operatorname{sign} : \mathbb{R} \to \mathbb{R} , \quad x \mapsto \begin{cases} -1 , & x < 0 , \\ 0 , & x = 0 , \\ 1 , & x > 0 , \end{cases}$$

is neither left nor right continuous at 0. ∎

The next proposition generalizes the sequence criterion of Theorem 1.4 to one-sided continuous functions.

1.12 Proposition *Let Y be a metric space, $X \subseteq \mathbb{R}$, and $f : X \to Y$. Then the following are equivalent:*

(i) *f is left (or right) continuous at $x \in X$.*

(ii) *For each sequence (x_n) in X such that $x_n \to x$ and $x_n \le x$ (or $x_n \ge x$), the sequence $\bigl(f(x_n)\bigr)$ converges to $f(x)$.*

Proof The proof of this claim is similar to the proof of Theorem 1.4. ∎

One-sided continuity can also be used to characterize continuity.

1.13 Proposition *Let Y be a metric space, $X \subseteq \mathbb{R}$, and $f : X \to Y$. Then the following are equivalent:*

(i) *f is continuous at x_0.*

(ii) *f is left and right continuous at x_0.*

Proof The implication '$\Rightarrow$' is clear.

'$\Leftarrow$' Let $\varepsilon > 0$. By the left and right continuity of f at x_0, there are positive numbers δ^- and δ^+ such that $d\big(f(x), f(x_0)\big) < \varepsilon$ for all $x \in X \cap (x_0 - \delta^-, x_0]$ and $x \in X \cap [x_0, x_0 + \delta^+)$. Set $\delta := \min\{\delta^-, \delta^+\}$. Then $d\big(f(x), f(x_0)\big) < \varepsilon$ for all $x \in X \cap (x_0 - \delta, x_0 + \delta)$. Therefore f is continuous at x_0. ■

Exercises

1 The function $\operatorname{zigzag} : \mathbb{R} \to \mathbb{R}$ is defined by

$$\operatorname{zigzag}(x) := \big|\lfloor x + 1/2 \rfloor - x\big| \ , \qquad x \in \mathbb{R} \ ,$$

where $\lfloor \cdot \rfloor$ is the floor function. Sketch the graph of zigzag and show the following:

(a) $\operatorname{zigzag}(x) = |x|$ for all $|x| \leq 1/2$.

(b) $\operatorname{zigzag}(x + n) = \operatorname{zigzag}(x), \ \ x \in \mathbb{R}, \ \ n \in \mathbb{Z}$.

(c) zigzag is continuous.

2 Let $q \in \mathbb{Q}$. Prove that the function $(0, \infty) \to (0, \infty), \ \ x \mapsto x^q$ is continuous.[6] (Hint: See Exercise II.2.7.)

3 Let $\varphi : \mathbb{R} \to (-1, 1), \ \ x \mapsto x/(1 + |x|)$. Show that φ is bijective and that φ and φ^{-1} are continuous.

4 Prove or disprove that the function

$$f : \mathbb{Q} \to \mathbb{R} \ , \quad x \mapsto \begin{cases} 0 \ , & x < \sqrt{2} \ , \\ 1 \ , & x > \sqrt{2} \ , \end{cases}$$

is continuous.

5 Let d_1 and d_2 be metrics on X, and $X_j := (X, d_j), \ j = 1, 2$. Then d_1 is **stronger** than d_2 if $\mathcal{U}_{X_1}(x) \supseteq \mathcal{U}_{X_2}(x)$ for each $x \in X$, that is, if each point has more d_1 neighborhoods than d_2 neighborhoods. In this case, one says also that d_2 is **weaker** than d_1.

Show the following:

(a) d_1 is stronger than d_2 if and only if the identity function $i : X_1 \to X_2, \ \ x \mapsto x$ is continuous.

(b) d_1 and d_2 are equivalent if and only if d_1 is both stronger and weaker than d_2, that is, for each $x \in X, \ \ \mathcal{U}_{X_1}(x) = \mathcal{U}_{X_2}(x)$.

6 Let $f : \mathbb{R} \to \mathbb{R}$ be a continuous[7] homomorphism of the additive group $(\mathbb{R}, +)$. Show that f is linear, that is, there is some $a \in \mathbb{R}$ such that $f(x) = ax, \ \ x \in \mathbb{R}$.
(Hint: Show that $f(q) = qf(1)$ for all $q \in \mathbb{Q}$ and use Proposition I.10.8.)

---

[6]In Section 6 we investigate the function $x \mapsto x^q$ in more generality.

[7]It can be proved that discontinuous homomorphisms of $(\mathbb{R}, +)$ exist (see Volume III, Exercise IX.5.6).

7 Let $f:\mathbb{R}\to\mathbb{R}$ be defined by

$$f(x):=\begin{cases} -1\ , & x\geq 1\ ,\\ 1/n\ , & 1/(n+1)\leq x<1/n\ ,\quad n\in\mathbb{N}^\times\ ,\\ 0\ , & x\leq 0\ .\end{cases}$$

Where is f continuous? left continuous? right continuous?

8 Suppose that X is a metric space and $f,g\in\mathbb{R}^X$ are continuous at x_0. Prove or disprove that[8]

$$|f|\ ,\quad f^+:=0\vee f\ ,\quad f^-:=0\vee(-f)\ ,\quad f\vee g\ ,\quad f\wedge g \tag{1.4}$$

are continuous at x_0. (Hint: Example 1.3(j) and Exercise I.8.11.)

9 Let $f:\mathbb{R}\to\mathbb{R}$ and $g:\mathbb{R}\to\mathbb{R}$ be defined by

$$f(x):=\begin{cases} 1\ , & x \text{ rational}\ ,\\ -1\ , & x \text{ irrational}\ ,\end{cases}\qquad g(x):=\begin{cases} x\ , & x \text{ rational}\ ,\\ -x\ , & x \text{ irrational}\ .\end{cases}$$

Where are the functions f, g, $|f|$, $|g|$ and $f\cdot g$ continuous?

10 Let $f:\mathbb{R}\to\mathbb{R}$ be defined by

$$f(x):=\begin{cases} 1/n\ , & x\in\mathbb{Q} \text{ and } x=m/n \text{ in lowest terms}\ ,\\ 0\ , & x\in\mathbb{R}\backslash\mathbb{Q}\ .\end{cases}$$

Show that f is continuous at each irrational number and discontinuous at each rational number.[9] (Hint: For each $x\in\mathbb{Q}$ there is, by Proposition I.10.11, a sequence $x_n\in\mathbb{R}\backslash\mathbb{Q}$ such that $x_n\to x$. So f cannot be continuous at x.
Let $x\in\mathbb{R}\backslash\mathbb{Q}$ and $\varepsilon>0$. Then there are only finitely many $n\in\mathbb{N}$ such that $n\leq 1/\varepsilon$. Thus there is some $\delta>0$ such that no $q=m/n$ with $n\leq 1/\varepsilon$ is in $(x-\delta,x+\delta)$. That is, for $y=m/n\in(x-\delta,x+\delta)$, we have $f(y)=f(m/n)=1/n<\varepsilon$.)

11 Consider the function

$$f:\mathbb{R}^2\to\mathbb{R}\ ,\quad (x,y)\mapsto\begin{cases} xy/(x^2+y^2)\ , & (x,y)\neq(0,0)\ ,\\ 0\ , & (x,y)=(0,0)\ ,\end{cases}$$

and, for a fixed $x_0\in\mathbb{R}$, define

$$f_1:\mathbb{R}\to\mathbb{R}\ ,\quad x\mapsto f(x,x_0)\ ,\qquad f_2:\mathbb{R}\to\mathbb{R}\ ,\quad x\mapsto f(x_0,x)\ .$$

Prove the following:

(a) f_1 and f_2 are continuous.

(b) f is continuous on $\mathbb{R}^2\setminus\{(0,0)\}$ and discontinuous at $(0,0)$. (Hint: For a null sequence (x_n) consider $f(x_n,x_n)$.)

12 Show that any linear function from $\mathbb{K}^n$ to $\mathbb{K}^m$ is Lipschitz continuous. (Hint: Use Proposition II.3.12 with suitable norms.)

---

[8] See Example I.4.4(c).

[9] It can be shown that there is *no* function from $\mathbb{R}$ to $\mathbb{R}$ which is continuous at each rational number and discontinuous at each irrational number (see Exercise V.4.5).

13 Suppose that V and W are normed vector spaces and $f : V \to W$ is a continuous group homomorphism from $(V,+)$ to $(W,+)$. Prove that f is linear. (Hint: If $\mathbb{K} = \mathbb{R}$, $x \in V$ and $q \in \mathbb{Q}$, then $f(qx) = qf(x)$. See also Exercise 6.)

14 Let $\big(E, (\cdot\,|\,\cdot)\big)$ be an inner product space and $x_0 \in E$. Show that the functions

$$E \to \mathbb{K}\ , \quad x \mapsto (x\,|\,x_0)\ , \qquad E \to \mathbb{K}\ , \quad x \mapsto (x_0\,|\,x)$$

are continuous.

15 Let $A \in \mathrm{End}(\mathbb{K}^n)$. Prove that the function

$$\mathbb{K}^n \to \mathbb{K}\ , \quad x \mapsto (Ax\,|\,x)$$

is continuous. (Hint: Use Exercise 12 and the Cauchy-Schwarz inequality.)

16 Let $n \in \mathbb{N}^\times$. The **determinant** of a matrix $A = [a_{jk}] \in \mathbb{K}^{n\times n}$ is defined by (see Exercise I.9.6)

$$\det A = \sum_{\sigma\in \mathsf{S}_n} (\operatorname{sign}\sigma) a_{1\sigma(1)} \cdot \cdots \cdot a_{n\sigma(n)}\ .$$

Show that the function

$$\mathbb{K}^{n\times n} \to \mathbb{K}\ , \quad A \mapsto \det A$$

is continuous (see Exercise II.3.14). (Hint: Use the bijection

$$\mathbb{K}^{m\times n} \to \mathbb{K}^{mn}\ , \quad \begin{bmatrix} a_{11}, & \ldots, & a_{1n} \\ \vdots & & \vdots \\ a_{m1}, & \ldots, & a_{mn} \end{bmatrix} \mapsto (a_{11}, \ldots, a_{1n}, a_{21}, \ldots, a_{mn})$$

to define the natural topology on $\mathbb{K}^{m\times n}$.)

17 Let X and Y be metric spaces and $f : X \to Y$. For $x \in X$, the function

$$\omega_f(x,\cdot) : (0,\infty) \to \mathbb{R}\ , \quad \varepsilon \mapsto \sup_{y,z\in \mathbb{B}(x,\varepsilon)} d\big(f(y), f(z)\big)$$

is called the **modulus of continuity** of f. Set

$$\omega_f(x) := \inf_{\varepsilon>0} \omega_f(x,\varepsilon)\ .$$

Show that f is continuous at x if and only if $\omega_f(x) = 0$.

18 Show that the square root function $w : \mathbb{R}^+ \to \mathbb{R}$, $x \mapsto \sqrt{x}$ is continuous but not Lipschitz continuous. Show that $w\,|\,[a,\infty)$ is Lipschitz continuous for each $a > 0$.

2 The Fundamentals of Topology

For a deeper understanding of continuous functions, we introduce in this section some of the basic concepts of topological spaces. The main result is Theorem 2.20 which characterizes continuous functions as structure preserving functions between topological spaces.

Open Sets

In the following, $X := (X, d)$ is a metric space. An element a of a subset A of X is called an **interior point** of A if there is a neighborhood U of a such that $U \subseteq A$. The set A is called **open** if every point of A is an interior point.

2.1 Remarks (a) Clearly, a is an interior point of A if and only if there is some $\varepsilon > 0$ such that $\mathbb{B}(a, \varepsilon) \subseteq A$.

(b) A is open if and only if A is a neighborhood of each of its points.

2.2 Example The open ball $\mathbb{B}(a, r)$ is open.

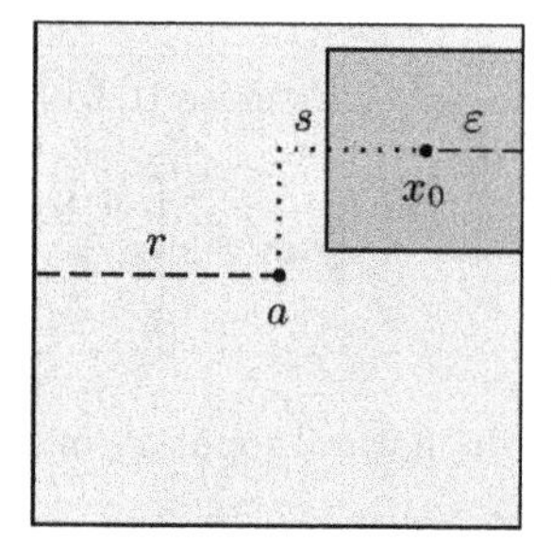

Proof For $x_0 \in \mathbb{B}(a, r)$, set $s := d(x_0, a)$. Then $\varepsilon := r - s$ is positive. For all $x \in \mathbb{B}(x_0, \varepsilon)$ we have

$$d(x, a) \leq d(x, x_0) + d(x_0, a) < \varepsilon + s = r ,$$

and so $\mathbb{B}(x_0, \varepsilon)$ is contained in $\mathbb{B}(a, r)$. This shows that x_0 is an interior point of $\mathbb{B}(a, r)$. ■

2.3 Remarks (a) The concepts 'interior point' and 'open set' depend on the surrounding metric space X. It is sometimes useful to make this explicit by saying 'a is an interior point of A with respect to X', or 'A is open in X'.

For example, an open ball in $\mathbb{R}$, that is, an open interval J, is open in $\mathbb{R}$ by the preceding example. However, if we consider $\mathbb{R}$ as embedded in $\mathbb{R}^2$, then J is not open in $\mathbb{R}^2$.

(b) Let $X = (X, \|\cdot\|)$ be a normed vector space and $\|\cdot\|_1$ and $\|\cdot\|$ equivalent norms on X. Then, by Remark II.3.13(d),

$$A \text{ is open in } (X, \|\cdot\|) \Longleftrightarrow A \text{ is open in } (X, \|\cdot\|_1) .$$

Thus if A is open with respect to a particular norm, it is open with respect to all equivalent norms.

(c) It follows from Example 2.2 that every point in a metric space has an *open* neighborhood. ■

2.4 Proposition *Let $\mathcal{T} := \{O \subseteq X \; ; \; O$ is open$\}$ be a family of open sets.*

(i) $\emptyset, X \in \mathcal{T}$.

(ii) *If $O_\alpha \in \mathcal{T}$ for all $\alpha \in \mathsf{A}$, then $\bigcup_\alpha O_\alpha \in \mathcal{T}$. That is,* arbitrary unions of open sets are open.

(iii) *If $O_0, \ldots, O_n \in \mathcal{T}$, then $\bigcap_{k=0}^n O_k \in \mathcal{T}$. That is,* finite intersections of open sets are open.

Proof (i) It is obvious that X is in $\mathcal{T}$, and, from Remark I.2.1(a), $\emptyset$ is also open.

(ii) Let A be an index set, $O_\alpha \in \mathcal{T}$ for all $\alpha \in \mathsf{A}$, and x_0 a point of $\bigcup_\alpha O_\alpha$. Then there is some $\alpha_0 \in \mathsf{A}$ such that $x_0 \in O_{\alpha_0}$. Since O_{α_0} is open, there is some neighborhood U of x_0 in X such that $U \subseteq O_{\alpha_0} \subseteq \bigcup_\alpha O_\alpha$. Hence $\bigcup_\alpha O_\alpha$ is open.

(iii) Let $O_0, \ldots, O_n \in \mathcal{T}$ and $x_0 \in \bigcap_{k=0}^n O_k$. Then there are positive numbers ε_k such that $\mathbb{B}(x_0, \varepsilon_k) \subseteq O_k$ for $k = 0, \ldots, n$. Set $\varepsilon := \min\{\varepsilon_0, \ldots, \varepsilon_n\} > 0$. Then $\mathbb{B}(x_0, \varepsilon)$ is contained in each O_k, and so $\mathbb{B}(x_0, \varepsilon) \subseteq \bigcap_{k=0}^n O_k$. ■

Properties (i)–(iii) of Proposition 2.4 involve the set operations $\bigcup$ and $\bigcap$, but do not involve the metric. This suggests the following generalization of the concept of a metric space: Let M be a set and $\mathcal{T} \subseteq \mathcal{P}(M)$, a set of subsets satisfying (i)–(iii). Then $\mathcal{T}$ is called a **topology** on M, and the elements of $\mathcal{T}$ are called the **open sets** with respect to $\mathcal{T}$. Finally the pair $(M, \mathcal{T})$ is a called a **topological space**.

2.5 Remarks **(a)** Let $\mathcal{T} \subseteq \mathcal{P}(X)$ be the family of sets of Proposition 2.4. Then $\mathcal{T}$ is called the **topology on** X **induced from the metric** d. If X is a normed vector space with metric induced from the norm, then $\mathcal{T}$ is called the **norm topology**.

(b) Let $(X, \|\cdot\|)$ be a normed vector space, and $\|\cdot\|_1$ a norm on X which is equivalent to $\|\cdot\|$. Let $\mathcal{T}_{\|\cdot\|}$ and $\mathcal{T}_{\|\cdot\|_1}$ be the norm topologies induced from $(X, \|\cdot\|)$ and $(X, \|\cdot\|_1)$. By Remark 2.3(b), $\mathcal{T}_{\|\cdot\|}$ and $\mathcal{T}_{\|\cdot\|_1}$ coincide, that is, *equivalent norms induce the same topology* on X. ■

Closed Sets

A subset A of the metric space X is called **closed** in X if A^c is open[1] in X.

2.6 Proposition

(i) $\emptyset$ and X *are closed.*

(ii) *Arbitrary intersections of closed sets are closed.*

(iii) *Finite unions of closed sets are closed.*

Proof These claims follow easily from Proposition 2.4 and Proposition I.2.7(iii). ■

[1] Note that A not being open does not imply that A is closed. For example, let $X := \mathbb{R}$ and $A := [0, 1)$. Then A is neither open nor closed in $\mathbb{R}$.

2.7 Remarks (a) Infinite intersections of open sets need not be open.

Proof In $\mathbb{R}$ we have, for example, $\bigcap_{n=1}^{\infty} \mathbb{B}(0,1/n) = \{0\}$. ∎

(b) Infinite unions of closed sets need not be closed.

Proof For example, $\bigcup_{n=1}^{\infty} \big[\mathbb{B}(0,1/n)\big]^c = \mathbb{R}^\times$ in $\mathbb{R}$. ∎

Let $A \subseteq X$ and $x \in X$. We call x an **accumulation point** of A if every neighborhood of x in X has a nonempty intersection with A. The element $x \in X$ is called a **limit point** of A if every neighborhood of x in X contains a point of A other than x. Finally we set

$$\overline{A} := \{\, x \in X \ ;\ x \text{ is an accumulation point of } A \,\} \ .$$

Clearly any element of A and any limit point of A is an accumulation point of A. Indeed $\overline{A}$ is the union of A and the set of limit points of A.

2.8 Proposition *Let A be a subset of a metric space X.*

(i) $A \subseteq \overline{A}$.

(ii) $A = \overline{A} \Longleftrightarrow A$ *is closed.*

Proof Claim (i) is clear.

(ii) '$\Longrightarrow$' Let $x \in A^c = (\overline{A})^c$. Since x is not an accumulation point of A, there is some $U \in \mathcal{U}(x)$ such that $U \cap A = \emptyset$. Thus $U \subseteq A^c$, that is, x is an interior point of A^c. Consequently A^c is open and A is closed in X.

'$\Longleftarrow$' Let A be closed in X. Then A^c is open in X. For any $x \in A^c$, there is some $U \in \mathcal{U}(x)$ such that $U \subseteq A^c$. This means that U and A are disjoint, and so x is not an accumulation point of A, that is, $x \in (\overline{A})^c$. Hence we have proved the inclusion $A^c \subseteq (\overline{A})^c$, which is equivalent to $\overline{A} \subseteq A$. With (i), this implies $\overline{A} = A$. ∎

The limit points of a set A are the limits of certain sequences in A.

2.9 Proposition *An element x of X is a limit point of A if and only if there is a sequence (x_k) in $A\backslash\{x\}$ which converges to x.*

Proof Let x be a limit point of A. For each $k \in \mathbb{N}^\times$, choose some element $x_k \neq x$ in $\mathbb{B}(x,1/k)$. Then (x_k) is a sequence in $A\backslash\{x\}$ such that $x_k \to x$.

Conversely, let (x_k) be a sequence in $A\backslash\{x\}$ such that $x_k \to x$. Then, for each neighborhood U of x, there is some $k \in \mathbb{N}$ such that $x_k \in U$. This means that $x_k \in U \cap \big(A\backslash\{x\}\big)$. Hence each neighborhood of x contains an element of A other than x. ∎

2.10 Corollary *An element x of X is an accumulation point of A if and only if there is a sequence (x_k) in A such that $x_k \to x$.*

Proof If x is a limit point, then the claim follows from Proposition 2.9. Otherwise, if x is an accumulation point, but not a limit point of A, then there is a neighborhood U of x such that $U \cap A = \{x\}$. Thus x is in A, and the constant sequence (x_k) with $x_k = x$ for all $k \in \mathbb{N}$ has the desired property. ∎

We can now characterize closed sets using convergent sequences.

2.11 Proposition *For $A \subseteq X$, the following are equivalent:*

(i) *A is closed.*

(ii) *A contains all its limit points.*

(iii) *Every sequence in A which converges in X, has its limit in A.*

Proof '(i)⇒(ii)' Any limit point of A is also an accumulation point and so is contained in $\overline{A}$. By (i) and Proposition 2.8, $A = \overline{A}$, and so all limit points are in A.

'(ii)⇒(iii)' Let (x_k) be a sequence in A such that $x_k \to x$ in X. Then, by Corollary 2.10, x is an accumulation point of A. This means that, either x is in A, or x is a limit point of A, so, by assumption, x is in A.

'(iii)⇒(i)' This implication follows from Proposition 2.8 and Corollary 2.10. ∎

The Closure of a Set

Let A be a subset of a metric space X. Define the **closure of** A by

$$\mathrm{cl}(A) := \mathrm{cl}_X(A) := \bigcap_{B \in M} B$$

with

$$M := \{\, B \subseteq X \ ; \ B \supseteq A \text{ and } B \text{ is closed in } X \,\} .$$

Since X is closed and contains A, the set M is nonempty and the definition makes sense. By Proposition 2.6(ii), $\mathrm{cl}(A)$ is closed. Since $A \subseteq \mathrm{cl}(A)$, the closure of A is precisely the smallest closed set which contains A, that is, any closed set which contains A, also contains $\mathrm{cl}(A)$.

In the next proposition we show that the closure of A is simply the set of all accumulation points of A, that is, $\overline{A} = \mathrm{cl}(A)$.

2.12 Proposition *Let A be a subset of a metric space X. Then $\overline{A} = \mathrm{cl}(A)$.*

Proof (i) First we prove that $\overline{A} \subseteq \mathrm{cl}(A)$. If $\mathrm{cl}(A) = X$, the statement is clearly true. Suppose otherwise that $\mathrm{cl}(A) \neq X$ and $x \in U := \big(\mathrm{cl}(A)\big)^c$. Since $\mathrm{cl}(A)$ is closed, U is open and hence is a neighborhood of x. It follows from $A \subseteq \mathrm{cl}(A)$ that A and U are disjoint, that is, x is not an accumulation point of A. This implies that $\big(\mathrm{cl}(A)\big)^c \subseteq (\overline{A})^c$ and so $\overline{A} \subseteq \mathrm{cl}(A)$.

(ii) We now prove the opposite inclusion, $\mathrm{cl}(A) \subseteq \overline{A}$. Once again the case $\overline{A} = X$ is trivial. If $x \notin \overline{A}$, then there is an open neighborhood U of x such that $U \cap A = \emptyset$, that is, A is contained in the closed set U^c. Thus $x \in U \subseteq \big(\mathrm{cl}(A)\big)^c$ and we have proved that $(\overline{A})^c \subseteq \big(\mathrm{cl}(A)\big)^c$, and equivalently $\mathrm{cl}(A) \subseteq \overline{A}$. ■

The following corollary collects some easy consequences of the fact that $\overline{A}$ is the smallest closed set which contains A.

2.13 Corollary *Let A and B be subsets of X.*

(i) $A \subseteq B \Rightarrow \overline{A} \subseteq \overline{B}$.

(ii) $\overline{(\overline{A})} = \overline{A}$.

(iii) $\overline{A \cup B} = \overline{A} \cup \overline{B}$.

Proof Claims (i) and (ii) follow directly from Proposition 2.12.

To prove (iii), we note first that, by Propositions 2.6(iii) and 2.12, $\overline{A} \cup \overline{B}$ is closed. Since $\overline{A} \cup \overline{B}$ contains $A \cup B$, Proposition 2.12 implies that $\overline{A \cup B} \subseteq \overline{A} \cup \overline{B}$. On the other hand $\overline{A \cup B}$ is also closed. Since $A \subseteq \overline{A \cup B}$ and $B \subseteq \overline{A \cup B}$, we get the inclusions $\overline{A} \subseteq \overline{A \cup B}$ and $\overline{B} \subseteq \overline{A \cup B}$. Together, these imply $\overline{A} \cup \overline{B} \subseteq \overline{A \cup B}$. ■

This corollary implies that the function $h \colon \mathcal{P}(X) \to \mathcal{P}(X)$, $A \mapsto \overline{A}$ is increasing and **idempotent**, that is, $h \circ h = h$.

The Interior of a Set

The relationship between closed sets, accumulation points and the closure has a parallel for open sets which we describe in this section. Taking the role of the closure is the **interior** of A, defined by

$$\mathrm{int}(A) := \mathrm{int}_X(A) := \bigcup \{\, O \subseteq A \ ; \ O \text{ is open in } X \,\} .$$

Clearly $\mathrm{int}(A)$ is a subset of A, and, by Proposition 2.4(ii), $\mathrm{int}(A)$ is open. Thus $\mathrm{int}(A)$ is the largest open subset of A. The role of accumulation points is taken by interior points and we define

$$\mathring{A} := \{\, a \in A \ ; \ a \text{ is an interior point of } A \,\} .$$

Then, corresponding to Proposition 2.12, we have the following:

2.14 Proposition *Let A be a subset of a metric space X. Then $\mathring{A} = \mathrm{int}(A)$.*

Proof (i) For each $a \in \mathring{A}$, there is an open neighborhood U of a such that $U \subseteq A$. Thus $a \in U \subseteq \mathrm{int}(A)$, and so we have proved that $\mathring{A} \subseteq \mathrm{int}(A)$.

(ii) Conversely, let $a \in \mathrm{int}(A)$. Then there is an open subset O of A such that $a \in O$. Thus O is a neighborhood of a which is contained in A, that is, a is an interior point of A. Thus we have the inclusion $\mathrm{int}(A) \subseteq \mathring{A}$. ■

The following corollary is an immediate consequence of this proposition.

2.15 Corollary *Let A and B be subsets of X.*

(i) $A \subseteq B \Rightarrow \mathring{A} \subseteq \mathring{B}$.

(ii) $(\mathring{A})^\circ = \mathring{A}$.

(iii) *A is open* $\Leftrightarrow A = \mathring{A}$.

Similar to the case of the closure, the function $\mathcal{P}(X) \to \mathcal{P}(X),\ A \mapsto \mathring{A}$ is increasing and idempotent.

The Boundary of a Set

Intuitively, we expect that the boundary of a disk in the plane is the circle which encloses it. This notion of what the boundary should be can be made precise using the concepts of open and closed sets. Specifically, for a subset A of a metric space X, the (topological) **boundary of** A is defined by $\partial A := \overline{A} \backslash \mathring{A}$. For example, the boundary of X is empty, that is, $\partial X = \emptyset$.

2.16 Proposition *Let A be a subset of X.*

(i) *∂A is closed.*

(ii) *x is in ∂A if and only if every neighborhood of x has nonempty intersection with both A and A^c.*

Proof These claims follow immediately from $\partial A = \overline{A} \cap (\mathring{A})^c$. ■

The Hausdorff Condition

The following proposition shows that, in metric spaces, any two distinct points have disjoint neighborhoods.

2.17 Proposition *Let $x, y \in X$ be such that $x \neq y$. Then there are a neighborhood U of x and a neighborhood V of y such that $U \cap V = \emptyset$.*

Proof Since $x \neq y$, we have $\varepsilon := d(x,y)/2 > 0$. Set $U := \mathbb{B}(x,\varepsilon)$ and $V := \mathbb{B}(y,\varepsilon)$. Suppose that $U \cap V \neq \emptyset$ so that there is some $z \in U \cap V$. Then, by the triangle inequality,

$$2\varepsilon = d(x,y) \leq d(x,z) + d(z,y) < \varepsilon + \varepsilon = 2\varepsilon \ ,$$

a contradiction. Thus U and V are disjoint. ■

The claim of Proposition 2.17 is called the **Hausdorff condition**. To prove this condition, we have made essential use of the existence of a metric. Indeed there are (non-metric) topological spaces for which Proposition 2.17 fails. A simple example of such a topological space appears in Exercise 10.

One easy consequence of the Hausdorff condition is

$$\bigcap \{\, U \ ; \ U \in \mathcal{U}_X(x) \,\} = \{x\} \ , \qquad x \in X \ ,$$

meaning that there are sufficiently many neighborhoods to distinguish the points of a metric space.

2.18 Corollary *Any one element subset of a metric space is closed.*

Proof[2] Fix $x \in X$. If $X = \{x\}$, then the claim follows from Proposition 2.6(i). Otherwise, if $y \in \{x\}^c$, then, by Proposition 2.17, there are neighborhoods U of x and V of y such that $U \cap V = \emptyset$. In particular, $\{x\} \cap V \subseteq U \cap V = \emptyset$ and so $V \subseteq \{x\}^c$. Thus $\{x\}^c$ is open. ■

Examples

We illustrate these new concepts with examples which, in particular, show that the previously defined notions, 'open interval', 'closed interval', 'open ball' and 'closed ball', are consistent with the topological concepts.

2.19 Examples **(a)** The open interval $(a,b) \subseteq \mathbb{R}$ is open in $\mathbb{R}$.

(b) The closed interval $[a,b] \subseteq \mathbb{R}$ is closed in $\mathbb{R}$.

(c) Let $I \subseteq \mathbb{R}$ be an interval, $a := \inf I$ and $b := \sup I$. Then

$$\partial I = \begin{cases} \emptyset \ , & I = \mathbb{R} \text{ or } I = \emptyset \ , \\ \{a\} \ , & a \in \mathbb{R} \text{ and } b = \infty \ , \\ \{b\} \ , & b \in \mathbb{R} \text{ and } a = -\infty \ , \\ \{a,b\} \ , & -\infty < a < b < \infty \ , \\ \{a\} \ , & a = b \in \mathbb{R} \ . \end{cases}$$

[2] This is also an easy consequence of Proposition 2.11(iii) (see also Remark 2.29(d)).

(d) The closed ball $\bar{\mathbb{B}}(x,r)$ is closed.

Proof If $X = \bar{\mathbb{B}}(x,r)$ there is nothing to show. So we suppose that $\bar{\mathbb{B}}(x,r) \neq X$ and y is not in $\bar{\mathbb{B}}(x,r)$, that is, $\varepsilon := d(x,y) - r > 0$. Then, for $z \in \mathbb{B}(y,\varepsilon)$, it follows from the reversed triangle inequality that

$$d(x,z) \geq d(x,y) - d(y,z) > d(x,y) - \varepsilon = r \ .$$

Hence the ball $\mathbb{B}(y,\varepsilon)$ is contained in $\big(\bar{\mathbb{B}}(x,r)\big)^c$. Since this holds for all $y \in \big(\bar{\mathbb{B}}(x,r)\big)^c$, $\big(\bar{\mathbb{B}}(x,r)\big)^c$ is open. ■

(e) In any metric space, $\overline{\mathbb{B}(x,r)} \subseteq \bar{\mathbb{B}}(x,r)$ for $r \geq 0$. If X is a normed vector space[3] and $r > 0$, then $\overline{\mathbb{B}(x,r)} = \bar{\mathbb{B}}(x,r)$.

Proof The first claim is a consequence of (d) and Proposition 2.12.

For the second claim, suppose that X is a normed vector space and $r > 0$. It suffices to show the inclusion $\bar{\mathbb{B}}(x,r) \subseteq \overline{\mathbb{B}(x,r)}$. Suppose, to the contrary, that $\overline{\mathbb{B}(x,r)} \subset \bar{\mathbb{B}}(x,r)$. Choose some $y \in \bar{\mathbb{B}}(x,r) \backslash \overline{\mathbb{B}(x,r)}$ and note that $d(y,x) = \|y - x\| = r > 0$, and therefore $x \neq y$. For $\varepsilon \in (0,1)$, define

$$x_\varepsilon := x + (1-\varepsilon)(y-x) = \varepsilon x + (1-\varepsilon)y \ .$$

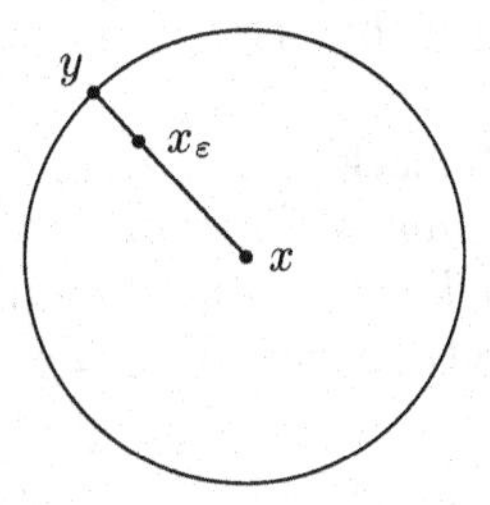

Then $\|x - x_\varepsilon\| = (1-\varepsilon)\,\|y - x\| = (1-\varepsilon)r < r$ and $\|y - x_\varepsilon\| = \varepsilon\,\|x - y\| = \varepsilon r > 0$. Now let (ε_k) be a null sequence in $(0,1)$ and $x_k := x_{\varepsilon_k}$ for all $k \in \mathbb{N}$. Then (x_k) is a sequence in $\mathbb{B}(x,r)$ such that $x_k \to y$. By Proposition 2.10, y is an accumulation point of $\mathbb{B}(x,r)$, that is, $y \in \overline{\mathbb{B}(x,r)}$. But this contradicts our choice of y. ■

(f) In any normed vector space X,

$$\partial\mathbb{B}(x,r) = \partial\bar{\mathbb{B}}(x,r) = \{\, y \in X \ ; \ \|x - y\| = r \,\} \ .$$

Proof This follows from (e). ■

(g) The **n-sphere** $S^n := \{\, x \in \mathbb{R}^{n+1} \ ; \ |x| = 1 \,\}$ is closed in $\mathbb{R}^{n+1}$.

Proof Since $S^n = \partial\mathbb{B}^{n+1}$, the claim follows from Proposition 2.16(i). ■

A Characterization of Continuous Functions

We now present the previously announced main result of this section.

2.20 Theorem *Let $f : X \to Y$ be a function between metric spaces X and Y. Then the following are equivalent:*

(i) *f is continuous.*

(ii) *$f^{-1}(O)$ is open in X for each open set O in Y.*

(iii) *$f^{-1}(A)$ is closed in X for each closed set A in Y.*

---

[3]There are metric spaces in which $\overline{\mathbb{B}(x,r)}$ is a *proper* subset of $\bar{\mathbb{B}}(x,r)$, as Exercise 3 shows.

Proof '(i)⇒(ii)' Let $O \subseteq Y$ be open. If $f^{-1}(O) = \emptyset$, then the claim follows from Proposition 2.4(i). Thus we suppose that $f^{-1}(O) \neq \emptyset$. Since f is continuous, for each $x \in f^{-1}(O)$, there is an open neighborhood U_x of x in X such that $f(U_x) \subseteq O$. This implies

$$x \in U_x \subseteq f^{-1}(O) \ , \qquad x \in f^{-1}(O) \ ,$$

from which we get

$$\bigcup_{x \in f^{-1}(O)} U_x = f^{-1}(O) \ .$$

By Example 2.2 and Proposition 2.4(iii), $f^{-1}(O)$ is open in X.

'(ii)⇒(iii)' Let $A \subseteq Y$ be closed. Then A^c is open in Y. By (ii) and Proposition I.3.8(iv′), $f^{-1}(A^c) = \bigl(f^{-1}(A)\bigr)^c$ is open in X. Thus $f^{-1}(A)$ is closed in X.

'(iii)⇒(i)' Let $x \in X$. If V is an open neighborhood of $f(x)$ in Y, then V^c is closed in Y. By Proposition I.3.8(iv′) and our hypothesis, $\bigl(f^{-1}(V)\bigr)^c = f^{-1}(V^c)$ is closed in X, that is, $U := f^{-1}(V)$ is open in X. Since $x \in U$, U is a neighborhood of x such that $f(U) \subseteq V$. This means that f is continuous at x. ■

2.21 Remark According to this theorem, *a function is continuous if and only if the preimage of any open set is open, if and only if the preimage of any closed set is closed.* For another formulation of this important result, we denote the topology of a metric space X by $\mathcal{T}_X$, that is,

$$\mathcal{T}_X := \{\, O \subseteq X \ ; \ O \text{ is open in } X \,\} \ .$$

Then

$$f : X \to Y \text{ is continuous} \Longleftrightarrow f^{-1} : \mathcal{T}_Y \to \mathcal{T}_X \ ,$$

that is, $f : X \to Y$ is continuous if and only if the image of $\mathcal{T}_Y$ under the set valued function $f^{-1} : \mathcal{P}(Y) \to \mathcal{P}(X)$ is contained in $\mathcal{T}_X$. ■

The following examples show how Theorem 2.20 can be used to prove that certain sets are open or closed.

2.22 Examples **(a)** Let X and Y be metric spaces, and $f : X \to Y$ continuous. Then, for each $y \in Y$, the fiber $f^{-1}(y)$ of f is closed in X, that is, *the solution set of the equation $f(x) = y$ is closed.*

Proof This follows from Corollary 2.18 and Theorem 2.20. ■

(b) Let $k, n \in \mathbb{N}^\times$ be such that $k \leq n$. Then $\mathbb{K}^k$ is closed in $\mathbb{K}^n$.

Proof If $k = n$ the claim is clear. For $k < n$, consider the projection

$$\mathrm{pr} : \mathbb{K}^n \to \mathbb{K}^{n-k} \ , \quad (x_1, \ldots, x_n) \mapsto (x_{k+1}, \ldots, x_n) \ .$$

Then Example 1.3(h) shows that this function is continuous. Moreover $\mathbb{K}^k = \mathrm{pr}^{-1}(0)$. Hence the claim follows from (a). ■

(c) *Solution sets of inequalities* Let $f : X \to \mathbb{R}$ be continuous and $r \in \mathbb{R}$. Then $\{\, x \in X \,;\, f(x) \le r \,\}$ is closed in X and $\{\, x \in X \,;\, f(x) < r \,\}$ is open in X.

Proof Clearly

$$\{\, x \in X \,;\, f(x) \le r \,\} = f^{-1}\big((-\infty, r]\big) \quad \text{and} \quad \{\, x \in X \,;\, f(x) < r \,\} = f^{-1}\big((-\infty, r)\big) \,.$$

Hence the claims follow from Examples 2.19(a), (b) and Theorem 2.20. ■

(d) The **closed n-dimensional unit cube**

$$I^n := \{\, x \in \mathbb{R}^n \,;\, 0 \le x_k \le 1,\ 1 \le k \le n \,\}$$

is closed in $\mathbb{R}^n$.

Proof Let $\mathrm{pr}_k : \mathbb{R}^n \to \mathbb{R},\ (x_1, \ldots, x_n) \mapsto x_k$ be the kth projection. Then

$$I^n = \bigcap_{k=1}^{n} \big(\{\, x \in \mathbb{R}^n \,;\, \mathrm{pr}_k(x) \le 1 \,\} \cap \{\, x \in \mathbb{R}^n \,;\, \mathrm{pr}_k(x) \ge 0 \,\}\big) \,.$$

By (c), I^n is a finite intersection of closed sets, and hence, by Proposition 2.6, is itself closed. ■

(e) Continuous images of closed (or open) sets need not be closed (or open).

Proof (i) Let $X := \mathbb{R}^2$ and $A := \{\, (x, y) \in \mathbb{R}^2 \,;\, xy = 1 \,\}$. Since the function $\mathbb{R}^2 \to \mathbb{R}$, $(x, y) \mapsto xy$ is continuous (see Proposition 1.5(ii)), it follows from (a) that the set A is closed in X. Even though the projection $\mathrm{pr}_1 : \mathbb{R}^2 \to \mathbb{R}$ is continuous, $\mathrm{pr}_1(A) = \mathbb{R}^\times$ is not closed in $\mathbb{R}$.

(ii) For the second claim, let $X := Y := \mathbb{R},\ O := (-1, 1)$ and $f : \mathbb{R} \to \mathbb{R},\ x \mapsto x^2$. Then O is open in $\mathbb{R}$ and f is continuous, but $f(O) = [0, 1)$ is not open in $\mathbb{R}$. ■

Continuous Extensions

Let X and Y be metric spaces. Suppose that $D \subseteq X,\ f : D \to Y$ is continuous and $a \in X$ is a limit point of D. If D is not closed, then a may not be in D and so f is not defined at a. In this section we consider whether $f(a)$ can be defined so that f is continuous on $D \cup \{a\}$. If such an extension exists, then, for any sequence (x_n) in D which converges to a, $\big(f(x_n)\big)$ converges to $f(a)$. Thus, for a (not necessarily continuous) function $f : D \to Y$ and a limit point a of D, we define

$$\lim_{x \to a} f(x) = y \tag{2.1}$$

if $y \in Y$ is such that, for *each* sequence (x_n) in D which converges to a, the sequence $\big(f(x_n)\big)$ converges to y in Y.

2.23 Remarks (a) The following are equivalent:

(i) $\lim_{x\to a} f(x) = y$.

(ii) For each neighborhood V of y in Y, there is a neighborhood U of a in X such that $f(U \cap D) \subseteq V$.

Proof '(i)⇒(ii)' We prove the contrapositive. Suppose that there is a neighborhood V of y in Y such that $f(U \cap D) \not\subseteq V$ for each neighborhood U of a in X. In particular,

$$f\big(\mathbb{B}_X(a, 1/n) \cap D\big) \cap V^c \neq \emptyset\ , \qquad n \in \mathbb{N}^\times\ .$$

Thus, for each $n \in \mathbb{N}^\times$, we can choose some $x_n \in \mathbb{B}_X(a, 1/n) \cap D$ such that $f(x_n) \in V^c$. In particular, the sequence (x_n) is in D and converges to a. Since $f(x_n) \notin V$ for each n, $\big(f(x_n)\big)$ cannot converge to y.

'(ii)⇒(i)' Let (x_n) be a sequence in D such that $x_n \to a$ in X, and V a neighborhood of y in Y. By hypothesis, there is some neighborhood U of a such that $f(U \cap D) \subseteq V$. Since (x_n) converges to a, there is some $N \in \mathbb{N}$ such that $x_n \in U$ for all $n \geq N$. Thus the image $\big(f(x_n)\big)$ is contained in V for all $n \geq N$. This means that $f(x_n) \to y$. ■

(b) If $a \in D$ is a limit point of D, then

$$\lim_{x\to a} f(x) = f(a) \Longleftrightarrow f \text{ is continuous at } a\ .$$

Proof This follows from (a). ■

2.24 Proposition *Let X and Y be metric spaces, $D \subseteq X$, and $f : D \to Y$ continuous. Suppose that $a \in D^c$ is a limit point of D and there is some $y \in Y$ such that $\lim_{x\to a} f(x) = y$. Then*

$$\overline{f} : D \cup \{a\} \to Y\ , \quad x \mapsto \begin{cases} f(x)\ , & x \in D\ , \\ y\ , & x = a\ , \end{cases}$$

is a continuous extension of f to $D \cup \{a\}$.

Proof We need to prove only that $\overline{f} : D \cup \{a\} \to Y$ is continuous at a. But this follows directly from Remarks 2.23. ■

For the special case $X \subseteq \mathbb{R}$, we can define one-sided limits as follows. Suppose that $D \subseteq X$, $f : D \to Y$ is a function and $a \in X$ is a limit point of $D \cap (-\infty, a]$ (or $D \cap [a, \infty)$). Then we define[4] the **left** (or **right**) **limit**

$$\lim_{x\to a-} f(x) \quad (\text{or } \lim_{x\to a+} f(x))$$

similarly to $\lim_{x\to a} f(x)$, by allowing only sequences such that $x_n < a$ (or $x_n > a$). Analogously, we write $y = \lim_{x\to\infty} f(x)$ (or $y = \lim_{x\to-\infty} f(x)$) if, for every sequence (x_n) such that $x_n \to \infty$ (or $x_n \to -\infty$), we have $f(x_n) \to y$.

[4]We write also $f(a-) := \lim_{x\to a-} f(x)$ and $f(a+) := \lim_{x\to a+} f(x)$ when no confusion is possible.

2.25 Examples (a) Suppose that $X := \mathbb{R}$, $D := \mathbb{R}\backslash\{1\}$, $n \in \mathbb{N}^\times$ and $f : D \to \mathbb{R}$ is defined by $f(x) := (x^n - 1)/(x - 1)$. Then

$$\lim_{x\to 1} f(x) = \lim_{x\to 1} \frac{x^n - 1}{x - 1} = n .$$

Proof By Exercise I.8.1(b) we have

$$\frac{x^n - 1}{x - 1} = 1 + x + x^2 + \cdots + x^{n-1} .$$

The claim follows from this and continuity of polynomials in $\mathbb{R}$. ■

(b) For $X := \mathbb{C}$ and $D := \mathbb{C}^\times$,

$$\lim_{z\to 0} \frac{\exp(z) - 1}{z} = 1 .$$

Proof From $\exp(z) = \sum z^k/k!$ we get

$$\frac{\exp(z) - 1}{z} - 1 = \frac{z}{2}\Big[1 + \frac{z}{3} + \frac{z^2}{3\cdot 4} + \frac{z^3}{3\cdot 4\cdot 5} + \cdots\Big] .$$

Hence, for all $z \in \mathbb{C}^\times$ such that $|z| < 1$, we have the inequality

$$\Big|\frac{\exp(z) - 1}{z} - 1\Big| \le \frac{|z|}{2}\big[1 + |z| + |z^2| + |z^3| + \cdots\big] = \frac{|z|}{2(1 - |z|)} .$$

The claim then follows from

$$\lim_{z\to 0}\Big(\frac{|z|}{2(1 - |z|)}\Big) = 0$$

which is a consequence of Remark 2.23(b) and the continuity of $|z|/(1 - |z|)$ at $z = 0$. ■

(c) Let $X := D := Y := \mathbb{R}$ and $f(x) := x^n$ for $n \in \mathbb{N}$. Then

$$\lim_{x\to\infty} x^n = \begin{cases} 1 , & n = 0 , \\ \infty , & n \in \mathbb{N}^\times , \end{cases}$$

and

$$\lim_{x\to -\infty} x^n = \begin{cases} 1 , & n = 0 , \\ \infty , & n \in 2\mathbb{N}^\times , \\ -\infty , & n \in 2\mathbb{N} + 1 . \end{cases}$$

(d) Because $\lim_{x\to 0-} 1/x = -\infty$ and $\lim_{x\to 0+} 1/x = \infty$, the function $\mathbb{R}^\times \to \mathbb{R}$, $x \mapsto 1/x$ cannot be extended to a continuous function on $\mathbb{R}$. ■

Relative Topology

Let X be a metric space and Y a subset of X. Then Y is itself a metric space with respect to the metric $d_Y := d|Y \times Y$ induced from X, and so 'open in (Y, d_Y)' and 'closed in (Y, d_Y)' are well defined concepts.

There is another way of defining the open subsets of Y which completely avoids the use of a metric. This definition requires only that X be a topological space. Specifically, a subset M of Y is **open** (or **closed**) **in** Y, if there is an open set O in X (or a closed set A in X) such that $M = O \cap Y$ (or $M = A \cap Y$). If $M \subseteq Y$ is open (or closed) in Y, we say also that M is **relatively open** (or **relatively closed**) **in** Y. Using these definitions, it is easy to see that the topological structure of X induces a topological structure on Y.

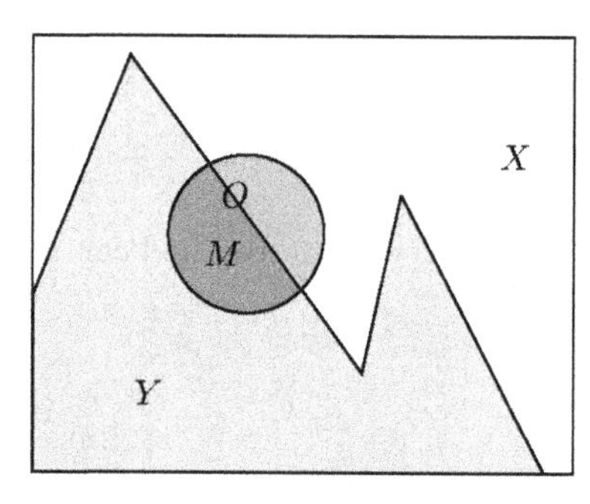

Thus we have two ways of defining the open subsets of Y. The next proposition shows that these definitions are equivalent.

2.26 Proposition *Let X be a metric space and $M \subseteq Y \subseteq X$. Then M is open (or closed) in Y if and only if M is open (or closed) in (Y, d_Y).*

Proof Without loss of generality we can assume that M is nonempty.

(i) Let M be open in Y. Then there is some open set O in X such that $M = O \cap Y$. Thus, for each $x \in M$, there is some $r > 0$ such that $\mathbb{B}_X(x, r) \subseteq O$. Since

$$\mathbb{B}_Y(x, r) = \mathbb{B}_X(x, r) \cap Y \subseteq O \cap Y = M ,$$

x is an interior point of M with respect to (Y, d_Y). Consequently M is open in (Y, d_Y).

(ii) Now let M be open in (Y, d_Y). For each $x \in M$, there is some $r_x > 0$ such that $\mathbb{B}_Y(x, r_x) \subseteq M$. Set $O := \bigcup_{x \in M} \mathbb{B}_X(x, r_x)$. Then, by Example 2.2 and Proposition 2.4(ii), O is an open subset of X. Moreover, from Proposition I.2.7(ii),

$$O \cap Y = \Bigl(\bigcup_{x \in M} \mathbb{B}_X(x, r_x)\Bigr) \cap Y = \bigcup_{x \in M} \bigl(\mathbb{B}_X(x, r_x) \cap Y\bigr) = \bigcup_{x \in M} \mathbb{B}_Y(x, r_x) = M .$$

Thus M is open in X.

(iii) Next we suppose that M is closed in Y, that is, there is a closed set A in X such that $M = Y \cap A$. Because $Y \backslash M = Y \cap A^c$, it follows from (i) that $Y \backslash M$ is open in (Y, d_Y). Hence M is closed in (Y, d_Y).

(iv) Finally, if M is closed in (Y, d_Y), then $Y \backslash M$ is open in (Y, d_Y). By (ii), $Y \backslash M$ is open in Y, and so there is an open set O in X such that $O \cap Y = Y \backslash M$. This implies $M = Y \cap O^c$, and so M is closed in Y. ■

2.27 Corollary *If $M \subseteq Y \subseteq X$, then M is open in Y if and only if $Y \backslash M$ is closed in Y.*

2.28 Examples (a) Let $X := \mathbb{R}^2$, $Y := \mathbb{R} \times \{0\}$ and $M := (0,1) \times \{0\}$. Then M is open in Y, but not in X.

(b) Let $X := \mathbb{R}$ and $Y := (0,2]$. Then $(1,2]$ is open in Y but not in X, and $(0,1]$ is closed in Y but not in X. ■

General Topological Spaces

Even though metric spaces are the natural framework for most of our discussion, in later chapters — and in other books — general topological spaces are also important. For this reason, it is useful to analyze the definitions and propositions of this section to find out which are true in any topological space. This we do in the following remarks.

2.29 Remarks Let $X = (X, \mathcal{T})$ be a topological space.

(a) As above, $A \subseteq X$ is called **closed** if A^c is open, that is, if $A^c \in \mathcal{T}$. The definitions of **accumulation point**, **limit point** and $\overline{A}$ remain unchanged. Then it is clear that Propositions 2.6 and 2.8 remain valid.

(b) A subset $U \subseteq X$ is called a **neighborhood** of a subset A of X if there is an open set O such that $A \subseteq O \subseteq U$. If $A = \{x\}$, then U is called a **neighborhood of** x. The set of all neighborhoods of x we again denote by $\mathcal{U}(x)$, or more precisely, by $\mathcal{U}_X(x)$. Clearly every point has an open neighborhood. A point x is called an **interior point** of $A \subseteq X$ if some neighborhood of x is contained in A. It is clear that these definitions are consistent with those introduced already for metric spaces.

Finally, the **interior** $\mathring{A}$ and **boundary** ∂A of $A \subseteq X$ are defined exactly as for metric spaces. It is then easy to check that Propositions 2.12 and 2.14, as well as Corollaries 2.13 and 2.15 remain true. Thus we have $\overline{A} = \mathrm{cl}(A)$ and $\mathring{A} = \mathrm{int}(A)$.

(c) Propositions 2.9 and 2.11, and Corollary 2.10 are not true in general topological spaces. Even so, the following is always true: If A is closed and (x_k) is a convergent sequence in A with $\lim x_k = x$, then x is in A. Of course, here the **convergence** of a sequence and the **limit** of a convergent sequence are defined just as in Section II.1. An analysis of the proof of Proposition 2.9 shows that the following property of metric spaces is used:

$$\left.\begin{array}{l}\text{For each point } x \in X, \text{ there is a sequence } (U_k) \text{ of neighborhoods of } x \text{ such}\\ \text{that, for any neighborhood } U \text{ of } x, \text{ there is some } k \in \mathbb{N} \text{ such that } U_k \subseteq U.\end{array}\right\} \quad (2.2)$$

For metric spaces it suffices to choose $U_k := \mathbb{B}(x, 1/k)$.

A sequence of neighborhoods (U_k) as above is called a **countable neighborhood basis** for x. A topological space for which (2.2) holds is said to satisfy the **first countability axiom**.

(d) We have already noted that Proposition 2.17 does not hold in general topological spaces. A topological space satisfying the Hausdorff condition is called a **Hausdorff space**. Proposition 2.17 shows that any metric space is a Hausdorff space.

In a Hausdorff space, Corollary 2.18 holds with exactly the same proof, and so every one element set is closed. Moreover a convergent sequence in a Hausdorff space has a unique limit.

(e) The continuity of a function between topological spaces is defined exactly as in Section 1. Thus Theorem 1.8, about the continuity of compositions, remains true. Propositions 1.5 and 1.10 are true when X is an arbitrary topological space, though the proofs must be changed so as to make a more direct use of the definition of continuity (see Exercise 19).

Finally, Theorem 2.20, the most important in this section, is true for arbitrary topological spaces. Thus a function between topological spaces is continuous if and only if the preimages of open (or closed) sets are open (or closed). Examples 2.22(a) and (c) remain true if X is a topological space and Y is a Hausdorff space (Why?).

(f) If X and Y are arbitrary topological spaces, then the first part of the proof of Theorem 1.4 shows that any continuous function from X to Y is also sequentially continuous. The second part of this same proof shows that the converse is true if X satisfies the first countability axiom.

(g) Let X and Y be topological spaces and $a \in X$ a limit point of $D \subseteq X$. Then, for $f: D \rightarrow Y$ the limit

$$\lim_{x \to a} f(x) \tag{2.3}$$

can be defined as in (2.1) only if X satisfies the first countability axiom (more precisely, if a has a countable neighborhood basis). In this case, Remark 2.23(a) remains true. If X is an arbitrary topological space, then (ii) of Remark 2.23(a) is used as the definition of (2.3). In either case, Remark 2.23(b) and Proposition 2.24 hold.

(h) If Y is a subset of a topological space X, the concepts **relatively open** (that is, **open in** Y) and **relatively closed** (that is, **closed in** Y) are defined as previously. Then

$$\mathcal{T}_Y := \{\, B \subseteq Y \ ; \ B \text{ is open in } Y \,\}$$

is a topology on Y called the **relative (or induced) topology** of Y with respect to X. Thus $(Y, \mathcal{T}_Y)$ is a topological space itself and so is a **topological subspace** of X. It is easy to see that $A \subseteq Y$ is relatively closed if and only if A is closed in $(Y, \mathcal{T}_Y)$, that is, if $A^c \in \mathcal{T}_Y$ (see Corollary 2.27). Moreover, $(Y, \mathcal{T}_Y)$ is a Hausdorff space (or satisfies the first countability axiom) if the same is true of X. If $i := i_Y : Y \rightarrow X,\ \ y \mapsto y$ is the inclusion of Example I.3.2(b), then $i^{-1}(A) = A \cap Y$ for all $A \subseteq X$. Hence, if Y has some other topology $\mathcal{T}'_Y$, then $i \colon (Y, \mathcal{T}'_Y) \rightarrow X$ is continuous if and only if $\mathcal{T}'_Y$ is stronger than the relative topology $\mathcal{T}_Y$.

(i) Let X and Y be topological spaces and A a subset of X with the relative topology. If $f: X \rightarrow Y$ is continuous at $x_0 \in A$, then $f \,|\, A \colon A \rightarrow Y$ is continuous at x_0 (see Example 1.3(k)).

Proof This follows from $f \,|\, A = f \circ i_A$ and (h). ■

Exercises

1 For the following subsets M of a metric space X, determine $\overline{M}$, $\mathring{M}$, ∂M and the set M' of all limit points of M:

(a) $M = (0,1]$, $X = \mathbb{R}$.

(b) $M = (0,1] \times \{0\}$, $X = \mathbb{R}^2$.

(c) $M = \{\, 1/n \ ; \ n \in \mathbb{N}^\times \,\}$, $X = \mathbb{R}$.

(d) $M = \mathbb{Q}$, $X = \mathbb{R}$.

(e) $M = \mathbb{R}\backslash\mathbb{Q}$, $X = \mathbb{R}$.

2 Let $\mathbb{Q}$ have the natural metric and $S := \bigl\{\, x \in \mathbb{Q} \ ; \ -\sqrt{2} < x < \sqrt{2} \,\bigr\}$. Prove or disprove the following:

(a) S is open in $\mathbb{Q}$.

(b) S is closed in $\mathbb{Q}$.

3 Let X be a nonempty set and d the discrete metric on X. Show the following:

(a) Every subset of X is open, that is, $\mathcal{P}(X)$ is the topology of (X,d).

(b) It is not true, in general, that $\bar{\mathbb{B}}(x,r) = \overline{\mathbb{B}(x,r)}$.

4 For $S := \{\, (x,y) \in \mathbb{R}^2 \ ; \ x^2 + y^2 < 1 \,\} \backslash \bigl([0,1) \times \{0\}\bigr)$, determine $(\overline{S})^0$. Is $(\overline{S})^0 = S$?

5 Let X be a metric space and $A \subseteq X$. Prove that $\mathring{A} = X \backslash \overline{(X \backslash A)}$.

6 Let X_j, $j = 1,\ldots,n$, be metric spaces and $X := X_1 \times \cdots \times X_n$. Show the following:

(a) If O_j is open in X_j for all j, then $O_1 \times \cdots \times O_n$ is open in X.

(b) If A_j is closed in X_j for all j, then $A_1 \times \cdots \times A_n$ is closed in X.

7 Let $h \colon \mathcal{P}(X) \to \mathcal{P}(X)$ be a function with the properties

(i) $h(\emptyset) = \emptyset$,

(ii) $h(A) \supseteq A$, $A \in \mathcal{P}(X)$,

(iii) $h(A \cup B) = h(A) \cup h(B)$, $A, B \in \mathcal{P}(X)$,

(iv) $h \circ h = h$.

(a) Set $\mathcal{T}_h := \{\, A^c \in \mathcal{P}(X) \ ; \ h(A) = A \,\}$ and show that $(X, \mathcal{T}_h)$ is a topological space.

(b) Given a topological space $(X, \mathcal{T})$, find a function $h \colon \mathcal{P}(X) \to \mathcal{P}(X)$ satisfying (i)–(iv) and $\mathcal{T}_h = \mathcal{T}$.

8 Let X be a metric space and $A, B \subseteq X$. Prove or disprove that $(A \cup B)^\circ = \mathring{A} \cup \mathring{B}$ and $(A \cap B)^\circ = \mathring{A} \cap \mathring{B}$.

9 Consider the metric on $\mathbb{R}$ given by $\delta(x,y) := |x-y|/(1+|x-y|)$ (see Exercise II.1.9). Show that the sets $A_n := [n,\infty)$, $n \in \mathbb{N}$, are closed and bounded in $(\mathbb{R},\delta)$, and that[5] $\bigcap_{n=0}^{k} A_n \neq \emptyset$ for each $k \in \mathbb{N}$ and $\bigcap A_n = \emptyset$.

10 Let $X := \{1,2,3,4,5\}$ and

$$\mathcal{T} := \bigl\{\, \emptyset, X, \{1\}, \{3,4\}, \{1,3,4\}, \{2,3,4,5\} \bigr\} .$$

Show that $(X, \mathcal{T})$ is a topological space and determine the closure of $\{2,4,5\}$.

[5] Compare also Exercise 3.5.

11 Let $\mathcal{T}_1$ and $\mathcal{T}_2$ be topologies on a set X. Prove or disprove that $\mathcal{T}_1 \cup \mathcal{T}_2$ and $\mathcal{T}_1 \cap \mathcal{T}_2$ are topologies on X.

12 Let X and Y be metric spaces. Prove that

$$f : X \to Y \text{ is continuous} \iff f(\overline{A}) \subseteq \overline{f(A)}, \ A \subseteq X .$$

13 Let A and B be closed subsets of a metric space X. Suppose that Y is a metric space and $g : A \to Y$ and $h : B \to Y$ are continuous functions such that

$$g \,|\, A \cap B = h \,|\, A \cap B \quad \text{if} \quad A \cap B \neq \emptyset .$$

Show that the function

$$f : A \cup B \to Y , \quad x \mapsto \begin{cases} g(x) , & x \in A , \\ h(x) , & x \in B , \end{cases}$$

is continuous.

14 A function $f : X \to Y$ between metric spaces (X, d) and (Y, δ) is called **open** if $f(\mathcal{T}_d) \subseteq \mathcal{T}_\delta$, that is, if the images of open sets are open. The function f is called **closed** if $f(A)$ is closed for any closed set A. Let d denote the natural metric and δ the discrete metric on $\mathbb{R}$. Prove the following:

(a) $\mathrm{id} : (\mathbb{R}, d) \to (\mathbb{R}, \delta)$ is open and closed, but not continuous.

(b) $\mathrm{id} : (\mathbb{R}, \delta) \to (\mathbb{R}, d)$ is continuous, but neither open nor closed.

15 Let $f : \mathbb{R} \to \mathbb{R}$, $x \mapsto \exp(x)\,\mathrm{zigzag}(x)$ (see Exercise 1.1). Then f is continuous, but neither open nor closed. (Hint: Consider Exercise II.8.10 and determine $f\bigl((-\infty, 0)\bigr)$ and $f\bigl(\{\, -(2n+1)/2 \ ; \ n \in \mathbb{N} \,\}\bigr)$.)

16 Prove that the function

$$f : [0, 2] \to [0, 2] , \quad x \mapsto \begin{cases} 0 , & x \in [0, 1] , \\ x - 1 , & x \in (1, 2] , \end{cases}$$

is continuous and closed, but not open.

17 Let $S^1 := \{\, (x, y) \in \mathbb{R}^2 \ ; \ x^2 + y^2 = 1 \,\}$, the unit circle in $\mathbb{R}^2$, with the natural metric. Show that the function

$$f : S^1 \to [0, 2) , \quad (x, y) \mapsto \begin{cases} 0 , & y \geq 0 , \\ 1 + x , & y \leq 0 , \end{cases}$$

is closed, but neither continuous nor open.

18 Let X and Y be metric spaces and

$$p : X \times Y \to X , \quad (x, y) \mapsto x$$

the canonical projection onto X. Then p is continuous and open, but not, in general, closed.

19 Prove Propositions 1.5 and 1.10 for an arbitrary topological space X.

20 Let X and Y be metric spaces and $f : X \to Y$. Show that (see Exercise 1.17)

$$A_n := \{\, x \in X \ ; \ \omega_f(x) \geq 1/n \,\}$$

is closed for each $n \in \mathbb{N}^\times$.

21 Let X be a metric space and $A \subseteq X$. Show the following:

(i) If A is complete, then A is closed in X. The converse is, in general, false.

(ii) If X is complete, then A is complete if and only if A is closed in X.

3 Compactness

We have seen that continuous images of open sets may not be open, and continuous images of closed sets may not be closed. In the next two sections we investigate certain properties of topological spaces which, in contrast, are preserved by continuous functions. These properties are of far reaching importance and are especially useful for the study of real valued functions.

Covers

In the following, $X := (X, d)$ is a metric space.

A family of sets $\{ A_\alpha \subseteq X \ ; \ \alpha \in \mathsf{A} \}$ is called a **cover** of the subset $K \subseteq X$ if $K \subseteq \bigcup_\alpha A_\alpha$. A cover is called **open** if each A_α is open in X. A subset $K \subseteq X$ is called **compact** if *every* open cover of K has a finite subfamily which is also a cover of K. In other words, $K \subseteq X$ is compact if *every* open cover of K has a *finite* **subcover**.

3.1 Examples **(a)** Let (x_k) be a convergent sequence in X with limit a. Then the set $K := \{a\} \cup \{x_k \ ; \ k \in \mathbb{N}\}$ is compact.

Proof Let $\{ O_\alpha \ ; \ \alpha \in \mathsf{A} \}$ be an open cover of K. Then there are α and $\alpha_k \in \mathsf{A}$ such that $a \in O_\alpha$ and $x_k \in O_{\alpha_k}$ for all $k \in \mathbb{N}$. Because $\lim x_k = a$, there is some $N \in \mathbb{N}$ such that $x_k \in O_\alpha$ for all $k > N$. Then $\{ O_{\alpha_k} \ ; \ 0 \leq k \leq N \} \cup \{O_\alpha\}$ is a finite subcover of the given cover of K. ■

(b) The statement of (a) is false, in general, if the limit a is not included in K.

Proof Let $X := \mathbb{R}$ and $A := \{ 1/k \ ; \ k \in \mathbb{N}^\times \}$. Set $O_1 := (1/2, 2)$ and, for all $k \geq 2$, $O_k := \big(1/(k+1), 1/(k-1)\big)$. Then $\{ O_k \ ; \ k \in \mathbb{N}^\times \}$ is an open cover of A with the property that each O_k contains exactly one element of A. Thus $\{ O_k \ ; \ k \in \mathbb{N} \}$ has no finite subcover of A. ■

(c) The set of natural numbers $\mathbb{N}$ is not compact in $\mathbb{R}$.

Proof It suffices once again to construct an open cover $\{ O_k \ ; \ k \in \mathbb{N} \}$ of $\mathbb{N}$ such that each O_k contains exactly one natural number, for example, $O_k := (k - 1/3, k + 1/3)$ for all $k \in \mathbb{N}$. ■

3.2 Proposition *Any compact set $K \subseteq X$ is closed and bounded in X.*

Proof Let $K \subseteq X$ be compact.

(i) We prove first that K is closed in X. It clearly suffices to consider the case $K \neq X$ since X is closed in X. Thus suppose that x_0 is in K^c. Because of the Hausdorff property, for each $y \in K$, there are open neighborhoods $U_y \in \mathcal{U}(y)$ and $V_y \in \mathcal{U}(x_0)$ such that $U_y \cap V_y = \emptyset$. Since $\{ U_y \ ; \ y \in K \}$ is an open cover of K, there are finitely many points $y_0, \ldots, y_m$ in K such that $K \subseteq \bigcup_{j=0}^m U_{y_j} =: U$.

By Proposition 2.4, U and $V := \bigcap_{j=0}^{m} V_{y_j}$ are open and disjoint. Thus V is a neighborhood of x_0 such that $V \subseteq K^c$, that is, x_0 is an interior point of K^c. Since this holds for each $x_0 \in K^c$, K^c is open and K is closed.

(ii) To verify the boundedness of K, fix some x_0 in X. Since, by Example 2.2, $\mathbb{B}(x_0, k)$ is open and $K \subseteq \bigcup_{k=1}^{\infty} \mathbb{B}(x_0, k) = X$, the compactness of K implies that there are $k_0, \ldots, k_m \in \mathbb{N}$ such that $K \subseteq \bigcup_{j=0}^{m} \mathbb{B}(x_0, k_j)$. In particular, $K \subseteq \mathbb{B}(x_0, N)$ where $N := \max\{k_0, \ldots, k_m\}$. Thus K is bounded. ∎

A Characterization of Compact Sets

The converse of Proposition 3.2 is false in general metric spaces (see Exercise 15) and so compact sets are *not* simply closed and bounded sets. Instead we have in the next theorem a characterization of compactness in terms of cluster points. For the proof we need the following concept which appears again in Theorem 3.10: A subset K of X is **totally bounded** if, for each $r > 0$, there are $m \in \mathbb{N}$ and $x_0, \ldots, x_m \in K$ such that $K \subseteq \bigcup_{k=0}^{m} \mathbb{B}(x_k, r)$. Obviously any totally bounded set is bounded.

3.3 Theorem *A subset $K \subseteq X$ is compact if and only if every sequence in K has a cluster point in K.*

Proof (i) First we suppose that K is compact and that there is a sequence in K with no cluster point in K. Thus, for each $x \in K$, there is an open neighborhood U_x of x which contains at most finitely many terms of the sequence. Because $\{\, U_x \,;\, x \in K \,\}$ is an open cover of K, there are $x_0, \ldots, x_m \in K$ such that $\{\, U_{x_k} \,;\, k = 0, \ldots, m \,\}$ is a cover of K. Hence K contains at most finitely many terms of the sequence. This contradiction shows that every sequence in K has a cluster point in K.

(ii) The proof of the converse is done in two steps:

(a) Let K be a subset of X with the property that each sequence in K has a cluster point in K. We claim that K is totally bounded.

Suppose, to the contrary, that K is not totally bounded. Then there is some $r > 0$ with the property that K is *not* contained in $\bigcup_{k=0}^{m} \mathbb{B}(x_k, r)$ for *any* finite set $x_0, \ldots, x_m \in K$. In particular, there is some $x_0 \in K$ such that K is not contained in $\mathbb{B}(x_0, r)$. Thus there is some $x_1 \in K \backslash \mathbb{B}(x_0, r)$. Since K is not contained in $\mathbb{B}(x_0, r) \cup \mathbb{B}(x_1, r)$, there is some $x_2 \in K \setminus \bigl[\mathbb{B}(x_0, r) \cup \mathbb{B}(x_1, r)\bigr]$. Iterating this process, we construct a sequence (x_k) in K such that x_{n+1} is not in $\bigcup_{k=0}^{n} \mathbb{B}(x_k, r)$ for all n. By hypothesis, the sequence (x_k) has a cluster point x in K, and so, in particular, there are $m, N \in \mathbb{N}^{\times}$ such that $d(x_N, x) < r/2$ and $d(x_{N+m}, x) < r/2$. The triangle inequality implies that $d(x_N, x_{N+m}) < r$, that is, x_{N+m} is in $\mathbb{B}(x_N, r)$. This contradicts the above property of the sequence (x_k) and so we have proved that K is totally bounded.

(b) Now let $\{ O_\alpha \ ; \ \alpha \in \mathsf{A} \}$ be an open cover of K. Suppose, contrary to our claim, that there is no finite subcover of $\{ O_\alpha \ ; \ \alpha \in \mathsf{A} \}$. Since K is totally bounded, for each $k \in \mathbb{N}^\times$, there is a finite set of open balls of radius $1/k$ and center in K which forms a cover of K. Then one of these open balls, B_k say, has the property that no finite subset of $\{ O_\alpha \ ; \ \alpha \in \mathsf{A} \}$ is a cover of $K \cap B_k$. Let x_k be the center of B_k for $k \in \mathbb{N}^\times$. By hypothesis, the sequence (x_k) has a cluster point $\overline{x}$ in K.

Now let $\overline{\alpha} \in \mathsf{A}$ be such that $\overline{x} \in O_{\overline{\alpha}}$. Since $O_{\overline{\alpha}}$ is open, there is some $\varepsilon > 0$ such that $\mathbb{B}(\overline{x}, \varepsilon) \subseteq O_{\overline{\alpha}}$. Since $\overline{x}$ is a cluster point of the sequence (x_k), there is some $M > 2/\varepsilon$ such that $d(x_M, \overline{x}) < \varepsilon/2$. Thus, for each $x \in B_M$, we have

$$d(x, \overline{x}) \leq d(x, x_M) + d(x_M, \overline{x}) < \frac{1}{M} + \frac{\varepsilon}{2} < \frac{\varepsilon}{2} + \frac{\varepsilon}{2} = \varepsilon \ ,$$

that is, $B_M \subseteq \mathbb{B}(\overline{x}, \varepsilon) \subseteq O_{\overline{\alpha}}$. This contradicts our choice of B_M and so the cover $\{ O_\alpha \ ; \ \alpha \in \mathsf{A} \}$ must have a finite subcover. ■

Sequential Compactness

We say that a subset $K \subseteq X$ is **sequentially compact** if every sequence in K has a subsequence which converges to an element of K.

The relationship between the cluster points of a sequence and convergent subsequences (see Proposition II.1.17) makes possible a reformulation of Theorem 3.3 in terms of sequential compactness.

3.4 Theorem *A subset of a metric space is compact if and only if it is sequentially compact.*

As an important application of Theorem 3.3 we describe the compact subsets of $\mathbb{K}^n$.

3.5 Theorem (Heine-Borel) *A subset of $\mathbb{K}^n$ is compact if and only if it is closed and bounded. In particular, an interval is compact if and only if it is closed and bounded.*

Proof By Proposition 3.2, any compact set is closed and bounded. The converse follows from the Bolzano-Weierstrass theorem (see Theorem II.5.8), Proposition 2.11 and Theorem 3.4. ■

Continuous Functions on Compact Spaces

The following theorem shows that compactness is preserved under continuous functions.

3.6 Theorem *Let X and Y be metric spaces and $f : X \to Y$ continuous. If X is compact, then $f(X)$ is compact. That is,* continuous images of compact sets are compact.

Proof Let $\{O_\alpha \; ; \; \alpha \in \mathsf{A}\}$ be an open cover of $f(X)$ in Y. By Theorem 2.20, for each $\alpha \in \mathsf{A}$, $f^{-1}(O_\alpha)$ is an open subset of X. Hence $\{f^{-1}(O_\alpha) \; ; \; \alpha \in \mathsf{A}\}$ is an open cover of the compact space X and there are $\alpha_0, \ldots, \alpha_m \in \mathsf{A}$ such that $X = \bigcup_{k=0}^m f^{-1}(O_{\alpha_k})$. It follows that $f(X) \subseteq \bigcup_{k=0}^m O_{\alpha_k}$, that is, $\{O_{\alpha_0}, \ldots, O_{\alpha_m}\}$ is a finite subcover of $\{O_\alpha \; ; \; \alpha \in \mathsf{A}\}$. Hence $f(X)$ is compact. ■

3.7 Corollary *Let X and Y be metric spaces and $f : X \to Y$ continuous. If X is compact, then $f(X)$ is bounded.*

Proof This follows directly from Theorem 3.6 and Proposition 3.2. ■

The Extreme Value Theorem

For real valued functions, Theorem 3.6 has the important consequence that a real valued continuous function on a compact set attains its minimum and maximum values.

3.8 Corollary (extreme value theorem) *Let X be a compact metric space and $f : X \to \mathbb{R}$ continuous. Then there are $x_0, x_1 \in X$ such that*

$$f(x_0) = \min_{x \in X} f(x) \quad \text{and} \quad f(x_1) = \max_{x \in X} f(x) \; .$$

Proof From Theorem 3.6 and Proposition 3.2 we know that $f(X)$ is closed and bounded in $\mathbb{R}$. Thus $m := \inf\bigl(f(X)\bigr)$ and $M := \sup\bigl(f(X)\bigr)$ exist in $\mathbb{R}$. By Proposition I.10.5, there are sequences (y_n) and (z_n) in $f(X)$ which converge to m and M in $\mathbb{R}$. Since $f(X)$ is closed, Proposition 2.11 implies that m and M are in $f(X)$, that is, there are $x_0, x_1 \in X$ such that $f(x_0) = m$ and $f(x_1) = M$. ■

The importance of this result can be seen in the following examples.

3.9 Examples (a) *All norms on $\mathbb{K}^n$ are equivalent.*

Proof (i) Let $|\cdot|$ be the Euclidean norm and $\|\cdot\|$ an arbitrary norm on $\mathbb{K}^n$. Then it suffices to show the equivalence of these two norms, that is, the existence of a positive constant C such that

$$C^{-1}\,|x| \le \|x\| \le C\,|x| \; , \qquad x \in \mathbb{K}^n \; . \tag{3.1}$$

(ii) Set $S := \{x \in \mathbb{K}^n \; ; \; |x| = 1\}$. From Example 1.3(j) we know that the function $|\cdot| : \mathbb{K}^n \to \mathbb{R}$ is continuous, and so, by Example 2.22(a), S is closed in $\mathbb{K}^n$. Of course, S is also bounded in $\mathbb{K}^n$. By the Heine-Borel theorem, S is a compact subset of $\mathbb{K}^n$.

(iii) We next show that $f: S \to \mathbb{R}$, $x \mapsto \|x\|$ is continuous.[1] Let $\{ e_k ; 1 \le k \le n \}$ be the standard basis of $\mathbb{K}^n$. For each $x = (x_1, \ldots, x_n) \in \mathbb{K}^n$, we have $x = \sum_{k=1}^n x_k e_k$ (see Example I.12.4(a) and Remark I.12.5). From the triangle inequality for $\|\cdot\|$ we get

$$\|x\| = \Bigl\|\sum_{k=1}^n x_k e_k\Bigr\| \le \sum_{k=1}^n |x_k| \, \|e_k\| \le C_0 \, |x| \ , \qquad x \in \mathbb{K}^n \ , \tag{3.2}$$

where we have set $C_0 := \sum_{k=1}^n \|e_k\|$ and used the inequality $|x_k| \le |x|$. This proves the second inequality of (3.1). Moreover, from (3.2) and the reversed triangle inequality for $\|\cdot\|$, we get

$$|f(x) - f(y)| = \bigl| \, \|x\| - \|y\| \, \bigr| \le \|x - y\| \le C_0 \, |x - y| \ , \qquad x, y \in S \ ,$$

which proves the Lipschitz continuity of f.

(iv) For all $x \in S$, we have $f(x) > 0$, so, by the extreme value theorem, we know that $m := \min f(S)$ is positive, that is,

$$0 < m = \min f(S) \le f(x) = \|x\| \ , \qquad x \in S \ . \tag{3.3}$$

Finally let $x \in \mathbb{K}^n \backslash \{0\}$. Then $x/|x|$ is in S, and so, from (3.3), we have $m \le \bigl\| x/|x| \bigr\|$, that is,

$$m \, |x| \le \|x\| \ , \qquad x \in \mathbb{K}^n \ . \tag{3.4}$$

The claim now follows from (3.2) and (3.4) with $C := \max\{C_0, 1/m\}$. ■

(b) The fundamental theorem of algebra[2] *Any nonconstant polynomial* $p \in \mathbb{C}[X]$ *has a zero in* $\mathbb{C}$.

Proof (i) Let p be a such a polynomial. Without loss of generality we can assume that the leading coefficient of p is 1 and so write p in the form

$$p = X^n + a_{n-1} X^{n-1} + \cdots + a_1 X + a_0$$

with $n \in \mathbb{N}^\times$ and $a_k \in \mathbb{C}$. If $n = 1$, the claim is clear, so we suppose that $n \ge 2$. Set

$$R := 1 + \sum_{k=0}^{n-1} |a_k| \ .$$

Then, for each $z \in \mathbb{C}$ such that $|z| > R \ge 1$, we have

$$\begin{aligned} |p(z)| &\ge |z|^n - |a_{n-1}| \, |z|^{n-1} - \cdots - |a_1| \, |z| - |a_0| \\ &\ge |z|^n - \bigl(|a_{n-1}| + \cdots + |a_1| + |a_0|\bigr) \, |z|^{n-1} \\ &= |z|^{n-1} \bigl(|z| - (R-1)\bigr) \ge |z|^{n-1} > R^{n-1} \ge R \ . \end{aligned}$$

Hence the absolute value of p outside of the ball $\bar{\mathbb{B}}_{\mathbb{C}}(0, R)$ is greater than R. Because $|p(0)| = |a_0| < R$, this means that

$$\inf_{z \in \mathbb{C}} |p(z)| = \inf_{|z| \le R} |p(z)| \ .$$

[1] Example 1.3(j) cannot be used here. Why not?

[2] The fundamental theorem of algebra is not valid for the field of real numbers $\mathbb{R}$, as the example $p = 1 + X^2$ shows.

(ii) We next consider the function

$$|p| : \bar{\mathbb{B}}_{\mathbb{C}}(0,R) \to \mathbb{R}\ , \quad z \mapsto |p(z)|\ ,$$

which, being a restriction of the composition of the continuous functions $|\cdot|$ and p, is continuous (see Examples 1.3(k) and 1.9(a), as well as Corollary 1.6). By the Heine-Borel theorem and Example 2.19(d), the closed ball $\bar{\mathbb{B}}_{\mathbb{C}}(0,R)$ is compact. Thus, applying the extreme value theorem to $|p|$, there is some $z_0 \in \bar{\mathbb{B}}_{\mathbb{C}}(0,R)$ such that the function $|p|$ is minimum at z_0.

(iii) Suppose that p has no zeros in $\bar{\mathbb{B}}_{\mathbb{C}}(0,R)$. Then, in particular, $p(z_0) \neq 0$, and $q := p(X+z_0)/p(z_0)$ is a polynomial of degree n such that

$$|q(z)| \geq 1\ , \quad z \in \mathbb{C}\ , \qquad \text{and} \qquad q(0) = 1\ . \tag{3.5}$$

Hence we can write q in the form

$$q = 1 + \alpha X^k + X^{k+1} r$$

for suitable $\alpha \in \mathbb{C}^\times$, $k \in \{1,\ldots,n-1\}$ and $r \in \mathbb{C}[X]$.

(iv) At this point we make use of the existence of complex roots, a result which we prove later in Section 6 (of course, without using the fundamental theorem of algebra). This theorem says, in particular that some $z_1 \in \mathbb{C}$ exists[3] such that $z_1^k = -1/\alpha$. Thus

$$q(tz_1) = 1 - t^k + t^{k+1} z_1^{k+1} r(tz_1)\ , \qquad t \in [0,1]\ ,$$

and hence

$$|q(tz_1)| \leq 1 - t^k + t^k \cdot t\,|z_1^{k+1} r(tz_1)|\ , \qquad t \in [0,1]\ . \tag{3.6}$$

(v) Finally we consider the function

$$h : [0,1] \to \mathbb{R}\ , \quad t \mapsto |z_1^{k+1} r(tz_1)|\ .$$

It is not difficult to see that h is continuous (see Proposition 1.5(ii), Corollary 1.6, Theorem 1.8 and Example 1.9(a)). By the Heine-Borel theorem and Corollary 3.7, there is some $M \geq 1$ such that

$$h(t) = |z_1^{k+1} r(tz_1)| \leq M\ , \qquad t \in [0,1]\ .$$

If we use this bound in (3.6) we get

$$|q(tz_1)| \leq 1 - t^k(1 - tM) \leq 1 - t^k/2 < 1\ , \qquad t \in \bigl(0, 1/(2M)\bigr)\ ,$$

which contradicts the first statement of (3.5). Therefore p must have a zero in $\bar{\mathbb{B}}_{\mathbb{C}}(0,R)$. ■

Corollary *Let*

$$p = a_n X^n + a_{n-1} X^{n-1} + \cdots + a_1 X + a_0$$

with $a_0,\ldots,a_n \in \mathbb{C}$, $a_n \neq 0$ and $n \geq 1$. Then there are $z_1,\ldots,z_n \in \mathbb{C}$ such that

$$p = a_n \prod_{k=1}^{n} (X - z_k)\ .$$

Thus each polynomial p over $\mathbb{C}$ has exactly $\deg(p)$ (counted with multiplicities) zeros.

---

[3] Note that this claim is false for $\mathbb{R}$. Indeed, this is the only place in the proof where the special properties of $\mathbb{C}$ are used.

Proof By the fundamental theorem of algebra, $p(z_1) = 0$ for some $z_1 \in \mathbb{C}$. By Theorem I.8.17, there is some $p_1 \in \mathbb{C}[X]$ such that $p = (X - z_1)p_1$ and $\deg(p_1) = \deg(p) - 1$. A simple induction argument finishes the proof. ■

(c) *Let A and K be disjoint subsets of a metric space with K compact and A closed. Then the distance $d(K, A)$ from K to A is positive, that is,*

$$d(K, A) := \inf_{k \in K} d(k, A) > 0 \ .$$

Proof By Examples 1.3(k) and (l), the real valued function $d(\cdot, A)$ is continuous on K and so, by the extreme value theorem, there is some $k_0 \in K$ such that $d(k_0, A) = d(K, A)$. Suppose that

$$d(k_0, A) = \inf_{a \in A} d(k_0, a) = 0 \ .$$

Then there is a sequence (a_k) in A such that $d(k_0, a_k) \to 0$ for all $k \to \infty$. Hence the sequence (a_k) converges to k_0. Because A is closed, k_0 is in A, contradicting $A \cap K = \emptyset$. Therefore we have $d(k_0, A) = d(K, A) > 0$. ■

(d) The compactness of K is necessary in (c).

Proof The sets $A := \mathbb{R} \times \{0\}$ and $B := \{\, (x, y) \in \mathbb{R}^2 \ ; \ xy = 1 \,\}$ are closed but not compact in $\mathbb{R}^2$. Since $d\bigl((n, 0), (n, 1/n)\bigr) = 1/n$ for $n \in \mathbb{N}^\times$, we have $d(A, B) = 0$. ■

Total Boundedness

With the practical importance of the concept of compactness amply demonstrated by the above examples, we now present another characterization of compact sets which uses completeness and total boundedness.

3.10 Theorem *A subset of a metric space is compact if and only if it is complete and totally bounded.*

Proof '$\Longrightarrow$' Let $K \subseteq X$ be compact and (x_j) a Cauchy sequence in K. Since K is sequentially compact, (x_j) has a subsequence which converges in K. Thus, by Proposition II.6.4, the sequence (x_j) itself converges in K. This implies that K is complete.

For each $r > 0$, the set $\{\, \mathbb{B}(x, r) \ ; \ x \in K \,\}$ is an open cover of K. Since K is compact, this cover has a finite subcover. Thus we have shown that K is totally bounded.

'$\Longleftarrow$' Let K be complete and totally bounded. Let (x_j) be a sequence in K. Since K is totally bounded, for each $n \in \mathbb{N}^\times$, there is a finite set of open balls with centers in K and radius $1/n$ which forms a cover of K. In particular, there is a subsequence $(x_{1,j})_{j \in \mathbb{N}}$ of (x_j) which is contained in a ball of radius 1. Then there is a subsequence $(x_{2,j})_{j \in \mathbb{N}}$ of $(x_{1,j})_{j \in \mathbb{N}}$ which is contained in a ball of radius $1/2$. Further, there is a subsequence $(x_{3,j})_{j \in \mathbb{N}}$ of $(x_{2,j})_{j \in \mathbb{N}}$ which is contained in a ball of radius $1/3$.

Iterating this construction yields, for each $n \in \mathbb{N}^\times$, a subsequence $(x_{n+1,j})_{j\in\mathbb{N}}$ of $(x_{n,j})_{j\in\mathbb{N}}$ which is contained in a ball of radius $1/(n+1)$.

Now set $y_n := x_{n,n}$ for all $n \in \mathbb{N}^\times$. It is easy to check that (y_n) is a Cauchy sequence in K (see Remark 3.11(a)). Since K is complete, the sequence (y_n) converges in K.

Thus the sequence (x_j) has a subsequence, namely (y_n), which converges in K. This shows that K is sequentially compact and also, by Theorem 3.4, that K is compact. ■

3.11 Remarks **(a)** In the second part of the preceding proof we have used a trick which is useful in many other situations: From a given sequence $(x_{0,j})_{j\in\mathbb{N}}$, choose successive subsequences $(x_{n+1,j})_{j\in\mathbb{N}}$ so that, for all $n \in \mathbb{N}$, $(x_{n+1,j})_{j\in\mathbb{N}}$ is a subsequence of $(x_{n,j})$. Then form the **diagonal sequence** by choosing, for each $n \in \mathbb{N}$, the nth element from the nth subsequence.

$$\begin{array}{ccccc}
\boxed{x_{0,0}}, & x_{0,1}, & x_{0,2}, & x_{0,3}, & \cdots \\
x_{1,0}, & \boxed{x_{1,1}}, & x_{1,2}, & x_{1,3}, & \cdots \\
x_{2,0}, & x_{2,1}, & \boxed{x_{2,2}}, & x_{2,3}, & \cdots \\
x_{3,0}, & x_{3,1}, & x_{3,2}, & \boxed{x_{3,3}}, & \cdots \\
\vdots & \vdots & \vdots & \vdots & \ddots
\end{array}$$

The diagonal sequence $(y_n) := (x_{n,n})_{n\in\mathbb{N}}$ clearly has the property that $(y_n)_{n\ge N}$ is a subsequence of $(x_{N,j})_{j\in\mathbb{N}}$ for each $N \in \mathbb{N}$, and so it has the same properties 'at infinity' as each of the subsequences $(x_{n,j})_{j\in\mathbb{N}}$.

(b) A subset K of a metric space X is compact if and only if K with the induced metric is a compact metric space.

Proof This is a simple consequence of the definition of relative topology and Proposition 2.26. ■

Because of Remark 3.11(b), it would have sufficed to formulate Theorems 3.3 and 3.4 for X rather than for a subset K of X. However, in applications an 'underlying' metric space X is usually given, for example, X is often a Banach space, and then it is certain subsets of X which are to be studied. So the above somewhat longer formulations are 'closer to reality'.

Uniform Continuity

Let X and Y be metric spaces and $f : X \to Y$ continuous. Then, by Proposition 1.1, for each $x_0 \in X$ and each $\varepsilon > 0$, there is some $\delta(x_0, \varepsilon) > 0$ such that for each $x \in X$ with $d(x, x_0) < \delta$ we have $d\bigl(f(x_0), f(x)\bigr) < \varepsilon$. As we noted after Proposition 1.1 and saw explicitly in Example 1.3(a), the number $\delta(x_0, \varepsilon)$ depends, in general, on $x_0 \in X$. On the other hand, Example 1.3(e) shows that there are continuous functions for which the number δ can be chosen independently of $x_0 \in X$. Such functions are called uniformly continuous and are of great practical importance. Specifically, a function $f : X \to Y$ is called **uniformly continuous** if, for each $\varepsilon > 0$, there is some $\delta(\varepsilon) > 0$ such that

$$d\bigl(f(x), f(y)\bigr) < \varepsilon \quad \text{for all } x, y \in X \text{ such that } d(x, y) < \delta(\varepsilon) \ .$$

3.12 Examples **(a)** Lipschitz continuous functions are uniformly continuous (see Example 1.3(e)).

(b) The function $r : (0, \infty) \to \mathbb{R},\ \ x \mapsto 1/x$ is continuous, but not uniformly continuous.

Proof Since r is the restriction of a rational function, it is certainly continuous. Now let $\varepsilon > 0$. Suppose that there is some $\delta := \delta(\varepsilon) > 0$ such that $|r(x) - r(y)| < \varepsilon$ for all $x, y \in (0, 1)$ such that $|x - y| < \delta$. Choose $x := \delta/(1 + \delta\varepsilon)$ and $y := x/2$. Then $x, y \in (0, 1)$ and $|x - y| = \delta/\bigl[2(1 + \delta\varepsilon)\bigr] < \delta$ and $|r(x) - r(y)| = (1 + \delta\varepsilon)/\delta > \varepsilon$. This contradicts our choice of δ. ∎

The following important theorem shows that in many cases, continuous functions are automatically uniformly continuous.

3.13 Theorem *Suppose that X and Y are metric spaces with X compact. If $f : X \to Y$ is continuous, then f is uniformly continuous. That is,* continuous functions on compact sets are uniformly continuous.

Proof Suppose that f is continuous but not uniformly continuous. Then there exists some $\varepsilon > 0$ with the property that, for each $\delta > 0$, there are $x, y \in X$ such that $d(x, y) < \delta$ but $d\bigl(f(x), f(y)\bigr) \geq \varepsilon$. In particular, there are sequences (x_n) and (y_n) in X such that

$$d(x_n, y_n) < 1/n \quad \text{and} \quad d\bigl(f(x_n), f(y_n)\bigr) \geq \varepsilon \ , \qquad n \in \mathbb{N}^\times .$$

Since X is compact, by Theorem 3.4, there is a subsequence $(x_{n_k})_{k\in\mathbb{N}}$ of (x_n) such that $\lim_{k\to\infty} x_{n_k} = \overline{x} \in X$. For the corresponding subsequence $(y_{n_k})_{k\in\mathbb{N}}$ of (y_n) we have

$$d(\overline{x}, y_{n_k}) \leq d(\overline{x}, x_{n_k}) + d(x_{n_k}, y_{n_k}) \leq d(\overline{x}, x_{n_k}) + 1/n_k \ , \qquad k \in \mathbb{N}^\times \ .$$

Hence $(y_{n_k})_{k\in\mathbb{N}}$ also converges to $\overline{x}$. Since f is continuous the images of the two sequences converge to $f(\overline{x})$, in particular, there is some $K \in \mathbb{N}$ such that

$$d\big(f(x_{n_K}), f(\overline{x})\big) < \varepsilon/2 \quad \text{and} \quad d\big(f(y_{n_K}), f(\overline{x})\big) < \varepsilon/2 .$$

This leads to the contradiction

$$\varepsilon \leq d\big(f(x_{n_K}), f(y_{n_K})\big) \leq d\big(f(x_{n_K}), f(\overline{x})\big) + d\big(f(\overline{x}), f(y_{n_K})\big) < \varepsilon .$$

Thus f is uniformly continuous. ∎

Compactness in General Topological Spaces

Just as at the end of the previous section, we want to briefly consider the case of general topological spaces. Admittedly, the general situation is no longer simple and we must limit our discussion here to a description of the results. For the proofs and a deeper exploration of (set theoretical) topology, Dugundji's book [Dug66] is highly recommended.

3.14 Remarks (a) Let $X = (X, \mathcal{T})$ be a topological space. Then X is **compact** if X is a Hausdorff space and every open cover of X has a finite subcover. The space X is **sequentially compact** if it is a Hausdorff space and every sequence has a convergent subsequence. A subset $Y \subseteq X$ is **compact** (or **sequentially compact**) if the topological subspace $(Y, \mathcal{T}_Y)$ is compact (or sequentially compact). By Propositions 2.17 and 2.26 as well as Remark 3.11(b), these definitions generalize the concepts of compact and sequentially compact subsets of a metric space.

(b) Any compact subset K of a Hausdorff space X is closed. For each $x_0 \in K^c$ there are disjoint open sets U and V in X such that $K \subseteq U$ and $x_0 \in V$. In other words, a compact subset of a Hausdorff space and a point, not in that subset, can be separated by open neighborhoods.

Proof This follows from the first part of the proof of Proposition 3.2 ∎

(c) Any closed subset of a compact space is compact.

Proof See Exercise 2. ∎

(d) Let X be compact and Y Hausdorff. Then the image of any continuous function $f : X \to Y$ is compact.

Proof The proof of Theorem 3.6 and the definition of relative topology show that every open cover of $f(X)$ has a finite subcover. Since a subspace of a Hausdorff space is itself a Hausdorff space, the claim follows. ∎

(e) In general topological spaces, compactness and sequential compactness are distinct concepts. That is, a compact space need not be sequentially compact, and a sequentially compact space need not be compact.

(f) Uniform continuity is undefined in general topological spaces since the definition given above makes essential use of the metric. ∎

Exercises

1 Let X_j, $j = 1, \ldots, n$, be metric spaces. Prove that $X_1 \times \cdots \times X_n$ is compact if and only if each X_j is compact.

2 Let X be a compact metric space and Y a subset of X. Prove that Y is compact if and only if Y is closed.

3 Let X and Y be metric spaces. A bijection $f : X \to Y$ is called a **homeomorphism** if both f and f^{-1} are continuous. Show the following:

(a) If $f : X \to Y$ is a homeomorphism, then $\mathcal{U}(f(x)) = f(\mathcal{U}(x))$ for all $x \in X$, that is, 'f maps neighborhoods to neighborhoods'.

(b) Suppose that X is compact and $f : X \to Y$ is continuous.

(i) f is closed (see Exercise 2.14).

(ii) If f is bijective, it is a homeomorphism.

4 A family $\mathcal{M}$ of subsets of a nonempty set has the **finite intersection property** if each finite subset of $\mathcal{M}$ has nonempty intersection.
Prove that the following are equivalent:

(a) X is a compact metric space.

(b) Every family $\mathcal{A}$ of closed subsets of X which has the finite intersection property, has nonempty intersection, that is, $\bigcap \mathcal{A} \neq \emptyset$.

5 Let (A_j) be a sequence of nonempty closed subsets of X with $A_j \supseteq A_{j+1}$ for all $j \in \mathbb{N}$. Show that, if A_0 is compact, then $\bigcap A_j \neq \emptyset$.[4]

6 Let E and F be finite dimensional normed vector spaces and $A : E \to F$ linear. Prove that A is Lipschitz continuous. (Hint: Example 3.9(a).)

7 Show that the set $O(n)$ of all real orthogonal matrices is a compact subset of $\mathbb{R}^{(n^2)}$.

8 Let

$$C_0 := [0,1] , \quad C_1 := C_0 \backslash (1/3, 2/3) , \quad C_2 := C_1 \backslash \big((1/9, 2/9) \cup (7/9, 8/9)\big) , \quad \ldots$$

In general, C_{n+1} is formed by removing the open middle third from each of the 2^n intervals which make up C_n. The intersection $C := \bigcap C_n$ is called the **Cantor set**. Prove the following:

(a) C is compact and has empty interior.

(b) C consists of all numbers in $[0,1]$ whose ternary expansion is $\sum_{k=1}^{\infty} a_k 3^{-k}$ with $a_k \in \{0, 2\}$.

(c) Every point of C is a limit point of C, that is, C is perfect.

(d) For $x \in C$ with the ternary expansion $\sum_{k=1}^{\infty} a_k 3^{-k}$, define

$$\varphi(x) := \sum_{k=1}^{\infty} a_k 2^{-(k+1)} .$$

Then $\varphi : C \to [0,1]$ is increasing, surjective and continuous.

(e) C is uncountable.

(f) φ has a continuous extension $f : [0,1] \to [0,1]$ which is constant on each interval in $[0,1] \backslash C$. The function f is called the **Cantor function** of C.

---

[4] Compare Exercise 2.9.

9 Let X be a metric space. A function $f : X \to \mathbb{R}$ is called **lower continuous at** $a \in X$ if, for each sequence (x_n) in X such that $\lim x_n = a$, we have $f(a) \leq \underline{\lim}\, f(x_n)$. It is called **upper continuous at** a if $-f$ is lower continuous at a. Finally f is called **lower continuous** (or **upper continuous**) if f is lower continuous (or upper continuous) at each point of X.

(a) Show the equivalence of the following:

(i) f is lower continuous.

(ii) For each $a \in X$ and $\varepsilon > 0$, there is some $U \in \mathcal{U}(a)$ such that $f(x) > f(a) - \varepsilon$ for all $x \in U$.

(iii) For each $\alpha \in \mathbb{R}$, $f^{-1}\big((\alpha, \infty)\big)$ is open.

(iv) For each $\alpha \in \mathbb{R}$, $f^{-1}\big((-\infty, \alpha]\big)$ is closed.

(b) f is continuous if and only if f is lower and upper continuous.

(c) Let χ_A be the characteristic function of $A \subseteq X$. Then A is open if and only if χ_A is lower continuous.

(d) Let X be compact and $f : X \to \mathbb{R}$ lower continuous. Then f attains its minimum, that is, there is some $x \in X$ such that $f(x) \leq f(y)$ for all $y \in X$. (Hint: Consider a sequence (x_n) in X such that $f(x_n) \to \inf f(X)$.)

10 Let $f, g : [0,1] \to \mathbb{R}$ be defined by

$$f(x) := \begin{cases} 1/n\ , & x \in \mathbb{Q} \text{ where } x = m/n \text{ in lowest terms}\ , \\ 0\ , & x \notin \mathbb{Q}\ , \end{cases}$$

and

$$g(x) := \begin{cases} (-1)^n n/(n+1)\ , & x \in \mathbb{Q} \text{ where } x = m/n \text{ in lowest terms}\ , \\ 0\ , & x \notin \mathbb{Q}\ . \end{cases}$$

Prove or disprove the following:

(a) f is upper continuous.

(b) f is lower continuous.

(c) g is upper continuous.

(d) g is lower continuous.

11 Let X be a metric space and $f : [0,1) \to X$ continuous. Show that f is uniformly continuous if $\lim_{t \to 1} f(t)$ exists.

12 Which of the functions

$$f : (0, \infty) \to \mathbb{R}\ , \quad t \mapsto (1 + t^2)^{-1}\ , \qquad g : (0, \infty) \to \mathbb{R}\ , \quad t \mapsto t^{-2}$$

is uniformly continuous?

13 Prove that a finite dimensional subspace of a normed vector space is closed. (Hint: Let E be a normed vector space and F a subspace of E with finite dimension. Let (v_n) be a sequence in F and $v \in E$ such that $\lim v_n = v$ in E. Because of Remark I.12.5, Proposition 1.10 and the Bolzano-Weierstrass theorem, there are some subsequence $(v_{n_k})_{k \in \mathbb{N}}$ of (v_n) and $w \in F$ such that $\lim_k v_{n_k} = w$ in F. Now use Propositions 2.11 and 2.17 to show that $v = w \in F$.)

14 Suppose that X is a metric space and $f : X \to \mathbb{R}$ is bounded. Show that $\omega_f : X \to \mathbb{R}$ is upper continuous (see Exercises 1.17 and 2.20).

15 Show that the closed unit ball in ℓ_∞ (see Remark II.3.6(a)) is not compact. (Hint: Consider the sequence (e_n) of 'unit vectors' e_n given by $e_n(j) := \delta_{nj}$ for all $j \in \mathbb{N}$.)

4 Connectivity

It is intuitively clear that an open interval in $\mathbb{R}$ is 'connected', but that it becomes 'disconnected' if we remove a single point. In this section, we make this intuitive concept of connectivity more precise. In doing so, we discover once again that topology plays an essential role.

Definition and Basic Properties

A metric space X is called **connected** if X cannot be represented as the union of two disjoint nonempty open subsets. Thus X is connected if and only if

$$\nexists\, O_1, O_2 \subseteq X, \text{ open, nonempty, with } O_1 \cap O_2 = \emptyset \text{ and } O_1 \cup O_2 = X\ .$$

A subset M of X is called **connected** in X if M is connected with respect to the metric induced from X.

4.1 Examples **(a)** Clearly, the empty set and any one element set are connected.

(b) The set of the natural numbers $\mathbb{N}$ is not connected.

Proof By Example 2.19(a) and Theorem 2.26, the subsets $O_1 := \{0\} = \mathbb{N} \cap (-\infty, 1/2)$ and $O_2 := \{1, 2, 3, \ldots\} = \mathbb{N} \cap (1/2, \infty)$ are open in $\mathbb{N}$. Since, of course, $O_1 \cap O_2 = \emptyset$ and $O_1 \cup O_2 = \mathbb{N}$, this shows that $\mathbb{N}$ is not connected. ■

(c) The set of rational numbers $\mathbb{Q}$ is not connected in $\mathbb{R}$.

Proof The subsets $O_1 := \{\, x \in \mathbb{Q}\ ;\ x < \sqrt{2}\,\}$ and $O_2 := \{\, x \in \mathbb{Q}\ ;\ x > \sqrt{2}\,\}$ are open, nonempty and satisfy $O_1 \cap O_2 = \emptyset$ and $O_1 \cup O_2 = \mathbb{Q}$. ■

4.2 Proposition *For any metric space X, the following are equivalent:*

(i) *X is connected.*

(ii) *X is the only nonempty subset of X which is both open and closed.*

Proof '(i)$\Rightarrow$(ii)' Let O be a nonempty subset of X which is both open and closed. Then O^c is also open and closed in X, and, of course, $O \cap O^c = \emptyset$ and $X = O \cup O^c$. Since X is connected and O is nonempty by hypothesis, it follows that O^c must be empty. Hence $O = X$.

'(ii)$\Rightarrow$(i)' Suppose that O_1 and O_2 are nonempty open subsets of X such that $O_1 \cap O_2 = \emptyset$ and $O_1 \cup O_2 = X$. Then $O_1 = O_2^c$ is nonempty, open and closed in X so, by hypothesis, $O_1 = O_2^c = X$. This implies $O_2 = \emptyset$, a contradiction. ■

4.3 Remark This proposition is often used for proving statements about connected sets as follows: Suppose that we want to prove that each element x of a connected set X has property E, that is, $E(x)$ holds for all $x \in X$. Set

$$O := \{ x \in X \ ; \ E(x) \text{ is true} \} .$$

Then it suffices to show that the set O is nonempty, open and closed, since then, by Proposition 4.2, $O = X$. ■

Connectivity in $\mathbb{R}$

The next proposition describes all connected subsets of $\mathbb{R}$ and also provides our first concrete examples of nontrivial connected sets.

4.4 Theorem *A subset of $\mathbb{R}$ is connected if and only if it is an interval.*

Proof Because of Example 4.1(a) we can suppose that the subset contains more than one element.

'$\Rightarrow$' Let $X \subseteq \mathbb{R}$ be connected.

(i) Set $a := \inf(X) \in \bar{\mathbb{R}}$ and $b := \sup(X) \in \bar{\mathbb{R}}$. Since X has at least two elements, the interval (a, b) is nonempty and[1] $X \subseteq (a, b) \cup \{a, b\}$.

(ii) We prove first the inclusion $(a, b) \subseteq X$. Suppose, to the contrary, that (a, b) is not contained in X. Then there is some $c \in (a, b)$ which is not in X. Set $O_1 := X \cap (-\infty, c)$ and $O_2 := X \cap (c, \infty)$. Then O_1 and O_2 are, by Proposition 2.26, open in X. Of course, O_1 and O_2 are disjoint and their union is X. By our choice of a, b and c there are elements $x, y \in X$ such that $x < c$ and $y > c$. This means that x is in O_1 and y is in O_2, and so O_1 and O_2 are nonempty. Hence X is not connected, contradicting our hypothesis.

(iii) Since we have shown the inclusions $(a, b) \subseteq X \subseteq (a, b) \cup \{a, b\}$, X is an interval.

'$\Leftarrow$' (i) Suppose, to the contrary, that X is an interval and there are open, nonempty subsets O_1 and O_2 of X such that $O_1 \cap O_2 = \emptyset$ and $O_1 \cup O_2 = X$. Choose $x \in O_1$ and $y \in O_2$ and consider first the case $x < y$. Since $\mathbb{R}$ is order complete, $z := \sup\bigl(O_1 \cap [x, y]\bigr)$ is a well defined real number.

(ii) The element z cannot be in O_1 because O_1 is open in X and X is an interval and so there is some $\varepsilon > 0$ such that $[z, z + \varepsilon) \subseteq O_1 \cap [x, y]$. This contradicts the supremum property of z. Similarly, z cannot be in O_2 since otherwise there is some $\varepsilon > 0$ such that

$$(z - \varepsilon, z] \subseteq O_2 \cap [x, y] ,$$

which contradicts $O_1 \cap O_2 = \emptyset$ and the definition of z. Thus $z \notin O_1 \cup O_2 = X$. On the other hand, $[x, y]$ is contained in X because X is an interval. This leads to the contradiction $z \in [x, y] \subseteq X$ and $z \notin X$. The case $y < x$ can be proved similarly. ■

[1] If a and b are real numbers, then $(a, b) \cup \{a, b\} = [a, b]$.

The Generalized Intermediate Value Theorem

Connected sets have the property that their images under continuous functions are also connected. This important fact can be proved easily using the results of Section 2.

4.5 Theorem *Let X and Y be metric spaces and $f : X \to Y$ continuous. If X is connected, then so is $f(X)$. That is,* continuous images of connected sets are connected.

Proof Suppose, to the contrary, that $f(X)$ is not connected. Then there are nonempty subsets V_1 and V_2 of $f(X)$ such that V_1 and V_2 are open in $f(X)$, $V_1 \cap V_2 = \emptyset$ and $V_1 \cup V_2 = f(X)$. By Proposition 2.26, there are open sets O_j in Y such that $V_j = O_j \cap f(X)$ for $j = 1, 2$. Set $U_j := f^{-1}(O_j)$. Then, by Theorem 2.20, U_j is open in X for $j = 1, 2$. Moreover

$$U_1 \cup U_2 = X\ , \quad U_1 \cap U_2 = \emptyset \qquad \text{and} \qquad U_j \neq \emptyset\ , \quad j = 1, 2\ ,$$

which is not possible for the connected set X. ■

4.6 Corollary *Continuous images of intervals are connected.*

We will demonstrate in the next two sections that Theorems 4.4 and 4.5 are extremely useful tools for the investigation of real functions. Already we note the following easy consequence of these theorems.

4.7 Theorem (generalized intermediate value theorem) *Let X be a connected metric space and $f : X \to \mathbb{R}$ continuous. Then $f(X)$ is an interval. In particular, f takes on every value between any two given function values.*

Proof This follows directly from Theorems 4.4 and 4.5. ■

Path Connectivity

Let $\alpha, \beta \in \mathbb{R}$ with $\alpha < \beta$. A continuous function $w : [\alpha, \beta] \to X$ is called a **continuous path** connecting the **end points** $w(\alpha)$ and $w(\beta)$.

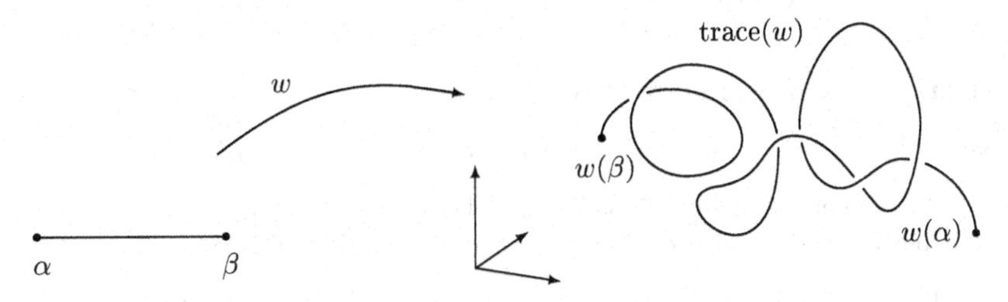

A metric space X is called **path connected** if, for each pair $(x, y) \in X \times X$, there is a continuous path in X connecting x and y. A subset of a metric space is called **path connected** if it is a path connected metric space with respect to the induced metric.

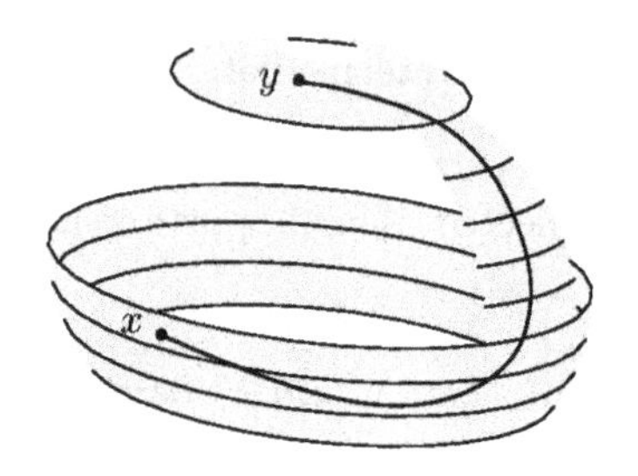

4.8 Proposition *Any path connected space is connected.*

Proof Suppose, to the contrary, that there is a metric space X which is path connected, but not connected. Then there are nonempty open sets O_1, O_2 in X such that $O_1 \cap O_2 = \emptyset$ and $O_1 \cup O_2 = X$. Choose $x \in O_1$ and $y \in O_2$. By hypothesis, there is a path $w\colon [\alpha, \beta] \to X$ such that $w(\alpha) = x$ and $w(\beta) = y$. Set $U_j := w^{-1}(O_j)$. Then, by Theorem 2.20, U_j is open in $[\alpha, \beta]$. We now have α in U_1 and β in U_2, as well as $U_1 \cap U_2 = \emptyset$ and $U_1 \cup U_2 = [\alpha, \beta]$, and so the interval $[\alpha, \beta]$ is not connected. This contradicts Theorem 4.4. ■

Let E be a normed vector space and $a, b \in E$. The linear structure of E allows us to consider 'straight' paths in E:

$$v\colon [0,1] \to E\ , \quad t \mapsto (1-t)a + tb\ . \tag{4.1}$$

We denote the image of the path v by $[\![a, b]\!]$.

A subset X of E is called **convex** if, for each pair $(a, b) \in X \times X$, $[\![a, b]\!]$ is contained in X.

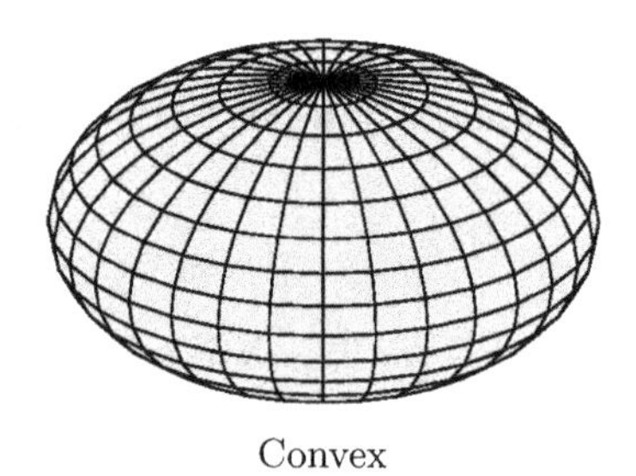
Convex

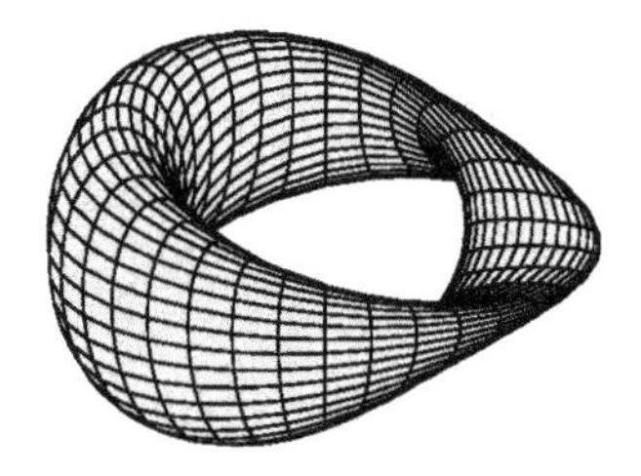
Not convex

4.9 Remarks Let E be a normed vector space.

(a) Every convex subset of E is path connected and connected.

Proof Let X be convex and $a, b \in X$. Then (4.1) defines a path in X connecting a and b. Thus X is path connected. Proposition 4.7 then implies that X is connected. ■

(b) For all $a \in E$ and $r > 0$, the balls $\mathbb{B}_E(a, r)$ and $\bar{\mathbb{B}}_E(a, r)$ are convex.

Proof For $x, y \in \mathbb{B}_E(a, r)$ and $t \in [0,1]$ we have

$$\begin{aligned} \|(1-t)x + ty - a\| &= \|(1-t)(x-a) + t(y-a)\| \\ &\leq (1-t)\,\|x-a\| + t\,\|y-a\| < (1-t)r + tr = r \ . \end{aligned}$$

This inequality implies that $[\![x, y]\!]$ is in $\mathbb{B}_E(a, r)$. The second claim can be proved similarly. ■

(c) A subset of $\mathbb{R}$ is convex if and only if it is an interval.

Proof Let $X \subseteq \mathbb{R}$ be convex. Then, by (a), X is connected and so, by Theorem 4.4, X is an interval. The claim that intervals are convex is clear. ■

In $\mathbb{R}^2$ there are simple examples of connected sets which are not convex. Even so, in such cases, it seems plausible that any pair of points in the set can be joined with a path which consists of finitely many straight line segments. The following theorem shows that this holds, not just in $\mathbb{R}^2$, but in any normed vector space, so long as the set is open.

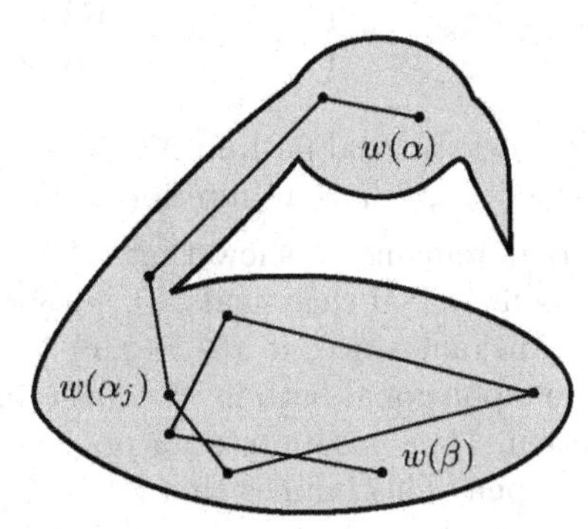

Let X be a subset of a normed vector space. A function $w\colon [\alpha, \beta] \to X$ is called a **polygonal path**[2] in X if there are $n \in \mathbb{N}$ and real numbers $\alpha_0, \ldots, \alpha_{n+1}$ such that $\alpha = \alpha_0 < \alpha_1 < \cdots < \alpha_{n+1} = \beta$ and

$$w\big((1-t)\alpha_j + t\alpha_{j+1}\big) = (1-t)w(\alpha_j) + tw(\alpha_{j+1})$$

for all $t \in [0,1]$ and $j = 0, \ldots, n$.

4.10 Theorem *Let X be a nonempty, open and connected subset of a normed vector space. Then any pair of points of X can be connected by a polygonal path in X.*

Proof Let $a \in X$ and

$$M := \big\{\, x \in X \ ; \ \text{there is a polygonal path in } X \text{ connecting } x \text{ and } a \big\}.$$

We now apply the proof technique described in Remark 4.3.

(i) Because $a \in M$, the set M is not empty.

(ii) We next prove that M is open in X. Let $x \in M$. Since X is open, there is some $r > 0$ such that $\mathbb{B}(x, r) \subseteq X$. By Remark 4.9(b), for each $y \in \mathbb{B}(x, r)$, the set $[\![x, y]\!]$ is contained in $\mathbb{B}(x, r)$ and so also in X. Since $x \in M$, there is a polygonal path $w\colon [\alpha, \beta] \to X$ such that $w(\alpha) = a$ and $w(\beta) = x$.

[2]The function $w\colon [\alpha, \beta] \to X$ is clearly left and right continuous at each point, and so, by Proposition 1.12, is continuous. Thus a polygonal path is, in particular, a path.

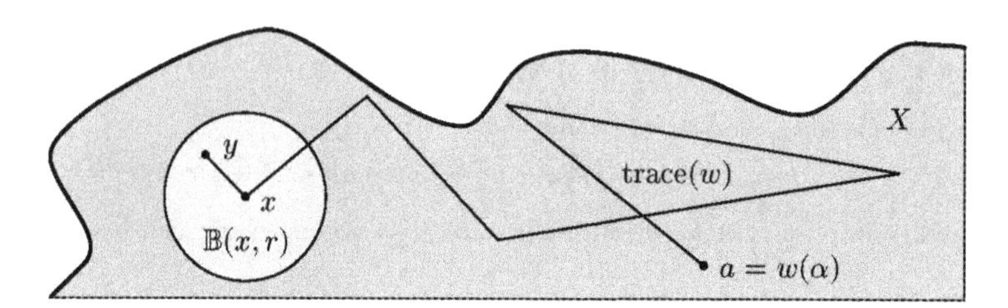

Now define $\widetilde{w}\colon [\alpha, \beta+1] \to X$ by

$$\widetilde{w}(t) := \begin{cases} w(t) \ , & t \in [\alpha, \beta] \ , \\ (t-\beta)y + (\beta + 1 - t)x \ , & t \in (\beta, \beta+1] \ . \end{cases}$$

Then $\widetilde{w}$ is a polygonal path in X which connects a and y. This shows that $\mathbb{B}(x,r)$ is contained in M, x is an interior point of M, and M is open in X.

(iii) It remains to show that M is closed. Let $y \in X \backslash M$. Since X is open, there is some $r > 0$ such that $\mathbb{B}(y,r)$ is contained in X. The sets $\mathbb{B}(y,r)$ and M must be disjoint since, if $x \in \mathbb{B}(y,r) \cap M$, then, by the argument of (ii), there would be a polygonal path in X connecting a and y, and so y is in M, contrary to assumption. Thus y is an interior point of $X \backslash M$ and, since $y \in X \backslash M$ is arbitrary, $X \backslash M$ is open. This implies that M is closed in X. ■

4.11 Corollary *An open subset of a normed vector space is connected if and only if it is path connected.*

Proof This follows from Proposition 4.8 and Theorem 4.10. ■

Connectivity in General Topological Spaces

To end this section we analyze the above proofs for their dependence on the existence of a metric.

4.12 Remarks (a) The definitions of 'connected' and 'path connected' depend on the topology only and do not make use of a metric. Hence these are valid in any topological space. The same is true for Propositions 4.2, 4.5 and 4.8. In particular, the generalized intermediate value theorem (Theorem 4.7) holds when X is an arbitrary topological space.

(b) There are examples of connected spaces which are not path connected. For this reason, Theorem 4.10 is particularly useful. ■

Exercises

In the following, X is a metric space.

1 Prove the equivalence of the following:

(a) X is connected.

(b) There is no continuous surjection $X \to \{0,1\}$.

2 Suppose that $C_\alpha \subseteq X$ is connected for each α in an index set A. Show that $\bigcup_\alpha C_\alpha$ is connected if $C_\alpha \cap C_\beta \neq \emptyset$ for all $\alpha, \beta \in \mathsf{A}$. That is, *arbitrary unions of connected pairwise nondisjoint sets are connected.* (Hint: Use Exercise 1 and prove by contradiction.)

3 Show by example that the intersection of connected sets is not, in general, connected.

4 Let X_j, $j = 1, \ldots, n$, be metric spaces. Prove that the product $X_1 \times \cdots \times X_n$ is connected if and only if each X_j is connected. (Hint: Write $X \times Y$ as a union of sets of the form $\bigl(X \times \{y\}\bigr) \cup \bigl(\{x\} \times Y\bigr)$.)

5 Show that the closure of a connected set is also connected. (Hint: Consider a continuous function $f : \overline{A} \to \{0,1\}$ and use $f(\overline{A}) \subseteq \overline{f(A)}$ (see Exercise 2.12).)

6 Given an element $x \in X$, the set

$$K(x) := \bigcup_{Y \in M} Y \quad \text{where} \quad M := \{\, Y \subseteq X \ ; \ Y \text{ is connected and } x \in Y \,\}$$

is, by Exercise 2, the largest connected subset of X which contains x, and hence is called the **connected component** of x in X. Prove the following:

(a) $\{\, K(x) \ ; \ x \in X \,\}$ is a partition of X, that is, each $x \in X$ is contained in exactly one connected component of X.

(b) Each connected component is closed.

7 Determine all the connected components of $\mathbb{Q}$ in $\mathbb{R}$.

8 Let $E = (E, \|\cdot\|)$ be a normed vector space with $\dim(E) \geq 2$. Prove that $E \backslash \{0\}$ and the unit sphere $S := \{\, x \in E \ ; \ \|x\| = 1 \,\}$ are connected.

9 Prove that the following metric spaces X and Y are not homeomorphic (see Exercise 3.3):

(a) $X := S^1$, $Y := [0,1]$.

(b) $X := \mathbb{R}$, $Y := \mathbb{R}^n$, $n \geq 2$.

(c) $X := (0,1) \cup (2,3)$, $Y := (0,1) \cup (2,3]$.

(Hint: In each case, remove one or two points from X.)

10 Show that the set $O(n)$ of all real orthogonal $n \times n$ matrices is not connected. (Hint: The function $O(n) \to \{-1, 1\}$, $A \mapsto \det A$ is continuous and surjective (see Exercise 1.16).)

11 For $b_{j,k} \in \mathbb{R}$, $1 \leq j, k \leq n$, consider the bilinear form

$$B : \mathbb{R}^n \times \mathbb{R}^n \to \mathbb{R} \ , \quad (x,y) \mapsto \sum_{j,k=1}^n b_{j,k} x_j y_k \ .$$

If $B(x,x) > 0$ (or $B(x,x) < 0$) for all $x \in \mathbb{R}^n \backslash \{0\}$, then B is called **positive** (or **negative**) **definite**. If B is neither positive nor negative definite, it is **indefinite**. Show the following:

(a) If B is indefinite, then there is some $x \in S^{n-1}$ such that $B(x,x) = 0$.

(b) If B is positive definite, then there is some $\beta > 0$ such that $B(x,x) \geq \beta \, |x|^2$, $x \in \mathbb{R}^n$.

(Hint: For (a), use the intermediate value theorem. For (b), use the extreme value theorem.)

12 Let E be a vector space. Suppose that $x_1, \ldots, x_n \in E$ and $\alpha_1, \ldots, \alpha_n \in \mathbb{R}^+$ are such that $\sum_{j=1}^n \alpha_j = 1$. Then $\sum_{j=1}^n \alpha_j x_j$ is called a **convex combination** of $x_1, \ldots, x_n$.

Prove the following:

(a) Arbitrary intersections of convex subsets of E are convex.

(b) A subset M of E is convex if and only if M is closed under convex combinations, that is, every convex combination of points of M is in M.

(c) If E is a normed vector space and $M \subseteq E$ is convex, then $\mathring{M}$ and $\overline{M}$ are also convex.

5 Functions on $\mathbb{R}$

Our abstract development of continuity is especially fruitful when applied to real valued functions on $\mathbb{R}$. This is, of course, a consequence of the rich structure of $\mathbb{R}$.

Bolzano's Intermediate Value Theorem

Applying the generalized intermediate value theorem to real valued functions gives Bolzano's original version of this important theorem.

5.1 Theorem (Bolzano's intermediate value theorem) *Suppose that $I \subseteq \mathbb{R}$ is an interval and $f : I \to \mathbb{R}$ is continuous. Then $f(I)$ is an interval. That is,* continuous images of intervals are intervals.

Proof This follows from Theorems 4.4 and 4.7. ∎

In the following I denotes a nonempty interval in $\mathbb{R}$.

5.2 Examples (a) The claim in Bolzano's intermediate value theorem is false if f is not continuous or is not defined on an interval. This is illustrated by the functions whose graphs are below:

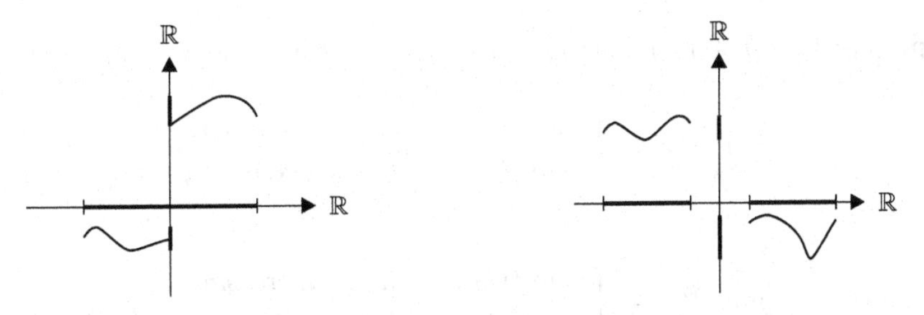

(b) *If $f : I \to \mathbb{R}$ is continuous and there are $a, b \in I$ such that $f(a) < 0 < f(b)$, then there is some ξ between a and b such that $f(\xi) = 0$.*

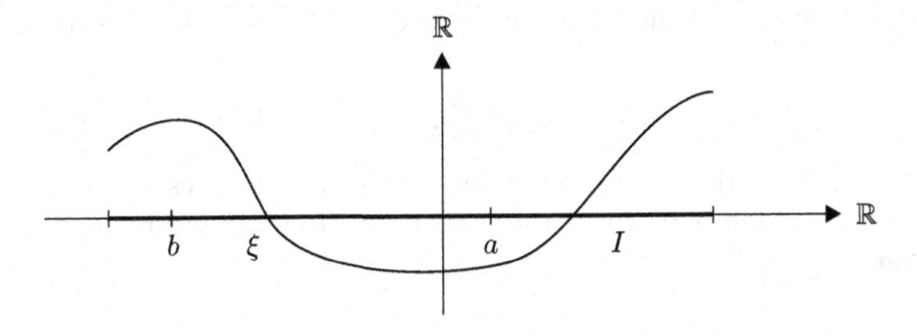

(c) *Every polynomial $p \in \mathbb{R}[X]$ with odd degree has a real zero.*

Proof Without loss of generality, we can write p in the form

$$p = X^{2n+1} + a_{2n}X^{2n} + \cdots + a_0$$

with $n \in \mathbb{N}$ and $a_k \in \mathbb{R}$. Then

$$p(x) = x^{2n+1}\Bigl(1 + \frac{a_{2n}}{x} + \cdots + \frac{a_0}{x^{2n+1}}\Bigr) , \qquad x \in \mathbb{R}^\times .$$

For a sufficiently large $R > 0$ we have

$$1 + \frac{a_{2n}}{R} + \cdots + \frac{a_0}{R^{2n+1}} \geq 1 - \frac{|a_{2n}|}{R} - \cdots - \frac{|a_0|}{R^{2n+1}} \geq \frac{1}{2} ,$$

and so $p(R) \geq R^{2n+1}/2 > 0$ and $p(-R) \leq -R^{2n+1}/2 < 0$. Since polynomial functions are continuous, the claim follows from (b). ■

Monotone Functions

The order completeness of $\mathbb{R}$ has far reaching consequences for monotone functions. As a first example, we show the existence of the left and right limits of a monotone, but not necessarily continuous, real function at the ends of an interval.

5.3 Proposition *Let $f : I \to \mathbb{R}$ be monotone, $\alpha := \inf I$ and $\beta := \sup I$. Then*

$$\lim_{x \to \alpha+} f(x) = \begin{cases} \inf f(I) , & \text{if } f \text{ is increasing} , \\ \sup f(I) , & \text{if } f \text{ is decreasing} , \end{cases}$$

and

$$\lim_{x \to \beta-} f(x) = \begin{cases} \sup f(I) , & \text{if } f \text{ is increasing} , \\ \inf f(I) , & \text{if } f \text{ is decreasing} . \end{cases}$$

Proof Suppose that f is increasing and $b := \sup f(I) \in \bar{\mathbb{R}}$. By the definition of b, for each $\gamma < b$, there is some $x_\gamma \in I$ such that $f(x_\gamma) > \gamma$. Since f is increasing we have

$$\gamma < f(x_\gamma) \leq f(x) \leq b , \qquad x \geq x_\gamma .$$

The analog of Remark 2.23 for left-sided limits then implies $\lim_{x \to \beta-} f(x) = b$. The claims for the left end of the interval and for decreasing functions are proved similarly. ■

To investigate discontinuities and continuous extensions of real functions, we need the following lemma.

5.4 Lemma *Let $D \subseteq \mathbb{R}$, $t \in \mathbb{R}$ and*

$$\overline{D}_t := \overline{D \cap (-\infty, t)} \cap \overline{D \cap (t, \infty)} \ .$$

If $\overline{D}_t$ is not empty, then $\overline{D}_t = \{t\}$ and there are sequences (r_n), (s_n) in D such that

$$r_n < t \ , \ s_n > t \ , \quad n \in \mathbb{N} \ , \qquad \text{and} \qquad \lim r_n = \lim s_n = t \ .$$

Proof Suppose that $\overline{D}_t \neq \emptyset$ and $\tau \in \overline{D}_t$. Then, by the definition of $\overline{D}_t$ and Proposition 2.9, there are sequences (r_n) and (s_n) in D such that
(i) $r_n < t$, $n \in \mathbb{N}$, and $\lim r_n = \tau$, (ii) $s_n > t$, $n \in \mathbb{N}$, and $\lim s_n = \tau$.
By Proposition II.2.7, (i) implies $\tau \le t$ and (ii) implies $\tau \ge t$. Thus $\tau = t$, and all claims are proved. ■

5.5 Examples **(a)** Let D be an interval. Then

$$\overline{D}_t = \begin{cases} \{t\} \ , & t \in \mathring{D} \ , \\ \emptyset \ , & t \notin \mathring{D} \ . \end{cases}$$

(b) If $D = \mathbb{R}^\times$, then $\overline{D}_t = \{t\}$ for each $t \in \mathbb{R}$. ■

We now consider a function $f : D \to X$ where $X = (X, d)$ is a metric space and D is a subset of $\mathbb{R}$. Let $t_0 \in \mathbb{R}$ be such that $\overline{D}_{t_0} \neq \emptyset$. If the one-sided limits $f(t_0+) = \lim_{t \to t_0+} f(t)$ and $f(t_0-) = \lim_{t \to t_0-} f(t)$ exist and are distinct, then t_0 is called a **jump discontinuity** of f and $d(f(t_0+), f(t_0-))$ is called the **size** of the jump discontinuity at t_0.

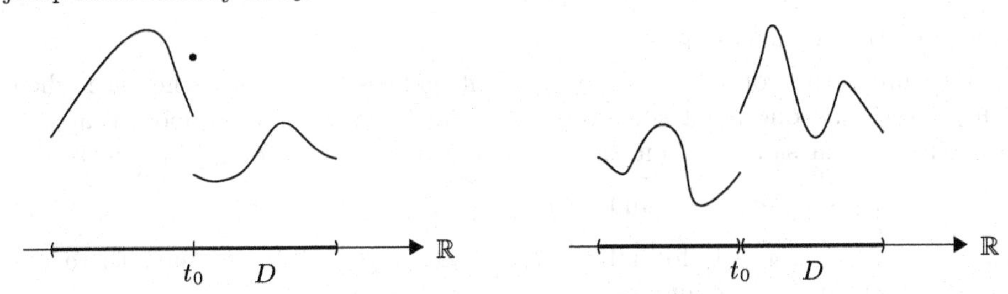

5.6 Proposition *If $f : I \to \mathbb{R}$ is monotone, then f is continuous except perhaps at countably many jump discontinuities.*

Proof It suffices to consider the case of an increasing function $f : I \to \mathbb{R}$. For $t_0 \in \mathring{I}$, Proposition 5.3 applied to each of the restricted functions $f \,|\, I \cap (-\infty, t_0)$ and $f \,|\, I \cap (t_0, \infty)$ implies that $\lim_{t \to t_0+} f(t)$ and $\lim_{t \to t_0-} f(t)$ exist. Because of Propositions 1.12 and 1.13, it suffices to show that the set

$$M := \big\{ \, t_0 \in \mathring{I} \ ; \ f(t_0-) \neq f(t_0+) \, \big\}$$

is countable. For each $t \in M$, we have $f(t-) < f(t+)$ and so we can choose some $r(t) \in \mathbb{Q} \cap \big(f(t-), f(t+)\big)$. This defines a function

$$r: M \to \mathbb{Q}, \quad t \mapsto r(t),$$

which must be injective because f is increasing. Thus M is equinumerous to a subset of $\mathbb{Q}$. In particular, by Propositions I.6.7 and I.9.4, M is countable. ∎

Continuous Monotone Functions

The important theorem which follows shows that any strictly monotone continuous function is injective and has a continuous monotone inverse function defined on its image.

5.7 Theorem (inverse function theorem for monotone functions) *Suppose that $I \subseteq \mathbb{R}$ is a nonempty interval and $f: I \to \mathbb{R}$ is continuous and strictly increasing (or strictly decreasing).*

(i) *$J := f(I)$ is an interval.*

(ii) *$f: I \to J$ is bijective.*

(iii) *$f^{-1}: J \to I$ is continuous and strictly increasing (or strictly decreasing).*

Proof Claim (i) follows from Theorem 5.1, and (ii) is a direct consequence of the strict monotonicity of f.

To prove (iii), suppose that f is strictly increasing and set $g := f^{-1}: J \to I$. If $s_1, s_2 \in J$ are such that $s_1 < s_2$, then $g(s_1) < g(s_2)$ since otherwise

$$s_1 = f\big(g(s_1)\big) \geq f\big(g(s_2)\big) = s_2 .$$

Thus g is strictly increasing.

To prove the continuity of $g: J \to I$ it suffices to consider the case when J has more than one point since otherwise the claim is clear. Suppose that g is not continuous at $s_0 \in J$. Then there are $\varepsilon > 0$ and a sequence (s_n) in J such that

$$|s_n - s_0| < 1/n \quad \text{and} \quad |g(s_n) - g(s_0)| \geq \varepsilon , \qquad n \in \mathbb{N}^\times . \tag{5.1}$$

Thus $s_n \in [s_0 - 1, s_0 + 1]$ for all $n \in \mathbb{N}^\times$, and, since g is increasing, there are $\alpha, \beta \in \mathbb{R}$ such that $\alpha < \beta$ and

$$t_n := g(s_n) \in [\alpha, \beta] .$$

By the Bolzano-Weierstrass theorem, the sequence (t_n) has a convergent subsequence $(t_{n_k})_{k \in \mathbb{N}}$. Let t_0 be the limit of this subsequence. Then the continuity of f implies that $f(t_{n_k}) \to f(t_0)$ as $k \to \infty$. But, from the first claim of (5.1), we also know that $f(t_{n_k}) = s_{n_k}$ converges to s_0. Thus $s_0 = f(t_0)$ and so

$$g(s_{n_k}) = t_{n_k} \to t_0 = g(s_0) \quad (k \to \infty) .$$

This contradicts the second claim of (5.1) and completes the proof. ∎

5.8 Examples (a) For each $n \in \mathbb{N}^{\times}$, the function

$$\mathbb{R}^{+} \to \mathbb{R}^{+}\ , \quad x \mapsto \sqrt[n]{x}$$

is continuous[1] and strictly increasing. In addition, $\lim_{x\to\infty} \sqrt[n]{x} = \infty$.

Proof For $n \in \mathbb{N}^{\times}$, let $f : \mathbb{R}^{+} \to \mathbb{R}^{+}$ be defined by $t \mapsto t^n$. Being the restriction of a polynomial function, f is continuous. If $0 \leq s < t$, then

$$f(t) - f(s) = t^n - s^n = t^n\Bigl(1 - \Bigl(\frac{s}{t}\Bigr)^n\Bigr) > 0\ ,$$

which shows that f is strictly increasing. Finally $\lim_{t\to\infty} f(t) = \infty$ and so all claims follow from Theorem 5.7. ■

(b) The continuity claim of Theorem 5.7(iii) is false, in general, if I is not an interval.

Proof The function $f : Z \to \mathbb{R}$ of Example 1.9(c) is continuous and strictly increasing, but the inverse function of f is not continuous. ■

Further important applications of the inverse function theorem for monotone functions appear in the following section.

Exercises

In the following, I is a compact interval containing more than one point.

1 Let $f : I \to I$ be continuous. Show that f has a **fixed point**, that is, there is some $\xi \in I$ such that $f(\xi) = \xi$.

2 Let $f : I \to \mathbb{R}$ be continuous and injective. Show that f is strictly monotone.

3 Let D be an open subset of $\mathbb{R}$ and $f : D \to \mathbb{R}$ continuous and injective. Prove that $f : D \to f(D)$ is a homeomorphism.[2]

4 Let $\alpha : \mathbb{N} \to \mathbb{Q}$ be a bijection and, for $x \in \mathbb{R}$, let N_x be the set $\{\, k \in \mathbb{N}\ ;\ \alpha(k) \leq x \,\}$. Let (y_n) be a sequence in $(0, \infty)$ such that $\sum y_n < \infty$. Define

$$f : \mathbb{R} \to \mathbb{R}\ , \quad x \mapsto \sum_{k \in N_x} y_k\ .$$

Prove the following:[3]

(a) f is strictly monotone.

(b) f is continuous at each irrational number.

(c) At each rational number q, there is a jump discontinuity of size y_n where $n = \alpha^{-1}(q)$.

---

[1] See also Exercise II.2.7.

[2] See Exercise 3.3.

[3] This exercise shows that Proposition 5.6 cannot be strengthened.

5 Consider the function

$$f:[0,1]\to[0,1]\ ,\qquad x\mapsto\begin{cases} x\ , & x \text{ rational}\ ,\\ 1-x\ , & x \text{ irrational}\ .\end{cases}$$

Show the following:

(a) f is bijective.

(b) f is not monotone on any subinterval of $[0,1]$.

(c) f is continuous only at $x=1/2$.

6 Let $f_0 :=$ zigzag (see Exercise 1.1) and

$$F(x) := \sum_{n=0}^{\infty} 4^{-n} f_0(4^n x)\ ,\qquad x\in\mathbb{R}\ .$$

Prove the following:

(a) F is well defined.

(b) F is not monotone on any interval.

(c) F is continuous.

(Hint: (a) For each $x\in\mathbb{R}$, find a convergent majorant for $\sum 4^{-n} f_0(4^n x)$.
(b) Let $f_n(x) := 4^{-n} f_0(4^n x)$ for all $x\in\mathbb{R}$ and $n\in\mathbb{N}$. Set $a := k\cdot 4^{-m}$ and $h := 4^{-2m-1}$ for $k\in\mathbb{Z}$ and $m\in\mathbb{N}^\times$. Then

$$f_n(a)=0\ ,\quad n\ge m\ ,\qquad\text{and}\qquad f_n(a\pm h)=0\ ,\quad n\ge 2m+1\ ,$$

and so $F(a\pm h)-F(a)\ge h$. Finally approximate an arbitrary $x\in\mathbb{R}$ by $k\cdot 4^{-m}$ with $k\in\mathbb{Z}$ and $m\in\mathbb{N}^\times$.
(c) For $x,y\in\mathbb{R}$ and $m\in\mathbb{N}^\times$, we have $|F(x)-F(y)|\le\sum_{k=0}^{m}|f_k(x)-f_k(y)|+4^{-m}/3$.)

7 Let $f: I\to\mathbb{R}$ be monotone. Prove that $\omega_f(x)=|f(x+)-f(x-)|$ where $\omega_f(x)$ is defined as in Exercise 1.17.

6 The Exponential and Related Functions

In this (rather long) section we study one of the most important functions of mathematics, the exponential function. Its importance is apparent already in its close relationship to the trigonometric and logarithm functions, which we also investigate.

Euler's Formula

In Chapter II, we defined the exponential function using the exponential series,

$$\exp(z) := e^z := \sum_{n=0}^{\infty} \frac{z^n}{n!} = 1 + z + \frac{z^2}{2!} + \frac{z^3}{3!} + \cdots \ , \qquad z \in \mathbb{C} \ .$$

The use of the notation e^z for $\exp(z)$ is justified by Example II.8.12(b). Associated with this series are the **cosine series**

$$\sum (-1)^n \frac{z^{2n}}{(2n)!} = 1 - \frac{z^2}{2!} + \frac{z^4}{4!} - + \cdots \ ,$$

and the **sine series**

$$\sum (-1)^n \frac{z^{2n+1}}{(2n+1)!} = z - \frac{z^3}{3!} + \frac{z^5}{5!} - + \cdots \ .$$

We will show that — analogous to the exponential series — the cosine and sine series converge absolutely everywhere. The functions defined by these series,

$$\cos : \mathbb{C} \to \mathbb{C} \ , \quad z \mapsto \sum_{n=0}^{\infty} (-1)^n \frac{z^{2n}}{(2n)!}$$

and

$$\sin : \mathbb{C} \to \mathbb{C} \ , \quad z \mapsto \sum_{n=0}^{\infty} (-1)^n \frac{z^{2n+1}}{(2n+1)!} \ ,$$

are called the **cosine** and **sine** functions.[1]

6.1 Theorem

(i) *The exponential, cosine and sine series have infinite radii of convergence.*

(ii) *The functions* exp, cos, sin *are real valued on real arguments.*

(iii) *The* **addition theorem for the exponential function** *holds*:

$$e^{w+z} = e^w e^z \ , \qquad w, z \in \mathbb{C} \ .$$

[1] We will later see that these definitions give the familiar trigonometric functions.

(iv) **Euler's formula** *holds*:

$$e^{iz} = \cos z + i \sin z \,, \qquad z \in \mathbb{C} \,. \tag{6.1}$$

(v) *The functions* exp, cos *and* sin *are continuous on* $\mathbb{C}$.

Proof (i) In Example II.8.7(c), we have already proved that the exponential series has radius of convergence ∞. Thus Hadamard's formula yields

$$\infty = \frac{1}{\overline{\lim\limits_{n\to\infty}} \sqrt[n]{1/n!}} = \lim_{n\to\infty} \sqrt[n]{n!} \,.$$

By Theorem II.5.7, the sequence $\left(\sqrt[n]{n!}\right)_{n\in\mathbb{N}}$ and all of its subsequences converge to ∞. Thus

$$\frac{1}{\overline{\lim\limits_{n\to\infty}} \sqrt[2n]{1/(2n)!}} = \lim_{n\to\infty} \sqrt[2n]{(2n)!} = \infty$$

and

$$\frac{1}{\overline{\lim\limits_{n\to\infty}} \sqrt[2n+1]{1/(2n+1)!}} = \lim_{n\to\infty} \sqrt[2n+1]{(2n+1)!} = \infty \,,$$

so that, by Hadamard's formula, the cosine and sine series have infinite radii of convergence.

(ii) Because $\mathbb{R}$ is a field, all partial sums of the above series are real if z is real. Since $\mathbb{R}$ is closed in $\mathbb{C}$, the claim follows.

(iii) This is proved in Example II.8.12(a).

(iv) For $n \in \mathbb{N}$, we have

$$i^{2n} = (i^2)^n = (-1)^n \quad \text{and} \quad i^{2n+1} = i \cdot i^{2n} = i \cdot (-1)^n \,.$$

Thus, by Proposition II.7.5,

$$e^{iz} = \sum_{n=0}^{\infty} \frac{(iz)^n}{n!} = \sum_{k=0}^{\infty} \frac{(iz)^{2k}}{(2k)!} + \sum_{k=0}^{\infty} \frac{(iz)^{2k+1}}{(2k+1)!} = \cos z + i \sin z$$

for all $z \in \mathbb{C}$.

(v) This follows from Proposition 1.7. ■

6.2 Remarks **(a)** Cosine is an even function and sine is an odd function, that is,[2]

$$\cos(z) = \cos(-z) \quad \text{and} \quad \sin(z) = -\sin(-z) \,, \qquad z \in \mathbb{C} \,. \tag{6.2}$$

[2]See Exercise II.9.7.

(b) From (a) and Euler's formula (6.1) we get

$$\cos(z) = \frac{e^{iz} + e^{-iz}}{2}\,, \qquad \sin(z) = \frac{e^{iz} - e^{-iz}}{2i}\,, \qquad z \in \mathbb{C}\,. \tag{6.3}$$

(c) For $w, z \in \mathbb{C}$, we have

$$e^z \neq 0\,, \quad e^{-z} = 1/e^z\,, \quad e^{z-w} = e^z/e^w\,, \quad \overline{e^z} = e^{\overline{z}}\,.$$

Proof From the addition theorem we get $e^z e^{-z} = e^{z-z} = e^0 = 1$, from which the first three claims follow.

By Example 1.3(i), the function $\mathbb{C} \to \mathbb{C},\ w \mapsto \overline{w}$ is continuous. Theorem 1.4 then implies that

$$\overline{e^z} = \overline{\lim_{n\to\infty} \sum_{k=0}^{n} \frac{z^k}{k!}} = \lim_{n\to\infty} \sum_{k=0}^{n} \frac{\overline{z}^k}{k!} = e^{\overline{z}}$$

for all $z \in \mathbb{C}$. ■

(d) For all $x \in \mathbb{R}$, $\cos(x) = \mathrm{Re}(e^{ix})$ and $\sin(x) = \mathrm{Im}(e^{ix})$.

Proof This follows from Euler's formula and Theorem 6.1(ii). ■

In the following proposition we use the name 'trigonometric function' for cosine and sine. This usage is justified after Remarks 6.18.

6.3 Proposition (addition theorem for trigonometric functions) *For all $z, w \in \mathbb{C}$ we have*[3]

(i) $\cos(z \pm w) = \cos z \cos w \mp \sin z \sin w$,
$\sin(z \pm w) = \sin z \cos w \pm \cos z \sin w$.

(ii) $\sin z - \sin w = 2\cos\dfrac{z+w}{2}\sin\dfrac{z-w}{2}$,
$\cos z - \cos w = -2\sin\dfrac{z+w}{2}\sin\dfrac{z-w}{2}$.

Proof (i) The formulas in (6.3) and the addition theorem for the exponential function yield

$$\begin{aligned}\cos z \cos w - \sin z \sin w &= \frac{1}{4}\bigl\{(e^{iz} + e^{-iz})(e^{iw} + e^{-iw}) + (e^{iz} - e^{-iz})(e^{iw} - e^{-iw})\bigr\}\\ &= \frac{1}{2}\bigl\{e^{i(z+w)} + e^{-i(z+w)}\bigr\} = \cos(z+w)\end{aligned}$$

for all $z, w \in \mathbb{C}$. Using also (6.2), we get

$$\cos(z - w) = \cos z \cos w + \sin z \sin w\,, \qquad z, w \in \mathbb{C}\,.$$

The second formula in (i) can be proved similarly.

[3] When no misunderstanding is possible, it is usual to write $\cos z$ and $\sin z$ instead of $\cos(z)$ and $\sin(z)$.

(ii) For $z, w \in \mathbb{C}$, set $u := (z+w)/2$ and $v := (z-w)/2$. Then $u+v = z$ and $u - v = w$, and so, using (i), we get

$$\begin{aligned}\sin z - \sin w &= \sin(u+v) - \sin(u-v) = 2\cos u \sin v \\ &= 2\cos\frac{z+w}{2}\sin\frac{z-w}{2} \ .\end{aligned}$$

The second formula in (ii) can be proved similarly. ∎

6.4 Corollary *For $z \in \mathbb{C}$, $\cos^2 z + \sin^2 z = 1$.*

Proof Setting $z = w$ in Proposition 6.3(i) we get

$$\cos^2 z + \sin^2 z = \cos(z - z) = \cos(0) = 1 \ ,$$

which proves the claim. ∎

If we write $z \in \mathbb{C}$ in the form $z = x + iy$ with $x, y \in \mathbb{R}$, then $e^z = e^x e^{iy}$. This simple observation shows that the exponential function is completely determined by the real exponential function $\exp_{\mathbb{R}} := \exp|\mathbb{R}$ and the restriction of exp to $i\mathbb{R}$, that is, by $\exp_{i\mathbb{R}} := \exp|i\mathbb{R}$. Hence, to understand the 'complex' exponential function $\exp : \mathbb{C} \to \mathbb{C}$, we begin by studying these two functions separately.

The Real Exponential Function

We collect in the next proposition the most important qualitative properties of the function $\exp_{\mathbb{R}}$.

6.5 Proposition

(i) *If $x < 0$, then $0 < e^x < 1$. If $x > 0$, then $1 < e^x < \infty$.*

(ii) *$\exp_{\mathbb{R}} : \mathbb{R} \to \mathbb{R}^+$ is strictly increasing.*

(iii) *For each $\alpha \in \mathbb{Q}$,*

$$\lim_{x\to\infty} \frac{e^x}{x^\alpha} = \infty \ ,$$

that is, the exponential function increases faster than any power function.

(iv) $\lim_{x\to-\infty} e^x = 0$.

Proof (i) From

$$e^x = 1 + \sum_{n=1}^{\infty} \frac{x^n}{n!} \ , \qquad x \in \mathbb{R} \ ,$$

we see that $e^x > 1$ for all $x > 0$. If $x < 0$, then $-x > 0$ and so $e^{-x} > 1$. This implies $e^x = e^{-(-x)} = 1/e^{-x} \in (0, 1)$.

(ii) Let $x, y \in \mathbb{R}$ be such that $x < y$. Since $e^x > 0$ and $e^{y-x} > 1$, it follows that

$$e^y = e^{x+(y-x)} = e^x e^{y-x} > e^x .$$

(iii) It suffices to consider the case $\alpha > 0$. Let $n := \lfloor \alpha \rfloor + 1$. It follows from the exponential series that $e^x > x^{n+1}/(n+1)!$ for all $x > 0$. Thus

$$\frac{e^x}{x^\alpha} > \frac{e^x}{x^n} > \frac{x}{(n+1)!} , \qquad x > 0 ,$$

which proves the claim.

(iv) If we set $\alpha = 0$ in (iii) we get $\lim_{x\to\infty} e^x = \infty$. Thus

$$\lim_{x\to-\infty} e^x = \lim_{y\to\infty} e^{-y} = \lim_{y\to\infty} \frac{1}{e^y} = 0 ,$$

and all the claims are proved. ■

The Logarithm and Power Functions

From Proposition 6.5 we have

$$\exp_{\mathbb{R}} : \mathbb{R} \to \mathbb{R}^+ \text{ is continuous and strictly increasing and } \exp(\mathbb{R}) = (0, \infty) .$$

Thus, by Theorem 5.7, the real exponential function has a continuous and strictly increasing inverse function defined on $(0, \infty)$. This inverse function is called the (natural) **logarithm** and is written log, that is,

$$\log := (\exp_{\mathbb{R}})^{-1} : (0, \infty) \to \mathbb{R} .$$

In particular, $\log 1 = 0$ and $\log e = 1$.

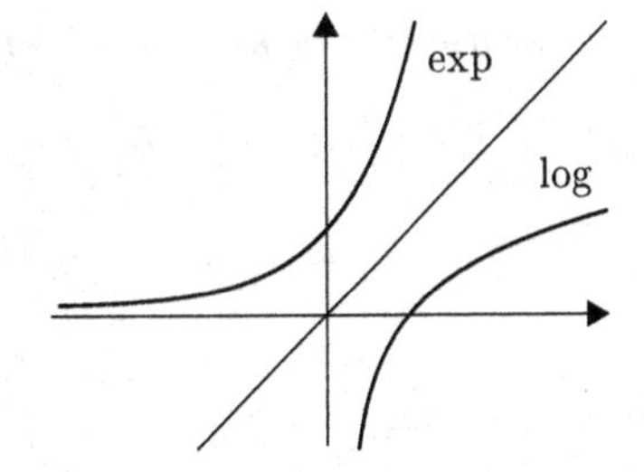

6.6 Theorem (addition theorem for the logarithm function) *For all $x, y \in (0, \infty)$,*

$$\log(xy) = \log x + \log y \quad \textit{and} \quad \log(x/y) = \log x - \log y .$$

Proof Let $x, y \in (0, \infty)$. For $a := \log x$ and $b := \log y$, we have $x = e^a$ and $y = e^b$. The addition theorem for the exponential function then implies $xy = e^a e^b = e^{a+b}$ and $x/y = e^a/e^b = e^{a-b}$, from which the claims follow. ■

6.7 Proposition *For all $a > 0$ and $r \in \mathbb{Q}$,*[4]

$$a^r = e^{r \log a} . \tag{6.4}$$

Proof By definition, we have $a = e^{\log a}$, and so Theorem 6.1(iii) implies that $a^n = (e^{\log a})^n = e^{n \log a}$ for all $n \in \mathbb{N}$. In addition,

$$a^{-n} = (e^{\log a})^{-n} = \frac{1}{(e^{\log a})^n} = \frac{1}{e^{n \log a}} = e^{-n \log a} , \qquad n \in \mathbb{N} .$$

Now set $x := e^{\frac{1}{n} \log a}$. Then $x^n = e^{n(\frac{1}{n} \log a)} = e^{\log a} = a$ and hence, by Proposition I.10.9, $e^{\frac{1}{n} \log a} = a^{\frac{1}{n}}$ for all $n \in \mathbb{N}^\times$.

Now let $r \in \mathbb{Q}$. Then there are $p \in \mathbb{Z}$ and $q \in \mathbb{N}^\times$ such that $r = p/q$. From above we have

$$a^r = a^{\frac{p}{q}} = \left(a^{\frac{1}{q}}\right)^p = \left(e^{\frac{1}{q} \log a}\right)^p = e^{\frac{p}{q} \log a} = e^{r \log a} ,$$

which completes the proof of (6.4). ■

Let $a > 0$. So far we have defined a^r only for rational exponents r, and for such exponents we have shown that $a^r = e^{r \log a}$. Since $e^{r \log a}$ is defined for any real number $r \in \mathbb{R}$, this suggests an obvious generalization. Specifically, we define

$$a^x := e^{x \log a} , \qquad x \in \mathbb{R} , \quad a > 0 .$$

6.8 Proposition *For all $a, b > 0$ and $x, y \in \mathbb{R}$,*

$$a^x a^y = a^{x+y} , \quad \frac{a^x}{a^y} = a^{x-y} , \quad a^x b^x = (ab)^x , \quad \frac{a^x}{b^x} = \left(\frac{a}{b}\right)^x ,$$
$$\log(a^x) = x \log a , \quad (a^x)^y = a^{xy} .$$

Proof For example,

$$a^x a^y = e^{x \log a} e^{y \log a} = e^{(x+y) \log a} = a^{x+y}$$

and

$$(a^x)^y = (e^{x \log a})^y = e^{xy \log a} = a^{xy} .$$

The remaining claims can be proved similarly. ■

[4]Note that the left side of (6.4) is the rth power of the positive number a as defined in Remark I.10.10(d), whereas, the right side is the value of the exponential function at $r \cdot \log a \in \mathbb{R}$. Note also that in the case $a = e$, (6.4) reduces to Example II.8.12(b).

6.9 Proposition *For all $\alpha > 0$,*

$$\lim_{x\to\infty} \frac{\log x}{x^\alpha} = 0 \quad \textit{and} \quad \lim_{x\to 0+} x^\alpha \log x = 0 \; .$$

In particular, the logarithm function increases more slowly than any (arbitrarily small) positive power function.

Proof Since the logarithm is increasing, it follows from Proposition 6.5(iii) that

$$\lim_{x\to\infty} \frac{\log x}{x^\alpha} = \lim_{x\to\infty} \frac{\log x}{e^{\alpha \log x}} = \lim_{y\to\infty} \frac{y}{e^{\alpha y}} = \frac{1}{\alpha} \lim_{t\to\infty} \frac{t}{e^t} = 0 \; .$$

For the second limit we have

$$\lim_{x\to 0+} x^\alpha \log x = \lim_{y\to\infty} \Bigl(\frac{1}{y}\Bigr)^\alpha \log \frac{1}{y} = -\lim_{y\to\infty} \frac{\log y}{y^\alpha} = 0 \; ,$$

which proves the claim. ■

Note that Proposition 6.5(iii) is also valid for $\alpha \in \mathbb{R}$.

The Exponential Function on $i\mathbb{R}$

The function $\exp_{i\mathbb{R}}$ has a completely different nature than the real exponential function $\exp_{\mathbb{R}}$. For example, while $\exp_{\mathbb{R}}$ is strictly increasing, we will prove that $\exp_{i\mathbb{R}}$ is a *periodic function.* In the process of determining its period we will define the constant π. To prove these claims, we first need a few lemmas.

6.10 Lemma $|e^{it}| = 1$ *for all* $t \in \mathbb{R}$.

Proof Since $\overline{e^z} = e^{\overline{z}}$ for all $z \in \mathbb{C}$, we have

$$|e^{it}|^2 = e^{it}\overline{(e^{it})} = e^{it}e^{-it} = e^0 = 1 \; , \qquad t \in \mathbb{R} \; ,$$

from which the claim follows. ■

Rather than $\exp_{i\mathbb{R}}$, it is sometimes useful to consider the function

$$\operatorname{cis} : \mathbb{R} \to \mathbb{C} \; , \quad t \mapsto e^{it} \; .$$

Lemma 6.10 says that the image of cis is contained in $S^1 := \{\, z \in \mathbb{C} \; ; \; |z| = 1 \,\}$. In the next lemma we strengthen this result and prove that the image of cis is all of S^1.

6.11 Lemma $\mathrm{cis}(\mathbb{R}) = S^1$.

Proof (i) In the first step we show that the image of the cosine function is

$$\cos(\mathbb{R}) = \mathrm{pr}_1\bigl[\mathrm{cis}(\mathbb{R})\bigr] = [-1,1] \ . \tag{6.5}$$

The first equality in (6.5) is a clear consequence of Euler's formula. To prove the second equality, set $I := \cos(\mathbb{R})$. Then it follows from Bolzano's intermediate value theorem (Theorem 5.1) that I is an interval. In addition, we know from Lemma 6.10 that

$$I = \mathrm{pr}_1\bigl(\mathrm{cis}(\mathbb{R})\bigr) \subseteq [-1,1] \ .$$

Of course, $1 = \cos(0)$ is in I, but $I = \{1\}$ is not possible since, if $\cos(z) = 1$ for all $z \in \mathbb{R}$, then, by Corollary II.9.9, the cosine series would be $1 + 0z + 0z^2 + \cdots$. Thus I has the form

$$I = [a,1] \quad \text{or} \quad I = (a,1]$$

for some suitable $a \in [-1,1)$.

Suppose that a is not equal to -1. Since $a_0 := (a+1)/2$ is in I, there is some $t_0 \in \mathbb{R}$ such that $a_0 = \cos t_0$. Set

$$z_0 := \mathrm{cis}(t_0) = \cos t_0 + i \sin t_0 \ .$$

Then, by Corollary 6.4,

$$\begin{aligned}\mathrm{pr}_1(z_0^2) &= \mathrm{Re}\bigl((\cos t_0 + i \sin t_0)^2\bigr) = \cos^2 t_0 - \sin^2 t_0 \\ &= 2\cos^2 t_0 - 1 = 2a_0^2 - 1 = a - \frac{1-a^2}{2} < a \ ,\end{aligned}$$

since, by assumption, $a^2 < 1$. The inequality $\mathrm{pr}_1(z_0^2) < a$ contradicts the fact that $\mathrm{pr}_1(z_0^2) = \mathrm{pr}_1(e^{2it_0})$ is in I. Thus we conclude that $a = -1$.

To complete the proof of (6.5), it remains to show that -1 is in I. We know that there is some $t_0 \in \mathbb{R}$ such that $\cos t_0 = 0$. Since $\sin^2 t_0 = 1 - \cos^2 t_0 = 1$, this implies $z_0 = e^{it_0} = i \sin t_0 = \pm i$. Thus

$$-1 = \mathrm{pr}_1(-1) = \mathrm{pr}_1(z_0^2) = \mathrm{pr}_1(e^{2it_0}) \in I \ ,$$

as claimed.

(ii) We prove next that $S^1 \subseteq \mathrm{cis}(\mathbb{R})$. If $z \in S^1$, then

$$\mathrm{Re}\, z \in [-1,1] = \mathrm{pr}_1\bigl(\mathrm{cis}(\mathbb{R})\bigr) \ ,$$

and so, by (i), there is some $t \in \mathbb{R}$ such that $\mathrm{Re}\, z = \mathrm{Re}\, e^{it}$. Moreover, it follows from $|z| = 1 = |e^{it}|$ that either $z = e^{it}$ or $\overline{z} = e^{it}$. In the first case, $z \in \mathrm{cis}(\mathbb{R})$ is clear. Otherwise $\overline{z} = e^{it}$ and, from $z = \overline{\overline{z}} = \overline{e^{it}} = e^{-it}$, it follows that $z \in \mathrm{cis}(\mathbb{R})$ in this case too. Thus we have proved $S^1 \subseteq \mathrm{cis}(\mathbb{R})$. This, together with Lemma 6.10, implies $\mathrm{cis}(\mathbb{R}) = S^1$. ■

6.12 Lemma *The set $M := \{ t > 0 \ ; \ e^{it} = 1 \}$ has a minimum element.*

Proof (i) First we show that M is nonempty. By Lemma 6.11, there is some $t \in \mathbb{R}^\times$ such that $e^{it} = -1$. Because

$$e^{-it} = \frac{1}{e^{it}} = \frac{1}{-1} = -1 \ ,$$

we can suppose that $t > 0$. Then $e^{2it} = (e^{it})^2 = (-1)^2 = 1$ and M is nonempty.

(ii) Next we show that M is closed in $\mathbb{R}$. To prove this, choose a sequence (t_n) in M which converges to $t^* \in \mathbb{R}$. Since t_n is positive for all n, we have $t^* \geq 0$. In addition, the continuity of cis implies

$$e^{it^*} = \operatorname{cis}(t^*) = \operatorname{cis}(\lim t_n) = \lim \operatorname{cis}(t_n) = 1 \ .$$

To prove that M is closed, it remains to show that t^* is positive. Suppose, to the contrary, that $t^* = 0$. Then there is some $m \in \mathbb{N}$ such that $t_m \in (0, 1)$. From Euler's formula we have $1 = e^{it_m} = \cos t_m + i \sin t_m$ and so $\sin t_m = 0$.

Applying Corollary II.7.9 to the sine series

$$\sin t = t - \frac{t^3}{6} + \frac{t^5}{5!} - + \cdots$$

we get

$$\sin t \geq t(1 - t^2/6) \ , \qquad 0 < t < 1 \ . \tag{6.6}$$

For t_m, this yields $0 = \sin t_m \geq t_m(1 - t_m^2/6) > 5t_m/6$, a contradiction. Thus M is closed.

(iii) Since M is a nonempty closed set which is bounded below, it has minimum element. ■

The Definition of π and its Consequences

The preceding lemma makes it possible to define a number π by

$$\pi := \frac{1}{2} \min\{ t > 0 \ ; \ e^{it} = 1 \} \ .$$

We will see in Section VI.5 that the number π defined this way has the usual geometrical meaning, for example, as the area of a unit circle. For the moment, however, π is simply the smallest positive real number such that $e^{2\pi i} = 1$.

Consider the number $e^{i\pi}$. We have $(e^{i\pi})^2 = e^{2\pi i} = 1$, and so $e^{i\pi} = \pm 1$. By the definition of π, the case $e^{i\pi} = 1$ is not possible, and so $e^{i\pi} = -1$. This implies also $e^{-i\pi} = 1/e^{i\pi} = -1$. These special cases can be used to determine all other $z \in \mathbb{C}$ such that $e^z = 1$ or $e^z = -1$.

6.13 Proposition

(i) $e^z = 1 \iff z \in 2\pi i\mathbb{Z}$.

(ii) $e^z = -1 \iff z \in \pi i + 2\pi i\mathbb{Z}$.

Proof (i) '$\Leftarrow$' For all $k \in \mathbb{Z}$, we have $e^{2\pi i k} = (e^{2\pi i})^k = 1$.
'$\Rightarrow$' Suppose that $z = x + iy$ with $x, y \in \mathbb{R}$ is such that $e^z = 1$. Then

$$1 = |e^z| = |e^x|\,|e^{iy}| = e^x ,$$

and so $x = 0$. If $k \in \mathbb{Z}$ and $r \in [0, 2\pi)$ are such that $y = 2\pi k + r$, then

$$1 = e^{iy} = e^{2\pi k i} e^{ir} = e^{ir} .$$

The definition of π implies that $r = 0$, and so $z = 2\pi i k$ with $k \in \mathbb{Z}$.

(ii) Since $e^{-i\pi} = -1$, we have $e^z = -1$ if and only if $e^{z-i\pi} = e^z e^{-i\pi} = 1$. By (i), $e^{z-i\pi} = 1$ if and only if $z - i\pi = 2\pi i k$, that is, $z = i\pi + 2\pi i k$ for some $k \in \mathbb{Z}$. ■

From Proposition 6.13(i) we have $e^{z+2\pi i k} = e^z e^{2\pi i k} = e^z$ for all $k \in \mathbb{Z}$, and hence the following corollary.

6.14 Corollary *The exponential function is periodic*[5] *with period* $2\pi i$, *that is,*

$$e^z = e^{z+2\pi i k} , \qquad z \in \mathbb{C} , \quad k \in \mathbb{Z} .$$

Using Proposition 6.13 we can show also that the function cis is bijective on half open intervals of length 2π.

6.15 Proposition *For each* $a \in \mathbb{R}$, *the functions*

$$\mathrm{cis} \,\big|\, [a, a+2\pi) : [a, a+2\pi) \to S^1 ,$$
$$\mathrm{cis} \,\big|\, (a, a+2\pi] : (a, a+2\pi] \to S^1$$

are bijective.

Proof (i) Suppose that $\mathrm{cis}\, t = \mathrm{cis}\, s$ for some $s, t \in \mathbb{R}$. Since $e^{i(t-s)} = 1$, there is, by Proposition 6.13, some $k \in \mathbb{Z}$ such that $t = s + 2\pi k$. This implies that each of the above functions is injective.

(ii) Let $z \in S^1$. By Lemma 6.11, there is some $t \in \mathbb{R}$ such that $\mathrm{cis}\, t = z$. Also there are $k_1, k_2 \in \mathbb{Z}$, $r_1 \in [0, 2\pi)$ and $r_2 \in (0, 2\pi]$ such that

$$t = a + 2\pi k_1 + r_1 = a + 2\pi k_2 + r_2 .$$

By Corollary 6.14, $\mathrm{cis}(a + r_1) = \mathrm{cis}(a + r_2) = \mathrm{cis}\, t = z$, and so these functions are also surjective. ■

[5]If E is a vector space and M a set, then $f : E \to M$ is **periodic** with **period** $p \in E \backslash \{0\}$ if $f(x + p) = f(x)$ for all $x \in E$.

6.16 Theorem

(i) $\cos z = \cos(z + 2k\pi)$, $\sin z = \sin(z + 2k\pi)$, $z \in \mathbb{C}$, $k \in \mathbb{Z}$, *that is,* cos and sin are periodic with period 2π.

(ii) *For all* $z \in \mathbb{C}$,

$$\cos z = 0 \Longleftrightarrow z \in \pi/2 + \pi\mathbb{Z} \ ,$$
$$\sin z = 0 \Longleftrightarrow z \in \pi\mathbb{Z} \ .$$

(iii) *The function* $\sin : \mathbb{R} \to \mathbb{R}$ *is positive on* $(0, \pi)$ *and is strictly increasing on the closed interval* $[0, \pi/2]$.

(iv) $\cos(z + \pi) = -\cos z$, $\sin(z + \pi) = -\sin z$, $z \in \mathbb{C}$.

(v) $\cos z = \sin(\pi/2 - z)$, $\sin z = \cos(\pi/2 - z)$, $z \in \mathbb{C}$.

(vi) $\cos(\mathbb{R}) = \sin(\mathbb{R}) = [-1, 1]$.

Proof Claim (i) follows from (6.3) and Corollary 6.14.

(ii) From (6.3) and Proposition 6.13 we have

$$\cos z = 0 \Longleftrightarrow e^{iz} + e^{-iz} = 0 \Longleftrightarrow e^{2iz} = -1 \Longleftrightarrow z \in \pi/2 + \pi\mathbb{Z} \ .$$

Similarly

$$\sin z = 0 \Longleftrightarrow e^{iz} - e^{-iz} = 0 \Longleftrightarrow e^{2iz} = 1 \Longleftrightarrow z \in \pi\mathbb{Z} \ .$$

(iii) From what we just proved, $\sin x \neq 0$ for $x \in (0, \pi)$. The inequality (6.6) shows that $\sin x$ is positive for all $x \in \bigl(0, \sqrt{6}\bigr)$. Because of the intermediate value theorem (Theorem 5.1) we must have, in fact,

$$\sin x > 0 \ , \qquad x \in (0, \pi) \ . \tag{6.7}$$

Similarly, since $\cos(0) = 1$ and $\cos t \neq 0$ for all $t \in (-\pi/2, \pi/2)$, the intermediate value theorem implies that $\cos t > 0$ for all t in $(-\pi/2, \pi/2)$.

For the second claim, suppose that $0 \leq x < y \leq \pi/2$. From Proposition 6.3(ii) we have

$$\sin y - \sin x = 2 \cos \frac{y + x}{2} \sin \frac{y - x}{2} \ . \tag{6.8}$$

Since $(y + x)/2$ and $(y - x)/2$ are in $(0, \pi/2)$, the right side of (6.8) is positive, and hence, $\sin y > \sin x$.

(iv) From (ii) we have $\sin \pi = 0$, and so, by Proposition 6.13(ii),

$$\cos \pi = \cos \pi + i \sin \pi = e^{i\pi} = -1 \ .$$

Now let $z \in \mathbb{C}$. From Proposition 6.3(i) we get

$$\cos(z + \pi) = \cos z \cos \pi - \sin z \sin \pi = -\cos z$$

and

$$\sin(z + \pi) = \sin z \cos \pi + \cos z \sin \pi = -\sin z \ .$$

(v) From (ii) we have $\cos(\pi/2) = 0$, and so, using (iii) and Corollary 6.4, we get

$$0 < \sin(\pi/2) = |\sin(\pi/2)| = \sqrt{1 - \cos^2(\pi/2)} = 1 \ .$$

From Proposition 6.3(i) we now have

$$\cos(\pi/2 - z) = \cos(\pi/2)\cos z + \sin(\pi/2)\sin z = \sin z$$

and

$$\sin(\pi/2 - z) = \sin(\pi/2)\cos z - \cos(\pi/2)\sin z = \cos z \ .$$

(vi) We have already shown in (6.5) that $\cos(\mathbb{R}) = [-1, 1]$. From (v) we get also $\sin(\mathbb{R}) = \cos(\mathbb{R})$. ∎

6.17 Remarks **(a)** Because of the equations

$$\sin(x + \pi) = -\sin x \ , \quad \cos x = \sin(\pi/2 - x) \ , \qquad x \in \mathbb{R} \ ,$$

and the fact that sine is an odd function, the *real* sine and cosine functions are completely determined by values of $\sin x$ on $[0, \pi/2]$.

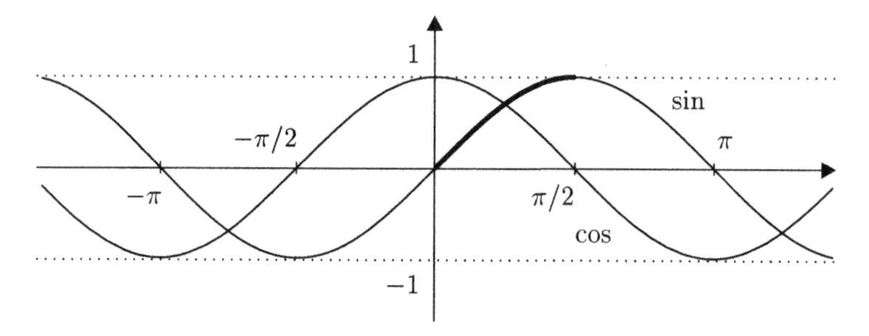

(b) *$\pi/2$ is the least positive zero of the cosine function.*

In principle, this observation, together with the cosine series, can be used to approximate the number π with arbitrary precision. For example, by Corollary II.7.9, we have the inequalities

$$1 - \frac{t^2}{2} < \cos t < 1 - \frac{t^2}{2} + \frac{t^4}{24} \ , \qquad t \in \mathbb{R}^\times \ ,$$

and so $\cos 2 < -1/3$ and $\cos t > 0$ for all $0 \le t < \sqrt{2}$. From the intermediate value theorem we know that the cosine function has a zero in the interval $[\sqrt{2}, 2)$. Indeed, since $\cos t > 0$ for all $0 \le t < \sqrt{2}$, the least positive zero, namely $\pi/2$, must be in this interval. Thus $2\sqrt{2} \le \pi < 4$. Since two distinct zeros are separated by a distance π or more, $\pi/2$ is the only zero in the interval $(0, 2)$.

For a better approximation, pick some t in the middle of the interval $[\sqrt{2}, 2)$, and then use the cosine series and Corollary II.7.9 to calculate the sign of $\cos t$.

This will determine whether $\pi/2$ is in $\bigl[\sqrt{2},t\bigr)$ or in $\bigl(t,2\bigr)$. By repeating this process, π can be determined with arbitrary precision. After *considerable* effort, one gets[6]

$$\pi = 3.14159\ 26535\ 89793\ 23846\ 26433\ 83279\ \ldots$$

We will later develop a far more efficient procedure for calculating π.

(c) A complex number is called **algebraic** if it is a zero of a nonconstant polynomial with integer coefficients. Complex numbers which are not algebraic are called **transcendental numbers**. In particular, real transcendental numbers are irrational.

In 1882, F. Lindemann proved that π is transcendental. This, together with classical results from algebra, provides a mathematical proof of the impossibility of 'squaring the circle'. That is, it is not possible, using only a rule and a compass, to construct a square whose area is equal to the area of a given circle. ■

The Tangent and Cotangent Functions

The **tangent** and **cotangent** functions are defined by

$$\tan z := \frac{\sin z}{\cos z}\,, \quad z \in \mathbb{C}\backslash\Bigl(\frac{\pi}{2}+\pi\mathbb{Z}\Bigr)\ , \qquad \cot z := \frac{\cos z}{\sin z}\,, \quad z \in \mathbb{C}\backslash\pi\mathbb{Z}\ .$$

Restricted to the real numbers, the tangent and cotangent functions have the following graphs:

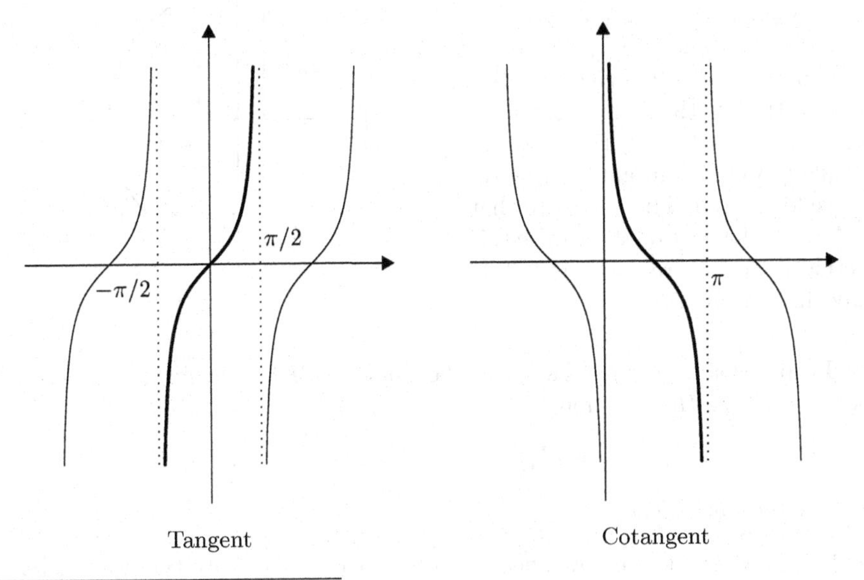

[6]A common mnemonic for the digits of π is

HOW I LIKE A DRINK, ALCOHOLIC OF COURSE,
AFTER THE HEAVY LECTURES INVOLVING QUANTUM MECHANICS.

The number of letters in each word gives the corresponding digit of π.

6.18 Remarks **(a)** The tangent and cotangent functions are continuous, periodic with period π, and odd.

(b) The **addition theorem for the tangent function** holds:

$$\tan(z \pm w) = \frac{\tan z \pm \tan w}{1 \mp \tan z \tan w}$$

for all $w, z \in \mathrm{dom}(\tan)$ such that $z \pm w \in \mathrm{dom}(\tan)$.

Proof This follows easily from Proposition 6.3(i). ■

(c) For all $z \in \mathbb{C} \backslash \pi\mathbb{Z}$, $\cot z = -\tan(z - \pi/2)$.

Proof This follows directly from Theorem 6.16(iv). ■

The Complex Exponential Function

In Propositions 6.1(v) and 6.15 we have seen that the function

$$\mathrm{cis} : [0, 2\pi) \to S^1 , \quad t \mapsto e^{it}$$

is continuous and bijective, and so, for each $z \in S^1$, there is a unique $\alpha \in [0, 2\pi)$ such that

$$z = e^{i\alpha} = \mathrm{cis}(\alpha) = \cos\alpha + i \sin\alpha .$$

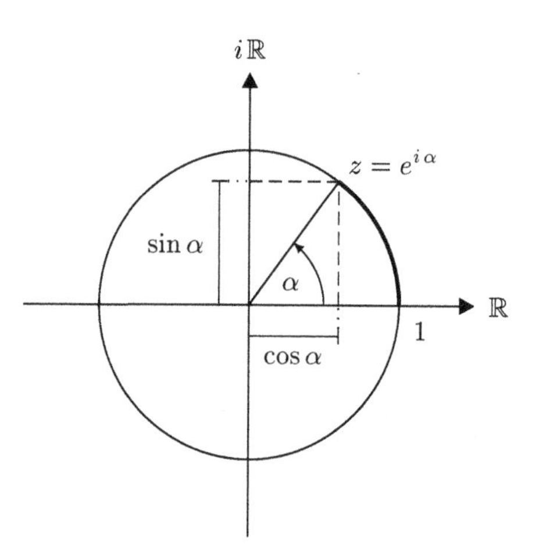

The number $\alpha \in [0, 2\pi)$ can be interpreted as the length of the circular arc from 1 to $z = e^{i\alpha}$ (see Exercise 12) or, equally well, as an angle. In addition we know from Theorem 6.16 that $\mathrm{cis} : \mathbb{R} \to S^1$ has period 2π. Hence this function wraps the real axis infinitely many times around S^1.

6.19 Proposition *For $a \in \mathbb{R}$, let I_a be an interval of the form $[a, a + 2\pi)$ or $(a, a + 2\pi]$. Then the function*

$$\exp(\mathbb{R} + i I_a) : \mathbb{R} + i I_a \to \mathbb{C}^\times , \quad z \mapsto e^z \tag{6.9}$$

is continuous and bijective.

Proof The continuity is, by Theorem 6.1(v), clear. To verify the injectivity, we suppose that there are $w, z \in \mathbb{R} + i I_a$ such that $e^z = e^w$. Write $z = x + iy$ and $w = \xi + i\eta$ with x, y, ξ and η real. Then Lemma 6.10 implies that

$$e^x = |e^x e^{iy}| = |e^\xi e^{i\eta}| = e^\xi , \tag{6.10}$$

and so, by Proposition 6.5, $x = \xi$. We now have

$$e^{i(y-\eta)} = e^{x+iy-(\xi+iy)} = e^z/e^w = 1 ,$$

which, by Proposition 6.13, implies $y - \eta \in 2\pi\mathbb{Z}$. By assumption, $|y - \eta| < 2\pi$ and so $y = \eta$. Thus we have shown that the function in (6.9) is injective.

Let $w \in \mathbb{C}^\times$. For $x := \log|w| \in \mathbb{R}$ we have $e^x = |w|$. By Proposition 6.15, there is a unique $y \in I_a$ such that $e^{iy} = w/|w| \in S^1$. Setting $z = x + iy \in \mathbb{R} + iI_a$, we have $e^z = e^x e^{iy} = |w|\,(w/|w|) = w$. ■

The function of the previous proposition can be represented graphically as below.

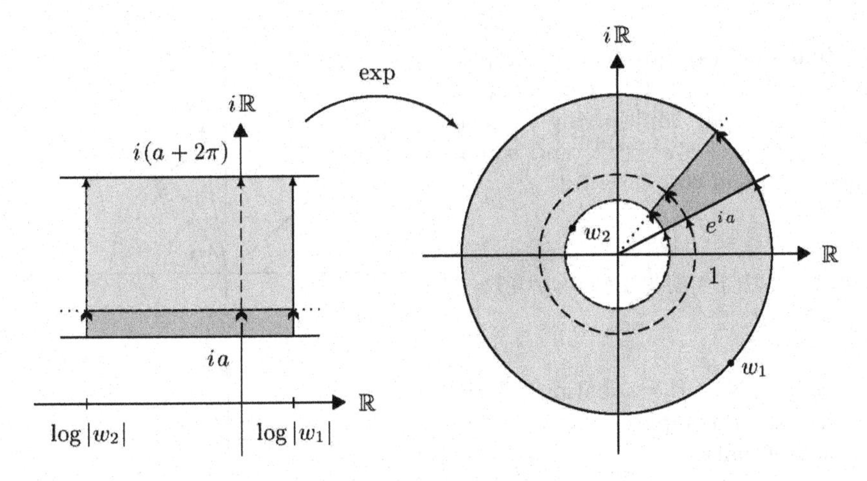

Finally we note that

$$\mathbb{C} = \bigcup_{k\in\mathbb{Z}} \{\, \mathbb{R} + i\,[a + 2k\pi, a + 2(k+1)\pi)\,\}$$

is a partition of the complex plane, so it follows from Proposition 6.19 that the exponential function $\exp : \mathbb{C} \to \mathbb{C}^\times$ wraps the complex plane infinitely many times around the origin, covering infinitely many times the punctured complex plane $\mathbb{C}^\times$.

Polar Coordinates

Using the exponential function, we can represent complex numbers using polar coordinates. In this representation the multiplication of two complex numbers has a simple geometrical interpretation.

6.20 Theorem (polar coordinate representation of complex numbers) *Each $z \in \mathbb{C}^\times$ has a unique representation in the form*

$$z = |z|\, e^{i\alpha}$$

with $\alpha \in [0, 2\pi)$.

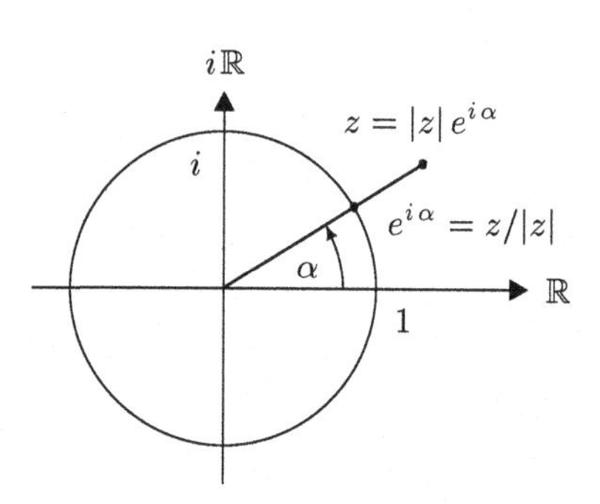

Proof This follows directly from Proposition 6.19. ∎

The real number $\alpha \in [0, 2\pi)$ from this theorem is called the **normalized argument of** $z \in \mathbb{C}^\times$ and is denoted $\arg_N(z)$.

6.21 Remarks **(a)** (product of complex numbers) Let $w, z \in \mathbb{C}^\times$, $\alpha := \arg_N(z)$ and $\beta := \arg_N(w)$. Multiplying z and w we get $zw = |z|\,|w|\, e^{i(\alpha+\beta)}$, and so, by Lemma 6.10 and Corollary 6.14,

$$|zw| = |z|\,|w| \;,$$
$$\arg_N(zw) \equiv \arg_N(z) + \arg_N(w)$$

modulo 2π.

(b) For $n \in \mathbb{N}^\times$, the equation $z^n = 1$ has exactly n complex solutions, the nth **roots of unity**,

$$z_k := e^{2\pi i k/n} \;, \qquad k = 0, \ldots, n-1 \;.$$

The points z_k are on the unit circle and are the vertices of a regular n-gon with one vertex at 1.

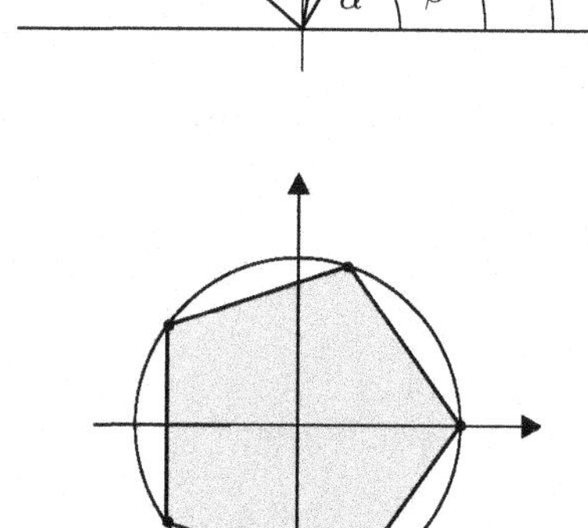

(c) For all $a \in \mathbb{C}$ and $k \in \mathbb{N}^\times$, the equation $z^k = a$ is solvable in $\mathbb{C}$.[7]

Proof If $a = 0$, the claim is clear. Otherwise, $a = |a|\, e^{i\alpha}$ with $\alpha := \arg_N(a) \in [0, 2\pi)$. Set $z := \sqrt[k]{|a|}e^{i\alpha/k}$. Then

$$z^k = (\sqrt[k]{|a|}e^{i\alpha/k})^k = (\sqrt[k]{|a|})^k (e^{i\alpha/k})^k = |a|\, e^{i\alpha} = a \;,$$

and we have found the desired solution. ∎

[7] This closes the gap in the proof of the fundamental theorem of algebra in Example 3.9(b). It also shows the validity of the assumption in Exercise I.11.15 about the solutions of cubic equations.

(d) (polar coordinate representation of the plane) For each $(x,y) \in \mathbb{R}^2 \setminus \{(0,0)\}$ there are unique real numbers $r > 0$ and $\alpha \in [0, 2\pi)$ such that

$$x = r\cos\alpha \quad \text{and} \quad y = r\sin\alpha \ .$$

Proof Let $x, y \in \mathbb{R}$ with $z := x + iy \in \mathbb{C}^\times$. Set

$$r := |z| = \sqrt{x^2+y^2} > 0 \quad \text{and} \quad \alpha := \arg_N(z) \in [0, 2\pi) \ .$$

Then, by Theorem 6.20 and Euler's formula, we have

$$x + iy = z = re^{i\alpha} = r\cos\alpha + ir\sin\alpha \ ,$$

from which the claim follows. ■

(e) For all $z \in \mathbb{C}$, $|e^z| = e^{\operatorname{Re} z}$.

Proof This follows from $|e^z| = |e^{\operatorname{Re} z} e^{i \operatorname{Im} z}| = e^{\operatorname{Re} z} \, |e^{i \operatorname{Im} z}| = e^{\operatorname{Re} z}$. ■

Complex Logarithms

For a given $w \in \mathbb{C}^\times$, we want to determine *all* solutions of the equation $e^z = w$. From Theorem 6.20 we know that this equation is solvable since

$$w = e^{\log|w| + i\arg_N(w)} \ .$$

Now let $z \in \mathbb{C}$ be an arbitrary solution of $e^z = w$. By Corollary 6.14 and Proposition 6.15 there is a unique $k \in \mathbb{Z}$ such that $z = \log|w| + i\arg_N(w) + 2\pi k i$. Hence

$$\bigl\{ \log|w| + i\bigl(\arg_N(w) + 2\pi k\bigr) \in \mathbb{C} \ ; \ k \in \mathbb{Z} \bigr\}$$

is the set of all solutions of the equation $e^z = w$. The *set*

$$\operatorname{Arg}(w) := \arg_N(w) + 2\pi\mathbb{Z}$$

is called the **argument of** w, and the *set*

$$\operatorname{Log}(w) := \log|w| + i\operatorname{Arg}(w)$$

is called the (**complex**) **logarithm of** w.

These equations define two *set valued* functions

$$\begin{aligned} \operatorname{Arg}\colon \mathbb{C}^\times &\to \mathcal{P}(\mathbb{C}) \ , \quad w \mapsto \operatorname{Arg}(w) \ , \\ \operatorname{Log}\colon \mathbb{C}^\times &\to \mathcal{P}(\mathbb{C}) \ , \quad w \mapsto \operatorname{Log}(w) \ , \end{aligned}$$

called the **argument** and **logarithm** functions.

Since set valued functions are, in general, rather cumbersome, we make use of the fact that, for each $w \in \mathbb{C}^\times$, there is a unique $\varphi =: \arg(w) \in (-\pi, \pi]$ such that $w = |w|\, e^{i\varphi}$. This defines a real valued function

$$\arg : \mathbb{C}^\times \to (-\pi, \pi] , \quad w \mapsto \arg(w)$$

called the **principal value of the argument**. The **principal value of the logarithm**[8] is defined by

$$\log : \mathbb{C}^\times \to \mathbb{R} + i\,(-\pi, \pi] , \quad w \mapsto \log |w| + i \arg(w) .$$

Propositions 6.5 and 6.15 imply that log is a bijection, and

$$\begin{aligned} e^{\log w} &= w , \qquad & w &\in \mathbb{C}^\times , \\ \log e^z &= z , \qquad & z &\in \mathbb{R} + i\,(-\pi, \pi] . \end{aligned} \tag{6.11}$$

In particular, $\log w$ is defined for $w < 0$, and, in this case, $\log w = \log |w| + i\pi$.

For the set valued complex logarithm we have

$$e^{\operatorname{Log} w} = w , \quad w \in \mathbb{C}^\times , \qquad \operatorname{Log} e^z = z + 2\pi i \mathbb{Z} , \quad z \in \mathbb{C} .$$

Finally,[9]

$$\operatorname{Log}(zw) = \operatorname{Log} z + \operatorname{Log} w , \quad \operatorname{Log}(z/w) = \operatorname{Log} z - \operatorname{Log} w \tag{6.12}$$

for all $w, z \in \mathbb{C}^\times$. This can be proved similarly to the addition theorem of the natural logarithm (Theorem 6.6).

Complex Powers

For $z \in \mathbb{C}^\times$ and $w \in \mathbb{C}$,

$$z^w := e^{w \operatorname{Log} z}$$

is called the (**complex**) **power** of z. Because Log is a set valued function, z^w is a set. Specifically,

$$z^w = \bigl\{ e^{w(\log|z| + i\,(\arg_N(z) + 2\pi k))} \ ; \ k \in \mathbb{Z} \bigr\} .$$

The **principal value of** z^w is, of course, defined using the principal value of the logarithm:

$$\mathbb{C}^\times \to \mathbb{C} , \qquad z \mapsto z^w := e^{w \log z} .$$

The rules in Proposition 6.8 generalize easily to the principal value of the power function:

$$z^a z^b = z^{a+b} \quad \text{and} \quad z^a \cdot w^a = (zw)^a \tag{6.13}$$

for all $w, z \in \mathbb{C}^\times$ and $a, b \in \mathbb{C}$.

[8]For $w \in (0, \infty)$, this definition is consistent with the real logarithm $(\exp | \mathbb{R})^{-1}$.

[9]See (I.4.1) for the meaning of $+$ and $-$ on sets.

6.22 Remarks (a) Theorem 6.1(iii) says that

$$\exp\colon (\mathbb{C},+) \to (\mathbb{C}^\times,\cdot) \tag{6.14}$$

is a group homomorphism between the Abelian groups $(\mathbb{C},+)$ and $(\mathbb{C}^\times,\cdot)$. Moreover, Propositions 6.13 and 6.15 imply that (6.14) is surjective and has kernel $2\pi i\,\mathbb{Z}$. By Example I.7.8(c), the quotient group $(\mathbb{C},+)/(2\pi i\mathbb{Z})$ is isomorphic to $(\mathbb{C}^\times,\cdot)$.

(b) The unit circle S^1 forms an Abelian group $(S^1,\cdot)$, the **circle group** under multiplication (see Exercise I.11.9). From Theorem 6.1(iii) and Propositions 6.13 and 6.15, it follows that

$$\operatorname{cis}\colon (\mathbb{R},+) \to (S^1,\cdot)$$

is a surjective group homomorphism with kernel $2\pi\mathbb{Z}$. Hence the groups $(S^1,\cdot)$ and $(\mathbb{R},+)/(2\pi\mathbb{Z})$ are isomorphic.

(c) The function

$$\exp_{\mathbb{R}}\colon (\mathbb{R},+) \to \big((0,\infty),\cdot\big)$$

is a group isomorphism with inverse $\log\colon (0,\infty) \to \mathbb{R}$. ∎

A Further Representation of the Exponential Function

In Exercise II.4.3 we saw that, for rational arguments, the exponential function is given by

$$e^r = \lim_{n\to\infty}\Big(1+\frac{r}{n}\Big)^n .$$

This result can be generalized to arbitrary complex numbers.

6.23 Theorem *For all $z \in \mathbb{C}$,*

$$e^z = \lim_{n\to\infty}\Big(1+\frac{z}{n}\Big)^n .$$

Proof Let $z \in \mathbb{C}$. From Exercise I.8.1 we have

$$a^n - b^n = (a-b)\sum_{k=0}^{n-1} a^k b^{n-k-1} , \qquad a,b \in \mathbb{C} ,$$

and so

$$\begin{aligned} e^z - (1+z/n)^n &= (e^{z/n})^n - (1+z/n)^n \\ &= \big[e^{z/n} - (1+z/n)\big]\sum_{k=0}^{n-1}(e^{z/n})^k(1+z/n)^{n-1-k} . \end{aligned} \tag{6.15}$$

From Example 2.25(b) we know that

$$r_n := \left[\frac{e^{z/n}-1}{z/n} - 1\right] \to 0 , \qquad n \to \infty . \tag{6.16}$$

To estimate

$$L_n := \sum_{k=0}^{n-1} (e^{z/n})^k (1+z/n)^{n-1-k} , \qquad n \in \mathbb{N}^\times , \tag{6.17}$$

we use the inequalities

$$|e^w| = e^{\operatorname{Re} w} \le e^{|w|} , \qquad |1+w| \le 1+|w| \le e^{|w|}$$

to get

$$|L_n| \le \sum_{k=0}^{n-1} (e^{|z|/n})^k (e^{|z|/n})^{n-1-k} = n(e^{|z|/n})^{n-1} \le n e^{|z|} , \qquad n \in \mathbb{N}^\times . \tag{6.18}$$

Combining (6.15), (6.17) and (6.18) we get

$$\left| e^z - \left(1 + \frac{z}{n}\right)^n \right| = \left| \frac{z}{n} r_n L_n \right| \le \left| \frac{z}{n} \right| |r_n| \, n e^{|z|} = |z| \, e^{|z|} \, |r_n| ,$$

which, with (6.16), proves the claim. ■

Exercises

1 Show that the functions $\operatorname{cis} : \mathbb{R} \to \mathbb{C}$ and $\cos, \sin : \mathbb{R} \to \mathbb{R}$ are Lipschitz continuous with Lipschitz constant 1. (Hint: See Example 2.25(b).)

2 For $z \in \mathbb{C}$ and $m \in \mathbb{N}$, prove **de Moivre's formula**,

$$(\cos z + i \sin z)^m = \cos(mz) + i \sin(mz) .$$

3 Prove the following trigonometric identities:

(a) $\cos^2(z/2) = (1+\cos z)/2, \ \sin^2(z/2) = (1-\cos z)/2, \ z \in \mathbb{C}$.

(b) $\tan(z/2) = (1-\cos z)/\sin z = \sin z/(1+\cos z), \ z \in \mathbb{C}\backslash(\pi\mathbb{Z})$.

4 The **hyperbolic cosine** and **hyperbolic sine** functions are defined by

$$\cosh(z) := \frac{e^z + e^{-z}}{2} \quad \text{and} \quad \sinh(z) := \frac{e^z - e^{-z}}{2} , \qquad z \in \mathbb{C} .$$

For $w, z \in \mathbb{C}$, show the following:

(a) $\cosh^2 z - \sinh^2 z = 1$.

(b) $\cosh(z+w) = \cosh z \cosh w + \sinh z \sinh w$.

(c) $\sinh(z+w) = \sinh z \cosh w + \cosh z \sinh w$.

(d) $\cosh z = \cos iz, \ \sinh z = -i \sin iz$.

(e) $\cosh z = \sum_{k=0}^{\infty} \frac{z^{2k}}{(2k)!}, \quad \sinh z = \sum_{k=0}^{\infty} \frac{z^{2k+1}}{(2k+1)!}$.

5 The **hyperbolic tangent** and **hyperbolic cotangent** functions are defined by

$$\tanh z := \frac{\sinh(z)}{\cosh(z)}\ , \quad z \in \mathbb{C}\backslash\pi i\,(\mathbb{Z}+1/2)\ , \qquad \text{and} \qquad \coth z := \frac{\cosh(z)}{\sinh(z)}\ , \quad z \in \mathbb{C}\backslash\pi i\mathbb{Z}\ .$$

The functions cosh, sinh, tanh and coth have real values for real arguments. Sketch the graphs of these real valued functions.

Show the following:

(a) The functions

$$\cosh\ , \quad \sinh\ , \quad \tanh\colon \mathbb{C}\backslash\pi i\,(\mathbb{Z}+1/2) \to \mathbb{C}\ , \quad \coth\colon \mathbb{C}^\times \to \mathbb{C}\backslash\pi i\mathbb{Z}$$

are continuous.

(b) $\lim_{x\to\pm\infty}\tanh(x) = \pm 1$, $\lim_{x\to\pm 0}\coth(x) = \pm\infty$.

(c) $\cosh\colon [0,\infty) \to \mathbb{R}$ is strictly increasing with $\cosh\bigl([0,\infty)\bigr) = [1,\infty)$.

(d) $\sinh\colon \mathbb{R} \to \mathbb{R}$ is strictly increasing and bijective.

(e) $\tanh\colon \mathbb{R} \to (-1,1)$ is strictly increasing and bijective.

(f) $\coth\colon (0,\infty) \to \mathbb{R}$ is strictly decreasing with $\coth\bigl((0,\infty)\bigr) = (1,\infty)$.

(g) $\tanh\colon \mathbb{R} \to (-1,1)$ is Lipschitz continuous with Lipschitz constant 1.

6 Determine the following limits:

$$\text{(a)}\ \lim_{x\to 0+} x^x\ , \quad \text{(b)}\ \lim_{x\to 0+} x^{1/x}\ , \quad \text{(c)}\ \lim_{z\to 0}\frac{\log(1+z)}{z}\ .$$

(Hint: (c) See Example 2.25(b).)

7 For $x, y > 0$, prove the inequality

$$\frac{\log x + \log y}{2} \le \log\Bigl(\frac{x+y}{2}\Bigr)\ .$$

8 Determine the following limits:

$$\text{(a)}\ \lim_{z\to 0}\frac{\sin z}{z}\ , \qquad \text{(b)}\ \lim_{z\to 0}\frac{a^z - 1}{z}\ , \quad a \in \mathbb{C}^\times\ .$$

9 Show that the functions

$$\arg\colon \mathbb{C}\backslash(-\infty,0] \to (-\pi,\pi)\ , \quad \log\colon \mathbb{C}\backslash(-\infty,0] \to \mathbb{R} + i(-\pi,\pi)\ ,$$

are continuous. (Hint: (i) $\arg = \arg\circ\nu$ with $\nu(z) := z/|z|$ for all $z \in \mathbb{C}^\times$. (ii) $\arg\bigm|\bigl(S^1\setminus\{-1\}\bigr) = \bigl[\mathrm{cis}\bigm|(-\pi,\pi)\bigr]^{-1}$. (iii) Use Exercise 3.3(b) for intervals of the form $[-a,a]$ with $a \in (0,\pi)$.)

10 Prove the following rules for the principal value of the power function:

$$z^a z^b = z^{a+b}\ , \quad z^a w^a = (zw)^a\ , \qquad z, w \in \mathbb{C}^\times\ , \quad a, b \in \mathbb{C}\ .$$

11 Calculate i^i and and its principal value.

12 Let $x \in \mathbb{R}$, $n \in \mathbb{N}^\times$ and $z_{n,k} := e^{ixk/n} \in S^1$ for all $k = 0, 1, \ldots, n$. Set

$$L_n := \sum_{k=1}^{n} |z_{n,k} - z_{n,k-1}| \ ,$$

the length of the polygonal path with vertices $z_{n,0}, z_{n,1}, \ldots, z_{n,n}$. Show that

$$L_n = 2n\,\bigl|\sin\bigl(x/(2n)\bigr)\bigr| \quad \text{and} \quad \lim_{n\to\infty} L_n = |x| \ .$$

Remark For large $n \in \mathbb{N}$ and $x \in [0, 2\pi]$, the image of $[0, x]$ under the function cis is approximated by the polygonal path with vertices $z_{n,0}, z_{n,1}, \ldots, z_{n,n}$. Thus L_n is an approximation of the length of the arc of the circle between 1 and $\operatorname{cis}(x) = e^{ix}$. This exercise shows that the function $\operatorname{cis} : \mathbb{R} \to S^1$ 'wraps' the line $\mathbb{R}$ around S^1 in such a way that length is preserved.

13 Investigate the behavior of the function $\mathbb{C} \to \mathbb{C}$, $z \mapsto z^2$. In particular, calculate the images of the hyperbolas $x^2 - y^2 = \text{const}$, $xy = \text{const}$, as well as the lines $x = \text{const}$, $y = \text{const}$ for $z = x + iy$.

14 Determine all solutions in $\mathbb{C}$ of the following equations:

(a) $z^4 = (\sqrt{2}/2)(1+i)$.

(b) $z^5 = i$.

(c) $z^3 + 6z + 2 = 0$.

(d) $z^3 + (1-2i)z^2 - (1+2i)z - 1 = 0$.

(Hint: For the cubic equations in (c) and (d), use Exercise I.11.15.)

15 For $x \in \mathbb{R}$ and $n \in \mathbb{N}$, let

$$f_n(x) := \lim_{k\to\infty} \bigl(\cos(n!\,\pi x)\bigr)^{2k} \ .$$

Determine $\lim_{n\to\infty} f_n(x)$. (Hint: Consider separately the case $x \in \mathbb{R}\backslash\mathbb{Q}$, and use the fact that $|\cos(m\pi)| = 1$ if and only if $m \in \mathbb{Z}$.)

16 Prove that $\cosh 1$ is irrational. (Hint: Exercise II.7.10.)

Chapter IV

Differentiation in One Variable

In Chapter II we explored the limit concept, one of the most fundamental and essential notions of analysis. We developed methods for calculating limits and presented many of its important applications. In Chapter III we considered in detail the topological foundations of analysis and the concept of continuity. In doing so we saw, in particular, the connection between continuity and the limit concept. In the last section of the previous chapter, by applying much of our accumulated understanding, we investigated several of the most important functions in mathematics.

Even though we seem to know a lot about the exponential function and its relatives, the cosine and sine functions, our understanding is, in fact, rudimentary and is limited largely to the global aspects of these functions. In this chapter we consider primarily the local properties of functions. In doing so, we encounter again a common theme of analysis, which, expressed simply, is the approximation of complicated 'continuous' behavior by simple (often discrete) structures. This approximation idea is, of course, at the foundation of the limit concept, and it appears throughout all of 'continuous' mathematics.

Guided at first by our intuitions, we consider the graphs of real valued functions of a real variable. One conceptually simple local approximation of a complicated appearing graph at a particular point is a tangent line. This is a line which passes through the point and which nearby 'fits' the graph as closely as possible. Then, near the point (as though seen through an arbitrarily powerful microscope), the function is almost indistinguishable from this linear approximation. We show that it is possible to describe the local properties of rather general functions using such linear approximations.

This notion of linear approximations is remarkably fruitful and not restricted to the intuitive one dimensional case. In fact, it is the foundation for practically all local investigations in analysis. We will see that finding a linear approximation is the same as differentiation. Indeed, differentiation, which is covered in the first three sections of this chapter, is nothing more than an efficient calculus of linear approximations. The importance of this idea is seen in its many beautiful and often surprising applications, some of which appear in the last section of this chapter.

In the first section we introduce the concept of differentiability and show its connection to linear approximations. We also derive the basic rules for calculating derivatives.

In Section 2, the geometric idea behind differentiation comes fully into play. By studying the tangent lines to a graph, we determine the local behavior of the corresponding function. The utility of this technique is made clear, in particular, in the study of convex functions. As a first simple application, we prove some of the fundamental inequalities of analysis.

Section 3 is dedicated to approximations of higher order. Instead of approximating a given function locally using a line, that is, by a polynomial of degree one, one looks for approximations by polynomials of higher degree. Of course, by doing so one gets further local information about the function. Such information is, in particular, useful to determine the nature of extrema.

In the last section we consider the approximate determination of the zeros of real functions. We prove the Banach fixed point theorem whose practical and theoretical importance cannot be overstated, and we use it to prove the convergence of Newton's method.

In the entire chapter, we limit ourselves to the study of functions from the real or complex numbers to arbitrary Banach spaces. The differentiation of functions of two or more variables is discussed in Chapter VII.

1 Differentiability

As already mentioned in the introduction to this chapter, our motivation for the development of differentiation is the desire to describe the local behavior of functions using linear approximations. Thus we are lead to the tangent line problem: Given a point on the graph of a real function, determine the tangent line to the graph at that point.

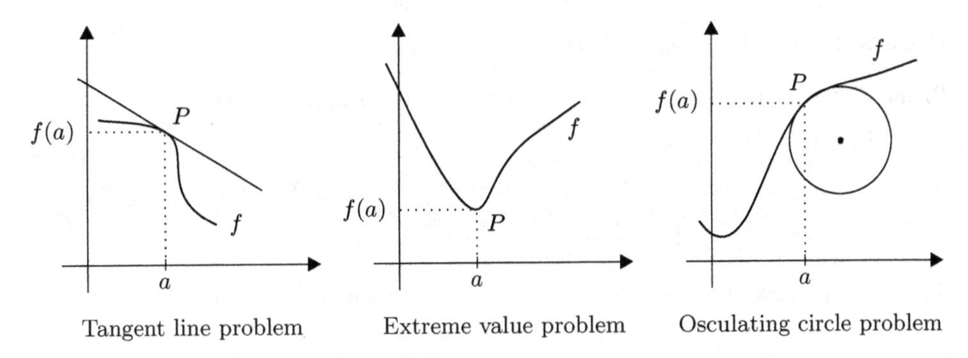

Tangent line problem — Extreme value problem — Osculating circle problem

The problem of finding the extreme values of the function or an osculating circle at a point, that is, a circle which best fits the graph, is closely related to the tangent line problem, and thus also to differentiation.

In the following, $X \subseteq \mathbb{K}$ is a set, $a \in X$ is a limit point of X and $E = (E, \|\cdot\|)$ is a normed vector space over $\mathbb{K}$.

The Derivative

A function $f : X \to E$ is called **differentiable at** a if the limit

$$f'(a) := \lim_{x \to a} \frac{f(x) - f(a)}{x - a}$$

exists in E. When this occurs, $f'(a) \in E$ is called the **derivative of** f **at** a. Besides the symbol $f'(a)$, many other notations for the derivative are used:

$$\dot{f}(a)\ , \quad \partial f(a)\ , \quad Df(a)\ , \quad \frac{df}{dx}(a)\ .$$

Before we systematically investigate differentiable functions, we provide some useful reformulations of the definition.

1.1 Theorem *For $f : X \to E$, the following are equivalent:*

(i) *f is differentiable at a.*

(ii) *There is some* $m_a \in E$ *such that*

$$\lim_{x \to a} \frac{f(x) - f(a) - m_a(x-a)}{x-a} = 0 .$$

(iii) *There are* $m_a \in E$ *and a function* $r : X \to E$ *which is continuous at* a *such that* $r(a) = 0$ *and*

$$f(x) = f(a) + m_a(x-a) + r(x)(x-a) , \qquad x \in X .$$

In cases (ii) *and* (iii), $m_a = f'(a)$.

Proof The implication '(i)$\Rightarrow$(ii)' is clear by setting $m_a := f'(a)$.

'(ii)$\Rightarrow$(iii)' Define

$$r(x) := \begin{cases} 0 , & x = a , \\ \dfrac{f(x) - f(a) - m_a(x-a)}{x-a} , & x \neq a . \end{cases}$$

Then, by Remark III.2.23(b) and (ii), r has the claimed properties.

'(iii)$\Rightarrow$(i)' This is also clear. ■

1.2 Corollary *If* $f : X \to E$ *is differentiable at* a, *then* f *is continuous at* a.

Proof This follows immediately from the implication '(i)$\Rightarrow$(iii)' of Theorem 1.1. ■

The converse of Corollary 1.2 is false: There are functions which are continuous but not differentiable (see Example 1.13(k)).

Linear Approximation

Let $f : X \to E$ be differentiable at a. Then the function

$$g : \mathbb{K} \to E , \quad x \mapsto f(a) + f'(a)(x-a)$$

is affine and $g(a) = f(a)$. Moreover, it follows from Theorem 1.1 that

$$\lim_{x \to a} \frac{\|f(x) - g(x)\|}{|x-a|} = 0 .$$

Thus f and g coincide at the point a and the 'error' $\|f(x) - g(x)\|$ approaches zero more quickly than $|x-a|$ as $x \to a$. This observation suggests the following definition: The function $f : X \to E$ is called **approximately linear** at a if there is an affine function $g : \mathbb{K} \to E$ such that

$$f(a) = g(a) \quad \text{and} \quad \lim_{x \to a} \frac{\|f(x) - g(x)\|}{|x-a|} = 0 .$$

The following corollary shows that this property and differentiability are, in fact, identical.

1.3 Corollary *A function $f\colon X \to E$ is differentiable at a if and only if it is approximately linear at a. In this case, the approximating affine function g is unique and given by*

$$g\colon \mathbb{K} \to E\ , \quad x \mapsto f(a) + f'(a)(x-a)\ .$$

Proof '$\Rightarrow$' This follows directly from Theorem 1.1.

'$\Leftarrow$' Let $g\colon \mathbb{K} \to E$ be an affine function which approximates f at a. By Proposition I.12.8, there are unique elements $b, m \in E$ such that $g(x) = b + mx$ for all $x \in \mathbb{K}$. Since $g(a) = f(a)$, we have, in fact, $g(x) = f(a) + m(x-a)$ for all $x \in \mathbb{K}$. The claim then follows from Theorem 1.1. ■

1.4 Remarks **(a)** Suppose that the function $f\colon X \to E$ is differentiable at a. As above, define $g(x) := f(a) + f'(a)(x-a)$ for all $x \in \mathbb{K}$. Then the graph of g is an affine line through $\big(a, f(a)\big)$ which approximates the graph of f near the point $\big(a, f(a)\big)$. This line is called the **tangent line** to f at $\big(a, f(a)\big)$. In the case $\mathbb{K} = \mathbb{R}$, this definition agrees with our intuitions from elementary geometry.

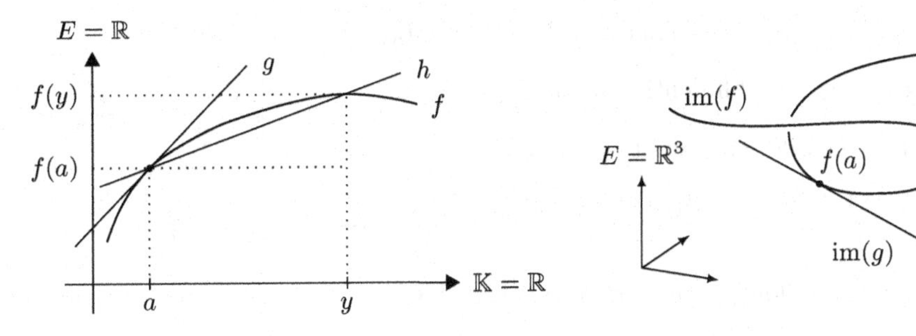

The expression

$$\frac{f(y) - f(a)}{y - a}\ , \qquad y \neq a\ ,$$

is called a **difference quotient** of f. The graph of the affine function

$$h(x) := f(a) + \frac{f(y) - f(a)}{y - a}(x - a)\ , \qquad x \in \mathbb{K}\ ,$$

is called the **secant line** through $\big(a, f(a)\big)$ and $\big(y, f(y)\big)$. In the case $\mathbb{K} = \mathbb{R} = E$, the differentiability of f at a means that, as $y \to a$, the **slope** $\big(f(y) - f(a)\big)/(y-a)$ of the secant line through $\big(a, f(a)\big)$ and $\big(y, f(y)\big)$ converges to the slope $f'(a)$ of the tangent line at $\big(a, f(a)\big)$.

(b) Let $X = J \subseteq \mathbb{R}$ be an interval and $E = \mathbb{R}^3$. Suppose that $f(t)$ gives the position of a point in space at time $t \in J$. Then $|f(t) - f(t_0)|/|t - t_0|$ is the absolute value of the 'average speed' between times t_0 and t, and $\dot{f}(t_0)$ represents the **instantaneous velocity** of the point at the time t_0.

(c) (i) Suppose that $\mathbb{K} = E = \mathbb{R}$ and $f : X \subseteq \mathbb{R} \to \mathbb{R}$ is a function which is differentiable at a. Consider f as a function from $\mathbb{C}$ to $\mathbb{C}$, that is, set

$$f_{\mathbb{C}} : X \subseteq \mathbb{C} \to \mathbb{C} , \qquad f_{\mathbb{C}}(x) := f(x) , \quad x \in X .$$

Then $f_{\mathbb{C}}$ is also differentiable at a and $f'_{\mathbb{C}}(a) = f'(a)$.

(ii) Now suppose that $\mathbb{K} = E = \mathbb{C}$ and $f : X \subseteq \mathbb{C} \to \mathbb{C}$ is a function which is differentiable at $a \in Y := X \cap \mathbb{R}$. Suppose also that a is a limit point of Y and $f(Y) \subseteq \mathbb{R}$. Then $f\,|\,Y : Y \to \mathbb{R}$ is differentiable at a and $(f\,|\,Y)'(a) = f'(a) \in \mathbb{R}$.

Proof This follows directly from the definition, the differentiability of f, and the fact that $\mathbb{R}$ is closed in $\mathbb{C}$. ■

Rules for Differentiation

1.5 Proposition *Let $E_1, \ldots, E_n$ be normed vector spaces and $E := E_1 \times \cdots \times E_n$. Then $f = (f_1, \ldots, f_n) : X \to E$ is differentiable at a if and only if each component function $f_j : X \to E_j$ is differentiable at a. In this case,*

$$\partial f(a) = \big(\partial f_1(a), \ldots, \partial f_n(a)\big) .$$

Thus vectors can be differentiated componentwise.

Proof For the difference quotient we have

$$\frac{f(x) - f(a)}{x - a} = \Big(\frac{f_1(x) - f_1(a)}{x - a}, \ldots, \frac{f_n(x) - f_n(a)}{x - a}\Big) , \qquad x \neq a .$$

Thus the claim follows from Example II.1.8(e). ■

In the next theorem we collect further rules for differentiation which make the calculation of the derivatives of functions rather easy.

1.6 Theorem

(i) (linearity) *Let $f, g : X \to E$ be differentiable at a and $\alpha, \beta \in \mathbb{K}$. Then the function $\alpha f + \beta g$ is also differentiable at a and*

$$(\alpha f + \beta g)'(a) = \alpha f'(a) + \beta g'(a) .$$

In other words, the set of functions which are differentiable at a forms a subspace V of E^X, and the function $V \to E$, $f \mapsto f'(a)$ is linear.

(ii) (product rule) *Let $f, g : X \to \mathbb{K}$ be differentiable at a. Then the function $f \cdot g$ is also differentiable at a and*

$$(f \cdot g)'(a) = f'(a)g(a) + f(a)g'(a) .$$

The set of functions which are differentiable at a forms a subalgebra of $\mathbb{K}^X$.

(iii) (quotient rule) *Let $f, g\colon X \to \mathbb{K}$ be differentiable at a with $g(a) \neq 0$. Then the function f/g is also differentiable at a and*

$$\Big(\frac{f}{g}\Big)'(a) = \frac{f'(a)g(a) - f(a)g'(a)}{\big[g(a)\big]^2} .$$

Proof All of these claims follow directly from the rules for convergent sequences which we proved in Section II.2.

For (i) this is particularly clear. For the proof of the product rule (ii), we write the difference quotient of $f \cdot g$ in the form

$$\frac{f(x)g(x) - f(a)g(a)}{x-a} = \frac{f(x) - f(a)}{x-a}g(x) + f(a)\frac{g(x) - g(a)}{x-a} , \qquad x \neq a .$$

By Corollary 1.2, g is continuous at a, and so the claim follows from Propositions II.2.2 and II.2.4, as well as Theorem III.1.4.

For (iii) we have $g(a) \neq 0$, and so, by Example III.1.3(d), there is a neighborhood U of a in X such that $g(x) \neq 0$ for all $x \in U$. Then, for each $x \in U\backslash\{a\}$ we have

$$\Big(\frac{f(x)}{g(x)} - \frac{f(a)}{g(a)}\Big)(x-a)^{-1} = \frac{1}{g(x)g(a)}\Big[\frac{f(x) - f(a)}{x-a}g(a) - f(a)\frac{g(x) - g(a)}{x-a}\Big] ,$$

from which the claim follows. ∎

The Chain Rule

It is often possible to express a complicated function as a composition of simpler functions. The following rule describes how such compositions can be differentiated.

1.7 Theorem (chain rule) *Suppose that $f\colon X \to \mathbb{K}$ is differentiable at a, and $f(a)$ is a limit point of Y with $f(X) \subseteq Y \subseteq \mathbb{K}$. If $g\colon Y \to E$ is differentiable at $f(a)$, then $g \circ f$ is differentiable at a and*

$$(g \circ f)'(a) = g'\big(f(a)\big)f'(a) .$$

Proof By hypothesis and Theorem 1.1, there is a function $r\colon X \to \mathbb{K}$ which is continuous at a such that $r(a) = 0$ and

$$f(x) = f(a) + f'(a)(x-a) + r(x)(x-a) , \qquad x \in X . \tag{1.1}$$

Similarly, there is a function $s\colon Y \to E$ which is continuous at $b := f(a)$ such that $s(b) = 0$ and

$$g(y) = g(b) + g'(b)(y-b) + s(y)(y-b) , \qquad y \in Y . \tag{1.2}$$

Now let $x \in X$ and set $y := f(x)$ in (1.2). Then, using (1.1),

$$\begin{aligned}(g \circ f)(x) &= g\big(f(a)\big) + g'\big(f(a)\big)\big(f(x) - f(a)\big) + s\big(f(x)\big)\big(f(x) - f(a)\big) \\ &= (g \circ f)(a) + g'\big(f(a)\big)f'(a)(x-a) + t(x)(x-a) ,\end{aligned}$$

where $t(x) := g'\big(f(a)\big)r(x) + s\big(f(x)\big)\big(f'(a) + r(x)\big)$ for all $x \in X$. By hypothesis, Corollary 1.2 and Theorem III.1.8, $t\colon X \to E$ is continuous at a. Moreover,

$$t(a) = g'\big(f(a)\big)r(a) + s(b)\big(f'(a) + r(a)\big) = 0 .$$

The claim now follows from Theorem 1.1. ■

Inverse Functions

Using the chain rule we can derive a criterion for the differentiability of inverse functions and calculate their derivatives.

1.8 Theorem (differentiability of inverse functions) *Let $f : X \to \mathbb{K}$ be injective and differentiable at a. In addition, suppose that $f^{-1} : f(X) \to X$ is continuous at $b := f(a)$. Then f^{-1} is differentiable at b if and only if $f'(a)$ is nonzero. In this case,*

$$(f^{-1})'(b) = \frac{1}{f'(a)} , \qquad b = f(a) .$$

Proof '$\Rightarrow$' Applying the chain rule to the identity $f^{-1} \circ f = \mathrm{id}_X$ we get

$$1 = (\mathrm{id}_X)'(a) = (f^{-1})'\big(f(a)\big)f'(a) ,$$

and hence, $(f^{-1})'(b) = 1/f'(a)$.

'$\Leftarrow$' We first confirm that b is a limit point of $Y := f(X)$. By hypothesis, a is a limit point of X, and so, by Proposition III.2.9, there is a sequence (x_k) in $X\backslash\{a\}$ such that $\lim x_k = a$. Since f is continuous, we have $\lim f(x_k) = f(a)$. Since f is injective, we also have $f(x_k) \neq f(a)$ for all $k \in \mathbb{N}$, which shows that $b = f(a)$ is a limit point of Y.

Now let (y_k) be a sequence in Y such that $y_k \neq b$ for all $k \in \mathbb{N}$ and $\lim y_k = b$. Set $x_k := f^{-1}(y_k)$. Then $x_k \neq a$ and $\lim x_k = a$, since f^{-1} is continuous at b. Because

$$0 \neq f'(a) = \lim_k \frac{f(x_k) - f(a)}{x_k - a} ,$$

there is some K such that

$$0 \neq \frac{f(x_k) - f(a)}{x_k - a} = \frac{y_k - b}{f^{-1}(y_k) - f^{-1}(b)} , \qquad k \geq K .$$

Hence, for the difference quotient of f^{-1}, we have

$$\frac{f^{-1}(y_k) - f^{-1}(b)}{y_k - b} = \frac{x_k - a}{f(x_k) - f(a)} = 1 \Big/ \frac{f(x_k) - f(a)}{x_k - a} , \qquad k \geq K ,$$

and the claim follows by taking the limit $k \to \infty$. ■

1.9 Corollary *Let I be an interval and $f : I \to \mathbb{R}$ strictly monotone and continuous. Suppose that f is differentiable at $a \in I$. Then f^{-1} is differentiable at $f(a)$ if and only if $f'(a)$ is nonzero and, in this case, $(f^{-1})'\bigl(f(a)\bigr) = 1/f'(a)$.*

Proof By Theorem III.5.7, f is injective and f^{-1} is continuous on the interval $J := f(I)$. Hence the claim follows from Theorem 1.8. ■

Differentiable Functions

So far we have considered the following situation: X is an arbitrary subset of $\mathbb{K}$ and $a \in X$ is a limit point of X. Under these conditions, we have studied the differentiability of $f : X \to E$ at a. The obvious next question is whether f is differentiable at *every* point of X. For this question to make sense it is necessary that each point of X is a limit point of X.

Let M be a metric space. A subset $A \subseteq M$ is called **perfect** if each $a \in A$ is a limit point of A.[1]

1.10 Examples **(a)** Any nonempty open subset of a normed vector space is perfect.

(b) A convex subset of a normed vector space (in particular, an interval in $\mathbb{R}$) is perfect if and only if it contains more than one point. ■

Let $X \subseteq \mathbb{K}$ be perfect. Then $f : X \to E$ is called **differentiable** on X if f is differentiable at each point of X. The function

$$f' : X \to E , \quad x \mapsto f'(x)$$

is called the **derivative** of f. It is also denoted by $\dot{f}$, ∂f, Df and df/dx.

Higher Derivatives

If $f : X \to E$ is differentiable, then it is natural to ask whether the derivative f' is itself differentiable. When this occurs f is said to be twice differentiable and we

[1] This definition agrees with the definition in Section 1.10 in the case that $M = \mathbb{R}$ and A is an interval (see Example 1.10(b)).

call $\partial^2 f := f'' := \partial(\partial f)$ the second derivative of f. Repeating this process we can define further *higher derivatives* of f. Specifically, we set

$$\partial^0 f := f^{(0)} := f \ , \quad \partial^1 f(a) := f^{(1)}(a) := f'(a) \ ,$$
$$\partial^{n+1} f(a) := f^{(n+1)}(a) := \partial(\partial^n f)(a)$$

for all $n \in \mathbb{N}$. The element $\partial^n f(a) \in E$ is called the nth **derivative** of f at a. The function f is called n**-times differentiable** on X if the nth derivative exists at each $a \in X$. If f is n-times differentiable and the nth derivative

$$\partial^n f : X \to E \ , \quad x \mapsto \partial^n f(x)$$

is continuous, then f is n**-times continuously differentiable.**

The **space of** n**-times continuously differentiable functions** from X to E is denoted by $C^n(X,E)$. In particular, $C^0(X,E) = C(X,E)$ is the space of continuous E-valued functions on X already introduced in Section III.1. Finally

$$C^\infty(X,E) := \bigcap_{n \in \mathbb{N}} C^n(X,E)$$

is the space of **infinitely differentiable** or **smooth** functions from X to E. We write

$$C^n(X) := C^n(X, \mathbb{K}) \ , \qquad n \in \bar{\mathbb{N}} \ ,$$

when no misunderstanding is possible.

1.11 Remarks Let $n \in \mathbb{N}$.

(a) For the $(n+1)^{\text{th}}$ derivative at a to be defined, a must be a limit point of the domain of the nth derivative. This is the case, in particular, if the nth derivative exists on some neighborhood of a.

(b) If a function $f : X \to E$ is $(n+1)$-times differentiable at $a \in X$, then, by Corollary 1.2, for each $j \in \{0, 1, \ldots, n\}$, the jth derivative of f is continuous at a.

(c) It is not difficult to see that the inclusions,

$$C^\infty(X,E) \subseteq C^{n+1}(X,E) \subseteq C^n(X,E) \subseteq C(X,E) \ , \qquad n \in \mathbb{N} \ ,$$

hold. ■

We collect in the next theorem some of the most important rules which hold in the space of n-times continuously differentiable functions $C^n(X,E)$.

1.12 Theorem *Let $X \subseteq \mathbb{K}$ be perfect, $k \in \mathbb{N}$ and $n \in \bar{\mathbb{N}} = \mathbb{N} \cup \{\infty\}$.*

(i) (linearity) *For all $f, g \in C^k(X, E)$ and $\alpha, \beta \in \mathbb{K}$,*

$$\alpha f + \beta g \in C^k(X, E) \quad \text{and} \quad \partial^k(\alpha f + \beta g) = \alpha \partial^k f + \beta \partial^k g .$$

Hence $C^n(X, E)$ is a subspace of $C(X, E)$ and the **differentiation operator**

$$\partial \colon C^{n+1}(X, E) \to C^n(X, E) , \quad f \mapsto \partial f$$

is linear.

(ii) (Leibniz' rule) *Let $f, g \in C^k(X)$. Then $f \cdot g$ is in $C^k(X)$ and*

$$\partial^k(fg) = \sum_{j=0}^{k} \binom{k}{j} (\partial^j f) \partial^{k-j} g . \tag{1.3}$$

Hence $C^n(X)$ is a subalgebra of $\mathbb{K}^X$.

Proof (i) The first statement follows from Theorem 1.6 and Proposition III.1.5.

(ii) Because of Theorem 1.6 and Proposition III.1.5, it suffices to confirm Leibniz' rule (1.3). This we do using induction on k. The case $k = 0$ is proved in Proposition III.1.5. For the induction step $k \to k + 1$, we use the equation

$$\binom{k+1}{j} = \binom{k}{j-1} + \binom{k}{j} , \qquad k \in \mathbb{N} , \quad 1 \leq j \leq k ,$$

from Exercise I.5.5. The induction hypothesis, the product rule and (i) imply

$$\begin{aligned}
\partial^{k+1}(fg) &= \partial\Bigl(\sum_{j=0}^{k} \binom{k}{j} (\partial^j f) \partial^{k-j} g\Bigr) \\
&= \sum_{j=0}^{k} \binom{k}{j} \bigl[(\partial^{j+1} f) \partial^{k-j} g + (\partial^j f) \partial^{k-j+1} g\bigr] \\
&= (\partial^{k+1} f) g + f \partial^{k+1} g + \sum_{j=1}^{k} \Bigl[\binom{k}{j-1} + \binom{k}{j}\Bigr] (\partial^j f) \partial^{k-j+1} g \\
&= \sum_{j=0}^{k+1} \binom{k+1}{j} (\partial^j f) \partial^{k+1-j} g .
\end{aligned}$$

Thus the induction is complete. ■

1.13 Examples **(a)** Let a be a limit point of $X \subseteq \mathbb{R}$. Then $f : X \to \mathbb{C}$ is differentiable at a if and only if $\operatorname{Re} f$ and $\operatorname{Im} f$ are differentiable at a. In this case,

$$f'(a) = (\operatorname{Re} f)'(a) + i\,(\operatorname{Im} f)'(a) .$$

Proof This follows from Proposition 1.5. ∎

(b) Let $p = \sum_{k=0}^{n} a_k X^k$ be a polynomial.[2] Then p is smooth and

$$p'(x) = \sum_{k=1}^{n} k a_k x^{k-1} , \qquad x \in \mathbb{C} .$$

Proof Let $\mathbf{1} := 1X^0$ be the unity element in the algebra $\mathbb{K}[X]$, which, by our conventions, is the same as the constant function defined by $\mathbf{1}(x) = 1$ for all $x \in \mathbb{K}$. Then clearly

$$\mathbf{1} \in C^\infty(\mathbb{K}) \quad \text{and} \quad \partial \mathbf{1} = 0 . \tag{1.4}$$

By induction, we now show that

$$X^n \in C^\infty(\mathbb{K}) \quad \text{and} \quad \partial(X^n) = nX^{n-1} , \qquad n \in \mathbb{N}^\times . \tag{1.5}$$

The case $n = 1$ is true since, trivially, $\partial X = \mathbf{1}$, and by (1.4), $\mathbf{1} \in C^\infty(\mathbb{K})$. For the induction step $n \to n+1$, we use the product rule:

$$\partial(X^{n+1}) = \partial(X^n X) = \partial(X^n)X + X^n \partial X = nX^{n-1}X + X^n \mathbf{1} = (n+1)X^n .$$

Hence (1.5) is true. For an arbitrary polynomial $\sum_{k=0}^{n} a_k X^k$, the claim now follows from Theorem 1.12(i). ∎

(c) A rational function is smooth on its domain.

Proof This follows from (b), Theorem 1.6 and Corollary III.1.6. ∎

(d) The exponential function is in $C^\infty(\mathbb{K})$ and satisfies $\partial(\exp) = \exp$.

Proof It suffices to prove the formula $\partial(\exp) = \exp$. For $z \in \mathbb{C}$, the difference quotient is given by

$$\frac{e^{z+h} - e^z}{h} = e^z \frac{e^h - 1}{h} , \qquad h \in \mathbb{C}^\times ,$$

and so the claim follows from Example III.2.25(b). ∎

(e) For the logarithm, we have

$$\log \in C^\infty\bigl(\mathbb{C}\backslash(-\infty, 0], \mathbb{C}\bigr) , \qquad (\log)'(z) = 1/z , \quad z \in \mathbb{C}\backslash(-\infty, 0] .$$

Proof From (III.6.11) we have $\log = \bigl[\exp \bigm| \mathbb{R} + i(-\pi, \pi]\bigr]^{-1}$, and the logarithm is continuous on $\mathbb{R} + i(-\pi, \pi)$ (see Exercise III.6.9). For each $z \in \mathbb{C}\backslash(-\infty, 0]$, there is a unique x in $\mathbb{R} + i(-\pi, \pi)$ such that $z = e^x$. From Theorem 1.8 and (d) we then have

$$(\log)'(z) = \frac{1}{(\exp)'(x)} = \frac{1}{\exp(x)} = \frac{1}{z} ,$$

and the claim follows from (c). ∎

[2] By the convention at the end of Section I.8, we consider polynomials to be also functions.

(f) Let $a \in \mathbb{C}\backslash(-\infty,0]$. Then[3]

$$[z \mapsto a^z] \in C^\infty(\mathbb{C}) \quad \text{and} \quad (a^z)' = a^z \log a \ , \quad z \in \mathbb{C} \ .$$

Proof Since $a^z = e^{z \log a}$ for all $z \in \mathbb{C}$,

$$(a^z)' = (e^{z \log a})' = (\log a) e^{z \log a} = a^z \log a$$

follows from the chain rule and (d). Since $[z \mapsto a^z] : \mathbb{C} \to \mathbb{C}$ is continuous (why?), an easy induction shows that $[z \mapsto a^z] \in C^\infty(\mathbb{C})$. ■

(g) Let $a \in \mathbb{C}$. Then, for the power function, we have

$$[z \mapsto z^a] \in C^\infty\bigl(\mathbb{C}\backslash(-\infty,0],\mathbb{C}\bigr) \quad \text{and} \quad (z^a)' = az^{a-1} \ .$$

Proof As in (f), we have $z^a = e^{a \log z}$ for all $z \in \mathbb{C}\backslash(-\infty,0]$, and so, from the chain rule and (e), we get

$$(z^a)' = (e^{a \log z})' = \frac{a}{z} e^{a \log z} = \frac{a}{z} z^a = az^{a-1} \ ,$$

where in the last step we have also used (III.6.13). ■

(h) $\operatorname{cis} \in C^\infty(\mathbb{R},\mathbb{C})$ and $\operatorname{cis}'(t) = i \operatorname{cis}(t)$ for $t \in \mathbb{R}$.

Proof From (d) and the chain rule we get $\operatorname{cis}'(t) = (e^{it})' = i e^{it} = i \operatorname{cis}(t)$ for all $t \in \mathbb{R}$. ■

(i) cos and sin are in $C^\infty(\mathbb{C})$ with $\cos' = -\sin$ and $\sin' = \cos$.

Proof By (III.6.3), cos and sin can be written using the exponential function:

$$\cos z = \frac{e^{iz} + e^{-iz}}{2} \ , \quad \sin z = \frac{e^{iz} - e^{-iz}}{2i} \ , \qquad z \in \mathbb{C} \ .$$

Using (d) and the chain rule we get

$$\cos' z = i \frac{e^{iz} - e^{-iz}}{2} = -\sin z \ , \quad \sin' z = i \frac{e^{iz} + e^{-iz}}{2i} = \cos z \ ,$$

and so cos and sin are smooth. ■

(j) The tangent and cotangent functions are smooth on their domains and

$$\tan' = \frac{1}{\cos^2} = 1 + \tan^2 \ , \quad \cot' = \frac{-1}{\sin^2} = -1 - \cot^2 \ .$$

Proof The quotient rule and (i) yield

$$\tan' z = \Bigl(\frac{\sin}{\cos}\Bigr)'(z) = \frac{\cos^2 z + \sin^2 z}{\cos^2 z} = \frac{1}{\cos^2 z} = 1 + \tan^2 z \ , \qquad z \in \mathbb{C}\backslash(\pi/2 + \pi\mathbb{Z}) \ .$$

The proof for the cotangent function is similar. ■

(k) The function $f : \mathbb{R} \to \mathbb{R}, \ x \mapsto |x|$ is continuous, but not differentiable, at 0.

Proof Set $h_n := (-1)^n/(n+1)$ for all $n \in \mathbb{N}$. Then (h_n) is a null sequence such that $\bigl(f(h_n) - f(0)\bigr)h_n^{-1} = (-1)^n$ for all $n \in \mathbb{N}$. Thus f cannot be differentiable at 0. ■

[3] To avoid introducing a new symbol for the function $z \mapsto a^z$, we write somewhat imprecisely $(a^z)'$ for $[z \mapsto a^z]'(z)$. This simplified notation will be used in similar situations since it does not lead to misunderstanding.

(l) Consider the function

$$f(x) := \begin{cases} x^2 \sin \dfrac{1}{x} , & x \in \mathbb{R}^\times , \\ 0 , & x = 0 . \end{cases}$$

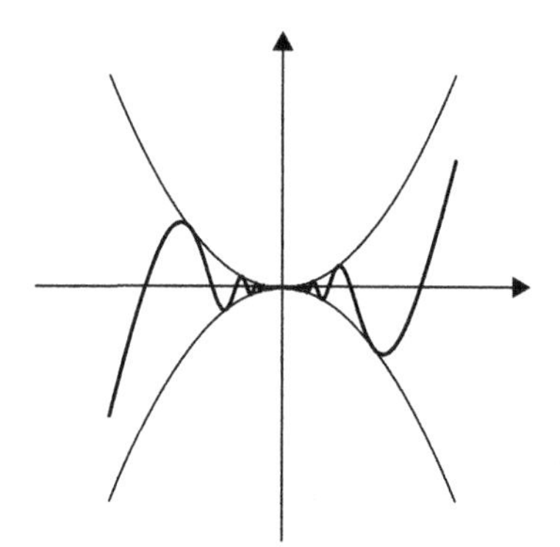

Then f is differentiable on $\mathbb{R}$, but the derivative f' is not continuous at 0. That is, $f \notin C^1(\mathbb{R})$.

Proof For the difference quotient of f at 0 we have

$$\frac{f(x) - f(0)}{x} = x \sin \frac{1}{x} , \qquad x \neq 0 ,$$

and so $f'(0) = 0$ by Proposition II.2.4. For all $x \in \mathbb{R}^\times$, $f'(x) = 2x \sin x^{-1} - \cos x^{-1}$ and hence

$$f'\Big(\frac{1}{2\pi n}\Big) = \frac{1}{\pi n} \sin(2\pi n) - \cos(2\pi n) = -1 , \qquad n \in \mathbb{N}^\times .$$

Thus f' is not continuous at 0. ■

(m) There are functions which are continuous on $\mathbb{R}$, but nowhere differentiable.

Proof Let f_0 be the function from Exercise III.1.1. For $n > 0$, define the function f_n by $f_n(x) := 4^{-n} f_0(4^n x)$ for all $x \in \mathbb{R}$. Clearly, f_n is piecewise affine with slope ± 1 and periodic with period 4^{-n}. From Exercise III.5.6 we know that the function $F := \sum_{n=0}^\infty f_n$ is continuous on $\mathbb{R}$.

Let $a \in \mathbb{R}$. Then, for each $n \in \mathbb{N}$, there is some $h_n \in \{\pm 4^{-(n+1)}\}$ such that, for $k \leq n$, f_k is affine between a and $a + h_n$. Thus $\big[f_k(a + h_n) - f_k(a)\big] / h_n = \pm 1$ for all $0 \leq k \leq n$. For $k > n$, we have $f_k(a + h_n) = f_k(a)$, since, in this case, f_k has period h_n. This implies

$$\frac{F(a + h_n) - F(a)}{h_n} = \sum_{k=0}^n \frac{f_k(a + h_n) - f_k(a)}{h_n} = \sum_{k=0}^n \pm 1 ,$$

and hence F is not differentiable at a. ■

(n) $C^\infty(X, E) \subset C^{n+1}(X, E) \subset C^n(X, E) \subset C(X, E)$, $n \in \mathbb{N}^\times$.

Proof In view of Remark 1.11(c), it suffices to show that these inclusions are proper. We consider only the case $X := \mathbb{R}$, $E := \mathbb{R}$ and leave the general case to the reader. For each $n \in \mathbb{N}$, define $f_n : \mathbb{R} \to \mathbb{R}$ by

$$f_n(x) := \begin{cases} x^{n+2} \sin(x^{-1}) , & x \neq 0 , \\ 0 , & x = 0 . \end{cases}$$

Then a simple induction argument shows that $f_n \in C^n(\mathbb{R}) \backslash C^{n+1}(\mathbb{R})$. The $n = 0$ case is proved in (l). ■

1.14 Remark The reader should note that Remark 1.4(c.ii) applied to Examples 1.13(b)–(g) and (i), (j) gives the usual rules for the derivatives of the *real* polynomial, rational, power, exponential, logarithm and trigonometric functions. ∎

One-Sided Differentiability

If $X \subseteq \mathbb{R}$, $a \in X$ is a limit point of $X \cap [a, \infty)$ and

$$\partial_+ f(a) := \lim_{x \to a+} \frac{f(x) - f(a)}{x - a}$$

exists, then $f : X \to E$ is **right differentiable** at a and $\partial_+ f(a) \in E$ is called the **right derivative** of f at a.

Similarly, if a is a limit point of $(-\infty, a] \cap X$ and

$$\partial_- f(a) := \lim_{x \to a-} \frac{f(x) - f(a)}{x - a}$$

exists, then f is **left differentiable** at a and $\partial_- f(a)$ is called the **left derivative** of f at a. If a is a limit point of both $X \cap [a, \infty)$ and $(-\infty, a] \cap X$, and f is differentiable at a, then clearly

$$\partial_+ f(a) = \partial_- f(a) = \partial f(a) .$$

1.15 Examples (a) For $f : \mathbb{R} \to \mathbb{R}$, $x \mapsto |x|$,

$$\partial_+ f(0) = 1 , \quad \partial_- f(0) = -1 , \qquad \partial_+ f(x) = \partial_- f(x) = \operatorname{sign}(x) , \quad x \neq 0 .$$

(b) Let $a < b$ and $f : [a, b] \to E$. Then f is differentiable at a (or b) if and only if f is right (or left) differentiable at a (or b). ∎

Example 1.15(a) shows that the existence of the right and left derivatives of a function $f : X \to E$ does not imply the existence of the derivative. The next proposition shows that the missing condition is that the one-sided derivatives must be equal.

1.16 Proposition *Let $X \subseteq \mathbb{R}$ and $f : X \to E$ be right and left differentiable at $a \in X$ with $\partial_+ f(a) = \partial_- f(a)$. Then f is differentiable at a and $\partial f(a) = \partial_+ f(a)$.*

Proof By hypothesis and Proposition 1.1(iii), there are functions

$$r_+ : X \cap [a, \infty) \to E \quad \text{and} \quad r_- : (-\infty, a] \cap X \to E$$

which are continuous at a and satisfy $r_+(a) = r_-(a) = 0$ and

$$f(x) = f(a) + \partial_\pm f(a)(x - a) + r_\pm(x)(x - a) , \qquad x \in X , \quad x \gtreqless a .$$

Now set $\partial f(a) := \partial_+ f(a) = \partial_- f(a)$ and

$$r(x) := \begin{cases} r_+(x) , & x \in X \cap [a, \infty) , \\ r_-(x) , & x \in (-\infty, a] \cap X . \end{cases}$$

Then $r\colon X \to E$ is, by Proposition III.1.12, continuous at a, $r(a) = 0$ and

$$f(x) = f(a) + \partial f(a)(x-a) + r(x)(x-a) , \qquad x \in X .$$

Thus the claim follows from Proposition 1.1(iii). ■

1.17 Example Let

$$f(x) := \begin{cases} e^{-1/x} , & x > 0 , \\ 0 , & x \le 0 . \end{cases}$$

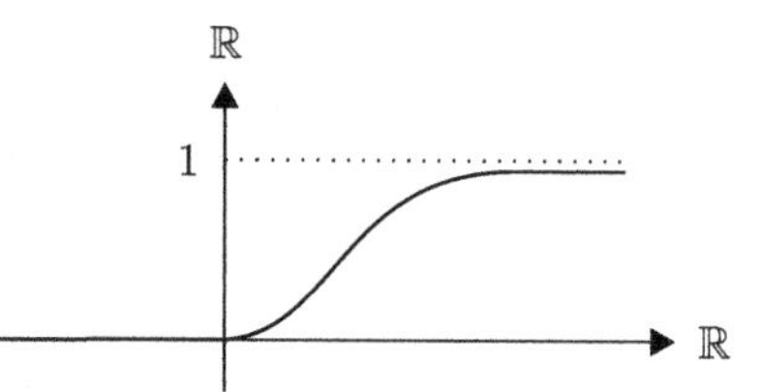

Then f is smooth and all its derivatives are zero at $x = 0$.

Proof It suffices to show that all the derivatives of f exist and satisfy

$$\partial^n f(x) = \begin{cases} p_{2n}(x^{-1}) e^{-x^{-1}} , & x > 0 , \\ 0 , & x \le 0 , \end{cases} \tag{1.6}$$

where p_{2n} denotes a polynomial of degree $\le 2n$ with real coefficients.

Clearly (1.6) holds for $x < 0$. In the case $x > 0$, (1.6) holds for $n = 0$. If the formula is true for some $n \in \mathbb{N}$, then

$$\begin{aligned} \partial^{n+1} f(x) &= \partial\big(p_{2n}(x^{-1}) e^{-x^{-1}}\big) \\ &= -\partial p_{2n}(x^{-1})(x^{-2}) e^{-x^{-1}} + p_{2n}(x^{-1}) e^{-x^{-1}} x^{-2} \\ &= p_{2(n+1)}(x^{-1}) e^{-x^{-1}} , \end{aligned}$$

with $p_{2(n+1)}(X) := \big(p_{2n}(X) - \partial p_{2n}(X)\big) X^2$. Because $\deg(p_{2n}) \le 2n$, the degree of ∂p_{2n} is at most $2n-1$ and (I.8.20) gives $\deg(p_{2(n+1)}) \le 2(n+1)$. Thus (1.6) holds for all $x > 0$.

It remains to consider the case $x = 0$. Once again we use a proof by induction. The $n = 0$ case is trivial. For the induction step $n \to n+1$, we calculate

$$\partial_+(\partial^n f)(0) = \lim_{x \to 0+} \frac{\partial^n f(x) - \partial^n f(0)}{x - 0} = \lim_{x \to 0+} \big[x^{-1} p_{2n}(x^{-1}) e^{-x^{-1}}\big] ,$$

where we used the induction hypothesis and (1.6) for the second equality. Further, by Propositions III.6.5(iii) and II.5.2(i), we have

$$\lim_{x \to 0+} \big[q(x^{-1}) e^{-x^{-1}}\big] = \lim_{y \to \infty} \frac{q(y)}{e^y} = 0 \tag{1.7}$$

for all $q \in \mathbb{R}[X]$. Thus $\partial^n f$ is right differentiable at 0 and $\partial_+(\partial^n f)(0) = 0$. Since $\partial^n f$ is obviously left differentiable at 0 with $\partial_-(\partial^n f)(0) = 0$, it follows from Proposition 1.16 that $\partial^{n+1} f(0) = 0$. This completes the proof of (1.6). ■

Exercises

1 Calculate the derivative of $f:(0,\infty)\to\mathbb{R}$ when $f(x)$ is

$$\text{(a) } (x^x)^x\ ,\quad \text{(b) } x^{(x^x)}\ ,\quad \text{(c) } x^{1/x}\ ,\quad \text{(d) } \log\log(1+x)\ ,$$
$$\text{(e) } x^{\sin x}\ ,\quad \text{(f) } \sqrt[3]{x^{3/5}+\sin^3(1/x)-\tan^2(x)}\ ,\quad \text{(g) } \frac{\cos x}{2+\sin\log x}\ .$$

2 For $m,n\in\mathbb{N}$, let $f_{m,n}:\mathbb{R}\to\mathbb{R}$ be defined by

$$f_{m,n}(x):=\begin{cases} x^n\sin(x^{-m})\ , & x\neq 0\ ,\\ 0\ , & x=0\ .\end{cases}$$

For what $k\in\bar{\mathbb{N}}$ is $f_{m,n}\in C^k(\mathbb{R})$?

3 Suppose that $f,g:\mathbb{K}\to\mathbb{K}$ satisfy $f'=f$, $f(x)\neq 0$ for all $x\in\mathbb{K}$, and $g'=g$. Show that f and g are in $C^\infty(\mathbb{K},\mathbb{K})$ and that there is some $c\in\mathbb{K}$ such that $g=cf$.

4 Show that $f:\mathbb{C}\to\mathbb{C}$, $z\mapsto\bar{z}$ is nowhere differentiable.

5 At what points is $f:\mathbb{C}\to\mathbb{C}$, $z\mapsto z\bar{z}$ differentiable?

6 Let U be a neighborhood of 0 in $\mathbb{K}$, E a normed vector space and $f:U\to E$.
(a) Suppose that there are numbers $K>0$ and $\alpha>1$ such that $|f(x)|\leq K\,|x|^\alpha$ for all $x\in U$. Show that f is differentiable at 0.
(b) Suppose that $f(0)=0$ and there are $K>0$ and $\alpha\in(0,1)$ such that $|f(x)|\geq K\,|x|^\alpha$ for all $x\in U$. Show that f is not differentiable at 0.
(c) What can be said if $|f(x)|=K\,|x|$ for all $x\in U$?

7 Calculate $\partial_\pm f(x)$ for the function $f:\mathbb{R}\to\mathbb{R}$, $x\mapsto\lfloor x\rfloor+\sqrt{x-\lfloor x\rfloor}$. Where is f differentiable?

8 Suppose that I is a perfect interval and $f,g:I\to\mathbb{R}$ are differentiable. Prove or disprove that the functions $|f|$, $f\vee g$ and $f\wedge g$ are (a) differentiable, (b) one-sided differentiable.

9 Let U be open in $\mathbb{K}$, $a\in U$ and $f:U\to E$. Prove or disprove the following:
(a) If f differentiable at a, then

$$f'(a)=\lim_{h\to 0}\frac{f(a+h)-f(a-h)}{2h}\ . \tag{1.8}$$

(b) If $\lim_{h\to 0}\bigl[f(a+h)-f(a-h)\bigr]/2h$ exists, then f is differentiable at a and (1.8) holds.

10 Let $n\in\mathbb{N}^\times$ and $f\in C^n(\mathbb{K})$. Prove that

$$\partial^n\bigl(xf(x)\bigr)=x\partial^n f(x)+n\partial^{(n-1)}f(x)\ .$$

11 For $n\in\mathbb{N}^\times$, show that

$$\partial^n\bigl(x^{n-1}e^{1/x}\bigr)=(-1)^n\frac{e^{1/x}}{x^{n+1}}\ ,\qquad x>0\ .$$

12 The **Legendre polynomial** P_n is defined by

$$P_n(x) := \frac{1}{2^n n!} \partial^n \left[(x^2-1)^n \right] , \qquad n \in \mathbb{N} .$$

(a) Calculate $P_0, P_1, \ldots, P_5$.

(b) Show that P_n is a polynomial of degree n which has n zeros in $(-1,1)$.

2 The Mean Value Theorem and its Applications

Let $f : \mathbb{R} \to \mathbb{R}$ be a differentiable function. If we view f' geometrically as the slope of tangent lines to the graph of f, it is intuitively clear that, with the help of f', not only the local properties, but also the global properties of f can be investigated. For example, if f has a local extremum at a, then the tangent line at $\bigl(a, f(a)\bigr)$ must be horizontal, that is, $f'(a) = 0$. If, on the other hand, the derivative f' is positive everywhere, then f has the global property of being increasing.

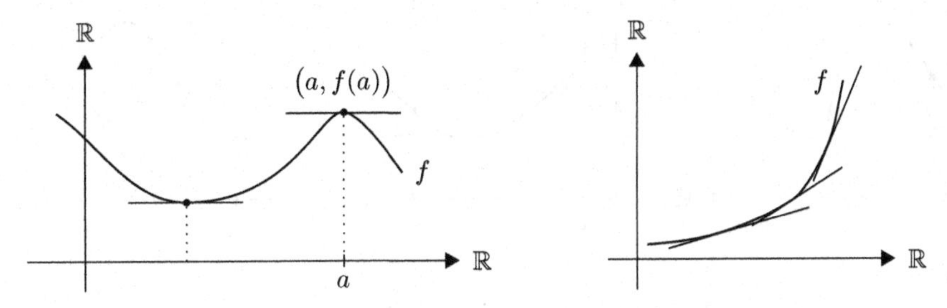

In the following, we generalize these ideas and make them more precise.

Extrema

Let X be a metric space and f a real valued function on X. Then f has a **local minimum** (or **local maximum**) at $x_0 \in X$ if there is a neighborhood U of x_0 such that $f(x_0) \le f(x)$ (or $f(x_0) \ge f(x)$) for all $x \in U$. The function f has a **global minimum** (or **global maximum**) at x_0 if $f(x_0) \le f(x)$ (or $f(x_0) \ge f(x)$) for all $x \in X$. Finally, we say that f has a **local** (or **global**) **extremum** at x_0 if f has a local (or global) minimum or maximum at x_0.

2.1 Theorem (necessary condition for local extrema) *Suppose that $X \subseteq \mathbb{R}$ and $f : X \to \mathbb{R}$ has a local extremum at $a \in \mathring{X}$. If f is differentiable at a, then $f'(a) = 0$.*

Proof Suppose that f has a local minimum at a. Then there is an open interval I with $a \in I \subseteq X$ and $f(x) \ge f(a)$ for all $x \in I$. Thus

$$\frac{f(x) - f(a)}{x - a} \begin{cases} \ge 0\,, & x \in I \cap (a, \infty)\,, \\ \le 0\,, & x \in (-\infty, a) \cap I\,. \end{cases}$$

In the limit $x \to a$, this implies $0 \le \partial_+ f(a) = \partial_- f(a) \le 0$, and so $f'(a) = 0$. If f has a local maximum at a, then $-f$ has a local minimum at a. Consequently, $f'(a) = 0$ in this case too. ■

If $X \subseteq \mathbb{K}$ and $f : X \to E$ is differentiable at $a \in X$ with $f'(a) = 0$, then a is called a **critical point** of f. Thus Theorem 2.1 says that *if f has a local extremum at $a \in \mathring{X}$ and is differentiable at a, then a is a critical point of f.*

2.2 Remarks Let $f\colon [a,b] \to \mathbb{R}$ with $-\infty < a < b < \infty$.

(a) If f is differentiable at a and has a local minimum (or maximum) at a, then $f'(a) \geq 0$ (or $f'(a) \leq 0$). Similarly, if f is differentiable at b and has a local minimum (or maximum) at b, then $f'(b) \leq 0$ (or $f'(b) \geq 0$).

Proof This follows directly from the proof of Theorem 2.1. ∎

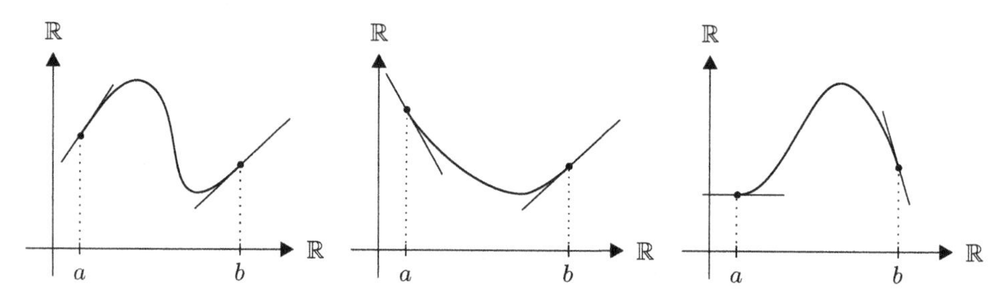

(b) Let f be continuous on $[a,b]$ and differentiable on (a,b). Then

$$\max_{x\in[a,b]} f(x) = f(a) \vee f(b) \vee \max\{\, f(x)\ ;\ x \in (a,b),\ f'(x) = 0 \,\}\ ,$$

that is, f *attains its maximum either at an end point of* $[a,b]$ *or at a critical point in* (a,b). Similarly

$$\min_{x\in[a,b]} f(x) = f(a) \wedge f(b) \wedge \min\{\, f(x)\ ;\ x \in (a,b),\ f'(x) = 0 \,\}\ .$$

Proof By the extreme value theorem (Corollary III.3.8), there is some $x_0 \in [a,b]$ such that $f(x_0) \geq f(x)$ for $x \in [a,b]$. If x_0 is not an end point of $[a,b]$, then, by Theorem 2.1, x_0 is a critical point of f. The second claim can be proved similarly. ∎

(c) If $x_0 \in (a,b)$ is a critical point of f it does *not* follow that f has an extremum at x_0.

Proof Consider the cubic polynomial $f(x) := x^3$ at $x_0 = 0$. ∎

The Mean Value Theorem

In the next two theorems a and b are real numbers such that $a < b$.

2.3 Theorem (Rolle's theorem) *Suppose that* $f \in C\bigl([a,b],\mathbb{R}\bigr)$ *is differentiable on* (a,b). *If* $f(a) = f(b)$, *then there is some* $\xi \in (a,b)$ *such that* $f'(\xi) = 0$.

Proof If f is constant on the interval $[a,b]$, then the claim is clear. Indeed, in this case, $f' = 0$. If f is not constant on $[a,b]$, then f has an extremum in (a,b) and the claim follows from Remark 2.2(b). ∎

2.4 Theorem (mean value theorem) *If $f \in C\big([a,b],\mathbb{R}\big)$ is differentiable on (a,b), then there is some $\xi \in (a,b)$ such that*

$$f(b) = f(a) + f'(\xi)(b-a) \ .$$

Proof Set

$$g(x) := f(x) - \frac{f(b)-f(a)}{b-a}x \ , \qquad x \in [a,b] \ .$$

Then $g\colon [a,b] \to \mathbb{R}$ satisfies the hypotheses of Rolle's theorem. Thus there is some $\xi \in (a,b)$ such that

$$0 = g'(\xi) = f'(\xi) - \frac{f(b)-f(a)}{b-a} \ ,$$

which proves the claim. ∎

Geometrically, the mean value theorem says that there is (at least) one point $\xi \in (a,b)$ such that the tangent line t to the graph of f at $\big(\xi, f(\xi)\big)$ is parallel to the secant line s through $\big(a, f(a)\big)$ and $\big(b, f(b)\big)$, that is, the slopes of these two lines are equal:

$$f'(\xi) = \frac{f(b)-f(a)}{b-a} \ .$$

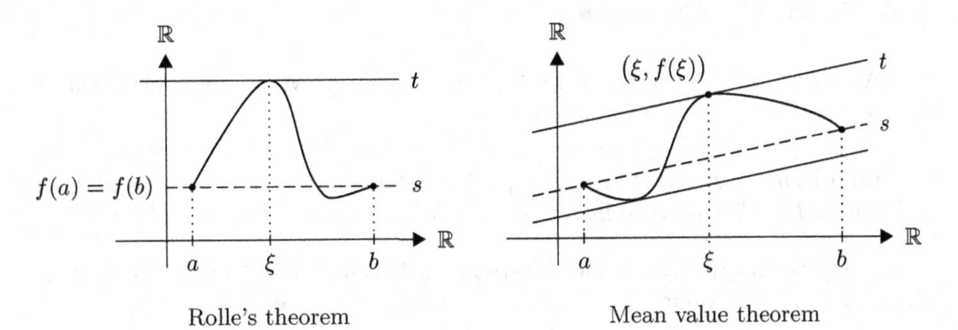

Rolle's theorem Mean value theorem

Monotonicity and Differentiability

2.5 Theorem (a characterization of monotone functions) *Suppose that I is a perfect interval and $f \in C(I,\mathbb{R})$ is differentiable on $\mathring{I}$.*

(i) *f is increasing (or decreasing) if and only if $f'(x) \geq 0$ (or $f'(x) \leq 0$) for all $x \in \mathring{I}$.*

(ii) *If $f'(x) > 0$ (or $f'(x) < 0$) for all $x \in \mathring{I}$, then f is strictly increasing (or strictly decreasing).*

Proof (i) '$\Rightarrow$' If f is increasing, then

$$\frac{f(y)-f(x)}{y-x} \geq 0 , \qquad x,y \in \mathring{I} , \quad x \neq y .$$

Taking the limit $y \to x$ we get $f'(x) \geq 0$ for all $x \in \mathring{I}$. The case of f decreasing is proved similarly.

'$\Leftarrow$' Let $x, y \in I$ with $x < y$. By the mean value theorem, there is some $\xi \in (x, y)$ such that

$$f(y) = f(x) + f'(\xi)(y - x) . \tag{2.1}$$

If $f'(z) \geq 0$ for all $z \in \mathring{I}$, then, in particular, $f'(\xi) \geq 0$, so it follows from (2.1) that $f(y) \geq f(x)$. Thus f is increasing. Similarly, if $f'(z) \leq 0$ for all $z \in \mathring{I}$, then f is decreasing.

Claim (ii) follows directly from (2.1). ■

2.6 Remarks **(a)** (a characterization of constant functions) With the hypotheses of Theorem 2.5, f is constant if and only if $f' = 0$.

Proof This follows from Theorem 2.5(i). ■

(b) The converse of Theorem 2.5(ii) is false. The function $f(x) := x^3$ is strictly increasing but its derivative is zero at 0. Moreover, in (a), it is essential that the domain be an interval (why?). ■

As a further application of Rolle's theorem we prove a simple criterion for the injectivity of real differentiable functions.

2.7 Proposition *Suppose that I is a perfect interval and $f \in C(I, \mathbb{R})$ is differentiable on $\mathring{I}$. If f' has no zero in $\mathring{I}$, then f is injective.*

Proof If f is not injective then there are $x, y \in I$ such that $x < y$ and $f(x) = f(y)$. Then, by Rolle's theorem, f' has a zero between x and y. ■

2.8 Theorem *Suppose that I is a perfect interval and $f : I \to \mathbb{R}$ is differentiable with $f'(x) \neq 0$ for all $x \in I$.*

(i) *f is strictly monotone.*

(ii) *$J := f(I)$ is a perfect interval.*

(iii) *$f^{-1} : J \to \mathbb{R}$ is differentiable and $(f^{-1})'(f(x)) = 1/f'(x)$ for all $x \in I$.*

Proof First we verify (ii). By Corollary 1.2 and Proposition 2.7, f is continuous and injective. So the intermediate value theorem and Example 1.10(b) imply that J is a perfect interval.

To prove (i), we suppose that f is not strictly monotone. Since, by Remark 2.6(a), f is not constant on any perfect subinterval, there are $x < y < z$ such that $f(x) > f(y) < f(z)$ or $f(x) < f(y) > f(z)$. By the intermediate value and the extreme value theorems, f has an extremum at some $\xi \in (x, z)$. By Theorem 2.1, we have $f'(\xi) = 0$, which contradicts our supposition. Finally, Claim (iii) follows from (i) and Corollaries 1.2 and 1.9. ■

2.9 Remarks **(a)** The function $\mathrm{cis}\colon \mathbb{R} \to \mathbb{C}$ has period 2π and so is certainly not injective. Nonetheless, $\mathrm{cis}'(t) = i\,e^{it} \neq 0$ for all $t \in \mathbb{R}$. This shows that Proposition 2.7 does not hold for complex valued (or vector valued) functions.

(b) If the hypothesis of Theorem 2.8 is satisfied, then it follows from (i) and Theorem 2.5 that either

$$f'(x) > 0\ , \quad x \in I\ , \qquad \text{or} \qquad f'(x) < 0\ , \quad x \in I\ . \tag{2.2}$$

Note that (2.2) does not follow from $f'(x) \neq 0$ for all $x \in I$ and the intermediate value theorem since f' may not be continuous. ■

2.10 Applications For the trigonometric functions we have

$$\begin{aligned} &\cos' x = -\sin x \neq 0\ , && \cot' x = -1/\sin^2 x \neq 0\ , && x \in (0, \pi)\ , \\ &\sin' x = \cos x \neq 0\ , && \tan' x = 1/\cos^2 x \neq 0\ , && x \in (-\pi/2, \pi/2)\ . \end{aligned}$$

Hence, by Theorem 2.8, the restrictions of these functions to the given intervals are injective and have differentiable inverse functions, the **inverse trigonometric functions**. The usual notation for these inverse functions is

$$\begin{aligned} \arcsin &:= \bigl(\sin \,\big|\, (-\pi/2, \pi/2)\bigr)^{-1} &&: (-1, 1) \to (-\pi/2, \pi/2)\ , \\ \arccos &:= \bigl(\cos \,\big|\, (0, \pi)\bigr)^{-1} &&: (-1, 1) \to (0, \pi)\ , \\ \arctan &:= \bigl(\tan \,\big|\, (-\pi/2, \pi/2)\bigr)^{-1} &&: \mathbb{R} \to (-\pi/2, \pi/2)\ , \\ \mathrm{arccot} &:= \bigl(\cot \,\big|\, (0, \pi)\bigr)^{-1} &&: \mathbb{R} \to (0, \pi)\ . \end{aligned}$$

To calculate the derivatives of the inverse trigonometric functions we use Theorem 2.8(iii). For the arcsine function this gives

$$\arcsin' x = \frac{1}{\sin' y} = \frac{1}{\cos y} = \frac{1}{\sqrt{1 - \sin^2 y}} = \frac{1}{\sqrt{1 - x^2}}\ , \qquad x \in (-1, 1)\ ,$$

where we have set $y := \arcsin x$ and used $x = \sin y$. Similarly, for the arctangent function,

$$\arctan' x = \frac{1}{\tan' y} = \frac{1}{1 + \tan^2 y} = \frac{1}{1 + x^2}\ , \qquad x \in \mathbb{R}\ ,$$

where $y \in (-\pi/2, \pi/2)$ is determined by $x = \tan y$.

The derivatives of the arccosine and arccotangent functions can be calculated the same way and, summarizing, we have

$$\begin{aligned} \arcsin' x &= \frac{1}{\sqrt{1-x^2}} , \quad & \arccos' x &= \frac{-1}{\sqrt{1-x^2}} , \quad & x &\in (-1,1) , \\ \arctan' x &= \frac{1}{1+x^2} , \quad & \operatorname{arccot}' x &= \frac{-1}{1+x^2} , \quad & x &\in \mathbb{R} . \end{aligned} \tag{2.3}$$

In particular, (2.3) shows that the inverse trigonometric functions are smooth.

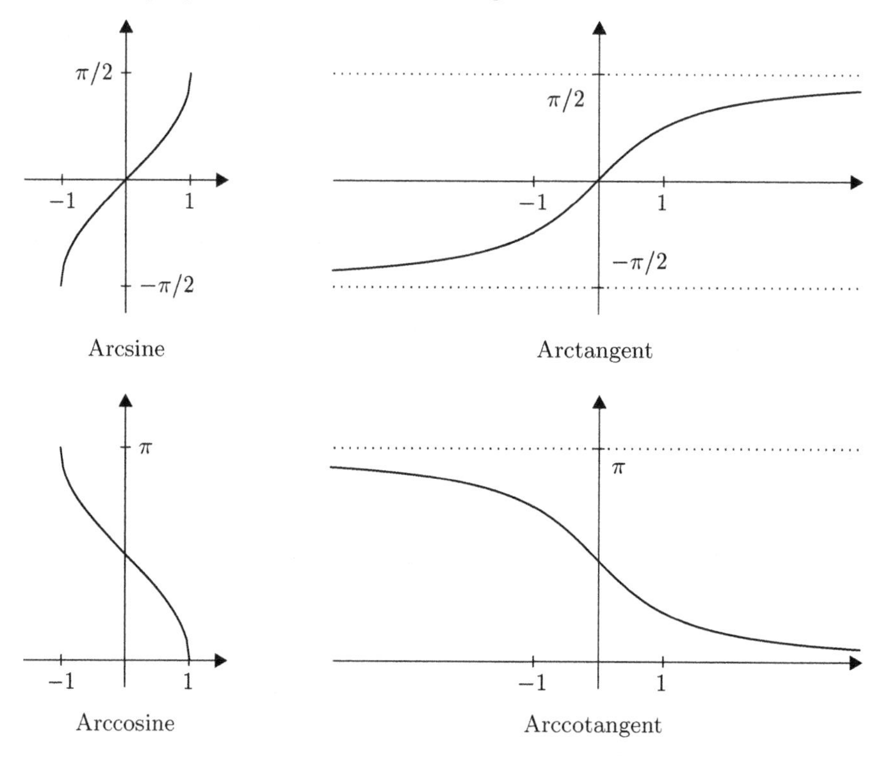

Convexity and Differentiability

We have already seen that monotonicity is a very useful concept for the investigation of real functions. It is therefore not surprising that differentiable functions with monotone derivatives have 'particularly nice' properties.

Let C be a convex subset of a vector space V. Then $f : C \to \mathbb{R}$ is **convex** if

$$f\big((1-t)x + ty\big) \le (1-t)f(x) + tf(y) , \qquad x, y \in C , \quad t \in (0,1) ,$$

and **strictly convex** if

$$f\big((1-t)x + ty\big) < (1-t)f(x) + tf(y) , \qquad x, y \in C , \quad x \ne y , \quad t \in (0,1) .$$

Finally we say f is **concave** (or **strictly concave**) if $-f$ is convex (or strictly convex).

2.11 Remarks (a) Clearly, f is concave (or strictly concave) if and only if

$$f\big((1-t)x+ty\big) \geq (1-t)f(x)+tf(y) \ ,$$

(or

$$f\big((1-t)x+ty\big) > (1-t)f(x)+tf(y) \ )$$

for all $x, y \in C$ such that $x \neq y$ and for all $t \in (0,1)$.

(b) Suppose that $I \subseteq \mathbb{R}$ is a perfect interval[1] and $f : I \to \mathbb{R}$. Then the following are equivalent:

(i) f is convex.

(ii) For all $a, b \in I$ such that $a < b$,

$$f(x) \leq f(a) + \frac{f(b)-f(a)}{b-a}(x-a) \ , \qquad a < x < b \ .$$

(iii) For all $a, b \in I$ such that $a < b$,

$$\frac{f(x)-f(a)}{x-a} \leq \frac{f(b)-f(a)}{b-a} \leq \frac{f(b)-f(x)}{b-x} \ , \qquad a < x < b \ .$$

(iv) For all $a, b \in I$ such that $a < b$,

$$\frac{f(x)-f(a)}{x-a} \leq \frac{f(b)-f(x)}{b-x} \ , \qquad a < x < b \ .$$

If, in (ii)–(iv), the symbol $\leq$ is replaced throughout by $<$, then these statements are equivalent to f being strictly convex. Analogous statements hold for concave and strictly concave functions.

Proof '(i)⇒(ii)' Let $a, b \in I$ with $a < b$ and $x \in (a, b)$. Set $t := (x-a)/(b-a)$. Then $t \in (0,1)$ and $(1-t)a + tb = x$, and so from the convexity of f we get

$$f(x) \leq \Big(1 - \frac{x-a}{b-a}\Big) f(a) + \frac{x-a}{b-a} f(b) = f(a) + \frac{f(b)-f(a)}{b-a}(x-a) \ .$$

'(ii)⇒(iii)' The first inequality in (iii) follows directly from (ii). From (ii) we also have

$$f(b) - f(x) \geq f(b) - f(a) - \frac{f(b)-f(a)}{b-a}(x-a) = \frac{f(b)-f(a)}{b-a}(b-x) \ ,$$

which implies

$$\frac{f(b)-f(a)}{b-a} \leq \frac{f(b)-f(x)}{b-x} \ .$$

'(iii)⇒(iv)' This implication is clear.

[1] By Remark III.4.9(c), a subset I of $\mathbb{R}$ is convex if and only if I is an interval.

'(iv)⇒(i)' Let $a, b \in I$ with $a < b$ and $t \in (0,1)$. Then $x := (1-t)a + tb$ is in (a,b), and so

$$\frac{f(x)-f(a)}{x-a} \leq \frac{f(b)-f(x)}{b-x} \ .$$

This inequality implies

$$f\big((1-t)a+tb\big) = f(x) \leq \frac{b-x}{b-a} f(a) + \frac{x-a}{b-a} f(b) = (1-t)f(a) + tf(b) \ .$$

Hence f is convex.

The remaining claims can be proved similarly. ■

(c) Viewed geometrically, (ii) of (b) says that the graph of $f \,|\, (a,b)$ lies below the secant line through $\big(a, f(a)\big)$ and $\big(b, f(b)\big)$.

The inequality (iii) of (b) says that the slope of the secant line through $\big(a, f(a)\big)$ and $\big(x, f(x)\big)$ is smaller than the slope of the secant line through $\big(a, f(a)\big)$ and $\big(b, f(b)\big)$, which is itself smaller than the slope of the secant line through $\big(x, f(x)\big)$ and $\big(b, f(b)\big)$.

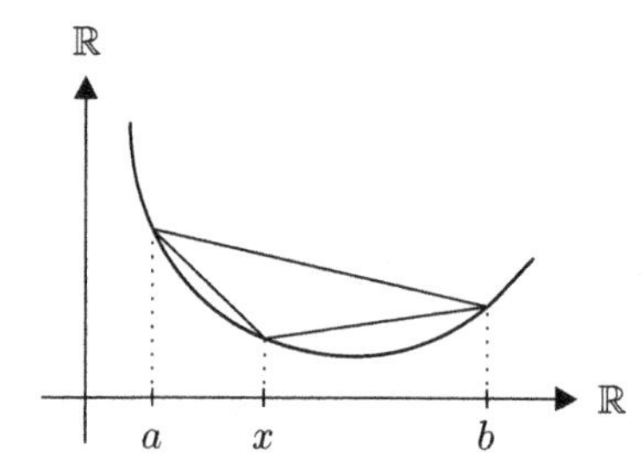

2.12 Theorem (a characterization of convex functions) *Suppose that I is a perfect interval and $f : I \to \mathbb{R}$ is differentiable. Then f is (strictly) convex if and only if f' is (strictly) increasing.*

Proof '⇒' Suppose that f is strictly convex and $a, b \in I$ are such that $a < b$. Then we can choose a strictly decreasing sequence (x_n) in (a,b) and a strictly increasing sequence (y_n) in (a,b) such that $\lim x_n = a$, $\lim y_n = b$ and $x_0 < y_0$. From Remark 2.11(b) we have

$$\frac{f(x_n)-f(a)}{x_n-a} < \frac{f(x_0)-f(a)}{x_0-a} < \frac{f(y_0)-f(x_0)}{y_0-x_0} < \frac{f(y_n)-f(y_0)}{y_n-y_0} < \frac{f(y_n)-f(b)}{y_n-b} \ .$$

Taking the limit $n \to \infty$ we get

$$f'(a) \leq \frac{f(x_0)-f(a)}{x_0-a} < \frac{f(y_0)-f(x_0)}{y_0-x_0} \leq f'(b) \ .$$

Thus f' is strictly increasing.

If f is convex, then the above discussion shows that the inequality $a < b$ implies $f'(a) \leq f'(b)$. That is, f' is increasing.

'⇐' Let $a, b, x \in I$ be such that $a < x < b$. By the mean value theorem there are $\xi \in (a,x)$ and $\eta \in (x,b)$ such that

$$\frac{f(x)-f(a)}{x-a} = f'(\xi) \quad \text{and} \quad \frac{f(b)-f(x)}{b-x} = f'(\eta) \ .$$

The claim then follows from Remark 2.11(b) and the (strict) monotonicity of f'. ■

2.13 Corollary *Suppose that I is a perfect interval and $f: I \to \mathbb{R}$ is twice differentiable.*

(i) *f is convex if and only if $f''(x) \geq 0$ for all $x \in I$.*

(ii) *If $f''(x) > 0$ for all $x \in I$, then f is strictly convex.*

Proof This follows directly from Theorems 2.5 and 2.12. ∎

2.14 Examples **(a)** $\exp: \mathbb{R} \to \mathbb{R}$ is strictly increasing and strictly convex.

(b) $\log: (0, \infty) \to \mathbb{R}$ is strictly increasing and strictly concave.

(c) For $\alpha \in \mathbb{R}$, let $f_\alpha: (0,\infty) \to \mathbb{R}$, $x \mapsto x^\alpha$ be the power function. Then f_α is

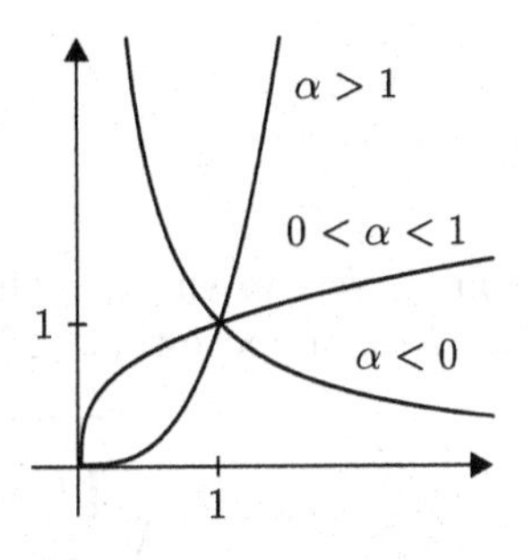

strictly increasing and strictly convex if $\alpha > 1$,

strictly increasing and strictly concave if $0 < \alpha < 1$,

strictly decreasing and strictly convex if $\alpha < 0$.

Proof All the claims follow from Theorem 2.5, Corollary 2.13 and the relationships

(α) $\exp = \exp' = \exp'' > 0$,

(β) $\log'(x) = x^{-1} > 0$, $\log''(x) = -x^{-2} < 0$ for all $x \in (0, \infty)$,

(γ) $f'_\alpha = \alpha f_{\alpha-1}$, $f''_\alpha = \alpha(\alpha-1) f_{\alpha-2}$, $f_\beta(x) > 0$ for all $x \in (0,\infty)$ and $\beta \in \mathbb{R}$. ∎

The Inequalities of Young, Hölder and Minkowski

The concavity of the logarithm function and the monotonicity of the exponential function make possible an elegant proof of one of the fundamental inequalities of analysis, the *Young inequality*. For this proof, it is useful to introduce the following notation: For $p \in (1, \infty)$, we say that $p' := p/(p-1)$ is the **Hölder conjugate** of p. It is determined by the equation[2]

$$\frac{1}{p} + \frac{1}{p'} = 1 . \tag{2.4}$$

2.15 Theorem (Young inequality) *For $p \in (1, \infty)$,*

$$\xi\eta \leq \frac{1}{p}\xi^p + \frac{1}{p'}\eta^{p'} , \qquad \xi, \eta \in \mathbb{R}^+ .$$

[2]From (2.4) it follows, in particular, that $(p')' = p$.

Proof It suffices to consider only the case $\xi, \eta \in (0, \infty)$. The concavity of the logarithm function and (2.4) imply the inequality

$$\log\Bigl(\frac{\xi^p}{p} + \frac{\eta^{p'}}{p'}\Bigr) \geq \frac{1}{p}\log \xi^p + \frac{1}{p'}\log \eta^{p'} = \log \xi + \log \eta = \log \xi\eta \ .$$

Since the exponential function is increasing and $\exp \log x = x$ for all x, the claimed inequality follows. ∎

2.16 Applications **(a)** (inequality of the geometric and arithmetic means[3]) *For* $n \in \mathbb{N}^\times$ *and* $x_j \in \mathbb{R}^+$, $1 \leq j \leq n$,

$$\sqrt[n]{\prod_{j=1}^n x_j} \leq \frac{1}{n}\sum_{j=1}^n x_j \ . \tag{2.5}$$

Proof We can suppose that all the x_j are positive. For $n = 1$, (2.5) is clearly true. Now suppose that (2.5) holds for some $n \in \mathbb{N}^\times$. Then

$$\sqrt[n+1]{\prod_{j=1}^{n+1} x_j} \leq \Bigl(\frac{1}{n}\sum_{j=1}^n x_j\Bigr)^{n/(n+1)} (x_{n+1})^{1/(n+1)} \ .$$

To the right side of this inequality we apply the Young inequality with

$$\xi := \Bigl(\frac{1}{n}\sum_{j=1}^n x_j\Bigr)^{n/(n+1)} \ , \qquad \eta := (x_{n+1})^{1/(n+1)} \ , \qquad p := 1 + \frac{1}{n} \ .$$

Then

$$\xi\eta \leq \frac{1}{p}\xi^p + \frac{1}{p'}\eta^{p'} = \frac{1}{n+1}\sum_{j=1}^n x_j + \frac{1}{n+1}x_{n+1} = \frac{1}{n+1}\sum_{j=1}^{n+1} x_j \ ,$$

which proves the claim. ∎

(b) (Hölder inequality) *For* $p \in (1, \infty)$ *and* $x = (x_1, \ldots, x_n) \in \mathbb{K}^n$, *define*

$$|x|_p := \Bigl(\sum_{j=1}^n |x_j|^p\Bigr)^{1/p} \ .$$

Then[4]

$$\sum_{j=1}^n |x_j y_j| \leq |x|_p \, |y|_{p'} \ , \qquad x, y \in \mathbb{K}^n \ .$$

[3] See Exercise I.10.10.

[4] In the case $p = p' = 2$ this reduces to the Cauchy-Schwarz inequality.

Proof It suffices to consider the case $x \neq 0$ and $y \neq 0$. From the Young inequality we have

$$\frac{|x_j|}{|x|_p}\frac{|y_j|}{|y|_{p'}} \le \frac{1}{p}\frac{|x_j|^p}{|x|_p^p} + \frac{1}{p'}\frac{|y_j|^{p'}}{|y|_{p'}^{p'}} , \qquad 1 \le j \le n .$$

Summing these inequalities over j yields

$$\frac{\sum_{j=1}^n |x_j y_j|}{|x|_p \, |y|_{p'}} \le \frac{1}{p} + \frac{1}{p'} = 1 ,$$

and so the claim follows. ■

(c) (Minkowski inequality) *For all* $p \in (1,\infty)$,

$$|x+y|_p \le |x|_p + |y|_p , \qquad x,y \in \mathbb{K}^n .$$

Proof From the triangle inequality we have

$$\begin{aligned}
|x+y|_p^p &= \sum_{j=1}^n |x_j+y_j|^{p-1}\,|x_j+y_j| \\
&\le \sum_{j=1}^n |x_j+y_j|^{p-1}\,|x_j| + \sum_{j=1}^n |x_j+y_j|^{p-1}\,|y_j| .
\end{aligned}$$

Thus the Hölder inequality implies

$$\begin{aligned}
|x+y|_p^p &\le |x|_p \Bigl(\sum_{j=1}^n |x_j+y_j|^p\Bigr)^{1/p'} + |y|_p \Bigl(\sum_{j=1}^n |x_j+y_j|^p\Bigr)^{1/p'} \\
&= \bigl(|x|_p + |y|_p\bigr)\,|x+y|_p^{p/p'} .
\end{aligned}$$

If $x+y=0$, then the claim is trivially true. Otherwise we can divide both sides of this inequality by $|x+y|_p^{p/p'}$ to get $|x+y|_p^{p-p/p'} \le |x|_p + |y|_p$. Since $p - p/p' = 1$, this proves the claim. ■

One immediate consequence of these inequalities is that $|\cdot|_p$ is a norm on $\mathbb{K}^n$:

2.17 Proposition *For each* $p \in [1,\infty]$, $|\cdot|_p$ *is a norm on* $\mathbb{K}^n$.

Proof We have already seen in Section II.3 that $|\cdot|_1$ and $|\cdot|_\infty$ are norms on $\mathbb{K}^n$. If $p \in (1,\infty)$, then the Minkowski inequality is exactly the triangle inequality for $|\cdot|_p$. The validity of the remaining norm axioms is clear. ■

The Mean Value Theorem for Vector Valued Functions

For the remainder of this section, a and b are real numbers such that $a < b$.

Let $f : [a, b] \to \mathbb{R}$ be differentiable. Then, by the mean value theorem, there is some $\xi \in (a, b)$ such that

$$f(b) - f(a) = f'(\xi)(b - a) . \tag{2.6}$$

Even when $\xi \in (a, b)$ is not known, (2.6) provides a relationship between the change of f on $[a, b]$ and the values of f' on the interval. For a differentiable function from $[a, b]$ into a normed vector space E, (2.6) is, in general, not true, as we know from Remark 2.9(a).

In applications it is often not necessary to know the exact change of f on $[a, b]$. Sometimes is suffices to know a suitable bound. For real valued functions, we get such a bound directly from (2.6):

$$|f(b) - f(a)| \leq \sup_{\xi \in (a,b)} |f'(\xi)| \, (b - a) .$$

The next theorem proves an analogous statement for vector valued functions.

2.18 Theorem (mean value theorem for vector valued functions) *Suppose that E is a normed vector space and $f \in C([a, b], E)$ is differentiable on (a, b). Then*

$$\|f(b) - f(a)\| \leq \sup_{t \in (a,b)} \|f'(t)\| \, (b - a) .$$

Proof It suffices to consider the case when f' is bounded, and so there is some $\alpha > 0$ such that $\alpha > \|f'(t)\|$ for all $t \in (a, b)$. Fix $\varepsilon \in (0, b - a)$ and set

$$S := \{ \sigma \in [a + \varepsilon, b] \ ; \ \|f(\sigma) - f(a + \varepsilon)\| \leq \alpha(\sigma - a - \varepsilon) \} .$$

The set S is not empty since $a + \varepsilon$ is in S. Because of the continuity of f, S is closed (see Example III.2.22(c)), and, by the Heine-Borel theorem, is compact. Hence $s := \max S$ exists and and is in the interval $[a + \varepsilon, b]$.

Suppose that $s < b$. Then, for all $t \in (s, b)$,

$$\|f(t) - f(a + \varepsilon)\| \leq \|f(t) - f(s)\| + \alpha(s - a - \varepsilon) . \tag{2.7}$$

Since f differentiable on $[a + \varepsilon, b)$, we have

$$\frac{\|f(t) - f(s)\|}{t - s} \to \|f'(s)\| \quad (t \to s) .$$

By the definition of α, there is some $\delta \in (0, b - s)$ such that

$$\|f(t) - f(s)\| \leq \alpha(t - s) , \qquad 0 < t - s < \delta .$$

Together with (2.7), this implies

$$\|f(t) - f(a+\varepsilon)\| \le \alpha(t - a - \varepsilon) , \qquad s < t < s + \delta ,$$

which contradicts the maximality of s. Thus we have $s = b$ and hence also

$$\|f(b) - f(a+\varepsilon)\| \le \alpha(b - a - \varepsilon)$$

for each upper bound α of $\{\, \|f'(t)\| \ ; \ t \in (a,b) \,\}$, that is,

$$\|f(b) - f(a+\varepsilon)\| \le \sup_{t\in(a,b)} \|f'(t)\| \, (b - a - \varepsilon) .$$

Since this holds for each $\varepsilon \in (0, b-a)$, the claim follows by taking the limit $\varepsilon \to 0$ and using the continuity of f. ■

2.19 Corollary *Suppose that I is a compact perfect interval, E is a normed vector space, and $f \in C(I, E)$ is differentiable on $\mathring{I}$. If f' is bounded on $\mathring{I}$, then f is Lipschitz continuous. In particular, any function in $C^1(I, E)$ is Lipschitz continuous.*

Proof The first claim follows directly from Theorem 2.18. If $f \in C^1(I, E)$, then, by Corollary III.3.7, the derivative f' is bounded on I. ■

The Second Mean Value Theorem

The following is often called the *second mean value theorem.*

2.20 Proposition *Suppose that $f, g \in C\big([a,b], \mathbb{R}\big)$ are differentiable on (a,b), and $g'(x) \neq 0$ for all $x \in (a,b)$. Then there is some ξ in (a,b) such that*

$$\frac{f(b) - f(a)}{g(b) - g(a)} = \frac{f'(\xi)}{g'(\xi)} .$$

Proof Rolle's theorem implies $g(a) \neq g(b)$, and so

$$h(x) := f(x) - \frac{f(b) - f(a)}{g(b) - g(a)} \big(g(x) - g(a)\big)$$

is well defined for all $x \in [a,b]$. Moreover, h is continuous on $[a,b]$, differentiable on (a,b) and satisfies $h(a) = h(b)$. By Rolle's theorem, there is some $\xi \in (a,b)$ such that $h'(\xi) = 0$. Since

$$h'(x) = f'(x) - \frac{f(b) - f(a)}{g(b) - g(a)} g'(x) , \qquad x \in (a,b) ,$$

the claim follows. ■

L'Hospital's Rule

As an application of the second mean value theorem, we derive a rule which is useful for calculating the limit of a quotient of two functions when the limit has the form '0/0' or '∞/∞'.

2.21 Proposition *Suppose that $f, g\colon (a,b) \to \mathbb{R}$ are differentiable and $g(x) \neq 0$ for all $x \in (a,b)$. Suppose also that either*

(i) $\lim_{x\to a} f(x) = \lim_{x\to a} g(x) = 0$

or

(ii) $\lim_{x\to a} g(x) = \pm\infty$.

Then

$$\lim_{x\to a} \frac{f(x)}{g(x)} = \lim_{x\to a} \frac{f'(x)}{g'(x)} ,$$

if the limit on the right exists in $\bar{\mathbb{R}}$.

Proof Suppose that $\alpha := \lim_{x\to a} f'(x)/g'(x) < \infty$. Then for each pair α_0 and α_1 such that $\alpha < \alpha_1 < \alpha_0$, there is some $x_1 \in (a,b)$ such that $f'(x)/g'(x) < \alpha_1$ for all $a < x < x_1$. By the second mean value theorem, for all $x, y \in (a, x_1)$ such that $x < y$, there is some $\xi \in (x,y)$ such that

$$\frac{f(y)-f(x)}{g(y)-g(x)} = \frac{f'(\xi)}{g'(\xi)} .$$

Since $\xi < y < x_1$, it follows that

$$\frac{f(y)-f(x)}{g(y)-g(x)} < \alpha_1 < \alpha_0 , \qquad x, y \in (a, x_1) . \tag{2.8}$$

Suppose that (i) is satisfied. Then taking the limit $x \to a$ in (2.8) yields

$$f(y)/g(y) \leq \alpha_1 < \alpha_0 , \qquad a < y < x_1 . \tag{2.9}$$

If instead $\lim_{x\to a} g(x) = \infty$, then there is, for each $y \in (a, x_1)$, some $x_2 \in (a, y)$ such that $g(x) > 1 \vee g(y)$ for all $a < x < x_2$. From (2.8) we get

$$\frac{f(x)}{g(x)} < \alpha_1 - \alpha_1 \frac{g(y)}{g(x)} + \frac{f(y)}{g(x)} , \qquad a < x < x_2 .$$

As $x \to a$, the right side of this inequality converges to α_1. Thus there is some $x_3 \in (a, x_2)$ such that $f(x)/g(x) < \alpha_0$ for all $a < x < x_3$. Since α_0 was chosen arbitrarily close to α, it follows from this and (2.9) that, in either case,

$$\overline{\lim_{x\to a}} f(x)/g(x) \leq \alpha .$$

If $\alpha \in (-\infty, \infty]$, then, by a similar argument,

$$\varliminf_{x \to a} f(x)/g(x) \geq \alpha \ .$$

Thus we have proved the claim if either (i) or $\lim_{x \to a} g(x) = \infty$ holds. The case $\lim_{x \to a} g(x) = -\infty$ can be proved similarly and is left to the reader. ∎

2.22 Remark Of course, the corresponding statements for the left limit $x \to b$ also are true. In addition, the proof of Proposition 2.21 remains valid in the cases $a = -\infty$ and $b = \infty$. ∎

2.23 Examples (a) For all $m, n \in \mathbb{N}^\times$ and $a \in \mathbb{R}$,

$$\lim_{x \to a} \frac{x^n - a^n}{x^m - a^m} = \lim_{x \to a} \frac{nx^{n-1}}{mx^{m-1}} = \frac{n}{m} a^{n-m} \ .$$

(b) Let $n \geq 2$ and $a_k \in [0, \infty)$ for $1 \leq k \leq n$. Then, from Proposition 2.21, we have

$$\begin{aligned} \lim_{x \to \infty} \left(\sqrt[n]{x^n + a_1 x^{n-1} + \cdots + a_n} - x \right) &= \lim_{y \to 0+} \frac{\sqrt[n]{1 + a_1 y + \cdots + a_n y^n} - 1}{y} \\ &= \lim_{y \to 0+} \frac{1}{n} \frac{a_1 + 2a_2 y + \cdots + n a_n y^{n-1}}{(1 + a_1 y + \cdots + a_n y^n)^{1-1/n}} \\ &= \frac{a_1}{n} \ . \end{aligned}$$

(c) For all $a \in \mathbb{R}^\times$,

$$\lim_{x \to 0} \frac{1 - \cos(ax)}{1 - \cos x} = a^2 \ .$$

Proof Using l'Hospital's rule twice we get

$$\lim_{x \to 0} \frac{1 - \cos(ax)}{1 - \cos x} = \lim_{x \to 0} \frac{a \sin(ax)}{\sin x} = \lim_{x \to 0} \frac{a^2 \cos(ax)}{\cos x} = a^2 \ ,$$

and so the claim is proved. ∎

Exercises

1 Let $f : \mathbb{R} \to \mathbb{R}$ be defined by

$$f(x) := \begin{cases} e^{-1/x^2} \ , & x \neq 0 \ , \\ 0 \ , & x = 0 \ . \end{cases}$$

Show that f is in $C^\infty(\mathbb{R})$, that f has an isolated[5] global minimum at $x = 0$, and that $f^{(k)}(0) = 0$ for $k \in \mathbb{N}$.

[5]A function f has an **isolated minimum** at x_0 if there is a neighborhood U of x_0 such that $f(x) > f(x_0)$ for all $x \in U \backslash \{x_0\}$.

2 Let f be the function of Example 1.17 and $F(x) := e^e f\big(f(1) - f(1-x)\big)$, $x \in \mathbb{R}$. Show that

$$F(x) = \begin{cases} 0 , & x \le 0 , \\ 1 , & x \ge 1 , \end{cases}$$

and F is strictly increasing on $[0,1]$.

3 Let $-\infty < a < b < \infty$ and $f \in C\big([a,b],\mathbb{R}\big)$ be differentiable on $(a,b]$. Show that, if $\lim_{x\to a} f'(x)$ exists, then f is in $C^1\big([a,b],\mathbb{R}\big)$ and $f'(a) = \lim_{x\to a} f'(x)$. (Hint: Use the mean value theorem.)

4 Let $a > 0$ and $f \in C^2\big([-a,a],\mathbb{R}\big)$ be even. Show that there is some $g \in C^1\big([0,a^2],\mathbb{R}\big)$ such that $f(x) = g(x^2)$ for all $x \in [-a,a]$. In particular, $f'(0) = 0$. (Hint: Exercise 3.)

5 The functions

$$\cosh^{-1} : [1,\infty) \to \mathbb{R}^+ \quad \text{and} \quad \sinh^{-1} : \mathbb{R} \to \mathbb{R}$$

are called the **inverse hyperbolic cosine** and **inverse hyperbolic sine** functions.

(a) Show that $\cosh^{-1}$ and $\sinh^{-1}$ are well defined and that

$$\begin{aligned} \cosh^{-1}(x) &= \log\big(x + \sqrt{x^2 - 1}\big) , \qquad & x \ge 1 , \\ \sinh^{-1}(x) &= \log\big(x + \sqrt{x^2 + 1}\big) , \qquad & x \in \mathbb{R} . \end{aligned}$$

(b) Calculate the first two derivatives of these functions.

(c) Discuss the convexity and concavity of cosh, sinh, $\cosh^{-1}$ and $\sinh^{-1}$. Sketch the graphs of these functions.

6 Let $n \in \mathbb{N}^\times$ and $f(x) := 1 + x + x^2/2! + \cdots + x^n/n!$ for all $x \in \mathbb{R}$. Show that the equation $f(x) = 0$ has exactly one real solution if n is odd, and no real solutions if n is even.

7 Suppose that $-\infty \le a < b \le \infty$ and $f : (a,b) \to \mathbb{R}$ is continuous. A point $x_0 \in (a,b)$ is called an **inflection point** of f if there are a_0, b_0 such that $a \le a_0 < x_o < b_0 \le b$ and $f\,|\,(a_0,x_0)$ is convex and $f\,|\,(x_0,b_0)$ is concave, or $f\,|\,(a_0,x_0)$ is concave and $f\,|\,(x_0,b_0)$ is convex.

(a) Let $f : \mathbb{R} \to \mathbb{R}$ be defined by

$$f(x) := \begin{cases} \sqrt{x} , & x \ge 0 , \\ -\sqrt{-x} , & x < 0 . \end{cases}$$

Show that f has an inflection point at 0.

(b) Suppose that $f : (a,b) \to \mathbb{R}$ is twice differentiable and has an inflection point at x_0. Show that $f''(x_0) = 0$.

(c) Show that the function $f : \mathbb{R} \to \mathbb{R}$, $x \mapsto x^4$ has no inflection points.

(d) Suppose that $f \in C^3\big((a,b),\mathbb{R}\big)$, $f''(x_0) = 0$ and $f'''(x_0) \ne 0$. Show that f has an inflection point at x_0.

8 Determine all the inflection points of f when $f(x)$ is given by

$$\text{(a) } x^2 - 1/x\ , \quad x > 0\ , \qquad \text{(b) } \sin x + \cos x\ , \quad x \in \mathbb{R}\ , \qquad \text{(c) } x^x\ , \quad x > 0\ .$$

9 Show that $f\colon (0,\infty) \to \mathbb{R},\ x \mapsto (1+1/x)^x$ is strictly increasing.

10 Suppose that $f \in C([a,b],\mathbb{R})$ is differentiable on (a,b) and satisfies $f(a) \geq 0$ and $f'(x) \geq 0,\ x \in (a,b)$. Prove that $f(x) \geq 0$ for all $x \in [a,b]$.

11 Show that

$$1 - 1/x \leq \log x \leq x - 1\ , \qquad x > 0\ .$$

12 Suppose that I is a perfect interval and $f, g\colon I \to \mathbb{R}$ is convex. Prove or disprove the following:

(a) $f \vee g$ is convex.

(b) $\alpha f + \beta g$ is convex for $\alpha, \beta \in \mathbb{R}$.

(c) fg is convex.

13 Suppose that I is a perfect interval, $f \in C(I,\mathbb{R})$ is convex, and $g\colon f(I) \to \mathbb{R}$ is convex and increasing. Show that $g \circ f\colon I \to \mathbb{R}$ is also convex. Find conditions on f and g which ensure that $g \circ f$ is strictly convex.

14 Suppose that $-\infty < a < b < \infty$ and $f\colon [a,b] \to \mathbb{R}$ is convex. Prove or disprove the following:

(a) For each $x \in (a,b)$, the limits $\partial_\pm f(x)$ exist and $\partial_- f(x) \leq \partial_+ f(x)$.

(b) $f\,|\,(a,b)$ is continuous.

(c) f is continuous.

15 Let I be a perfect interval, $a \in I$ and $n \in \mathbb{N}$. Suppose that $\varphi, \psi \in C^n(I,\mathbb{R})$ are such that

$$\varphi^{(k)}(a) = \psi^{(k)}(a) = 0\ , \qquad 0 \leq k \leq n\ ,$$

$\varphi^{(n+1)}$ and $\psi^{(n+1)}$ exist on $\mathring{I}$, and

$$\psi^{(k)}(x) \neq 0\ , \qquad x \in \mathring{I}\backslash\{a\}\ , \quad 0 \leq k \leq n+1\ .$$

Show that, for each $x \in I\backslash\{a\}$, there is some $\xi \in (x \wedge a, x \vee a)$ such that

$$\frac{\varphi(x)}{\psi(x)} = \frac{\varphi^{(n+1)}(\xi)}{\psi^{(n+1)}(\xi)}\ .$$

16 Suppose that I is an interval, $f \in C^{n-1}(I,\mathbb{R})$ and $f^{(n)}$ exists on $\mathring{I}$ for some $n \geq 2$. Let $x_0 < x_1 < \cdots < x_n$ be zeros of f. Show that there is some $\xi \in (x_0, x_n)$ such that $f^{(n)}(\xi) = 0$ (generalized Rolle's theorem).

17 Calculate the following limits:

$$\text{(a) } \lim_{x\to\infty}(1+2x)^{1/3x}\ , \quad \text{(b) } \lim_{x\to 1}\frac{1+\cos\pi x}{x^2-2x+1}\ , \quad \text{(c) } \lim_{x\to 0}\frac{\log\cos 3x}{\log\cos 2x}\ , \quad \text{(d) } \lim_{x\to 0}\Bigl(\frac{1}{\sin^2 x}-\frac{1}{x^2}\Bigr)\ .$$

18 Suppose that (x_k) and (y_k) are sequences in $\mathbb{K}^{\mathbb{N}}$, $1 < p < \infty$, and p' is the Hölder conjugate of p. Prove the **Hölder inequality for series**,

$$\Bigl|\sum_{k=0}^{\infty} x_k y_k\Bigr| \le \sum_{k=0}^{\infty} |x_k y_k| \le \Bigl(\sum_{k=0}^{\infty} |x_k|^p\Bigr)^{1/p} \Bigl(\sum_{k=0}^{\infty} |y_k|^{p'}\Bigr)^{1/p'} ,$$

and the **Minkowski inequality for series**,

$$\Bigl(\sum_{k=0}^{\infty} |x_k + y_k|^p\Bigr)^{1/p} \le \Bigl(\sum_{k=0}^{\infty} |x_k|^p\Bigr)^{1/p} + \Bigl(\sum_{k=0}^{\infty} |y_k|^p\Bigr)^{1/p} .$$

19 For $x = (x_k) \in \mathbb{K}^{\mathbb{N}}$, define

$$\|x\|_p := \begin{cases} \bigl(\sum_{k=0}^{\infty} |x_k|^p\bigr)^{1/p} , & 1 \le p < \infty , \\ \sup_{k\in\mathbb{N}} |x_k| , & p = \infty , \end{cases}$$

and

$$\ell_p := \{ x \in \mathbb{K}^{\mathbb{N}} \ ; \ \|x\|_p < \infty \} , \qquad 1 \le p \le \infty .$$

Show the following:

(a) $\ell_p := (\ell_p, \|\cdot\|_p)$ is a normed subspace of $\mathbb{K}^{\mathbb{N}}$.

(b) $\ell_\infty = B(\mathbb{N}, \mathbb{K})$.

(c) For $1 \le p \le q \le \infty$, we have $\ell_p \subseteq \ell_q$ and $\|x\|_q \le \|x\|_p$, $x \in \ell_p$.

20 Suppose that I is a perfect interval and $f : I \to \mathbb{R}$ is convex. Show that

$$f(\lambda_1 x_1 + \cdots + \lambda_n x_n) \le \lambda_1 f(x_1) + \cdots + \lambda_n f(x_n)$$

for all $x_1, \ldots, x_n \in I$ and $\lambda_1, \ldots, \lambda_n \in \mathbb{R}^+$ satisfying $\lambda_1 + \cdots + \lambda_n = 1$.

21 Suppose that I is an interval and $f : I \to \mathbb{R}$ is convex. Show that, if $x \in I$ and $h > 0$ satisfy $x + 2h \in I$, then $\triangle_h^2 f(x) \ge 0$. Here $\triangle_h$ is the divided difference operator of length h (see Section I.12). (Hint: Think geometrically.)

3 Taylor's Theorem

In this chapter we have already seen that, for a function f, being differentiable at a point a and being approximately linear at a are the same property. In addition, using the mean value theorem, we showed how f' determines certain local and global properties of f.

This suggests an obvious question: Can any smooth function $f: D \to E$ be approximated by a polynomial near a point $a \in D$? If so, what does this approximation say about the local and global properties of f? What can we say about f if we know sufficiently many, or even all, of its derivatives at a?

The Landau Symbol

Let X and E be normed vector spaces, D a nonempty subset of X and $f: D \to E$. In order to describe the behavior of f at a point $a \in \overline{D}$, we use the **Landau symbol** o. If $\alpha \geq 0$, we say 'f has a zero of order α at a' and write

$$f(x) = o(\|x - a\|^\alpha) \quad (x \to a) ,$$

if

$$\lim_{x \to a} \frac{f(x)}{\|x - a\|^\alpha} = 0 .$$

3.1 Remarks (a) A function f has a zero of order α at a if and only if, for each $\varepsilon > 0$, there is a neighborhood U of a in D such that

$$\|f(x)\| \leq \varepsilon \|x - a\|^\alpha , \qquad x \in U .$$

Proof This follows from Remark III.2.23(a). ■

(b) Suppose that $X = \mathbb{K}$ and $r: D \to E$ is continuous at $a \in D$. Then

$$f: D \to E , \quad x \mapsto \bigl(r(x) - r(a)\bigr)(x - a)$$

has a zero of order 1 at a, that is, $f(x) = o(|x - a|)$ $(x \to a)$.

(c) Let $X = \mathbb{K}$. Then $f: D \to E$ is differentiable at $a \in D$ if and only if there is some (unique) $m_a \in E$ such that

$$f(x) - f(a) - m_a(x - a) = o(|x - a|) \quad (x \to a) .$$

Proof This is a consequence of (b) and Theorem 1.1(iii) ■

(d) The function $f: (0, \infty) \to \mathbb{R}$, $x \mapsto e^{-1/x}$ has a zero of infinite order at 0, that is,

$$f(x) = o(|x|^\alpha) \quad (x \to 0) , \qquad \alpha > 0 .$$

Proof Let $\alpha > 0$. Then

$$\lim_{x \to 0} e^{-1/x}/x^\alpha = \lim_{y \to \infty} y^\alpha e^{-y} = 0$$

by Proposition III.6.5(iii). ■

The function $g \colon D \to E$ **approximates** the function $f \colon D \to E$ **with order** α at a if

$$f(x) = g(x) + o(\|x - a\|^\alpha) \quad (x \to a) ,$$

that is, if $f - g$ has a zero of order α at a.

As well as the symbol o, we will occasionally use the **Landau symbol** O. If $a \in \overline{D}$ and $\alpha \geq 0$, then we write

$$f(x) = O(\|x - a\|^\alpha) \quad (x \to a) ,$$

if there are $r > 0$ and $K > 0$ such that

$$\|f(x)\| \leq K \|x - a\|^\alpha , \qquad x \in \mathbb{B}(a, r) \cap D .$$

In this case we say that 'f increases with order at most α at a'. In particular, $f(x) = O(1)$ $(x \to a)$ implies that f is bounded in some neighborhood of a.

For the remainder of this section, $E := (E, \|\cdot\|)$ is a Banach space, D is a perfect subset of $\mathbb{K}$ and f is a function from D to E.

Taylor's Formula

We first investigate a necessary condition so that a 'polynomial' $p = \sum_{k=0}^n c_k X^k$ with coefficients[1] c_k in E can be chosen so that p approximates the function f with order n at $a \in D$.

Consider first the special case when $f = \sum_{k=0}^n b_k X^k$ is itself a polynomial with coefficients in E. From the binomial theorem we get

$$f(x) = \sum_{k=0}^n b_k (x - a + a)^k = \sum_{k=0}^n b_k \sum_{j=0}^k \binom{k}{j} (x - a)^j a^{k-j} , \qquad x \in \mathbb{K} .$$

Thus f can be written as a polynomial in $x - a$ in the form

$$f(x) = \sum_{k=0}^n c_k (x - a)^k , \qquad x \in \mathbb{K} ,$$

[1]A polynomial with coefficients in E is a formal expression of the form $\sum_{k=0}^n c_k X^k$ with $c_k \in E$. If the 'indeterminate' X is replaced by a field element $x \in \mathbb{K}$, we get a well determined element $p(x) := \sum_{k=0}^n c_k x^k \in E$. Thus the 'polynomial function' $\mathbb{K} \to E$, $x \mapsto p(x)$ is well defined. The set of polynomials with coefficients in E does not, in general, form a ring! See Section I.8.

where

$$c_k := \sum_{\ell=k}^{n} b_\ell \binom{\ell}{k} a^{\ell-k} , \qquad k = 0, \ldots, n .$$

Clearly we have

$$f(a) = \sum_{\ell=0}^{n} b_\ell a^\ell = c_0 , \qquad f'(a) = \sum_{\ell=1}^{n} b_\ell \ell a^{\ell-1} = c_1 ,$$

$$f''(a) = \sum_{\ell=2}^{n} b_\ell \ell(\ell-1) a^{\ell-2} = 2c_2 , \quad f'''(a) = \sum_{\ell=3}^{n} b_\ell \ell(\ell-1)(\ell-2) a^{\ell-3} = 6c_3 .$$

A simple induction argument shows that $f^{(k)}(a) = k!\, c_k$ for $k = 0, \ldots, n$ and so

$$f(x) = \sum_{k=0}^{n} \frac{f^{(k)}(a)}{k!}(x-a)^k , \qquad x \in \mathbb{K} . \tag{3.1}$$

Thus we have a simple expression for the coefficients of f when it is written as a polynomial in $x - a$ (see Proposition I.8.16).

The following fundamental theorem shows that any function $f \in C^n(D, E)$ has a polynomial approximation with order n at any point $a \in D$.

3.2 Theorem (Taylor's theorem) *Let D be convex and $n \in \mathbb{N}^\times$. Then, for each $f \in C^n(D, E)$ and $a \in D$, there is a function $R_n(f, a) \in C(D, E)$ such that*

$$f(x) = \sum_{k=0}^{n} \frac{f^{(k)}(a)}{k!}(x-a)^k + R_n(f,a)(x) , \qquad x \in D .$$

The **remainder function** *$R_n(f, a)$ satisfies*

$$\|R_n(f,a)(x)\| \le \frac{1}{(n-1)!} \sup_{0<t<1} \left\| f^{(n)}\bigl(a + t(x-a)\bigr) - f^{(n)}(a) \right\| |x-a|^n$$

for all $x \in D$.

Proof For $f \in C^n(D, E)$ and $a \in D$, define

$$R_n(f,a)(x) := f(x) - \sum_{k=0}^{n} \frac{f^{(k)}(a)}{k!}(x-a)^k , \qquad x \in D .$$

Then it suffices to prove the claimed bound on $R_n(f, a)$. For $t \in [0, 1]$ and a fixed $x \in D$, set

$$h(t) := f(x) - \sum_{k=0}^{n-1} \frac{f^{(k)}\bigl(a + t(x-a)\bigr)}{k!}(x-a)^k(1-t)^k - \frac{f^{(n)}(a)}{n!}(x-a)^n(1-t)^n .$$

Then $h(0) = R_n(f,a)(x)$ and $h(1) = 0$, and also

$$h'(t) = \left[f^{(n)}(a) - f^{(n)}\big(a + t(x-a)\big)\right]\frac{(1-t)^{n-1}}{(n-1)!}(x-a)^n \;, \qquad t \in (0,1) \;.$$

The mean value theorem for vector valued functions (Theorem 2.18) implies

$$\begin{aligned}\|R_n(f,a)(x)\| = \|h(1) - h(0)\| &\le \sup_{0<t<1} \|h'(t)\| \\ &\le \sup_{0<t<1} \frac{\left\|f^{(n)}\big(a + t(x-a)\big) - f^{(n)}(a)\right\|}{(n-1)!}\,|x-a|^n \;,\end{aligned}$$

which was to be proved. ■

3.3 Corollary (qualitative version of Taylor's theorem) *With the hypotheses of Theorem* 3.2,

$$f(x) = \sum_{k=0}^{n} \frac{f^{(k)}(a)}{k!}(x-a)^k + o(|x-a|^n) \quad (x \to a) \;.$$

Taylor Polynomials and Taylor Series

For $n \in \mathbb{N}$, $f \in C^n(D,E)$ and $a \in D$,

$$\mathcal{T}_n(f,a) := \sum_{k=0}^{n} \frac{f^{(k)}(a)}{k!}(X-a)^k$$

is a polynomial[2] of degree $\le n$ with coefficients in E, the nth **Taylor polynomial of f at a**, and

$$R_n(f,a) := f - \mathcal{T}_n(f,a)$$

is the nth **remainder function of f at a**. Corollary 3.3 shows that the Taylor polynomial $\mathcal{T}_n(f,a)$ approximates the function f with order higher than n at a.

Now let $E := \mathbb{K}$ and $f \in C^\infty(D) := C^\infty(D,\mathbb{K})$. Then the formal expression

$$\mathcal{T}(f,a) := \sum_k \frac{f^{(k)}(a)}{k!}(X-a)^k$$

is called the **Taylor series of f at a**, and by the **radius of convergence** of $\mathcal{T}(f,a)$ we mean the radius of convergence of the power series

$$\sum_{k=0}^{\infty} \frac{f^{(k)}(a)}{k!} X^k \;.$$

[2]In agreement with the conventions of Section I.8, we identify polynomials with coefficients in E with the corresponding E-valued polynomial functions.

If $\mathcal{T}(f,a)$ has positive radius of convergence ρ, then

$$\mathcal{T}(f,a) : \mathbb{B}(a,\rho) \to \mathbb{K} , \quad x \mapsto \sum_{k=0}^{\infty} \frac{f^{(k)}(a)}{k!}(x-a)^k \tag{3.2}$$

is a well defined function.

Just as for other power series, we identify the Taylor series $\mathcal{T}(f,a)$ with the function (3.2). Note that this identification is meaningful only on $\mathbb{B}(a,\rho)$.

3.4 Remarks Let D be open in $\mathbb{K}$ and $a \in D$.

(a) The Taylor polynomial $\mathcal{T}_n(f,a)$ is the nth 'partial sum' of the Taylor series $\mathcal{T}(f,a)$. If the radius of convergence of the Taylor series is positive, then f is approximated by $\mathcal{T}_n(f,a)$ with order n at a. This does not mean that f equals its Taylor series in some neighborhood U of a.

Proof The function f from Example 1.17 is smooth and satisfies $f^{(k)}(0) = 0$ for $k \in \mathbb{N}$. Consequently $\mathcal{T}(f,0) = 0 \neq f$. ■

(b) Suppose that the Taylor series $\mathcal{T}(f,a)$ for a function f has positive radius of convergence ρ. Then the function f is equal to its Taylor series in some neighborhood $U \subseteq \mathbb{B}(a,\rho) \cap D$ of a if and only if $\lim_{n\to\infty} R_n(f,a)(x) = 0$ for all $x \in U$. ■

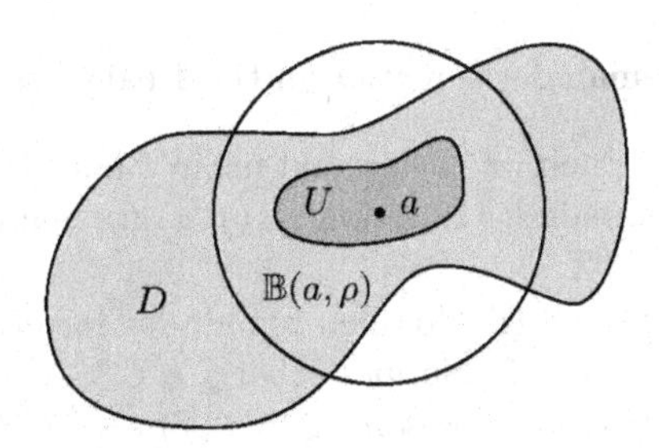

3.5 Example (series representation of the logarithm) For $|z| < 1/2$,

$$\log(1+z) = \sum_{k=1}^{\infty} \frac{(-1)^{k-1}}{k} z^k = z - \frac{z^2}{2} + \frac{z^3}{3} - \frac{z^4}{4} + - \cdots$$

Proof Let $f(z) := \log(1+z)$ for $z \in \mathbb{C}\backslash\{-1\}$. Then, by induction,

$$f^{(n)}(z) = (-1)^{n-1} \frac{(n-1)!}{(1+z)^n} , \qquad n \in \mathbb{N}^\times , \quad z \in \mathbb{C}\backslash(-\infty,-1] .$$

From Theorem 3.2 we get the formula

$$\log(1+z) = \sum_{k=1}^{n} \frac{(-1)^{k-1}}{k} z^k + R_n(f,0)(z) , \qquad n \in \mathbb{N}^\times , \quad z \in \mathbb{C}\backslash(-\infty,-1] ,$$

where the remainder function $R_n(f,0)$ satisfies

$$|R_n(f,0)(z)| \leq \sup_{0<t<1} \left| \frac{1}{(1+tz)^n} - 1 \right| |z|^n , \qquad n \in \mathbb{N}^\times , \quad z \in \mathbb{C}\backslash(-\infty,-1] .$$

For $|z| \leq 1/2$ and $t \in [0,1]$, we have the inequality $|1+tz| \geq 1-|z| \geq 1/2$, and hence

$$\left|\frac{1}{(1+tz)^n} - 1\right| \leq \frac{1}{(1-|z|)^n} + 1 \leq 2^n + 1 \leq 2^{n+1} , \qquad n \in \mathbb{N}^\times .$$

Thus

$$|R_n(f,0)(z)| \leq 2(2\,|z|)^n \to 0 \quad (n \to \infty)$$

for all $|z| < 1/2$, and the claim follows. ∎

3.6 Remark For $a_k := (-1)^{k-1}/k$, $k \in \mathbb{N}^\times$,

$$\lim_{k\to\infty} \frac{|a_{k+1}|}{|a_k|} = \lim_{k\to\infty} \frac{k}{k+1} = 1 .$$

So the power series $\sum(-1)^{k-1}X^k/k$ has radius of convergence 1. Thus the question arises whether this series equals the function $z \mapsto \log(1+z)$ on all of $\mathbb{B}_{\mathbb{C}}$. For complex z, we answer this question in Section V.3. For the real case, see Application 3.9(d). ∎

The Remainder Function in the Real Case

With the help of the second mean value theorem we can derive a further estimate of the remainder function $R_n(f,a)$ for the case $\mathbb{K} = \mathbb{R}$ and $E = \mathbb{R}$.

3.7 Theorem (Schlömilch remainder formula) *Let I be a perfect interval, $a \in I$, $p > 0$ and $n \in \mathbb{N}$. Suppose that $f \in C^n(I,\mathbb{R})$ and $f^{(n+1)}$ exists on $\mathring{I}$. Then, for each $x \in I \backslash \{a\}$, there is some $\xi := \xi(x) \in (x \wedge a, x \vee a)$ such that*

$$R_n(f,a)(x) = \frac{f^{(n+1)}(\xi)}{pn!}\Bigl(\frac{x-\xi}{x-a}\Bigr)^{n-p+1}(x-a)^{n+1} .$$

Proof Fix $x \in I$ and set $J := (x \wedge a, x \vee a)$. Define

$$g(t) := \sum_{k=0}^{n} \frac{f^{(k)}(t)}{k!}(x-t)^k , \quad h(t) := (x-t)^p , \qquad t \in J .$$

Obviously $g, h \in C(\overline{J},\mathbb{R})$ and both functions are differentiable on J with

$$g'(t) = f^{(n+1)}(t)\frac{(x-t)^n}{n!} , \quad h'(t) = -p(x-t)^{p-1} , \qquad t \in J .$$

By the second mean value theorem (Proposition 2.20), there is some ξ in J such that

$$g(x) - g(a) = \frac{g'(\xi)}{h'(\xi)}\bigl(h(x) - h(a)\bigr) .$$

Since $R_n(f,a)(x) = g(x) - g(a)$ and $h(x) - h(a) = -(x-a)^p$, the claim follows. ∎

3.8 Corollary (Lagrange and Cauchy remainder formulas) *With the hypotheses of Theorem 3.7,*

$$R_n(f,a)(x) = \frac{f^{(n+1)}(\xi)}{(n+1)!}(x-a)^{n+1} \qquad \text{(Lagrange)}$$

and

$$R_n(f,a)(x) = \frac{f^{(n+1)}(\xi)}{n!}\Big(\frac{x-\xi}{x-a}\Big)^n (x-a)^{n+1} \qquad \text{(Cauchy)}\ .$$

Proof Set $p = n+1$ and $p = 1$ respectively in Theorem 3.7. ■

3.9 Applications **(a)** (sufficient condition for local extrema) *Let I be a perfect interval and $f \in C^n(I,\mathbb{R})$ for some $n \geq 1$. Suppose that there is some $a \in \mathring{I}$ such that*

$$f'(a) = f''(a) = \cdots = f^{(n-1)}(a) = 0 \quad \text{and} \quad f^{(n)}(a) \neq 0\ .$$

(i) *If n is odd, then f has no extremum at a.*

(ii) *If n is even, then f has a local minimum at a if $f^{(n)}(a) > 0$, and f has a local maximum at a if $f^{(n)}(a) < 0$.*

Proof The hypotheses and Taylor's theorem (Corollary 3.3) imply

$$f(x) = f(a) + \Big[\frac{f^{(n)}(a)}{n!} + \frac{o(|x-a|^n)}{(x-a)^n}\Big](x-a)^n \quad (x \to a)\ . \tag{3.3}$$

Set $\gamma := |f^{(n)}(a)|/(2n!) > 0$. Then, by Remark 3.1, there is some $\delta > 0$ such that

$$\frac{|o(|x-a|^n)|}{|x-a|^n} \leq \gamma\ , \qquad x \in I \cap (a-\delta, a+\delta)\ . \tag{3.4}$$

We now distinguish the following cases:

(α) Let n be odd and $f^{(n)}(a) > 0$. Then, from (3.3) and (3.4), we have

$$f(x) \geq f(a) + \gamma(x-a)^n\ , \qquad x \in (a, a+\delta) \cap I\ ,$$

and

$$f(x) \leq f(a) - \gamma(a-x)^n\ , \qquad x \in (a-\delta, a) \cap I\ .$$

Thus f cannot have an extremum at a.

(β) If n is odd and $f^{(n)}(a) < 0$, then, from (3.3) and (3.4), we have

$$f(x) \leq f(a) - \gamma(x-a)^n\ , \qquad x \in (a, a+\delta) \cap I\ ,$$

and

$$f(x) \geq f(a) + \gamma(a-x)^n\ , \qquad x \in (a-\delta, a) \cap I\ .$$

So, in this case too, f cannot have an extremum at a.

(γ) Let n be even and $f^{(n)}(a) > 0$. Then

$$f(x) \geq f(a) + \gamma(x-a)^n \;, \qquad x \in (a-\delta, a+\delta) \cap I \;.$$

Hence f has a local minimum at a.

(δ) Finally, if n is even and $f^{(n)}(a) < 0$, then

$$f(x) \leq f(a) - \gamma(x-a)^n \;, \qquad x \in (a-\delta, a+\delta) \cap I \;,$$

that is, f has a local maximum at a. ■

Remark The above conditions are sufficient, but not necessary. For example, the function

$$f(x) := \begin{cases} e^{-1/x} \;, & x > 0 \;, \\ 0 \;, & x \leq 0 \;, \end{cases}$$

has a global minimum at 0, even though, by Example 1.17, f is smooth with $f^{(n)}(0) = 0$ for all $n \in \mathbb{N}$. ■

(b) (a characterization of the exponential function[3]) *Suppose that $a, b \in \mathbb{C}$, the function $f : \mathbb{C} \to \mathbb{C}$ is differentiable, and*

$$f'(z) = bf(z) \;, \quad z \in \mathbb{C} \;, \qquad f(0) = a \;. \tag{3.5}$$

Then $f(z) = ae^{bz}$ for all $z \in \mathbb{C}$.

Proof From $f' = bf$ and Corollary 1.2 we see that $f \in C^\infty(\mathbb{C})$ and $f^{(k)} = b^k f$ for all $k \in \mathbb{N}$. If, in addition, $f(0) = a$, then

$$\sum_k \frac{f^{(k)}(0)}{k!} X^k = f(0) \sum_k \frac{b^k}{k!} X^k = a \sum_k \frac{b^k}{k!} X^k \;.$$

Since this power series has infinite radius of convergence by Proposition II.9.4, we have

$$\mathcal{T}(f,0)(z) = ae^{bz} \;, \qquad z \in \mathbb{C} \;.$$

To complete the proof, we need to prove that this Taylor series equals f on $\mathbb{C}$, that is, we must show that the remainder converges to 0. For $z \in \mathbb{C}$, we estimate $R_n(f,0)(z)$ using Theorem 3.2 as follows:

$$\begin{aligned} |R_n(f,0)(z)| &\leq \sup_{0<t<1} \left| f^{(n)}(tz) - f^{(n)}(0) \right| \frac{|z|^n}{(n-1)!} = \frac{|b|^n \, |z|^n}{(n-1)!} \sup_{0<t<1} |f(tz) - a| \\ &\leq M \, |bz| \, \frac{|bz|^{n-1}}{(n-1)!} \;, \end{aligned}$$

where $M > 0$ has been chosen so that $|f(w) - a| \leq M$ for all $w \in \bar{\mathbb{B}}(0, |z|)$. From Example II.4.2(c), it now follows that $R_n(f,0)(z) \to 0$ for all $n \to \infty$. ■

[3]This says that $z \mapsto ae^{bz}$ is the unique solution of the differential equation $f' = bf$ satisfying the initial condition $f(0) = a$. Differential equations are studied in detail in Chapter IX.

(c) (a characterization of the exponential function by its functional equation) *If $f : \mathbb{C} \to \mathbb{C}$ satisfies*

$$f(z+w) = f(z)f(w) , \qquad z, w \in \mathbb{C} , \tag{3.6}$$

and

$$\lim_{z\to 0} \frac{f(z)-1}{z} = b \quad \text{for some } b \in \mathbb{C} , \tag{3.7}$$

then $f(z) = e^{bz}$ for all $z \in \mathbb{C}$.

Proof From (3.6), we have $f(0) = f(0)^2$, and so $f(0) \in \{0,1\}$. But, if $f(0) = 0$, then, by (3.6), $f(z) = f(z)f(0) = 0$ for all $z \in \mathbb{C}$, that is, $f = 0$. This contradicts (3.7), and so we must have $f(0) = 1$.

For each $z \in \mathbb{C}$, (3.6) implies

$$\frac{f(z+h) - f(z)}{h} = f(z)\frac{f(h)-1}{h} , \qquad h \in \mathbb{C}^\times .$$

Thus, by (3.7), f is differentiable and satisfies $f' = bf$. The claim now follows from (b). ∎

(d) (Taylor series for the real logarithm function) *For all $x \in (-1,1]$,*[4]

$$\log(1+x) = \sum_{k=1}^\infty \frac{(-1)^{k-1}}{k} x^k = x - \frac{x^2}{2} + \frac{x^3}{3} - \frac{x^4}{4} + - \cdots$$

In particular, the alternating harmonic series has the value $\log 2$.

Proof As in the proof of Example 3.5, let $f(x) := \log(1+x)$ for $x > -1$. Then

$$f^{(n+1)}(x) = (-1)^n \frac{n!}{(1+x)^{n+1}} , \qquad x > -1 ,$$

and

$$\log(1+x) = \sum_{k=1}^n \frac{(-1)^{k-1}}{k} x^k + R_n(f,0)(x) , \qquad x > -1 .$$

To estimate the remainder on $[0,1]$ we use the Lagrange formula (Corollary 3.8) and find, for each $x \in [0,1]$, some $\xi_n \in (0,x)$ such that

$$|R_n(f,0)(x)| = \left| \frac{x^{n+1}}{(n+1)(1+\xi_n)^{n+1}} \right| \le \frac{1}{n+1} , \qquad n \in \mathbb{N} .$$

Thus the Taylor series equals $\log(1+x)$ on $[0,1]$.

---

[4]See also Example 3.5.

For the case $x \in (-1,0)$, we use the Cauchy formula for $R_n(f,0)$ (Corollary 3.8). Thus, for each $n \in \mathbb{N}$, there is some $\eta_n \in (x,0)$ such that

$$|R_n(f,0)(x)| \le \left|\frac{1}{1+\eta_n}\right| \left|\frac{x-\eta_n}{1+\eta_n}\right|^n .$$

For $\eta \in (x,0)$, we have $\eta - x = \eta + 1 - (x+1)$ and so

$$\left|\frac{x-\eta}{1+\eta}\right| = \frac{\eta - x}{1+\eta} = 1 - \frac{1+x}{1+\eta} < -x < 1 .$$

Thus $\lim_n R_n(f,0)(x) = 0$ for all $x \in (-1,0)$. The second claim is obtained by setting $x = 1$ in the Taylor series. ■

(e) (a characterization of convex functions) *Let I be a perfect interval and $f \in C^2(I,\mathbb{R})$. Then f is convex if and only if the graph of f is above all of its tangent lines, that is, if*

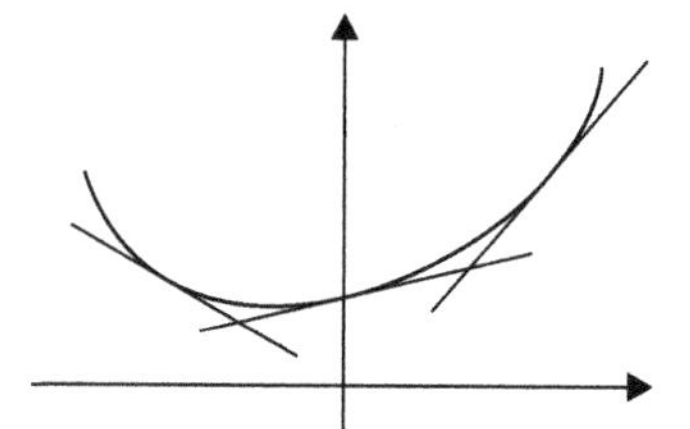

$$f(y) \ge f(x) + f'(x)(y-x)$$

for all $x, y \in I$.

Proof Let $x, y \in I$. Then, by Theorem 3.2 and the Lagrange formula for $R_1(f,x)$, there is some $\xi \in I$ such that

$$f(y) = f(x) + f'(x)(y-x) + \frac{f''(\xi)}{2}(y-x)^2 .$$

Since we know from Corollary 2.13 that f is convex if and only if $f''(\xi) \ge 0$ for all $\xi \in I$, this proves the claim. ■

Polynomial Interpolation

Let $-\infty < a \le x_0 < x_1 < \cdots < x_m \le b < \infty$ and $f : [a,b] \to \mathbb{R}$. In Proposition I.12.9 we showed that there is a unique interpolation polynomial $p = p_m[f; x_0, \ldots, x_m]$ of degree $\le m$ such that $f(x_j) = p(x_j)$ for all $1 \le j \le m$. We are now in a position to estimate the error function

$$r_m[f; x_0, \ldots, x_m] := f - p_m[f; x_0, \ldots, x_m]$$

on the interval $I := [a,b]$, assuming that f is sufficiently smooth.

3.10 Proposition *Let $m \in \mathbb{N}$ and $f \in C^m(I)$ be such that $f^{(m+1)}$ exists on $\mathring{I}$. Then there is some $\xi := \xi(x, x_0, \ldots, x_m) \in (x \wedge x_0, x \vee x_m)$ such that*

$$r_m[f; x_0, \ldots, x_m](x) = \frac{1}{(m+1)!} f^{(m+1)}(\xi) \prod_{j=0}^{m} (x - x_j) , \qquad x \in I .$$

Proof The claim is clearly true if x is equal to one of the x_j. So we suppose that $x \neq x_j$ for all $0 \leq j \leq m$ and define

$$g(x) := \frac{f(x) - p_m[f; x_0, \ldots, x_m](x)}{\prod_{j=0}^m (x - x_j)} \tag{3.8}$$

and

$$\varphi(t) := f(t) - p_m[f; x_0, \ldots, x_m](t) - g(x) \prod_{j=0}^m (t - x_j) , \qquad t \in I .$$

Then φ is in $C^m(I)$, $\varphi^{(m+1)}$ exists on $\mathring{I}$ and

$$\varphi^{(m+1)}(t) = f^{(m+1)}(t) - (m+1)!\, g(x) , \qquad t \in \mathring{I} . \tag{3.9}$$

Moreover, φ has the $m+2$ distinct zeros, $x, x_0, \ldots, x_m$. By the generalized Rolle's theorem (Exercise 2.16), there is some $\xi \in (x \wedge x_0, x \vee x_m)$ such that $\varphi^{(m+1)}(\xi) = 0$. Thus, by (3.9), $g(x) = f^{(m+1)}(\xi)/(m+1)!$. The claim now follows from (3.8). ■

3.11 Corollary *For $f \in C^{m+1}(I, \mathbb{R})$,*

$$\left| r_m[f; x_0, \ldots, x_m](x) \right| \leq \frac{\|f^{(m+1)}\|_\infty}{(m+1)!} \prod_{j=0}^m |x - x_j| , \qquad x \in I .$$

Higher Order Difference Quotients

By Remark I.12.10(b), we can also express the interpolation polynomial $p_m[f; x_0, \ldots, x_m]$ in the Newtonian form

$$p_m[f; x_0, \ldots, x_m] = \sum_{j=0}^m f[x_0, \ldots, x_j] \prod_{k=0}^{j-1} (X - x_k) . \tag{3.10}$$

Here $f[x_0, \ldots, x_n]$ are the divided differences of f. These can be calculated recursively using the formula

$$f[x_0, \ldots, x_n] = \frac{f[x_0, \ldots, x_{n-1}] - f[x_1, \ldots, x_n]}{x_0 - x_n} , \qquad 1 \leq n \leq m , \tag{3.11}$$

(see Exercise I.12.10). From (3.11) (with $n = 1$) and the mean value theorem, it follows that $f[x_0, x_1] = f'(\xi)$ for some suitable $\xi \in (x_0, x_1)$. The next proposition shows that a similar result holds for divided differences of higher order.

3.12 Proposition *Suppose that $f \in C^m(I, \mathbb{R})$ and $f^{(m+1)}$ exists on $\mathring{I}$. Then there is some $\xi \in (x \wedge x_0, x \vee x_m)$, depending on $x, x_0, \ldots, x_m$, such that*

$$f[x_0, \ldots, x_m, x] = \frac{1}{(m+1)!} f^{(m+1)}(\xi) , \qquad x \in I , \quad x \neq x_j , \quad 0 \leq j \leq m .$$

Proof From (3.10) (with m replaced by $m+1$), we have

$$p_{m+1}[f; x_0, \ldots, x_{m+1}] = p_m[f; x_0, \ldots, x_m] + f[x_0, \ldots, x_{m+1}] \prod_{j=0}^m (X - x_j) .$$

Evaluation at $x = x_{m+1}$ yields

$$f(x_{m+1}) = p_m[f; x_0, \ldots, x_m](x_{m+1}) + f[x_0, \ldots, x_{m+1}] \prod_{j=0}^{m} (x_{m+1} - x_j) .$$

Replacing x_{m+1} in this equation by x yields

$$f(x) - p_m[f; x_0, \ldots, x_m](x) = f[x_0, \ldots, x_m, x] \prod_{j=0}^{m} (x - x_j) . \tag{3.12}$$

By construction, this equation holds for all $x_m < x \leq b$, and it is clearly true for $x = x_m$. Exercise I.12.10(b) shows that the divided differences are symmetric functions of their arguments and so equation (3.12) holds, in fact, for all $x \in I$.

The left side of (3.12) is the error function $r_m[f; x_0, \ldots, x_m]$, so it follows from Proposition 3.10 that

$$\frac{1}{(m+1)!} f^{(m+1)}(\xi) \prod_{j=0}^{m} (x - x_j) = f[x_0, \ldots, x_m, x] \prod_{j=0}^{m} (x - x_j) , \qquad x \in I ,$$

for some $\xi := \xi(x, x_0, \ldots, x_m) \in (x \wedge x_0, x \vee x_m)$. ■

3.13 Corollary *Let $f \in C^{m+1}(I, \mathbb{R})$. Then, for all $x \in I$,*

$$\lim_{(x_0, \ldots, x_m) \to (x, \ldots, x)} f[x_0, \ldots, x_m, x] = \frac{1}{(m+1)!} f^{(m+1)}(x) ,$$

so long as the limit is taken so that no x_j ever equals x.

This corollary shows that the higher order divided differences can be used to approximate higher order derivatives in the same way that the usual difference quotient approximates the first derivative.

A particularly simple situation occurs if the points $x_0, x_1, \ldots, x_n$ are equally spaced, that is,

$$x_j := x_0 + jh , \qquad 0 \leq j \leq n ,$$

for some $h > 0$.

3.14 Proposition *Suppose that $f \in C^{n-1}(I, \mathbb{R})$, $f^{(n)}$ exists on $\mathring{I}$ and $0 < h \leq (b-a)/n$. Then there is some $\xi \in (a, a + nh)$ such that*

$$\triangle_h^n f(a) = f^{(n)}(\xi) .$$

Proof From (3.10), the uniqueness of the interpolation polynomial (Proposition I.12.9) and (I.12.15) we have

$$\frac{1}{n!} \triangle_h^n f(a) = f[x_0, x_1, \ldots, x_n] , \qquad x_0 := a . \tag{3.13}$$

The claim then follows from Proposition 3.12. ■

3.15 Corollary *For all $f \in C^n(I,\mathbb{R})$,*

$$\lim_{h\to 0+} \triangle_h^n f(x) = f^{(n)}(x) , \qquad x \in I .$$

Proof This follows directly from (3.13) and Corollary 3.13. ∎

3.16 Remarks **(a)** Let $f \in C^n(I,\mathbb{R})$. By Proposition I.12.13, the Newton interpolation polynomial for f with equally spaced points $x_j := x_0 + jh \in I$, $0 \le j \le n$, with $h > 0$, has the form

$$N_n[f;x_0;h] = \sum_{j=0}^{n} \frac{\triangle_h^j f(x_0)}{j!} \prod_{k=0}^{j-1} (X - x_k) .$$

From Corollary 3.15, we get

$$\lim_{h\to 0+} N_n[f;x_0;h] = \sum_{j=0}^{n} \frac{f^{(j)}(x_0)}{j!} (X - x_0)^j = \mathcal{T}_n(f,x_0) .$$

This shows that, in the limit $h \to 0+$, the Newton interpolation polynomial becomes the Taylor polynomial.

(b) Corollaries 3.13 and 3.15 are the theoretical foundation of numerical differentiation. For details and further development see, for example, [WS79] and [IK66], as well the literature on numerical analysis. ∎

Exercises

1 Suppose that $\alpha, \beta, R > 0$ and $p \in C^2\big([0,R),\mathbb{R}\big)$ satisfy

$$p(x) \ge \alpha , \quad (1+\beta)\big[p'(x)\big]^2 \le p''(x)p(x) , \qquad x \ge 0 .$$

Show that $R < \infty$ and $p(x) \to \infty$ as $x \to R-$.
(Hint: The function $p^{-\beta}$ is concave. Use a tangent line to $p^{-\beta}$ to provide a lower bound for p (see Application 3.9(e)).

2 Let $a, b \in \mathbb{C}$, $\omega \in \mathbb{R}$, and $f : \mathbb{C} \to \mathbb{C}$ be a twice differentiable function which satisfies

$$f(z) + \omega^2 f''(z) = 0 , \quad z \in \mathbb{C} , \qquad f(0) = a , \qquad f'(0) = \omega b . \tag{3.14}$$

(a) Show that f is in $C^\infty(\mathbb{C})$ and that f is uniquely determined by (3.14). Determine f.

(b) What is f if (3.14) is replaced by

$$f(z) = \omega^2 f''(z) , \quad z \in \mathbb{C} , \qquad f(0) = a , \qquad f'(0) = \omega b \ ?$$

3 Determine the Taylor series of $f : \mathbb{C} \to \mathbb{C}$ at the point 1 when
(a) $f(z) = 3z^3 - 7z^2 + 2z + 4$, (b) $f(z) = e^z$.

4 Calculate the nth Taylor polynomial at 0 of $\log\big((1+x)/(1-x)\big)$, $x \in (-1,1)$.

5 Determine the domains, the extrema and the inflection points of the following real functions:
(a) $x^3/(x-1)^2$, (b) $e^{\sin x}$, (c) $x^n e^{-x^2}$, (d) $x^2/\log x$, (e) $\sqrt[3]{(x-1)^2(x+1)}$,
(f) $\bigl(\log(3x)\bigr)^2/x$.

6 For $a > 1$, show that

$$\frac{1}{1+x} - \frac{1}{1+ax} \le \frac{\sqrt{a}-1}{\sqrt{a}+1} , \qquad x \ge 1 .$$

7 Suppose that $s \in \mathbb{R}$ and $n \in \mathbb{N}$. Show that, for each $x > -1$, there is some $\tau \in (0,1)$ such that

$$(1+x)^s = \sum_{k=0}^{n} \binom{s}{k} x^k + \binom{s}{n+1} \frac{x^{n+1}}{(1+\tau x)^{n+1-s}} . \tag{3.15}$$

Here

$$\binom{\alpha}{k} := \begin{cases} \dfrac{\alpha(\alpha-1)\cdot\cdots\cdot(\alpha-k+1)}{k!} , & k \in \mathbb{N}^\times , \\ 1 , & k = 0 , \end{cases}$$

denotes the **(general) binomial coefficient**[5] for $\alpha \in \mathbb{C}$.

8 Use (3.15) to approximate $\sqrt[5]{30}$. Estimate the error. (Hint: $\sqrt[5]{30} = 2\sqrt[5]{1-(1/16)}$.)

9 Prove the following Taylor series expansion for the general power function:[6]

$$(1+x)^s = \sum_{k=0}^{\infty} \binom{s}{k} x^k , \qquad x \in (-1,1) .$$

(Hint: To estimate the remainder, distinguish the cases $x \in (0,1)$ and $x \in (-1,0)$ (see Application 3.9(d)).

10 Let $X \subseteq \mathbb{K}$ be perfect and $f \in C^n(X,\mathbb{K})$ for some $n \in \mathbb{N}^\times$. A number $x_0 \in X$ is called a **zero of multiplicity** n of f if $f(x_0) = \cdots = f^{(n-1)}(x_0) = 0$ and $f^{(n)}(x_0) \ne 0$. Show that, if X is convex, then f has a zero of multiplicity $\ge n$ at x_0 if and only if there is some $g \in C(X,\mathbb{K})$ such that $f(x) = (x-x_0)^n g(x)$ for all $x \in X$.

11 Let $p = X^n + a_{n-1}X^{n-1} + \cdots + a_0$ be a polynomial with coefficients in $\mathbb{R}$. Prove or disprove that the function $p + \exp$ has a zero of multiplicity $\le n$ in $\mathbb{R}$.

12 Prove the following:

(a) For each $n \in \mathbb{N}$, $T_n(x) := \cos(n \arccos x)$, $x \in \mathbb{R}$, is a polynomial of degree n and

$$T_n(x) = x^n + \binom{n}{2} x^{n-2}(x^2-1) + \binom{n}{4} x^{n-4}(x^2-1)^2 + \cdots .$$

T_n is called the **Chebyschev polynomial** of degree n.

(b) These polynomials satisfy the recursion formula

$$T_{n+1} = 2XT_n - T_{n-1} , \qquad n \in \mathbb{N}^\times .$$

(c) For each $n \in \mathbb{N}^\times$, $T_n = 2^{n-1}X^n + p_n$ for some polynomial p_n with $\deg(p_n) < n$.

[5]See Section V.3.

[6]See also Theorem V.3.10.

(d) T_n has a simple zero, that is, a zero of multiplicity 1, at each of the points

$$x_k := \cos\frac{(2k-1)\pi}{2n}\ , \qquad k = 1,2,\ldots,n\ .$$

(e) T_n has an extremum at each of the points

$$y_k := \cos\frac{k\pi}{n}\ , \qquad k = 0,1,\ldots,n\ ,$$

in $[-1,1]$, and $T_n(y_k) = (-1)^k$.
(Hint: (a) For $\alpha \in [0,\pi]$ and $x := \cos\alpha$, $\cos n\alpha + i\sin n\alpha = \bigl(x + i\sqrt{1-x^2}\bigr)^n$.
(b) Addition theorem for the cosine function.)

13 Define the **normalized Chebyschev polynomials** by $\widetilde{T}_n := 2^{1-n}T_n$ for $n \in \mathbb{N}^\times$ and $\widetilde{T}_0 := T_0$. For $n \in \mathbb{N}$, let $\mathcal{P}_n$ be the set of all polynomials $X^n + a_1X^{n-1} + \cdots + a_n$ with $a_1,\ldots,a_n \in \mathbb{R}$. Let $\|\cdot\|_\infty$ be the maximum norm on $[-1,1]$. Prove the following:[7]

(a) In the set $\mathcal{P}_n$, the normalized Chebyschev polynomial of degree n is the best approximation of zero on the interval $[-1,1]$, that is, for each $n \in \mathbb{N}$,

$$\bigl\|\widetilde{T}_n\bigr\|_\infty \le \|p\|_\infty\ , \qquad p \in \mathcal{P}_n\ .$$

(b) For $-\infty < a < b < \infty$,

$$\max_{a\le x\le b}|p(x)| \ge 2^{1-2n}(b-a)^n\ , \qquad p \in \mathcal{P}_n\ .$$

(c) Let $x_0,\ldots,x_n$ be the zeros of T_{n+1}. Suppose that $f \in C^{n+1}\bigl([-1,1],\mathbb{R}\bigr)$ and p_n is the interpolation polynomial of degree $\le n$ such that $f(x_j) = p(x_j)$ for $j = 0,1,\ldots,n$. Then

$$\|r_n[f;x_0,\ldots,x_n]\|_\infty \le \frac{\|f^{(n+1)}\|_\infty}{2^n(n+1)!}\ .$$

Show that this bound on the error is optimal.

[7] Statement (a) is often called **Chebyschev's theorem**.

4 Iterative Procedures

We have already derived various theorems about the zeros of functions. The most prominent of these are the fundamental theorem of algebra, the intermediate value theorem and Rolle's theorem. These important and deep results have in common that they predict the existence of zeros, but say nothing about how to find these zeros. So we know, for example, that the real function

$$x \mapsto x^5 e^{|x|} - \frac{1}{\pi} x^2 \sin\bigl(\log(x^2)\bigr) + 1998$$

has at least one zero (why?), but we have, so far, no algorithm for finding this zero.[1]

In this section we develop methods to find zeros of functions and to solve equations — at least approximately. The central result of this section, the Banach fixed point theorem, is, in fact, of considerable importance beyond the needs of this section, as we will see in later chapters.

Fixed Points and Contractions

Let $f\colon X \to Y$ be a function between sets X and Y with $X \subseteq Y$. An element $a \in X$ such that $f(a) = a$ is called a **fixed point** of f.

4.1 Remarks (a) Suppose that E is a vector space, $X \subseteq E$ and $f : X \to E$. Set $g(x) := f(x) + x$ for all $x \in X$. Then $a \in X$ is a zero of f if and only if a is a fixed point of g. Thus determining the zeros of f is the same as determining the fixed points of g.

(b) Given a function $f : X \to E$, there are, in general, many possibilities for the function g as in (a). Suppose, for example, that $E = \mathbb{R}$ and 0 is the unique zero of the function $h\colon \mathbb{R} \to \mathbb{R}$. Set $g(x) = h\bigl(f(x)\bigr) + x$ for $x \in X$. Then $a \in X$ is a zero of f if and only if a is a fixed point of g.

(c) Let X be a metric space and a a fixed point of $f\colon X \to Y$. Suppose that $x_0 \in X$ and that the sequence (x_k) can be defined recursively by the 'iteration' $x_{k+1} := f(x_k)$. This means, of course, that $f(x_k)$ is in X for each k. If $x_k \to a$, then we say that 'a can be calculated by the **method of successive approximations**', or 'the method of successive approximations **converges** to a'.

The following graphs illustrate this method in the simplest cases. They show, in particular, that, even if f has only one fixed point, the sequence generated using this method may fail to converge.

---

[1]See Exercise 9.

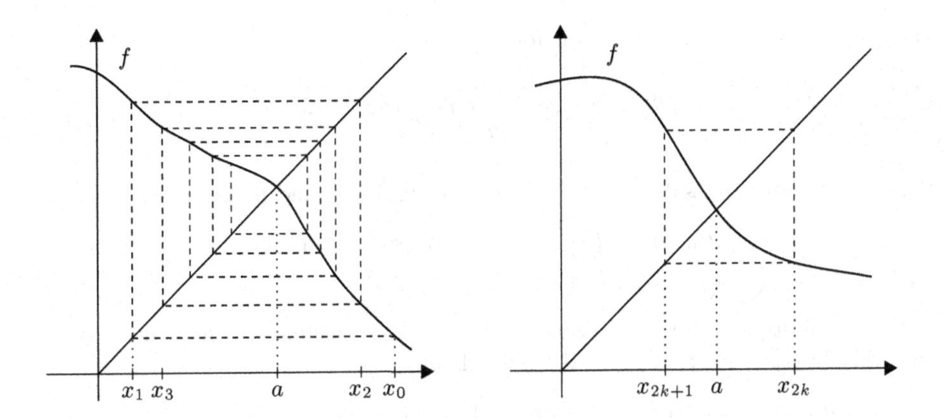

Consider, for example, the function $f: [0,1] \to [0,1]$ defined by $f(x) := 1 - x$. It has exactly one fixed point, namely $a = 1/2$. For the sequence (x_k) defined by $x_{k+1} := f(x_k)$ for all $k \in \mathbb{N}$, we have $x_{2k} = x_0$ and $x_{2k+1} = 1 - x_0$ for all $k \in \mathbb{N}$. Thus (x_k) diverges if $x_0 \neq 1/2$. ■

A function $f: X \to Y$ between two metric spaces X and Y is called a **contraction** if there is some $q \in (0,1)$ such that

$$d\big(f(x), f(x')\big) \le q d(x, x') , \qquad x, x' \in X .$$

In this case, q is called a **contraction constant** of f.

4.2 Remarks (a) A function $f: X \to Y$ is a contraction if and only if f is Lipschitz continuous with Lipschitz constant less than 1.

(b) Let E be a normed vector space and $X \subseteq \mathbb{K}$ convex and perfect. Suppose that $f: X \to E$ is differentiable and $\sup_X \|f'(x)\| < 1$. Then it follows from the mean value theorem for vector valued functions (Theorem 2.18) that f is a contraction. ■

The Banach Fixed Point Theorem

The following theorem is the main result of this section and has innumerable applications, especially in applied mathematics.

4.3 Theorem (contraction theorem, Banach fixed point theorem) *Suppose that X is a complete metric space and $f: X \to X$ is a contraction.*

(i) *f has a unique fixed point a.*

(ii) *For any initial value x_0, the method of successive approximations converges to a.*

(iii) *If q is a contraction constant for f, then*

$$d(x_k, a) \le \frac{q^k}{1-q} d(x_1, x_0) , \qquad k \in \mathbb{N} .$$

Proof (a) (uniqueness) If $a, b \in X$ are two distinct fixed points of f, then

$$d(a,b) = d\big(f(a), f(b)\big) \le q d(a,b) < d(a,b) ,$$

which is not possible.

(b) (existence and convergence) Let $x_0 \in X$. Define the sequence (x_k) recursively by $x_{k+1} := f(x_k)$ for all $k \in \mathbb{N}$. Then

$$d(x_{n+1}, x_n) = d\big(f(x_n), f(x_{n-1})\big) \le q d(x_n, x_{n-1}) , \qquad n \in \mathbb{N}^\times ,$$

and, by induction,

$$d(x_{n+1}, x_n) \le q^{n-k} d(x_{k+1}, x_k) \tag{4.1}$$

for all $n > k \ge 0$. This inequality implies

$$\begin{aligned} d(x_n, x_k) &\le d(x_n, x_{n-1}) + d(x_{n-1}, x_{n-2}) + \cdots + d(x_{k+1}, x_k) \\ &\le (q^{n-k-1} + q^{n-k-2} + \cdots + 1) d(x_{k+1}, x_k) \\ &= \frac{1-q^{n-k}}{1-q} d(x_{k+1}, x_k) \end{aligned} \tag{4.2}$$

for $n > k \ge 0$. Since, by (4.1), $d(x_{k+1}, x_k) \le q^k d(x_1, x_0)$, it follows from (4.2) that

$$d(x_n, x_k) \le \frac{q^k - q^n}{1-q} d(x_1, x_0) \le \frac{q^k}{1-q} d(x_1, x_0) , \qquad n > k \ge 0 . \tag{4.3}$$

This shows that (x_k) is a Cauchy sequence. Since X is a complete metric space, there is some $a \in X$ such that $\lim x_k = a$. By the continuity of f and the definition of the sequence (x_k), a is a fixed point of f.

(c) (error estimate) Since the sequence (x_n) converges to a, we can take the limit $n \to \infty$ in (4.3) to get the claimed estimate of the error (see Example III.1.3(l)). ■

4.4 Remarks **(a)** As well as the a priori error estimate of Theorem 4.3(iii) we have the a posteriori bound

$$d(x_k, a) \le \frac{q}{1-q} d(x_k, x_{k-1}) , \qquad k \in \mathbb{N} .$$

Proof Taking the limit $n \to \infty$ in (4.2) yields

$$d(x_k, a) \le \frac{1}{1-q} d(x_{k+1}, x_k) \le \frac{q}{1-q} d(x_k, x_{k-1}) ,$$

where we have also used (4.1). ■

(b) Suppose that $f : X \to X$ is a contraction with contraction constant q and a is a fixed point of f. Then, for the method of successive approximations, we have a further error estimate:

$$d(x_{k+1}, a) = d\big(f(x_k), f(a)\big) \le q d(x_k, a) , \qquad k \in \mathbb{N} .$$

Thus one says that this iterative process **converges linearly**.

In general, one says that a sequence (x_n) **converges with order** α to a if $\alpha \ge 1$ and there are constants n_0 and c such that

$$d(x_{n+1}, a) \le c\big[d(x_n, a)\big]^\alpha , \qquad n \ge n_0 .$$

If $\alpha = 1$, that is, the convergence is linear, we also require that $c < 1$. In general, a sequence converges faster the higher its order of convergence. For example, for **quadratic convergence**, if $d(x_{n_0}, a) < 1$ and $c \le 1$, then each step doubles the number of correct decimal places in the approximation. In practice, c is often larger than 1 and so this effect is partly diminished.

(c) In applications the following situation often occurs: Suppose that E is a Banach space, X is a *closed* subset of E and $f : X \to E$ is a contraction such that $f(X) \subseteq X$. Then, since X is a complete metric space (see Exercise II.6.4), all the statements of the contraction theorem hold for f.

(d) The hypothesis of (b), that $f(X) \subseteq X$, can be weakened. If there is some initial value $x_0 \in X$ such that the iteration $x_{k+1} = f(x_k)$ can be carried out for all k, then the claims of the contraction theorem hold for *this particular* x_0. ■

With the help of the previous remark we can derive a useful 'local version' of the Banach fixed point theorem.

4.5 Proposition *Let E be a Banach space and $X := \bar{\mathbb{B}}_E(x_0, r)$ with $x_0 \in E$ and $r > 0$. Suppose that $f : X \to E$ is a contraction with contraction constant q which satisfies $\|f(x_0) - x_0\| \le (1-q)r$. Then f has a unique fixed point and the method of successive approximations converges if x_0 is the initial value.*

Proof Since X is a closed subset of a Banach space, X is a complete metric space. Thus, by Remark 4.4(d), it suffices to show that $x_{k+1} = f(x_k)$ remains in X at each iteration. For $x_1 = f(x_0)$ this holds because of the hypothesis $\|f(x_0) - x_0\| = \|x_1 - x_0\| \le (1-q)r$.

Suppose that $x_1, \ldots, x_k \in X$. From (4.3) it follows that

$$\|x_{k+1} - x_0\| \le \frac{1 - q^{k+1}}{1 - q} \|x_1 - x_0\| \le (1 - q^{k+1})r < r .$$

Consequently, x_{k+1} is also in X and the iteration $x_{k+1} = f(x_k)$ is defined for all k. ■

4.6 Examples **(a)** Consider the problem of finding the solution ξ of the equation $\tan x = x$ in the interval $\pi/2 < \xi < 3\pi/2$. Set $I := (\pi/2, 3\pi/2)$ and $f(x) := \tan x$ for $x \in I$ so that $f'(x) = 1 + f^2(x)$. It follows from the mean value theorem that f is not a contraction on any neighborhood of ξ.

To use the contraction theorem we consider instead the inverse function of f, that is, the function

$$g\colon \left[\tan \mid (\pi/2, 3\pi/2)\right]^{-1} : \mathbb{R} \to (\pi/2, 3\pi/2) \ .$$

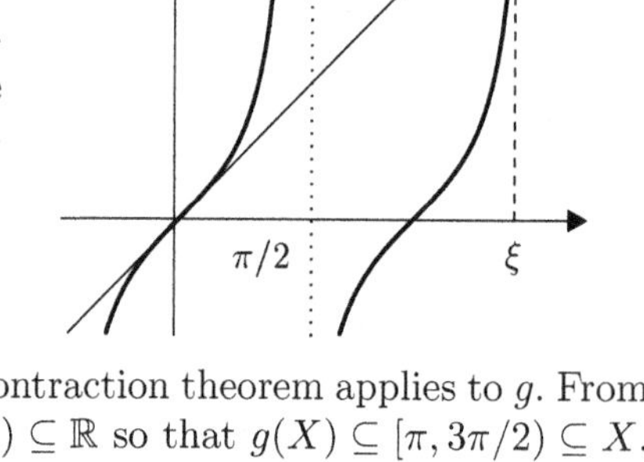

Since the tangent function is strictly increasing on $(\pi/2, 3\pi/2)$, the function g is well defined and $g(x) = \arctan(x) + \pi$. Moreover the fixed point problems for f and g are equivalent, that is, for all $a \in (\pi/2, 3\pi/2)$,

$$a = \tan a \Longleftrightarrow a = \arctan(a) + \pi \ .$$

Since $g'(x) = 1/(1 + x^2)$ (see (IV.2.3)), the contraction theorem applies to g. From the graph we see that $\xi > \pi$. Set $X := [\pi, \infty) \subseteq \mathbb{R}$ so that $g(X) \subseteq [\pi, 3\pi/2) \subseteq X$. Because $|g'(x)| \le 1/(1 + \pi^2) < 1$ for all $x \in X$, g is a contraction on X. Thus it follows from Theorem 4.3 that there is a unique $\xi \in [\pi, 3\pi/2)$ such that $\xi = g(\xi)$, and that, with initial value $x_0 := \pi$, the method of successive approximations converges to ξ.

(b) Let $-\infty < a < b < \infty$ and $f \in C^1\big([a,b], \mathbb{R}\big)$ be a contraction. Suppose that the iterative procedure $x_{k+1} = f(x_k)$ for $x_0 \in [a,b]$ defines an infinite sequence. By Remark 4.4(d), there is a unique $\xi \in [a,b]$ such that $x_k \to \xi$. The convergence is monotone if $f'(x) > 0$ for all $x \in [a,b]$, and *alternating*, that is, ξ is between each pair x_k and x_{k+1}, if $f'(x) < 0$ for all $x \in [a,b]$.

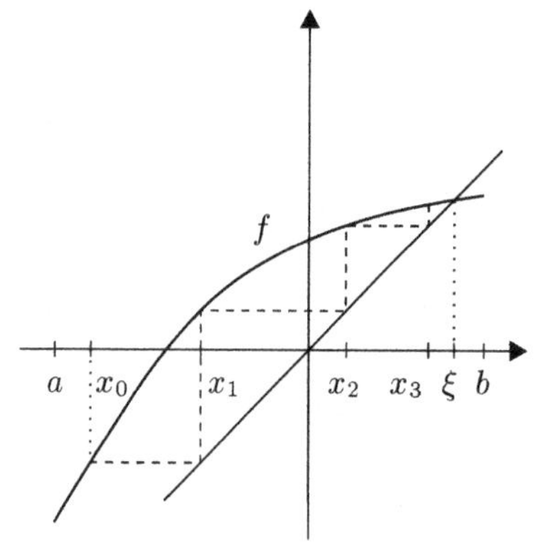

Monotone convergence

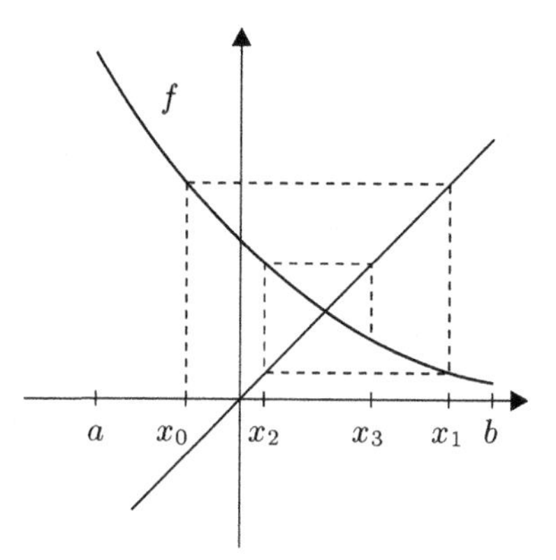

Alternating convergence

Proof By the mean value theorem, for each $k \in \mathbb{N}$, there is some $\eta_k \in (a, b)$ such that

$$x_{k+1} - x_k = f(x_k) - f(x_{k-1}) = f'(\eta_k)(x_k - x_{k-1}) .$$

If $f'(\eta_k) \geq 0$ for all $k \in \mathbb{N}^\times$, then

$$\operatorname{sign}(x_{k+1} - x_k) \in \{\operatorname{sign}(x_k - x_{k-1}), 0\} , \qquad k \in \mathbb{N}^\times ,$$

and so (x_k) is a monotone sequence. If $f'(\eta_k) \leq 0$ for all $k \in \mathbb{N}^\times$, then

$$\operatorname{sign}(x_{k+1} - x_k) \in \{-\operatorname{sign}(x_k - x_{k-1}), 0\} , \qquad k \in \mathbb{N}^\times ,$$

that is, the convergence is alternating. ■

Example 4.6(a) shows most importantly that, for concrete applications, it is important to analyze the problem theoretically first and, if needed, to put the problem in a new form so that the method of successive approximations can be used effectively.

Newton's Method

In the remainder of this section, we consider the following situation:

$$\begin{aligned}&\text{Let } -\infty < a < b < \infty \text{ and } f \in C^2([a,b],\mathbb{R}) \text{ be such that}\\ &f'(x) \neq 0 \text{ for all } x \in [a,b]. \text{ We suppose further that there}\\ &\text{is some } \xi \in (a,b) \text{ such that } f(\xi) = 0.\end{aligned} \tag{4.4}$$

Using linear approximations of f, we will develop a method to approximate the zero ξ of f. Geometrically, ξ is the intersection of the graph of f and the x-axis.

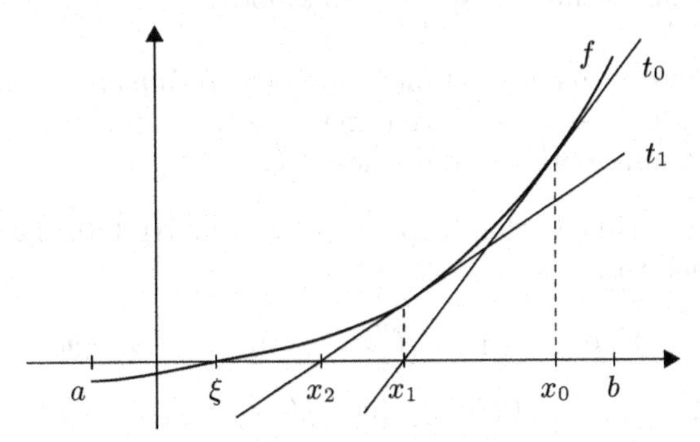

Starting with an initial approximation x_0 of ξ, we replace the graph of f by its tangent line t_0 at the point $(x_0, f(x_0))$. By hypothesis (4.4), f' is nonzero on $[a, b]$, and so the tangent line t_0 intersects the x-axis at a point x_1 which is a new approximation of ξ. The tangent line at the point $(x_0, f(x_0))$ is given by the equation

$$x \mapsto f(x_0) + f'(x_0)(x - x_0) ,$$

and so x_1 can be calculated from the equation $f(x_0) + f'(x_0)(x_1 - x_0) = 0$:

$$x_1 = x_0 - \frac{f(x_0)}{f'(x_0)} .$$

Iteration of this procedure is called **Newton's method**:

$$x_{n+1} = x_n - \frac{f(x_n)}{f'(x_n)} , \quad n \in \mathbb{N} , \qquad x_0 \in [a,b] .$$

The hypotheses in (4.4) do not suffice to ensure the convergence $x_n \to \xi$, as the following graph illustrates:

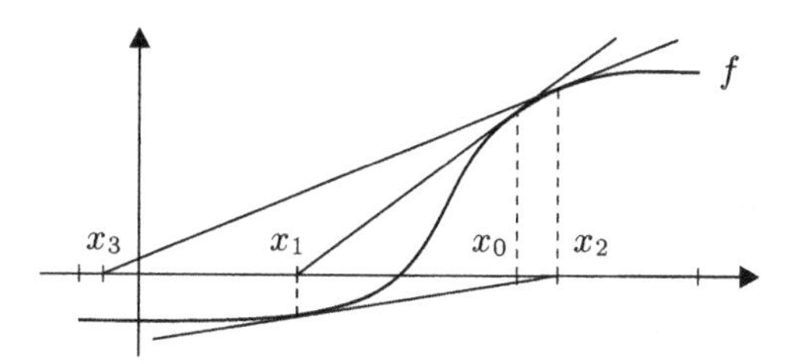

Define $g\colon [a,b] \to \mathbb{R}$ by

$$g(x) := x - f(x)/f'(x) . \tag{4.5}$$

Then ξ is clearly a fixed point of g, and Newton's method is simply the method of successive approximations for the function g. This suggests applying the Banach fixed point theorem, and indeed, this theorem is at the center of the proof of the following convergence result for Newton's method.

4.7 Theorem *There is some $\delta > 0$ such that Newton's method converges to ξ for any x_0 in the interval $[\xi - \delta, \xi + \delta]$. In other words:* Newton's method converges if the initial value is sufficiently close to the zero ξ.

Proof (i) By the extreme value theorem (Corollary III.3.8), there are constants $M_1, M_2, m > 0$ such that

$$m \le |f'(x)| \le M_1 , \quad |f''(x)| \le M_2 , \qquad x \in [a,b] . \tag{4.6}$$

For the function g defined in (4.5), we have $g' = ff''/[f']^2$, and so

$$|g'(x)| \le \frac{M_2}{m^2} |f(x)| , \qquad x \in [a,b] .$$

Since $f(\xi) = 0$, the absolute value of f can be estimated using the mean value theorem as follows:

$$|f(x)| = |f(x) - f(\xi)| \le M_1 |x - \xi| , \qquad x \in [a,b] . \tag{4.7}$$

Thus

$$|g'(x)| \leq \frac{M_1 M_2}{m^2} |x - \xi| \ , \qquad x \in [a,b] \ .$$

(ii) Choose $\delta_1 > 0$ such that

$$I := [\xi - \delta_1, \xi + \delta_1] \subseteq [a,b] \quad \text{and} \quad \frac{M_1 M_2}{m^2} \delta_1 \leq \frac{1}{2} \ .$$

Then g is a contraction on I with the contraction constant $1/2$. Now set $r := \delta_1/2$ and choose $\delta > 0$ such that $M_1\delta/m \leq r/2$. Because $M_1 \geq m$, we have $\delta \leq \delta_1/4$. Thus, for each $x_0 \in [\xi - \delta, \xi + \delta]$ and $x \in [x_0 - r, x_0 + r]$, we have

$$|x - \xi| \leq |x - x_0| + |x_0 - \xi| \leq r + \delta \leq \frac{\delta_1}{2} + \frac{\delta_1}{4} < \delta_1 \ .$$

This shows the inclusion $\bar{\mathbb{B}}(x_0, r) \subseteq I$ for each $x_0 \in [\xi - \delta, \xi + \delta]$. Thus g is a contraction on $\bar{\mathbb{B}}(x_0, r)$ with contraction constant $1/2$.

Finally, it follows from (4.6) and (4.7) that

$$|x_0 - g(x_0)| = \left| \frac{f(x_0)}{f'(x_0)} \right| \leq \frac{M_1}{m} |x_0 - \xi| \leq \frac{M_1 \delta}{m} \leq \frac{r}{2} \ .$$

Hence g satisfies the hypotheses of Proposition 4.5, and there is a unique fixed point η of g in $[\xi - \delta, \xi + \delta]$. Since η is a zero of f, and, by Rolle's theorem, f has only one zero in $[a,b]$, we have $\eta = \xi$. The claimed convergence property now follows from the Banach fixed point theorem. ■

4.8 Remarks **(a)** Newton's method converges quadratically, that is, there is some $c > 0$ such that

$$|x_{n+1} - \xi| \leq c\,|x_n - \xi|^2 \ , \qquad n \in \mathbb{N} \ .$$

Proof For each $n \in \mathbb{N}$, the Lagrange remainder formula for the Taylor series ensures the existence of some $\eta_n \in (\xi \wedge x_n, \xi \vee x_n)$ such that

$$0 = f(\xi) = f(x_n) + f'(x_n)(\xi - x_n) + \frac{1}{2} f''(\eta_n)(\xi - x_n)^2 \ .$$

Thus from Newton's method, we have

$$\xi - x_{n+1} = \xi - x_n + \frac{f(x_n)}{f'(x_n)} = -\frac{1}{2} \frac{f''(\eta_n)}{f'(x_n)} (\xi - x_n)^2 \ .$$

With the notation of (4.6) and $c := M_2/(2m)$, the claim now follows. ■

(b) Newton's method converges monotonically if f is convex and $f(x_0)$ is positive (or if f is concave and $f(x_0)$ is negative).

Proof This follows directly from Application 3.9(e) and the characterization of convex and concave functions in Theorem 2.12. ■

4.9 Example (calculating roots) For $a > 0$ and $n \geq 2$, we consider how $\sqrt[n]{a}$ can be determined using Newton's method. Setting $f(x) = x^n - a$ for all $x \geq 0$ we have the iteration

$$x_{k+1} = x_k - \frac{x_k^n - a}{n x_k^{n-1}} = \Big(1 - \frac{1}{n}\Big) x_k + \frac{a}{n x_k^{n-1}} , \qquad k \in \mathbb{N} . \tag{4.8}$$

Let $x_0 > \max\{1, a\}$. Since $f(x_0) = x_0^n - a > 0$ and f is convex, by Remark 4.8(b), (x_k) converges monotonically to $\sqrt[n]{a}$. In the special case $n = 2$, (4.8) becomes

$$x_{k+1} = \frac{1}{2}\Big(x_k + \frac{a}{x_k}\Big) , \qquad k \in \mathbb{N} , \qquad x_0 = \max\{1, a\} ,$$

which is the Babylonian algorithm of Exercise II.4.4. ∎

Exercises

1 Let X be a complete metric space and, for $f : X \to X$, let f^n denote the nth **iterate** of f, that is, $f^0 := \mathrm{id}_X$ and $f^n := f \circ f^{n-1}$, $n \in \mathbb{N}^\times$. Suppose that, for each $n \in \mathbb{N}$, there is some $q_n \geq 0$ such that

$$d\big(f^n(x), f^n(y)\big) \leq q_n d(x, y) , \qquad x, y \in X .$$

Show that, if (q_n) is a null sequence, then f has a fixed point in X.

2 Let X and Λ be metric spaces with X complete, and $f \in C(X \times \Lambda, X)$. Suppose that there is some $\alpha \in [0, 1)$ and, for each $\lambda \in \Lambda$, some $q(\lambda) \in [0, \alpha]$ such that

$$d\big(f(x, \lambda), f(y, \lambda)\big) \leq q(\lambda) d(x, y) , \qquad x, y \in X .$$

By the Banach fixed point theorem, for each $\lambda \in \Lambda$, $f(\cdot, \lambda)$ has a unique fixed point $x(\lambda)$ in X. Prove that $\big[\lambda \mapsto x(\lambda)\big] \in C(\Lambda, X)$.

3 Verify that the function $f : \mathbb{R} \to \mathbb{R}$, $x \mapsto e^{x-1} - e^{1-x}$ has a unique fixed point x^*. Calculate x^* approximately.

4 Using Newton's method, approximate the real zeros of $X^3 - 2X - 5$.

5 Determine numerically the least positive solutions of the following equations:

$$x \tan x = 1 , \quad x^3 + e^{-x} = 2 , \quad x - \cos^2 x = 0 , \quad 2 \cos x = x^2 .$$

6 By Exercise 2.6, the function $f(x) = 1 + x + x^2/2! + \cdots + x^n/n!$, $x \in \mathbb{R}$, has a unique zero for odd $n \in \mathbb{N}^\times$. Determine these zeros approximately.

7 Suppose that $-\infty < a < b < \infty$ and $f : [a, b] \to \mathbb{R}$ is a differentiable convex function such that either

$$f(a) < 0 < f(b) \quad \text{or} \quad f(a) > 0 > f(b) .$$

Show that the recursively defined sequence

$$x_{n+1} := x_n - \frac{f(x_n)}{f'(x_0)} , \qquad n \in \mathbb{N}^\times , \tag{4.9}$$

converges to the zero of f in $[a, b]$ for any initial value x_0 such that $f(x_0) > 0$.[2] For which initial values does this method converge if f is concave?

[2]The iterative procedure given in (4.9) is called the **simplified Newton's method**.

8 Suppose that $-\infty < a < b < \infty$ and $f \in C\big([a,b],\mathbb{R}\big)$ satisfies $f(a) < 0 < f(b)$. Set $a_0 := a,\ b_0 := b$ and recursively define

$$c_{n+1} := a_n - \frac{b_n - a_n}{f(b_n) - f(a_n)} f(a_n)\ , \qquad n \in \mathbb{N}\ , \tag{4.10}$$

and

$$a_{n+1} := \begin{cases} c_{n+1}\ , & f(c_{n+1}) \le 0\ , \\ a_n & \text{otherwise}\ , \end{cases} \quad , \quad b_{n+1} := \begin{cases} b_n\ , & f(c_{n+1}) \le 0\ , \\ c_{n+1} & \text{otherwise}\ . \end{cases} \tag{4.11}$$

Show that (c_n) converges to some zero of f. What is the graphical interpretation of this procedure (called the **regula falsi** or the **method of false position**)? How should the formulas be modified if $f(a) > 0 > f(b)$?

9 Determine approximately a zero of

$$x^5 e^{|x|} - \frac{1}{\pi} x^2 \sin\big(\log(x^2)\big) + 1998\ .$$

10 Let I be a compact perfect interval and $f \in C^1(I, I)$ a contraction such that $f'(x) \neq 0$ for all $x \in I$. Let $x_0 \in I$ and denote by $x^* := \lim f^n(x_0)$ the unique fixed point of f in I. Finally, suppose that $x_0 \neq x^*$. Prove the following:

(a) $f^n(x_0) \neq x^*$ for each $n \in \mathbb{N}^\times$.

(b) $\displaystyle\lim_{n\to\infty} \frac{f^{n+1}(x_0) - x^*}{f^n(x_0) - x^*} = f'(x^*)$.

Chapter V

Sequences of Functions

In this chapter, approximations are once again the center of our interest. Just as in Chapter II, we study sequences and series. The difference is that we consider here the more complex situation of sequences whose terms are functions. In this circumstance there are two viewpoints: We can consider such sequences locally, that is, at each point, or globally. In the second case it is natural to consider the terms of the sequence as elements of a function space so that we are again in the situation of Chapter II. If the functions in the sequence are all bounded, then we have a sequence in the Banach space of bounded functions, and we can apply all the results about sequences and series which we developed in the second chapter. This approach is particularly fruitful, allows short and elegant proofs, and, for the first time, demonstrates the advantages of the abstract framework in which we developed the fundamentals of analysis.

In the first section we analyze the various concepts of convergence which appear in the study of sequences of functions. The most important of these is uniform convergence which is simply convergence in the space of bounded functions. The main result of this section is the Weierstrass majorant criterion which is nothing more than the majorant criterion from the second chapter applied to the Banach space of bounded functions.

Section 2 is devoted to the connections between continuity, differentiability and convergence for sequences of functions. To our supply of concrete Banach spaces, we add one extremely important and natural example: the space of continuous functions on a compact metric space.

In the following section we continue our earlier investigations into power series and study those functions, the analytic functions, which can be represented locally by power series. In particular, we analyze Taylor series again and derive several classical power series representations. A deeper penetration into the beautiful and important theory of analytic functions must be postponed until we have the concept of the integral.

The last section considers the approximation of continuous functions by polynomials. Whereas the Taylor polynomial provides a local approximation, here we are interested in uniform approximations. The main result is the Stone-Weierstrass theorem. In addition, we take a first look at the behavior of periodic functions, and prove that the Banach algebra of continuous 2π-periodic functions is isomorphic to the Banach algebra of continuous functions on the unit circle. Directly from this fact we get the Weierstrass approximation theorem for periodic functions.

1 Uniform Convergence

For sequences of functions, several different kinds of convergence are possible depending on whether we are interested in the pointwise behavior, or the 'global' behavior, of the functions involved. In this section, we introduce both pointwise and uniform convergence and study the relations between them. The results we derive in this section form the foundation on which all deeper investigations into analysis are built.

Throughout this section, X is a set and $E := (E, |\cdot|)$ is a Banach space over $\mathbb{K}$.

Pointwise Convergence

An E-**valued sequence of functions on** X is simply a sequence (f_n) in E^X. If the choice of X and E is clear from the context (or irrelevant) we say simply that (f_n) is a **sequence of functions.**

The sequence of functions (f_n) **converges pointwise** to $f \in E^X$ if, for each $x \in X$, the sequence $\big(f_n(x)\big)$ converges to $f(x)$ in E. In this circumstance we write $f_n \xrightarrow[\text{pointw}]{} f$ or $f_n \to f$ (pointw) and call f the (**pointwise**) **limit** or the (**pointwise**) **limit function** of (f_n).

1.1 Remarks (a) Suppose that (f_n) converges pointwise. Then the limit function is unique.

Proof This follows directly from Corollary II.1.13. ∎

(b) The following are equivalent:

(i) $f_n \to f$ (pointw).

(ii) For each $x \in X$ and $\varepsilon > 0$, there is a natural number $N = N(x, \varepsilon)$ such that $|f_n(x) - f(x)| < \varepsilon$ for $n \geq N$.

(iii) For each $x \in X$, $\big(f_n(x)\big)$ is a Cauchy sequence in E.

Proof The implications '(i)⇒(ii)⇒(iii)' are clear. The claim '(iii)⇒(i)' holds because E is complete. ∎

(c) The above definitions are also meaningful if E is replaced by an arbitrary metric space. ∎

1.2 Examples (a) Let $X := [0,1]$, $E := \mathbb{R}$ and $f_n(x) := x^{n+1}$. Then (f_n) converges pointwise to the function $f\colon [0,1] \to \mathbb{R}$ defined by

$$f(x) := \begin{cases} 0 , & x \in [0,1) , \\ 1 , & x = 1 . \end{cases}$$

(b) Let $X := [0,1]$, $E := \mathbb{R}$ and[1]

$$f_n(x) := \begin{cases} 2nx\ , & x \in [0, 1/2n]\ , \\ 2 - 2nx\ , & x \in [1/2n, 1/n]\ , \\ 0\ , & x \in (1/n, 1]\ . \end{cases}$$

Then converges (f_n) pointwise to 0.

(c) Let $X := \mathbb{R}$, $E := \mathbb{R}$ and

$$f_n(x) := \begin{cases} 1/(n+1)\ , & x \in [n, n+1)\ , \\ 0 & \text{otherwise}\ . \end{cases}$$

In this case too, (f_n) converges pointwise to 0. ■

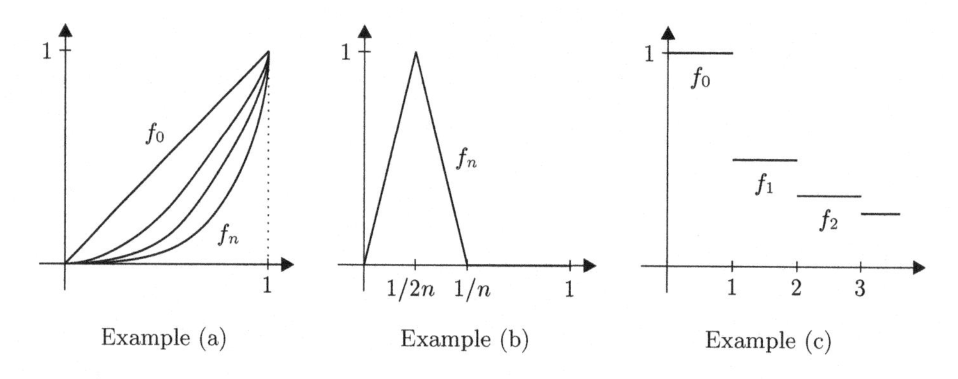

Example (a) Example (b) Example (c)

In Example 1.2(a), we see that, even though all terms of the sequence are infinitely differentiable, the limit function is not even continuous. Thus, for many purposes, pointwise convergence is too weak, and we need to define a stronger kind of convergence which ensures that the properties of the functions in the sequence are shared by the limit function.

Uniform Convergence

A sequence of functions (f_n) **converges uniformly** to f if, for each $\varepsilon > 0$, there is some $N = N(\varepsilon) \in \mathbb{N}$ such that

$$|f_n(x) - f(x)| < \varepsilon\ , \qquad n \geq N\ , \quad x \in X\ . \tag{1.1}$$

In this case we write $f_n \underset{\text{unf}}{\longrightarrow} f$ or $f_n \to f$ (unf).

[1] Here, and in similar situations, $1/ab$ means $1/(ab)$ and not $(1/a)b = b/a$.

The essential difference between pointwise and uniform convergence is that, for uniform convergence, N depends on ε but not on $x \in X$, whereas, for pointwise convergence, for a given ε, $N(\varepsilon, x)$ varies, in general, from point to point. For uniform convergence, the inequality (1.1) holds *uniformly* with respect to $x \in X$.

1.3 Remarks and Examples **(a)** Any uniformly convergent sequence of functions converges pointwise, that is, $f_n \to f$ (unf) implies $f_n \to f$ (pointw).

(b) The converse of (a) is false, that is, there are pointwise convergent sequences of functions which do not converge uniformly.

Proof Let (f_n) be the sequence of Example 1.2(b). Set $x_n := 1/2n$ for all $n \in \mathbb{N}^\times$. Then $|f_n(x_n) - f(x_n)| = 1$. Thus (f_n) cannot converge uniformly. ■

(c) The sequence of functions (f_n) of Example 1.2(c) converges uniformly to 0.

(d) Let $X := (0, \infty)$, $E := \mathbb{R}$ and $f_n(x) := 1/nx$ for all $n \in \mathbb{N}^\times$.

(i) $f_n \to 0$ (pointw).

(ii) For each $a > 0$, (f_n) converges uniformly to 0 on $[a, \infty)$.

(iii) The sequence of functions (f_n) does not converge uniformly to 0.

Proof The first claim is clear.

(ii) Let $a > 0$. Then

$$|f_n(x)| = 1/nx \le 1/na , \qquad n \in \mathbb{N}^\times , \quad x \ge a .$$

Thus (f_n) converges uniformly to 0 on $[a, \infty)$.

(iii) For $\varepsilon > 0$ and $x > 0$ we have $|f_n(x)| = 1/nx < \varepsilon$ if and only if $n > 1/x\varepsilon$. Hence (f_n) cannot converge uniformly to 0 on $(0, \infty)$. ■

(e) The following are equivalent:

(i) $f_n \to f$ (unf).

(ii) $(f_n - f) \to 0$ in $B(X, E)$.

(iii) $\|f_n - f\|_\infty \to 0$ in $\mathbb{R}$.

Note that it is possible for f_n to converge uniformly to f even if f_n and f are not in $B(X, E)$. For example, let $X := \mathbb{R}$, $E := \mathbb{R}$, $f_n(x) := x + 1/n$ for all $n \in \mathbb{N}^\times$ and $f(x) := x$. Then (f_n) converges uniformly to f, but neither f nor f_n is in $B(\mathbb{R}, \mathbb{R})$.

(f) If f_n and f are in $B(X, E)$, then (f_n) converges uniformly to f if and only if (f_n) converges to f in $B(X, E)$. ■

1.4 Proposition (Cauchy criterion for uniform convergence) *The following are equivalent:*

(i) *The sequence of functions (f_n) converges uniformly.*

(ii) *For each $\varepsilon > 0$, there is some $N := N(\varepsilon) \in \mathbb{N}$ such that*

$$\|f_n - f_m\|_\infty < \varepsilon , \qquad n, m \ge N .$$

Proof '(i)$\Rightarrow$(ii)' By hypothesis, there is some $f \in E^X$ such that $f_n \to f$ (unf). Thus, by Remark 1.3(e), $(f_n - f)$ converges to 0 in the space $B(X, E)$. The claim now follows from the triangle inequality

$$\|f_n - f_m\|_\infty \leq \|f_n - f\|_\infty + \|f - f_m\|_\infty .$$

'(ii)$\Rightarrow$(i)' For each $\varepsilon > 0$, there is some $N = N(\varepsilon)$ such that $\|f_n - f_m\|_\infty < \varepsilon$ for all $m, n \geq N$. Setting $\varepsilon := 1$ and $\widehat{f} := f_{N(1)}$, we see that, for all $n \geq N(1)$, $f_n - \widehat{f}$ is in $B(X, E)$. Thus $(f_n - \widehat{f})$ is a Cauchy sequence in $B(X, E)$. By Theorem II.6.6, $B(X, E)$ is complete and so there is some $\widetilde{f} \in B(X, E)$ such that $(f_n - \widehat{f}) \to \widetilde{f}$ in $B(X, E)$. By Remark 1.3(e), the sequence (f_n) converges uniformly to $\widetilde{f} + \widehat{f}$. ■

Series of Functions

Let (f_k) be an E-valued sequence of functions on X, that is, a sequence in E^X. Then

$$s_n := \sum_{k=0}^{n} f_k \in E^X , \qquad n \in \mathbb{N} ,$$

and so we have a well defined sequence (s_n) in E^X. As in Section II.7, this sequence is denoted $\sum f_k$ or $\sum_k f_k$ and is called a **series of E-valued functions** on X, or simply a **series of functions** (on X). In addition, s_n is called the nth partial sum and f_k is called the kth summand of this series.

The series $\sum f_k$ is called

pointwise convergent $:\Longleftrightarrow \sum f_k(x)$ converges in E for each $x \in X$,
absolutely convergent $:\Longleftrightarrow \sum |f_k(x)| < \infty$ for each $x \in X$,
uniformly convergent $:\Longleftrightarrow (s_n)$ converges uniformly,
norm convergent $:\Longleftrightarrow \sum \|f_k\|_\infty < \infty$.

1.5 Remarks (a) Let $\sum f_k$ be a pointwise convergent E-valued series of functions on X. Then

$$X \to E , \quad x \mapsto \sum_{k=0}^{\infty} f_k(x)$$

defines a function called the (**pointwise**) **sum** or (**pointwise**) **limit function** of the series $\sum f_k$.

(b) Let (f_k) be a sequence in $B(X, E)$. Then we can consider the series $\sum f_k$ as a series in $B(X, E)$ or as an E-valued series of functions on X. The norm convergence of the series of functions is then nothing other than the absolute convergence[2] of the series $\sum f_k$ in the Banach space $B(X,E)$.

(c) These convergence concepts are related as follows:[3]

(i) $\sum f_k$ absolutely convergent $\Rightarrow \sum f_k$ pointwise convergent.

(ii) $\sum f_k$ uniformly convergent $\underset{\not\Rightarrow}{\not\Leftarrow} \sum f_k$ absolutely convergent.

(iii) $\sum f_k$ norm convergent $\underset{\not\Leftarrow}{\Rightarrow} \sum f_k$ absolutely and uniformly convergent.

Proof The first claim follows from Proposition II.8.1.

(ii) Set $X := \mathbb{R}$, $E := \mathbb{R}$ and $f_k(x) := (-1)^k/k$ for all $k \in \mathbb{N}^\times$. Then $\sum f_k$ converges uniformly but not absolutely (see Remark II.8.2(a)).

To verify the second claim, consider $X := (0,1)$, $E := \mathbb{R}$ and $f_k(x) := x^k$, $k \in \mathbb{N}$. Then $\sum f_k$ is absolutely convergent and has the limit function

$$s(x) = \sum_{k=0}^{\infty} f_k(x) = 1/(1-x) , \qquad x \in (0,1) .$$

Since

$$s(x) - s_n(x) = \sum_{k=n+1}^{\infty} x^k = x^{n+1}/(1-x) , \qquad x \in (0,1) , \quad n \in \mathbb{N} ,$$

we have

$$s(x) - s_n(x) < \varepsilon \Leftrightarrow \frac{x^{n+1}}{1-x} < \varepsilon$$

for all $\varepsilon, x \in (0,1)$. Because the right inequality is not satisfied for x sufficiently near 1, the sequence of partial sums (s_n) does not converge uniformly.

(iii) Let $\sum f_k$ be norm convergent. Then, for each $x \in X$, we have the inequality $\sum |f_k(x)| \le \sum \|f_k\|_\infty < \infty$. Hence $\sum f_k$ is absolutely convergent. Further, it follows from (b) and Proposition II.8.1 that the series $\sum f_k$ converges in $B(X,E)$. Thus the uniform convergence of $\sum f_k$ follows from Remark 1.3(f).

Finally, let (f_k) be the sequence of functions of Example 1.2(c). Then $\sum f_k$ converges absolutely and uniformly, but because $\sum \|f_k\|_\infty = \sum 1/(k+1) = \infty$, $\sum f_k$ is not norm convergent. ■

The Weierstrass Majorant Criterion

A particularly simple situation occurs for a series of functions in the Banach space $B(X,E)$ since it is then possible to apply directly the results of Chapter II. For example, we get the following easy and important convergence theorem.

[2] It is important to distinguish the (pointwise) absolute convergence of a series of functions $\sum f_k$ and the absolute convergence of $\sum f_k$ in the Banach space $B(X,E)$. For this reason, the latter is called 'norm convergence'.

[3] $(A \not\Rightarrow B) := \neg(A \Rightarrow B)$.

1.6 Theorem (Weierstrass majorant criterion) *Suppose that $f_k \in B(X,E)$ for all $k \in \mathbb{N}$. If there is a convergent series $\sum \alpha_k$ in $\mathbb{R}$ such that $\|f_k\|_\infty \le \alpha_k$ for almost all $k \in \mathbb{N}$, then $\sum f_k$ is norm convergent. In particular, $\sum f_k$ converges absolutely and uniformly.*

Proof Since $\|f_k\|_\infty < \infty$ for all $k \in \mathbb{N}$, we can consider the series $\sum f_k$ to be in the Banach space $B(X,E)$. Then the claim follows directly from the majorant criterion (Theorem II.8.3) and Remark 1.5(c). ∎

1.7 Examples **(a)** The series of functions $\sum_k \cos(kx)/k^2$ is norm convergent on $\mathbb{R}$.

Proof For $x \in \mathbb{R}$ and $k \in \mathbb{N}^\times$,

$$|\cos(kx)/k^2| \le 1/k^2 \ .$$

Hence the claim follows from Theorem 1.6 and Example II.7.1(b). ∎

(b) For each $\alpha > 1$, the series[4] $\sum_k 1/k^z$ is norm convergent on

$$X_\alpha := \{\, z \in \mathbb{C} \ ; \ \operatorname{Re} z \ge \alpha \,\} \ .$$

Proof Clearly

$$|1/k^z| = 1/k^{\operatorname{Re} z} \le 1/k^\alpha \ , \qquad z \in X_\alpha \ , \quad k \in \mathbb{N}^\times \ .$$

Since the series $\sum 1/k^\alpha$ converges (see Exercise II.7.12), the claim follows from Theorem 1.6. ∎

(c) For each $m \in \mathbb{N}^\times$, the series $\sum_k x^{m+2} e^{-kx^2}$ is norm convergent on $\mathbb{R}$.

Proof Define $f_{m,k}(x) := |x^{m+2} e^{-kx^2}|$ for all $x \in \mathbb{R}$. Then $f_{m,k}$ attains its absolute maximum value $\bigl[(m+2)/2ek\bigr]^{(m+2)/2}$ at the point $x_M := \sqrt{(m+2)/2k}$. In other words, $\|f_{m,k}\|_\infty = c_m k^{-(m+2)/2}$ where $c_m := \bigl[(m+2)/2e\bigr]^{(m+2)/2}$. By Exercise II.7.12 the series $\sum_k k^{-(m+2)/2}$ converges, and so the claim follows once again from Theorem 1.6. ∎

As an important application of the Weierstrass majorant criterion we prove that a power series is norm convergent on any compact subset of its disk of convergence.

1.8 Theorem *Let $\sum a_k Y^k$ be a power series with positive radius of convergence ρ and $0 < r < \rho$. Then the series[5] $\sum a_k Y^k$ is norm convergent on $r\bar{\mathbb{B}}_{\mathbb{K}}$. In particular, it converges absolutely and uniformly.*

---

[4]The function $\zeta(z) := \sum_k 1/k^z$ is defined for all $\{\, z \in \mathbb{C} \ ; \ \operatorname{Re} z > 1 \,\}$ and is called the **(Riemann) zeta function**. We study this function in detail in Section VI.6.

[5]By the conventions of Section I.8, we identify the monomial $a_k Y^k$ with the corresponding 'monomial' function.

Proof By Theorem II.9.2, any power series converges absolutely in the interior of its disk of convergence, and so, setting $X := r\bar{\mathbb{B}}_{\mathbb{K}}$ and $f_k(x) := a_k x^k$ for all $x \in X$ and $k \in \mathbb{N}$, we have

$$\sum \|f_k\|_\infty = \sum |a_k| r^k < \infty .$$

The claim now follows from Theorem 1.6. ■

Exercises

1 Which of the following sequences of functions (f_n) converge uniformly on $X := (0,1)$?
(a) $f_n := \sqrt[n]{x}$, (b) $f_n := 1/(1+nx)$, (c) $f_n := x/(1+nx)$.

2 Show that (f_n), defined by $f_n(x) := \sqrt{(1/n^2) + |x|^2}$, converges uniformly on $\mathbb{K}$ to the absolute value function $x \mapsto |x|$.

3 Prove or disprove that $\sum x^n/n^2$ and $\sum x^n$ converge uniformly on $\mathbb{B}_{\mathbb{C}}$.

4 Prove or disprove that $\sum(-1)^n/nx$ converges pointwise (or uniformly, or absolutely) on $(0,1]$.

5 Let $X := \mathbb{B}_{\mathbb{K}}$. Investigate the norm convergence of the series $\sum f_n$ for the following cases:
(a) $f_n := x^n$, (b) $f_n := |x|^2/(1+|x|^2)^n$, (c) $f_n := x(1-x^2)^n$, (d) $f_n := \left[x(1-x^2)\right]^n$.

6 Verify that each of the series
(a) $\sum(1-\cos(x/n))$, (b) $\sum n(x/n - \sin(x/n))$,
converges uniformly on any compact subinterval of $\mathbb{R}$.
(Hint: Approximate the terms of these series using Taylor polynomials of first and second degree.)

7 Let (f_n) and (g_n) be uniformly convergent E-valued sequences of functions on X with limit functions f and g respectively. Show the following:

(a) $(f_n + g_n)$ converges uniformly to $f+g$.

(b) If f or g is in $B(X,\mathbb{K})$, then $(f_n g_n)$ converges uniformly to fg.

Show by example that, in (b), the boundedness of one of the limit functions is necessary.

8 Let (f_n) be a uniformly convergent sequence of $\mathbb{K}$-valued functions on X with limit function f. Suppose that there is some $\alpha > 0$ such that

$$|f_n(x)| \geq \alpha > 0 , \qquad n \in \mathbb{N} , \quad x \in X .$$

Show that $(1/f_n)$ converges uniformly to $1/f$.

9 Let (f_n) be a uniformly convergent sequence of E-valued functions on X, and F a Banach space. Suppose that $f_n(X) \subseteq D$ for all $n \in \mathbb{N}$ and $g : D \to F$ is uniformly continuous. Show that $(g \circ f_n)$ is uniformly convergent.

2 Continuity and Differentiability for Sequences of Functions

In this section we consider convergent sequences of functions whose terms are continuous or continuously differentiable, and investigate the conditions under which the limit function 'inherits' these same properties.

In the following $X := (X, d)$ is a metric space, $E := (E, |\cdot|)$ is a Banach space and (f_n) is a sequence of E-valued functions on X.

Continuity

Example 1.2(a) shows that the pointwise limit of a sequence of continuous (or even infinitely differentiable) functions may not be continuous. If the convergence is uniform however, then the continuity of the limit function is guaranteed, as the following theorem shows.

2.1 Theorem *If (f_n) converges uniformly to f and almost all f_n are continuous at $a \in X$, then f is also continuous at a.*

Proof Let $\varepsilon > 0$. Because f_n converges uniformly to f, there is, by Remark 1.3(e), some $N \in \mathbb{N}$ such that $\|f_n - f\|_\infty < \varepsilon/3$ for all $n \geq N$. Since almost all f_n are continuous at a, we can suppose that f_N is continuous at a. Thus there is a neighborhood U of a in X such that $|f_N(x) - f_N(a)| < \varepsilon/3$ for all $x \in U$. Then, for each $x \in U$, we have

$$\begin{aligned}|f(x) - f(a)| &\leq |f(x) - f_N(x)| + |f_N(x) - f_N(a)| + |f_N(a) - f(a)| \\ &\leq 2\,\|f - f_N\|_\infty + |f_N(x) - f_N(a)| < \varepsilon\ ,\end{aligned}$$

which shows the continuity of f at a. ■

2.2 Remark Clearly Theorem 2.1 and its proof remain valid if X is replaced by an arbitrary topological space and E by a metric space. This holds also for any statement of this section that involves continuity only. ■

Locally Uniform Convergence

An inspection of the proof of Theorem 2.1 shows that it remains true if there is a neighborhood U of a such that (f_n) converges uniformly on U. The behavior of (f_n) outside of U is irrelevant for the continuity of f at a, since continuity is a 'local' property. This motivates the definition of a 'local' version of uniform convergence.

A sequence of functions (f_n) is called **locally uniformly convergent** if each $x \in X$ has a neighborhood U such that $(f_n \,|\, U)$ converges uniformly. A series of functions $\sum f_n$ is called **locally uniformly convergent** if the sequence of partial sums (s_n) converges locally uniformly.

2.3 Remarks (a) Any uniformly convergent sequence of functions is locally uniformly convergent.

(b) Any locally uniformly convergent sequence of functions converges pointwise.

(c) If X is compact and (f_n) converges locally uniformly, then (f_n) converges uniformly.

Proof By (b), the (pointwise) limit function f of (f_n) is well defined. Let $\varepsilon > 0$. Because (f_n) converges locally uniformly, for each $x \in X$, there is an open neighborhood U_x of x and some $N(x) \in \mathbb{N}$ such that

$$|f_n(y) - f(y)| < \varepsilon , \qquad y \in U_x , \quad n \geq N(x) .$$

The family $\{ U_x ; x \in X \}$ is an open cover of the compact space X, and so there are finitely many points $x_0, \ldots, x_m \in X$ such that X is covered by U_{x_j}, $0 \leq j \leq m$. For $N := \max\{N(x_0), \ldots, N(x_m)\}$, we then have

$$|f_n(x) - f(x)| < \varepsilon , \qquad x \in X , \quad n \geq N .$$

This shows that (f_n) converges uniformly to f. ■

2.4 Theorem (continuity of the limits of sequences of functions) *If a sequence of continuous functions (f_n) converges locally uniformly to f, then f is also continuous. In other words,* locally uniform limits of continuous functions are continuous.

Proof Since the continuity of f is a local property, the claim follows directly from Theorem 2.1 ■

2.5 Remarks (a) If a sequence of functions (f_n) converges pointwise to f and all f_n and f are continuous, then it does *not* follow, in general, that (f_n) converges locally uniformly to f.

Proof For the sequence of functions (f_n) from Example 1.2(b) we have $f_n \in C(\mathbb{R})$ with $f_n \xrightarrow[\text{pointw}]{} 0$. Even so, there is no neighborhood of 0 on which (f_n) converges uniformly. ■

(b) Theorem 2.4 can be interpreted as a statement about exchanging limits: If the sequence of functions (f_n) converges locally uniformly to f, then, for all $a \in X$,

$$\lim_{x \to a} \lim_{n \to \infty} f_n(x) = \lim_{n \to \infty} \lim_{x \to a} f_n(x) = \lim_{n \to \infty} f_n(a) = f(a) .$$

Similarly, for a locally uniformly convergent series of functions we have

$$\lim_{x \to a} \sum_{k=0}^{\infty} f_k(x) = \sum_{k=0}^{\infty} \lim_{x \to a} f_k(x) = \sum_{k=0}^{\infty} f_k(a) , \qquad a \in X .$$

These facts can be expressed by saying that 'locally uniform convergence respects the taking of limits'.

Proof This is a consequence of the remark following Theorem III.1.4. ■

(c) A power series with positive radius of convergence represents a continuous function on its disk of convergence.[1]

Proof By Theorem 1.8, a power series converges locally uniformly on its disk of convergence. Thus the claim follows from Theorem 2.4. ■

The Banach Space of Bounded Continuous Functions

A particularly important subspace of the space $B(X, E)$ of bounded E-valued functions on X is the space

$$BC(X, E) := B(X, E) \cap C(X, E)$$

of **bounded continuous functions** from X to E. Clearly, $BC(X, E)$ is a subspace of $B(X, E)$ (and of $C(X, E)$), and is also a normed space with the supremum norm

$$\|\cdot\|_{BC} := \|\cdot\|_\infty ,$$

that is, with the subspace topology induced from $B(X, E)$. The following theorem shows that $BC(X, E)$ is a Banach space.

2.6 Theorem

(i) *$BC(X, E)$ is a closed subspace of $B(X, E)$ and hence a Banach space.*

(ii) *If X is compact, then*

$$BC(X, E) = C(X, E) ,$$

and the supremum norm $\|\cdot\|_\infty$ coincides with the **maximum norm**

$$f \mapsto \max_{x \in X} |f(x)| .$$

Proof (i) Let (f_n) be a sequence in $BC(X, E)$ which converges to f in $B(X, E)$. Then, by Remark 1.3(e), (f_n) converges uniformly to f, and, by Theorem 2.4, f is continuous, that is, f is in $BC(X, E)$. This shows that $BC(X, E)$ is a closed subspace of $B(X, E)$ and also that $BC(X, E)$ is complete (see Exercise II.6.4).

(ii) If X is compact, then, from the extreme value theorem (Corollary III.3.8), we have $C(X, E) \subseteq B(X, E)$ and

$$\max_{x \in X} |f(x)| = \sup_{x \in X} |f(x)| = \|f\|_\infty ,$$

which proves the claim. ■

[1] We show in the next section that such functions are, in fact, infinitely differentiable.

2.7 Remark If X is a metric space which is not compact, for example, an open subset of $\mathbb{K}^n$, then it is not possible to characterize locally uniform convergence in $C(X, E)$ using a norm. In other words, *if X is not compact, then $C(X, E)$ is not a normed vector space.* For a proof of this fact, we must refer the reader to the functional analysis literature. ∎

Differentiability

We now investigate the conditions under which the pointwise limit of a sequence of differentiable functions is itself differentiable.

2.8 Theorem (differentiability of the limits of sequences of functions) *Let X be an open (or convex) perfect subset of $\mathbb{K}$ and $f_n \in C^1(X, E)$ for all $n \in \mathbb{N}$. Suppose that there are $f, g \in E^X$ such that*

(i) *(f_n) converges pointwise to f, and*

(ii) *(f_n') converges locally uniformly to g.*

Then f is in $C^1(X, E)$, and $f' = g$. In addition, (f_n) converges locally uniformly to f.

Proof Let $a \in X$. Then there is some $r > 0$ such that (f_n') converges uniformly to g on $B_r := \mathbb{B}_{\mathbb{K}}(a, r) \cap X$. If X is open we can choose $r > 0$ so that $\mathbb{B}(a, r)$ is contained in X. Hence with either of our assumptions, B_r is convex and perfect. Thus, for each $x \in B_r$, we can apply the mean value theorem (Theorem IV.2.18) to the function

$$[0, 1] \to E , \quad t \mapsto f_n\big(a + t(x - a)\big) - t f_n'(a)(x - a)$$

to get

$$|f_n(x) - f_n(a) - f_n'(a)(x - a)| \le \sup_{0<t<1} \big|f_n'\big(a + t(x - a)\big) - f_n'(a)\big|\, |x - a| \ .$$

Taking the limit $n \to \infty$ we get

$$|f(x) - f(a) - g(a)(x - a)| \le \sup_{0<t<1} \big|g\big(a + t(x - a)\big) - g(a)\big|\, |x - a| \tag{2.1}$$

for each $x \in B_r$. Theorem 2.4 shows that g is in $C(X, E)$, so it follows from (2.1) that

$$f(x) - f(a) - g(a)(x - a) = o(|x - a|) \quad (x \to a) \ .$$

Hence f is differentiable at a and $f'(a) = g(a)$. Since this holds for all $a \in X$, we have shown that $f \in C^1(X, E)$.

It remains to prove that (f_n) converges locally uniformly to f. Applying the mean value theorem to the function

$$[0, 1] \to E , \quad t \mapsto (f_n - f)\big(a + t(x - a)\big)$$

we get the inequality

$$
\begin{aligned}
|f_n(x) - f(x)| &\le \bigl|f_n(x) - f(x) - \bigl(f_n(a) - f(a)\bigr)\bigr| + |f_n(a) - f(a)| \\
&\le r \sup_{0<t<1} \bigl|f_n'\bigl(a + t(x-a)\bigr) - f'\bigl(a + t(x-a)\bigr)\bigr| + |f_n(a) - f(a)| \\
&\le r \, \|f_n' - f'\|_{\infty, B_r} + |f_n(a) - f(a)|
\end{aligned}
$$

for each $x \in B_r$. The right side of this inequality is independent of $x \in B_r$ and converges to 0 as $n \to \infty$ because of (ii) and the fact that $f' = g$. Thus (f_n) converges uniformly to f on B_r. ■

2.9 Corollary (differentiability of the limit of a series of functions) *Suppose that $X \subseteq \mathbb{K}$ is open (or convex) and perfect, and (f_n) is a sequence in $C^1(X, E)$ for which $\sum f_n$ converges pointwise and $\sum f_n'$ converges locally uniformly. Then the sum $\sum_{n=0}^{\infty} f_n$ is in $C^1(X, E)$ and*

$$\Bigl(\sum_{n=0}^{\infty} f_n\Bigr)' = \sum_{n=0}^{\infty} f_n' .$$

In addition, $\sum f_n$ converges locally uniformly.

Proof This follows directly from Theorem 2.8. ■

2.10 Remarks **(a)** Let (f_n) be a sequence in $C^1(X, E)$ which converges uniformly to f. Even if f is continuously differentiable, (f_n') does not, in general, converge pointwise to f'.

Proof Let $X := \mathbb{R}$, $E := \mathbb{R}$ and $f_n(x) := (1/n)\sin(nx)$ for all $n \in \mathbb{N}^\times$. Because

$$|f_n(x)| = |\sin(nx)|/n \le 1/n , \qquad x \in X ,$$

(f_n) converges uniformly to 0. Since $\lim f_n'(0) = 1$, the sequence $\bigl(f_n'(0)\bigr)$ does not converge to the derivative of the limit function at the point 0. ■

(b) Let (f_n) be a sequence in $C^1(X, E)$ such that $\sum f_n$ converges uniformly. Then, in general, $\sum f_n'$ does not converge even pointwise.

Proof Suppose that $X := \mathbb{R}$, $E := \mathbb{R}$, and $f_n(x) := (1/n^2)\sin(nx)$ for all $n \in \mathbb{N}^\times$. Then $\|f_n\|_\infty = 1/n^2$ and so, by the Weierstrass majorant criterion, the series $\sum f_n$ converges uniformly. Since $f_n'(x) = (1/n)\cos(nx)$, $\sum f_n'(0)$ does not converge. ■

Exercises

1 Prove the following:

(a) If (f_n) converges uniformly and each f_n is uniformly continuous, then the limit function is also uniformly continuous.

(b) $BUC(X, E) := \bigl(\{\, f \in BC(X, E) \ ; \ f \text{ is uniformly continuous} \,\}, \ \|\cdot\|_\infty\bigr)$ is a Banach space.

(c) If X is compact, then $BUC(X, E) = C(X, E)$.

2 Consider a double sequence (x_{jk}) in E such that

(i) $(x_{jk})_{k\in\mathbb{N}}$ converges for each $j \in \mathbb{N}$.

(ii) For each $\varepsilon > 0$, there is some $N \in \mathbb{N}$ such that

$$|x_{mk} - x_{nk}| < \varepsilon\ , \qquad m,n \geq N\ , \quad k \in \mathbb{N}\ .$$

Show that $(x_{jk})_{j\in\mathbb{N}}$ converges for each $k \in \mathbb{N}$. Show that the sequences $(\lim_k x_{jk})_{j\in\mathbb{N}}$ and $(\lim_j x_{jk})_{k\in\mathbb{N}}$ converge and

$$\lim_j(\lim_k x_{jk}) = \lim_k(\lim_j x_{jk})\ .$$

3 Suppose that X is compact and (f_n) is a pointwise convergent sequence of real valued continuous functions on X. Prove that, if the limit function is continuous and (f_n) is monotone, then (f_n) converges uniformly (**Dini's theorem**).
(Hint: If (f_n) is increasing, then

$$0 \leq f(y) - f_{N_x}(y) = \big(f(y) - f(x)\big) + \big(f(x) - f_{N_x}(x)\big) + \big(f_{N_x}(x) - f_{N_x}(y)\big)$$

for all $x, y \in X$ and $N_x \in \mathbb{N}$.)

4 Show by example that, in Dini's theorem, the continuity of the limit function and the monotone convergence are necessary hypotheses.

5 Let (f_n) be a sequence of monotone functions on a compact interval I which converges pointwise to a continuous function f. Show that f is monotone and that (f_n) converges uniformly to f.

6 Consider a sequence of real valued functions (f_n) on X satisfying the following conditions:

(i) For each $x \in X$, $\big(f_n(x)\big)$ is decreasing.

(ii) (f_n) converges uniformly to 0.

Show that $\sum(-1)^n f_n$ converges uniformly.

7 Let (f_n) be a sequence of real valued functions on X, and (g_n) a sequence of $\mathbb{K}$-valued functions on X which satisfy the following conditions:

(i) For each $x \in X$, $\big(f_n(x)\big)$ is decreasing.

(ii) (f_n) converges uniformly to 0.

(iii) $\sup_n \big\|\sum_{k=0}^n g_k\big\|_\infty < \infty$.

Show that $\sum g_n f_n$ converges uniformly.
(Hint: Setting $\alpha_k := \sum_{j=0}^k g_j$ we have

$$\sum_{k=m+1}^{n} g_k f_k = \sum_{k=m}^{n-1} \alpha_k(f_k - f_{k+1}) + \alpha_n f_n - \alpha_m f_m$$

for all $m < n$. For a given $\varepsilon > 0$ and $M := \sup_k \|\alpha_k\|_\infty$, there is some $N \in \mathbb{N}$ such that $\|f_n\|_\infty < \varepsilon/2M$ for all $n \geq N$. It follows that

$$\Big|\sum_{k=m+1}^{n} g_k(x) f_k(x)\Big| \leq M \sum_{k=m}^{n-1} (f_k - f_{k+1})(x) + M(f_n + f_m)(x) < \varepsilon$$

for all $x \in X$ and $n > m \geq N$. Now use Proposition 1.4.)

8 With the help of the previous exercise, show that, for each $\alpha \in (0,\pi)$, the series $\sum_k e^{ikx}/k$ converges uniformly on $[\alpha, 2\pi-\alpha]$.
(Hint: We have

$$|e^{ix}-1| \geq \sqrt{2(1-\cos\alpha)}\ , \qquad x \in [\alpha, 2\pi-\alpha]\ ,$$

and so

$$\Bigl|\sum_{k=0}^{n} e^{ikx}\Bigr| = \frac{|e^{inx}-1|}{|e^{ix}-1|} \leq \sqrt{2/(1-\cos\alpha)}$$

for all $x \in [\alpha, 2\pi-\alpha]$.)

9 Suppose that $A\colon E \to E$ is linear and $\alpha \geq 0$ satisfies $\|Ax\| \leq \alpha\,\|x\|$ for all $x \in E$. Fix $x_0 \in E$ and define

$$u(z) := \sum_{k=0}^{\infty} \frac{z^k}{k!} A^k x_0\ , \qquad z \in \mathbb{K}\ .$$

Here A^k denotes the kth iterate of A. Show that $u \in C^\infty(\mathbb{K}, E)$ and determine $\partial^n u$ for all $n \in \mathbb{N}^\times$.
(Hint: The series $\sum (z^k/k!)A^k x_0$ has $\|x_0\|\,e^{|z|\,\alpha}$ as a convergent majorant. We also have $\sum A(z^k/k!)A^k x_0 = Au(z)$.)

10 Let X be open in $\mathbb{K}$, $n \in \mathbb{N}^\times$ and

$$BC^n(X,E) := \bigl(\{\, f \in C^n(X,E)\ ;\ \partial^j f \in B(X,E),\ j = 0,\ldots,n \,\},\ \|\cdot\|_{BC^n}\bigr)$$

with $\|f\|_{BC^n} := \max_{1\leq j\leq n} \|\partial^j f\|_\infty$. Prove the following:

(a) $BC^n(X,E)$ is not a closed subspace of $BC(X,E)$.

(b) $BC^n(X,E)$ is a Banach space.

11 Let $-\infty < a < b < \infty$ and $f_n \in C^1\bigl([a,b],E\bigr)$ for all $n \in \mathbb{N}$. Suppose that the sequence (f_n') converges uniformly and there is some $x_0 \in [a,b]$ for which $\bigl(f_n(x_0)\bigr)_{n\in\mathbb{N}}$ converges. Prove that (f_n) converges uniformly. (Hint: Use Theorem IV.2.18.)

3 Analytic Functions

In this section we study power series again. These are, of course, series of functions having a particularly simple form. We know already that a power series converges locally uniformly on its disk of convergence. We show in this section that such a series can be differentiated 'termwise' and that the result is again a power series with the same radius of convergence as the original series. It follows directly from this that a power series represents a smooth function on its disk of convergence.

These observations lead us to the study of *analytic functions*, functions which can be represented locally by power series. These functions have a very rich 'internal' structure whose beauty and importance we explore further in later chapters.

Differentiability of Power Series

Let $a = \sum_k a_k X^k \in \mathbb{K}[\![X]\!]$ be a power series with radius of convergence $\rho = \rho_a > 0$, and $\underline{a}$ the function on $\rho\mathbb{B}_{\mathbb{K}}$ represented by a. When no misunderstanding is possible, we write $\mathbb{B}$ for $\mathbb{B}_{\mathbb{K}}$.

3.1 Theorem (differentiability of power series) *Let $a = \sum_k a_k X^k$ be a power series. Then $\underline{a}$ is continuously differentiable on $\rho\mathbb{B}$. The 'termwise differentiated' series $\sum_{k\geq 1} k a_k X^{k-1}$ has radius of convergence ρ and*

$$\underline{a}'(x) = \Bigl(\sum_{k=0}^{\infty} a_k x^k\Bigr)' = \sum_{k=1}^{\infty} k a_k x^{k-1}\ , \qquad x \in \rho\mathbb{B}\ .$$

Proof Let ρ' be the radius of convergence of the power series $\sum k a_k X^{k-1}$. From Hadamard's formula (II.9.3), Example II.4.2(d) and Exercise II.5.2(d) we have

$$\rho' = \frac{1}{\overline{\lim}\, \sqrt[k]{k\,|a_k|}} = \frac{1}{\lim \sqrt[k]{k}\ \overline{\lim}\, \sqrt[k]{|a_k|}} = \frac{1}{\overline{\lim}\, \sqrt[k]{|a_k|}} = \rho\ .$$

By Theorem 1.8, the power series $\sum_{k\geq 1} k a_k X^{k-1}$ converges locally uniformly on $\rho\mathbb{B}$, so the claim follows from Corollary 2.9 ■

3.2 Corollary *If $a = \sum a_k X^k$ is a power series with positive radius of convergence ρ, then $\underline{a} \in C^\infty(\rho\mathbb{B}, \mathbb{K})$ and $\underline{a} = \mathcal{T}(\underline{a}, 0)$. In other words, $\sum a_k X^k$ is the* Taylor series of $\underline{a}$ at 0 and $a_k = \underline{a}^{(k)}(0)/k!$.

Proof By induction, it follows from Theorem 3.1, that $\underline{a}$ is smooth on $\rho\mathbb{B}$ and that, for all $x \in \rho\mathbb{B}$,

$$\underline{a}^{(k)}(x) = \sum_{n=k}^{\infty} n(n-1)\cdots(n-k+1) a_n x^{n-k}\ , \qquad k \in \mathbb{N}\ .$$

Hence $\underline{a}^{(k)}(0) = k!\, a_k$ for all $k \in \mathbb{N}$ and we have proved the claim. ■

Analyticity

Let D be open in $\mathbb{K}$. A function $f : D \to \mathbb{K}$ is called **analytic** (on D) if, for each $x_0 \in D$, there is some $r = r(x_0) > 0$ such that $\mathbb{B}(x_0, r) \subseteq D$ and a power series $\sum_k a_k X^k$ with radius of convergence $\rho \geq r$, such that

$$f(x) = \sum_{k=0}^{\infty} a_k (x - x_0)^k \ , \qquad x \in \mathbb{B}(x_0, r) \ .$$

In this case, we say that $\sum_k a_k (X - x_0)^k$ is the **power series expansion for** f **at** x_0. The set of all analytic functions on D is denoted by $C^\omega(D, \mathbb{K})$, or by $C^\omega(D)$ if no misunderstanding is possible. Further, $f \in C^\omega(D)$ is called **real (or complex) analytic** if $\mathbb{K} = \mathbb{R}$ (or $\mathbb{K} = \mathbb{C}$).

3.3 Examples (a) Polynomial functions are analytic on $\mathbb{K}$.

Proof This follows from (IV.3.1). ■

(b) The function $\mathbb{K}^\times \to \mathbb{K}^\times$, $x \mapsto 1/x$ is analytic.

Proof Let $x_0 \in \mathbb{K}^\times$. Then, by Example II.7.4, for each $x \in \mathbb{B}(x_0, |x_0|)$, we have

$$\frac{1}{x} = \frac{1}{x_0} \frac{1}{1 + (x - x_0)/x_0} = \frac{1}{x_0} \sum_{k=0}^{\infty} (-1)^k \Bigl(\frac{x - x_0}{x_0}\Bigr)^k = \sum_{k=0}^{\infty} \frac{(-1)^k}{x_0^{k+1}} (x - x_0)^k \ .$$

This proves that $x \mapsto 1/x$ is analytic on $\mathbb{K}^\times$. ■

3.4 Remarks Let D be open in $\mathbb{K}$ and $f \in \mathbb{K}^D$.

(a) If f is analytic, then the power series expansion of f at x_0 is unique.

Proof This follows from Corollary II.9.9. ■

(b) f is analytic if and only if f is in $C^\infty(D)$ and each $x_0 \in D$ has a neighborhood U in D such that

$$f(x) = \mathcal{T}(f, x_0)(x) \ , \qquad x \in U \ ,$$

that is, at each $x_0 \in D$, $f \in C^\infty(D)$ can be represented locally by its Taylor series.

Proof This follows directly from Corollary 3.2. ■

(c) Analyticity is a local property, that is, f is analytic on D if and only if each $x_0 \in D$ has a neighborhood U such that $f \,|\, U \in C^\omega(U)$.

(d) By Example IV.1.17, the function $f : \mathbb{R} \to \mathbb{R}$ defined by

$$f(x) := \begin{cases} e^{-1/x} \ , & x > 0 \ , \\ 0 \ , & x \leq 0 \ , \end{cases}$$

satisfies $f \in C^\infty(\mathbb{R})$ and $f(x) \neq \mathcal{T}(f, 0)(x) = 0$ for all $x > 0$. Hence there is no neighborhood of 0 on which the function f is represented by its Taylor series and f is not analytic.

(e) $C^\omega(D,\mathbb{K})$ is a subalgebra of $C^\infty(D,\mathbb{K})$ and $\mathbf{1} \in C^\omega(D,\mathbb{K})$.

Proof From Theorem IV.1.12 we know that $C^\infty(D,\mathbb{K})$ is a $\mathbb{K}$-algebra, and so the claim follows from Proposition II.9.7. ■

Next we prove that a power series represents an analytic function on its disk of convergence. In view of Remark 3.4(b) and Corollary 3.2, it suffices to show that a power series is locally representable by its Taylor series.

3.5 Proposition *Suppose that $a = \sum a_k X^k$ is a power series with radius of convergence $\rho > 0$. Then $\underline{a} \in C^\omega(\rho\mathbb{B},\mathbb{K})$ and*

$$\underline{a}(x) = \mathcal{T}(\underline{a}, x_0)(x) , \qquad x_0 \in \rho\mathbb{B} , \quad x \in \mathbb{B}(x_0, \rho - |x_0|) .$$

A power series represents an analytic function on its disk of convergence.

Proof (i) As in the proof of Corollary 3.2, we have

$$\underline{a}^{(k)}(x_0) = \sum_{n=k}^{\infty} n(n-1)\cdots(n-k+1)a_n x_0^{n-k} = k! \sum_{n=k}^{\infty} \binom{n}{k} a_n x_0^{n-k}$$

for all $x_0 \in \rho\mathbb{B}$. Noting that $\binom{n}{k} = 0$ for all $k > n$, we have

$$\mathcal{T}(\underline{a}, x_0) = \sum_{k=0}^{\infty}\Bigl(\sum_{n=0}^{\infty}\binom{n}{k} a_n x_0^{n-k}\Bigr)(X - x_0)^k . \tag{3.1}$$

(ii) With $r := \rho - |x_0| > 0$ and

$$b_{n,k}(x) := \binom{n}{k} a_n x_0^{n-k}(x - x_0)^k , \qquad n,k \in \mathbb{N} , \quad x \in \mathbb{B}(x_0, r) ,$$

it follows from the binomial theorem (Theorem I.8.4) that

$$\sum_{n,k=0}^{m} |b_{n,k}(x)| = \sum_{n=0}^{m} |a_n| \, (|x_0| + |x - x_0|)^n , \qquad m \in \mathbb{N} , \quad x \in \mathbb{B}(x_0, r) . \tag{3.2}$$

For $x \in \rho\mathbb{B}$, we have $|x_0| + |x - x_0| < \rho$, and so, since the power series a converges absolutely on $\rho\mathbb{B}$,

$$M(x) := \sum_{n=0}^{\infty} |a_n| \, (|x_0| + |x - x_0|)^n < \infty .$$

Together with (3.2), we now have

$$\sup_{m\in\mathbb{N}} \sum_{n,k=0}^{m} |b_{n,k}(x)| \le M(x) , \qquad x \in \mathbb{B}(x_0, r) .$$

This implies that the double series $\sum_{n,k}\binom{n}{k}a_n x_0^{n-k}(x-x_0)^k$ is summable for each $x \in \mathbb{B}(x_0,r)$. From Theorem II.8.10(ii) and (3.1) we now get

$$\begin{aligned}\mathcal{T}(\underline{a},x_0)(x) &= \sum_{k=0}^{\infty}\sum_{n=0}^{\infty}\binom{n}{k}a_n x_0^{n-k}(x-x_0)^k \\ &= \sum_{n=0}^{\infty}\Bigl(\sum_{k=0}^{n}\binom{n}{k}x_0^{n-k}(x-x_0)^k\Bigr)a_n = \sum_{n=0}^{\infty}a_n x^n = \underline{a}(x)\end{aligned}$$

for all $x \in \mathbb{B}(x_0,r)$, where we set $\binom{n}{k}=0$ for all $k>n$ and have used once again the binomial theorem. Because of Corollary 3.2 and Remark 3.4(b), this completes the proof. ■

3.6 Corollary

(i) *The functions* exp, cos *and* sin *are analytic on* $\mathbb{K}$.

(ii) *If* $f \in C^\omega(D,\mathbb{K})$, *then* $f' \in C^\omega(D,\mathbb{K})$.

Proof The first claim follows directly from Proposition 3.5. Because of Theorem 3.1, (ii) also follows from Proposition 3.5. ■

Antiderivatives of Analytic Functions

Suppose that D is open in $\mathbb{K}$, E is a normed vector space and $f\colon D \to E$. Then $F\colon D \to E$ is called an **antiderivative** of f if F is differentiable and $F'=f$.

A nonempty open and connected subset of a metric space is called a **domain**.

3.7 Remarks (a) Let $D \subseteq \mathbb{K}$ be a domain and $f\colon D \to E$. If $F_1, F_2 \in E^D$ are antiderivatives of f, then $F_2 - F_1$ is constant. That is, *antiderivatives are unique up to an additive constant.*

Proof (i) Let $F := F_2 - F_1$. Then F is differentiable with $F'=0$. We need to show that F is constant. Fix $x_0 \in D$ and define $Y := \{\, x \in D\ ;\ F(x) = F(x_0)\,\}$. This set is nonempty since it contains x_0.

(ii) We claim that Y is open in D. Let $y \in Y$. Since D is open, there is some $r>0$ such that $\mathbb{B}(y,r) \subseteq D$. For $x \in \mathbb{B}(y,r)$, define $\varphi(t) := F\bigl(y+t(x-y)\bigr)$, $t \in [0,1]$. Then $\varphi\colon [0,1] \to E$ is differentiable, and since $F'=0$, its derivative satisfies

$$\varphi'(t) = F'\bigl(y+t(x-y)\bigr)(x-y) = 0\ , \qquad t \in [0,1]\ .$$

By Remark IV.2.6(a), φ is a constant and so $F(x) = \varphi(1) = \varphi(0) = F(y) = F(x_0)$. This means that $\mathbb{B}(y,r)$ is contained in Y and Y is open in D.

(iii) The function F is differentiable and hence continuous. Since Y is the fiber of F at the point $F(x_0)$, that is, $Y = F^{-1}\bigl(F(x_0)\bigr)$, Y is closed in D (see Example III.2.22(a)).

(iv) Since D is connected, it follows from Remark III.4.3 that $Y = D$, that is, F is constant. ■

(b) Let $a = \sum a_k X^k$ be a power series with radius of convergence $\rho > 0$. Then $\underline{a}$ has an antiderivative on $\rho\mathbb{B}$ represented by the power series $\sum \bigl(a_k/(k+1)\bigr) X^{k+1}$, and this antiderivative is unique up to an additive constant.

Proof Since $\rho\mathbb{B}$ is connected it suffices, by (a), to show that the given power series represents an antiderivative of $\underline{a}$ on $\rho\mathbb{B}$. This follows directly from Theorem 3.1. ∎

3.8 Proposition *If $f \in C^\omega(D, \mathbb{K})$ has an antiderivative F, then F is also analytic.*

Proof Let $x_0 \in D$. Then there is some $r > 0$ such that

$$f(x) = \sum_{k=0}^{\infty} \frac{f^{(k)}(x_0)}{k!}(x - x_0)^k , \qquad x \in \mathbb{B}(x_0, r) \subseteq D .$$

By Remark 3.7(b), there is some $a \in \mathbb{K}$ such that

$$F(x) = a + \sum_{k=0}^{\infty} \frac{f^{(k)}(x_0)}{(k+1)!}(x - x_0)^{k+1} , \qquad x \in \mathbb{B}(x_0, r) . \tag{3.3}$$

It follows from Proposition 3.5, that F is analytic on $\mathbb{B}(x_0, r)$. Since analyticity is a local property, the claim follows. ∎

The Power Series Expansion of the Logarithm

In the next theorem we strengthen the results of Example IV.3.5 and Application IV.3.9(d).

3.9 Theorem *The logarithm function is analytic on $\mathbb{C}\backslash(-\infty, 0]$ and, for all $z \in \mathbb{B}_{\mathbb{C}}$,*

$$\log(1 + z) = \sum_{k=1}^{\infty} (-1)^{k-1} z^k / k .$$

Proof We know from Example IV.1.13(e) that the logarithm function is an antiderivative of $z \mapsto 1/z$ on $\mathbb{C}\backslash(-\infty, 0]$. Thus the first claim follows from Proposition 3.8 and Example 3.3(b).

From the power series expansion

$$\frac{1}{z} = \sum_{k=0}^{\infty} \frac{(-1)^k}{z_0^{k+1}}(z - z_0)^k , \qquad z_0 \in \mathbb{C}^\times , \quad z \in \mathbb{B}_{\mathbb{C}}(z_0, |z_0|) ,$$

and Remark 3.7(b), it follows that

$$\log z = c + \sum_{k=0}^{\infty} \frac{(-1)^k}{(k+1) z_0^{k+1}}(z - z_0)^{k+1} , \qquad z, z_0 \in \mathbb{C}\backslash(-\infty, 0] , \quad |z - z_0| < |z_0| ,$$

for some suitable constant c. By setting $z = z_0$, we find $c = \log z_0$, and so with $z_0 = 1$ we get the claimed power series expansion. ∎

The Binomial Series

The (**general**) **binomial coefficient** for $\alpha \in \mathbb{C}$ and $n \in \mathbb{N}$ is defined by

$$\binom{\alpha}{n} := \frac{\alpha(\alpha-1)\cdot\cdots\cdot(\alpha-n+1)}{n!}\ , \qquad n \in \mathbb{N}^\times\ , \qquad \binom{\alpha}{0} := 1\ .$$

This definition clearly agrees with the definition from Section I.5 if $\alpha \in \mathbb{N}$. Moreover the formulas

$$\binom{\alpha}{n} = \binom{\alpha-1}{n} + \binom{\alpha-1}{n-1} \quad \text{and} \quad \alpha\binom{\alpha-1}{n} = (n+1)\binom{\alpha}{n+1} \tag{3.4}$$

hold for all $\alpha \in \mathbb{C}$ and $n \in \mathbb{N}$ (see Exercise 7). The power series

$$\sum_k \binom{\alpha}{k} X^k \in \mathbb{C}[\![X]\!]$$

is called the **binomial series for the exponent** α. If $\alpha \in \mathbb{N}$, then $\binom{\alpha}{k} = 0$ for all $k > \alpha$ and the binomial series reduces to the polynomial

$$\sum_{k=0}^{\alpha} \binom{\alpha}{k} X^k = (1+X)^\alpha\ .$$

In the following theorem we generalize this statement to the case of arbitrary exponents.

3.10 Theorem *Let $\alpha \in \mathbb{C}\backslash\mathbb{N}$.*

(i) *The binomial series has radius of convergence 1 and*

$$\sum_{k=0}^{\infty} \binom{\alpha}{k} z^k = (1+z)^\alpha\ , \qquad z \in \mathbb{B}_{\mathbb{C}}\ . \tag{3.5}$$

(ii) *The power function $z \mapsto z^\alpha$ is analytic on $\mathbb{C}\backslash(-\infty,0]$ and*

$$z^\alpha = \sum_{k=0}^{\infty} \binom{\alpha}{k} z_0^{\alpha-k}(z-z_0)^k\ , \qquad z, z_0 \in \mathbb{C}\backslash(-\infty,0]\ , \quad |z-z_0| < |z_0|\ .$$

(iii) *For all $z, w \in \mathbb{C}\backslash(-\infty,0]$ such that $z+w \in \mathbb{C}\backslash(-\infty,0]$ and $|z| > |w|$,*

$$(z+w)^\alpha = \sum_{k=0}^{\infty} \binom{\alpha}{k} z^{\alpha-k} w^k\ .$$

(iv) *For all $\alpha \in (0,\infty)$, the binomial series is norm convergent on $\bar{\mathbb{B}}_{\mathbb{C}}$.*

Proof In the following, let $a_k := \binom{\alpha}{k}$.

(i) Since $\alpha \notin \mathbb{N}$ we have $\lim |a_k/a_{k+1}| = \lim_k \bigl((k+1)/|\alpha - k|\bigr) = 1$, and so, by Proposition II.9.4, the binomial series has radius of convergence 1.

Define $f(z) := \sum_{k=0}^{\infty} a_k z^k$ for all $z \in \mathbb{B}_{\mathbb{C}}$. From Theorem 3.1 and (3.4) it follows that

$$f'(z) = \sum_{k=1}^{\infty} k\binom{\alpha}{k} z^{k-1} = \sum_{k=0}^{\infty} (k+1)\binom{\alpha}{k+1} z^k = \alpha \sum_{k=0}^{\infty} \binom{\alpha-1}{k} z^k ,$$

and, using the first formula of (3.4),

$$\begin{aligned}(1+z)f'(z) &= \alpha\Bigl(\sum_{k=0}^{\infty}\binom{\alpha-1}{k}z^k + \sum_{k=0}^{\infty}\binom{\alpha-1}{k}z^{k+1}\Bigr)\\ &= \alpha\Bigl\{1 + \sum_{k=1}^{\infty}\Bigl(\binom{\alpha-1}{k} + \binom{\alpha-1}{k-1}\Bigr)z^k\Bigr\} = \alpha f(z)\end{aligned}$$

for all $z \in \mathbb{B}_{\mathbb{C}}$. Hence

$$(1+z)f'(z) - \alpha f(z) = 0 , \qquad z \in \mathbb{B}_{\mathbb{C}} ,$$

from which follows

$$\bigl[(1+z)^{-\alpha} f(z)\bigr]' = (1+z)^{-\alpha-1}\bigl[(1+z)f'(z) - \alpha f(z)\bigr] = 0 , \qquad z \in \mathbb{B}_{\mathbb{C}} .$$

Since $\mathbb{B}_{\mathbb{C}}$ is a domain, Remark 3.7(a) implies that $(1+z)^{-\alpha} f(z) = c$ for some constant $c \in \mathbb{C}$. Since $f(0) = 1$, we have $c = 1$, and so $f(z) = (1+z)^{\alpha}$ for all $z \in \mathbb{B}_{\mathbb{C}}$.

(ii) Let $z, z_0 \in \mathbb{C}\backslash(-\infty, 0]$ be such that $|z - z_0| < |z_0|$. Then, from (3.5), it follows that

$$\begin{aligned}z^{\alpha} &= \bigl(z_0 + (z - z_0)\bigr)^{\alpha} = z_0^{\alpha}\Bigl(1 + \frac{z - z_0}{z_0}\Bigr)^{\alpha}\\ &= z_0^{\alpha}\sum_{k=0}^{\infty}\binom{\alpha}{k}\frac{(z-z_0)^k}{z_0^k} = \sum_{k=0}^{\infty}\binom{\alpha}{k} z_0^{\alpha-k}(z-z_0)^k .\end{aligned}$$

In particular, $z \mapsto z^{\alpha}$ is analytic on $\mathbb{C}\backslash(-\infty, 0]$.

(iii) Since $|w/z| < 1$, (3.5) implies

$$(z+w)^{\alpha} = z^{\alpha}\Bigl(1 + \frac{w}{z}\Bigr)^{\alpha} = z^{\alpha}\sum_{k=0}^{\infty}\binom{\alpha}{k}\Bigl(\frac{w}{z}\Bigr)^k = \sum_{k=0}^{\infty}\binom{\alpha}{k} z^{\alpha-k} w^k .$$

(iv) Set $\alpha_k := \bigl|\binom{\alpha}{k}\bigr|$ for all $k \in \mathbb{N}$. Then

$$k\alpha_k - (k+1)\alpha_{k+1} = \alpha\alpha_k > 0 , \qquad k > \alpha > 0 . \tag{3.6}$$

Hence the sequence $(k\alpha_k)$ is decreasing for all $k > \alpha$ and there is some $\beta \geq 0$ such that $\lim k\alpha_k = \beta$. This implies

$$\lim_n \sum_{k=0}^{n} \bigl(k\alpha_k - (k+1)\alpha_{k+1}\bigr) = -\lim_n \bigl((n+1)\alpha_{n+1}\bigr) = -\beta \ .$$

From (3.6) we now get

$$\sum_{k>\alpha} \alpha_k = \frac{1}{\alpha} \sum_{k>\alpha} \bigl(k\alpha_k - (k+1)\alpha_{k+1}\bigr) < \infty \ .$$

Because $|a_k z^k| \leq \alpha_k$ for all $|z| \leq 1$, the claim is a consequence of the Weierstrass majorant criterion (Theorem 1.6). ■

3.11 Examples In the following we investigate further the binomial series for the special values $\alpha = 1/2$ and $\alpha = -1/2$.

(a) (The case $\alpha = 1/2$) First we calculate the binomial coefficients:

$$\begin{aligned} \binom{1/2}{k} &= \frac{1}{k!}\frac{1}{2}\Bigl(\frac{1}{2}-1\Bigr)\cdot\,\cdots\,\cdot\Bigl(\frac{1}{2}-k+1\Bigr) \\ &= \frac{(-1)^{k-1}}{k!}\,\frac{1\cdot 3\cdot\,\cdots\,\cdot(2k-3)}{2^k} \\ &= (-1)^{k-1}\frac{1\cdot 3\cdot\,\cdots\,\cdot(2k-3)}{2\cdot 4\cdot\,\cdots\,\cdot 2k} \end{aligned}$$

for all $k \geq 2$. From Theorem 3.10 we get the series expansion

$$\sqrt{1+z} = 1 + \frac{z}{2} + \sum_{k=2}^{\infty} (-1)^{k-1} \frac{1\cdot 3\cdot\,\cdots\,\cdot(2k-3)}{2\cdot 4\cdot\,\cdots\,\cdot 2k} z^k \ , \qquad z \in \bar{\mathbb{B}}_{\mathbb{C}} \ . \tag{3.7}$$

(b) (Calculation of square roots) Write (3.7) in the form

$$\sqrt{1+z} = 1 + \frac{z}{2} - z^2 \sum_{k=0}^{\infty} (-1)^k b_k z^k \ , \qquad z \in \bar{\mathbb{B}}_{\mathbb{C}} \ ,$$

with

$$b_0 := 1/8 \ , \quad b_{k+1} := b_k(2k+3)/(2k+6) \ , \qquad k \in \mathbb{N} \ ,$$

and consider this series on the interval $[0,1]$. From the error estimate for alternating series (Corollary II.7.9) it follows that

$$1 + \frac{x}{2} - x^2\Bigl(\sum_{k=0}^{2n} (-1)^k b_k x^k\Bigr) \leq \sqrt{1+x} \leq 1 + \frac{x}{2} - x^2\Bigl(\sum_{k=0}^{2n+1} (-1)^k b_k x^k\Bigr)$$

for all $n \in \mathbb{N}$ and $x \in [0,1]$.

This provides a further method of calculating numerical approximations of square roots. For example, for $n = 2$ and $x = 1$, we have

$$1 + \frac{1}{2} - \frac{1}{8} + \frac{1}{16} - \frac{5}{128} = 1.39843\ldots \le \sqrt{2} \le 1.39843\ldots + \frac{7}{256} = 1.42578\ldots$$

This method can be used to calculate approximations for the square roots of numbers in the interval $[0, 2]$.

A simple trick can be used to extend this method to numbers greater than 2: To determine the square root of $a > 2$, find $m \in \mathbb{N}$ such that $m^2 < a \le 2m^2$ and set $x := (a - m^2)/m^2$. Then $x \in (0, 1)$ and $a = m^2(1 + x)$. Hence

$$\sqrt{a} = m\sqrt{1+x} = m\Bigl(1 + \frac{x}{2} - \frac{x^2}{8} + \frac{x^3}{16} \mp \cdots\Bigr) ,$$

and

$$m\Bigl[1 + \frac{x}{2} - x^2 \sum_{k=0}^{2n} (-1)^k b_k x^k\Bigr] \le \sqrt{a} \le m\Bigl[1 + \frac{x}{2} - x^2 \sum_{k=0}^{2n+1} (-1)^k b_k x^k\Bigr] .$$

For example, $\sqrt{10}$ has the series expansion

$$\sqrt{10} = 3\Bigl(1 + \frac{1}{2\cdot 9} - \frac{1}{8\cdot 81} + \frac{1}{16\cdot 729} - \frac{5}{128\cdot 6561} \pm \cdots\Bigr)$$

which yields the inequalities

$$\begin{aligned} 3\Bigl(1 + \frac{1}{18} - \frac{1}{648} + \frac{1}{11664} - \frac{5}{839808}\Bigr) &= 3.16227637\ldots \le \sqrt{10} \\ &\le 3.16227637\ldots + \frac{21}{15116544} \\ &= 3.16227776\ldots \end{aligned}$$

For comparison, the exact decimal expansion of $\sqrt{10}$ begins $3.162277660\ldots$

(c) (The case $\alpha = -1/2$) Here we have

$$\binom{-1/2}{k} = (-1)^k \frac{1\cdot 3\cdot \cdots \cdot (2k-1)}{2\cdot 4\cdot \cdots \cdot 2k} , \qquad k \ge 2 .$$

From Theorem 3.10 we get

$$\frac{1}{\sqrt{1+z}} = 1 - \frac{z}{2} + \sum_{k=2}^{\infty} (-1)^k \frac{1\cdot 3\cdot \cdots \cdot (2k-1)}{2\cdot 4\cdot \cdots \cdot 2k} z^k , \qquad z \in \bar{\mathbb{B}}_{\mathbb{C}} .$$

If $|z| < 1$ then $|-z^2| < 1$ and so we can substitute $-z^2$ for z to get

$$\frac{1}{\sqrt{1-z^2}} = 1 + \frac{z^2}{2} + \sum_{k=2}^{\infty} \frac{1\cdot 3\cdot \cdots \cdot (2k-1)}{2\cdot 4\cdot \cdots \cdot 2k} z^{2k} , \qquad z \in \bar{\mathbb{B}}_{\mathbb{C}} . \tag{3.8}$$

In particular, it follows from Proposition 3.5 that the function $z \mapsto 1/\sqrt{1-z^2}$ is analytic on $\mathbb{B}_{\mathbb{C}}$. ■

For real arguments, (3.8) provides a power series expansion for the arcsine function.

3.12 Corollary *The arcsine function is real analytic on* $(-1,1)$ *and*

$$\arcsin(x) = x + \sum_{k=1}^{\infty} \frac{1 \cdot 3 \cdot \cdots \cdot (2k-1)}{2 \cdot 4 \cdot \cdots \cdot 2k} \frac{x^{2k+1}}{2k+1} , \qquad x \in (-1,1) .$$

Proof By Remark 3.7(b) and (3.8),

$$F(x) := x + \frac{x^3}{2 \cdot 3} + \sum_{k=2}^{\infty} \frac{1 \cdot 3 \cdot \cdots \cdot (2k-1)}{2 \cdot 4 \cdot \cdots \cdot 2k} \frac{x^{2k+1}}{2k+1} , \qquad x \in (-1,1) ,$$

is an antiderivative of $f : (-1,1) \to \mathbb{R},\ x \mapsto 1/\sqrt{1-x^2}$. Since the arcsine function is another antiderivative of f (Application IV.2.10) and $F(0) = 0 = \arcsin(0)$, it follows from Remark 3.4(a) that $F = \arcsin$. Finally, Proposition 3.5 shows that arcsin is analytic on $(-1,1)$. ■

The Identity Theorem for Analytic Functions

To close this section we prove an important *global* property of analytic functions: If an analytic function is zero on an open subset of its domain D, then it is zero on all of D.

3.13 Theorem (identity theorem for analytic functions) *Let D be a domain in $\mathbb{K}$ and $f \in C^\omega(D, \mathbb{K})$. If the set of zeros of f has a limit point in D, then f is zero on D.*

Proof Set

$$Y := \bigl\{ x \in D \ ;\ \exists\, (x_n) \text{ in } D \backslash \{x\} \text{ such that } \lim x_n = x \text{ and } f(x_n) = 0 \text{ for } n \in \mathbb{N} \bigr\} .$$

By supposition, Y is nonempty. Since f is continuous, we have $f(y) = 0$ for all $y \in Y$. Hence every limit point of Y is contained in Y, and, by Proposition III.2.11, Y is closed in D. Let $x_0 \in Y$. Since f is analytic, there is some neighborhood V of x_0 in D and a power series $\sum a_k X^k$ such that $f(x) = \sum a_k (x - x_0)^k$ for all $x \in V$. Since x_0 is in Y, there is a sequence (y_n) in $V \backslash \{x_0\}$ such that $y_n \to x_0$ and $f(y_n) = 0$ for all $n \in \mathbb{N}$. It then follows from the identity theorem for power series (Corollary II.9.9) that $a_k = 0$ for all $k \in \mathbb{N}$, that is, f is zero on V, and also that V is contained in Y. We have therefore shown that Y is open in D.

Since Y is a nonempty, open and closed subset of the domain D and D is connected, we have $Y = D$ (see Remark III.4.3). ■

3.14 Remarks (a) Let D be a domain in $\mathbb{K}$ and $f, g \in C^\omega(D,\mathbb{K})$. If there is a sequence (x_n) which converges in D such that $x_n \neq x_{n+1}$ and $f(x_n) = g(x_n)$ for all $n \in \mathbb{N}$, then $f = g$.

Proof The function $h := f - g$ is analytic on D and $\lim x_n$ is a limit point in D of the set of zeros of h, so the claim follows from Theorem 3.13. ■

(b) If D is open in $\mathbb{R}$, then $C^\omega(D,\mathbb{R})$ is a proper subalgebra of $C^\infty(D,\mathbb{R})$.

Proof Since both differentiability and analyticity are local properties, we can suppose that D is a bounded open interval. It is easy to see that, for all $x_0 \in D$ and $f \in C^\omega(D,\mathbb{R})$, the function $x \mapsto f(x - x_0)$ is analytic on $x_0 + D$. Thus it suffices to consider the case $D := (-a, a)$ for some $a > 0$. Let f be the restriction to D of the function of Example IV.1.17. Then $f \in C^\infty(D,\mathbb{R})$ and $f\,|\,(-a,0) = 0$, but $f \neq 0$. Hence it follows from (a) that f is not analytic. ■

(c) A nonzero analytic function may have infinitely many zeros, as the cosine function shows. Theorem 3.13 simply says that these zeros cannot have a limit point in the domain of the function.

(d) The proof of (b) shows that, in the real case, the analyticity of f is necessary in Theorem 3.13. In the complex case, the situation is completely different. We will see later that the concepts of 'complex differentiability' and 'complex analyticity' are the same, so that, for each open subset D of $\mathbb{C}$, $C^1(D,\mathbb{C})$ and $C^\omega(D,\mathbb{C})$ coincide. ■

3.15 Remark Suppose that D is open in $\mathbb{R}$ and $f : D \to \mathbb{R}$ is (real) analytic. Then, for each $x \in D$, there is some $r_x > 0$ such that

$$f(y) = \sum_{k=0}^{\infty} \frac{f^{(k)}(x)}{k!}(y - x)^k \;, \qquad y \in \mathbb{B}_{\mathbb{R}}(x, r_x) \cap D \;.$$

The set

$$D_{\mathbb{C}} := \bigcup_{x \in D} \mathbb{B}_{\mathbb{C}}(x, r_x)$$

is an open neighborhood of D in $\mathbb{C}$. By Proposition 3.5, for each $x \in D$,

$$f_{\mathbb{C},x}(z) := \sum_{k=0}^{\infty} \frac{f^{(k)}(x)}{k!}(z - x)^k \;, \qquad z \in \mathbb{B}_{\mathbb{C}}(x, r_x) \;,$$

defines an analytic function on $\mathbb{B}_{\mathbb{C}}(x, r_x)$. The identity theorem for analytic functions implies that any two such functions, $f_{\mathbb{C},x}$ and $f_{\mathbb{C},y}$, coincide on the intersection of their domains. This means that

$$f_{\mathbb{C}}(z) := f_{\mathbb{C},x}(z) \;, \qquad z \in \mathbb{B}_{\mathbb{C}}(x, r_x) \;, \quad x \in D \;,$$

defines an analytic function $f_{\mathbb{C}} : D_{\mathbb{C}} \to \mathbb{C}$ such that $f_{\mathbb{C}} \supseteq f$. The function $f_{\mathbb{C}}$ is called the **analytic continuation** of f on $D_{\mathbb{C}}$.

Now suppose that D is open in $\mathbb{C}$. Set $D_{\mathbb{R}} := D \cap \mathbb{R} \neq \emptyset$. If $f \in C^{\omega}(D, \mathbb{C})$ and $f(D_{\mathbb{R}}) \subseteq \mathbb{R}$, then $f \,|\, D_{\mathbb{R}}$ is real analytic.

These considerations show that in our further investigation of analytic functions, we can limit ourselves to the complex case. ■

Exercises

1 Let D be open in $\mathbb{C}$ with $D_{\mathbb{R}} := D \cap \mathbb{R} \neq \emptyset$ and $f \in C^{\omega}(D, \mathbb{C})$. Show the following:

(a) $(\operatorname{Re} f) \,|\, D_{\mathbb{R}}$ and $(\operatorname{Im} f) \,|\, D_{\mathbb{R}}$ are real analytic.

(b) Let $f = \sum a_k (X - x_0)^k$ be a power series expansion of f at $x_0 \in D_{\mathbb{R}}$ with radius of convergence $\rho > 0$. Set $\widetilde{D} := D_{\mathbb{R}} \cap (x_0 - \rho, x_0 + \rho)$. Then the following are equivalent:

(i) $f \,|\, \widetilde{D} \in C^{\omega}(\widetilde{D}, \mathbb{R})$.

(ii) $a_k \in \mathbb{R}$ for each $k \in \mathbb{N}$.

2 Suppose that $f \in C^{\omega}(D, \mathbb{K})$ has no zeros. Show that $1/f$ is also analytic.
(Hint: Use the division algorithm of Exercise II.9.9.)

3 Define $h \colon \mathbb{C} \to \mathbb{C}$ by

$$h(z) := \begin{cases} (e^z - 1)/z \,, & z \in \mathbb{C}^{\times} \,, \\ 1 \,, & z = 0 \,. \end{cases}$$

Show that $h \in C^{\omega}(\mathbb{C}, \mathbb{C})$ and $h(z) \neq 0$ if $|z| < 1/(e-1)$.
(Hint: For analyticity, consider the series $\sum X^k/(k+1)!$. From Remark II.8.2(c), we get the inequality

$$|h(z)| = \left| \frac{e^z - 1}{z} \right| \geq 1 - \sum_{k=1}^{\infty} \frac{|z|^k}{(k+1)!}$$

for all $z \in \mathbb{C}$.)

4 Let $h \colon \mathbb{C} \to \mathbb{C}$ be as in Exercise 3. By Exercises 2 and 3, the function $1/h$ is analytic on $\mathbb{B}\big(0, 1/(e-1)\big)$, and so there are $\rho > 0$ and $B_k \in \mathbb{C}$ such that

$$\frac{1}{h} = \sum_{k=0}^{\infty} \frac{B_k}{k!} z^k \,, \qquad z \in \rho\mathbb{B} \,.$$

Calculate $B_0, \ldots, B_{10}$ and show that all B_k are rational.

5 Suppose that D is a domain in $\mathbb{C}$ and $f \in C^{\omega}(D, \mathbb{C})$ satisfies one of the following conditions:

(i) $\operatorname{Re} f = \text{const}$.

(ii) $\operatorname{Im} f = \text{const}$.

(iii) $\overline{f} \in C^{\omega}(D, \mathbb{C})$.

(iv) $|f| = \text{const}$.

Show that f is constant. (Hint: (i) Using a suitable difference quotient, show that $f'(z) \in i\mathbb{R} \cap \mathbb{R}$. (iii) $2 \operatorname{Re} f = f + \overline{f}$ and (i). (iv) $|f|^2 = f\overline{f}$ and Exercise 2.)

6 Let $f \in C^\omega(\rho\mathbb{B})$ be represented by $\sum a_k X^k$ on $\rho\mathbb{B}$ for some $\rho > 0$. Suppose that (x_n) is a null sequence in $(\rho\mathbb{B})\backslash\{0\}$. Show that the following are equivalent:

(i) f is even.

(ii) $f(x_n) = f(-x_n),\ n \in \mathbb{N}$.

(iii) $a_{2m+1} = 0,\ m \in \mathbb{N}$.

Formulate an analogous characterization of odd analytic functions on $\rho\mathbb{B}$.

7 Prove the formulas (3.4).

8 Show that

$$\binom{\alpha+\beta}{k} = \sum_{\ell=0}^{k} \binom{\alpha}{\ell}\binom{\beta}{k-\ell}$$

for all $\alpha, \beta \in \mathbb{C}$ and $k \in \mathbb{N}$.

9 Verify that the functions

(a) $\sinh : \mathbb{C} \to \mathbb{C}$, $\cosh : \mathbb{C} \to \mathbb{C}$, $\tanh : \mathbb{C}\backslash i\pi(\mathbb{Z}+1/2) \to \mathbb{C}$;

(b) $\tan : \mathbb{C}\backslash\pi(\mathbb{Z}+1/2) \to \mathbb{C}$, $\cot : \mathbb{C}\backslash\pi\mathbb{Z} \to \mathbb{C}$;

are analytic. (Hint: Use Proposition 3.8.)

10 Show that the functions

$$\ln(\cos)\ , \quad \ln(\cosh)\ , \quad x \mapsto \ln^2(1+x)$$

are analytic in a neighborhood of 0. What are the corresponding power series expansions at 0? (Hint: First find power series expansions for the derivatives.)

11 Prove that, for $x \in [-1, 1]$,

$$\arctan x = \sum_{k=0}^{\infty} (-1)^k \frac{x^{2k+1}}{2k+1} = x - \frac{x^3}{3} + \frac{x^5}{5} - \frac{x^7}{7} + - \cdots\ ,$$

and hence (**Leibniz formula**)

$$\frac{\pi}{4} = \sum_{k=0}^{\infty} \frac{(-1)^k}{2k+1} = 1 - \frac{1}{3} + \frac{1}{5} - \frac{1}{7} + - \cdots\ .$$

(Hint: $\arctan' x = 1/(1+x^2)$. For $x = \pm 1$, convergence follows from the Leibniz criterion (Theorem II.7.8).)

4 Polynomial Approximation

An analytic function is represented locally by power series and so, near a given point x_0, it can be approximated with arbitrary precision by polynomials, that is, the error can be made arbitrarily small by allowing polynomials of sufficiently high degrees and by limiting the approximation to sufficiently small neighborhoods of the point x_0. Here the approximating polynomial, the Taylor polynomial, is given explicitly in terms of the values of the function to be approximated and its derivatives at the point x_0. In addition, the error in the approximation can be controlled using the various formulas for the remainder of Taylor series. This fact lies behind the great importance of Taylor's theorem, particularly for numerical mathematics which considers the derivation of efficient algorithms for the approximation of functions and solutions of equations.

In this section we investigate the problem of the *global* approximation of functions by polynomials. The main result of this section, the Stone-Weierstrass theorem, guarantees the existence of such polynomials for arbitrary continuous functions on compact subsets of $\mathbb{R}^n$.

Banach Algebras

An algebra A which is also is a Banach space satisfying

$$\|ab\| \leq \|a\| \, \|b\| \ , \qquad a,b \in A \ ,$$

is called a **Banach algebra**. If A contains a unity element e, we also require that $\|e\| = 1$.

4.1 Examples **(a)** Let X be a nonempty set. Then $B(X,\mathbb{K})$ is a Banach algebra with unity element $\mathbf{1}$.

Proof From Theorem II.6.6 we know that $B(X,\mathbb{K})$ is a Banach space. Moreover

$$\|fg\|_\infty = \sup_{x\in X} |f(x)g(x)| \leq \sup_{x\in X} |f(x)| \sup_{x\in X} |g(x)| = \|f\|_\infty \, \|g\|_\infty \ , \qquad f,g \in B(X,\mathbb{K}) \ .$$

This shows that $B(X,\mathbb{K})$ is a subalgebra of $\mathbb{K}^X$. For the unity element $\mathbf{1}$ of $\mathbb{K}^X$, we have $\mathbf{1} \in B(X,E)$ and $\|\mathbf{1}\|_\infty = 1$. ■

(b) Let X be a metric space. Then $BC(X,\mathbb{K})$ is a closed subalgebra of $B(X,\mathbb{K})$ which contains $\mathbf{1}$, and so is a Banach algebra with unity.

Proof By Theorem 2.6, $BC(X,\mathbb{K})$ is a closed subspace of $B(X,\mathbb{K})$, and so is itself a Banach space. The claim then follows from (a) and Proposition III.1.5. ■

(c) Let X be a compact metric space. Then $C(X,\mathbb{K})$ is a Banach algebra with unity element $\mathbf{1}$.

Proof In Theorem 2.6 we showed that the Banach spaces $C(X,\mathbb{K})$ and $BC(X,\mathbb{K})$ coincide in this circumstance. ■

(d) In a Banach algebra A, the multiplication operation $A \times A \to A,\ (a,b) \mapsto ab$ is continuous.

Proof For all (a,b) and (a_0,b_0) in $A \times A$ we have

$$\|ab - a_0b_0\| \le \|a - a_0\|\,\|b\| + \|a_0\|\,\|b - b_0\| \ ,$$

from which the claim follows easily (see the proof of Example III.1.3(m)). ■

(e) If B is a subalgebra of a Banach algebra A, then $\overline{B}$ is a Banach algebra.

Proof For $a, b \in \overline{B}$, there are sequences (a_n) and (b_n) such that $a_n \to a$ and $b_n \to b$ in A. From Proposition II.2.2 and Remark II.3.1(c) it follows that

$$a + \lambda b = \lim a_n + \lambda \lim b_n = \lim(a_n + \lambda b_n) \in \overline{B}$$

for all $\lambda \in \mathbb{K}$. Thus $\overline{B}$ is a closed subspace of A and hence also a Banach space. Because of (d), we also have $a_n b_n \to ab$, so that ab is in $\overline{B}$. Consequently $\overline{B}$ is a subalgebra of A and hence a Banach algebra. ■

Density and Separability

A subset D of a metric space X is **dense** in X if $\overline{D} = X$. A metric space is called **separable** if it contains a countable dense subset.

4.2 Remarks **(a)** The following are equivalent:

(i) D is dense in X.

(ii) For each $x \in X$ and neighborhood U of x, we have $U \cap D \neq \emptyset$.

(iii) For each $x \in X$, there is a sequence (d_n) in D such that $d_n \to x$.

(b) Suppose that $X_1, \ldots, X_m$ are metric spaces, and, for $1 \le j \le m$, D_j is dense in X_j. Then $D_1 \times \cdots \times D_m$ is dense in $X_1 \times \cdots \times X_m$.

Proof This is a direct consequence of (a) and Example II.1.8(e). ■

(c) The definitions of density and separability are clearly valid also for general topological spaces. Statements (i) and (ii) of (a) are equivalent to each other in general topological spaces, but not to (iii).

(d) Let X and Y be metric spaces and $h\colon X \to Y$ a homeomorphism. Then D is dense in X if and only if $h(D)$ is dense in Y.

Proof This follows directly from the characterization (ii) of (a) and the fact that homeomorphisms map neighborhoods to neighborhoods (see Exercise III.3.3). ■

4.3 Examples **(a)** $\mathbb{Q}$ is dense in $\mathbb{R}$. In particular, $\mathbb{R}$ is separable.

Proof This follows from Propositions I.10.8 and I.9.4. ■

(b) The irrational numbers $\mathbb{R}\backslash\mathbb{Q}$ form a dense subset of $\mathbb{R}$.

Proof This we proved in Proposition I.10.11. ■

(c) For any subset A of X, A is dense in $\overline{A}$.

(d) $\mathbb{Q}+i\mathbb{Q}$ is dense in $\mathbb{C}$. In particular, $\mathbb{C}$ is separable.

Proof This follows from (a) and Remark 4.2(b) (see also Remark II.3.13(e)). ■

(e) Any finite dimensional normed vector space is separable. In particular, $\mathbb{K}^n$ is separable.

Proof Let V be a normed vector space over $\mathbb{K}$ and $(b_1,\ldots,b_n)$ a basis for V. By (a) and (d), $\mathbb{K}$ is separable. Let D be a countable dense subset of $\mathbb{K}$ and

$$V_D := \Bigl\{\, \textstyle\sum_{k=1}^n \alpha_k b_k \ ; \ \alpha_k \in D \Bigr\} .$$

Then V_D is countable and dense in V (see Exercise 6). ■

In the following proposition we collect several useful equivalent formulations of density.

4.4 Proposition *Let X be a metric space and $D \subseteq X$. Then the following are equivalent:*

(i) *D is dense in X.*

(ii) *If A is closed and $D \subseteq A \subseteq X$, then $A = X$. Thus X is the unique closed dense subset of X.*

(iii) *For each $x \in X$ and $\varepsilon > 0$, there is some $y \in D$ such that $d(x,y) < \varepsilon$.*

(iv) *The complement of D has empty interior, that is, $(D^c)^\circ = \emptyset$.*

Proof '(i)⇒(ii)' Let A be closed with $D \subseteq A \subseteq X$. From Corollary III.2.13 it follows that $X = \overline{D} \subseteq \overline{A} = A$, that is, $A = X$.

'(ii)⇒(iii)' We argue by contradiction. If $x \in X$ and $\varepsilon > 0$ are such that $D \cap \mathbb{B}(x,\varepsilon) = \emptyset$, then $D \subseteq \bigl[\mathbb{B}(x,\varepsilon)\bigr]^c$. This contradicts (ii), since $\bigl[\mathbb{B}(x,\varepsilon)\bigr]^c$ is a closed subset of X such that $\bigl[\mathbb{B}(x,\varepsilon)\bigr]^c \neq X$.

'(iii)⇒(iv)' Suppose that $(D^c)^\circ$ is not empty. Since $(D^c)^\circ$ is open, there are $x \in (D^c)^\circ$ and $\varepsilon > 0$ such that $\mathbb{B}(x,\varepsilon) \subseteq (D^c)^\circ \subseteq D^c$. This implies $D \cap \mathbb{B}(x,\varepsilon) = \emptyset$, contradicting (iii).

'(iv)⇒(i)' From Exercise III.2.5, we have $\mathring{V} = X \setminus \overline{(X \setminus V)}$ for any subset V of X. From (iv) it follows that

$$\emptyset = (D^c)^\circ = X \setminus \overline{(D^c)^c} = X \setminus \overline{D} ,$$

that is, $\overline{D} = X$. This completes the proof. ■

Of course, condition (iii) is also equivalent to condition (ii) of Remark 4.2(a).

The Stone-Weierstrass Theorem

As preparation for the proof of the Stone-Weierstrass theorem, we prove the following two lemmas.

4.5 Lemma

$$|t| = \sum_{k=0}^{\infty} \binom{1/2}{k} (t^2-1)^k , \qquad t \in [-1,1] ,$$

and this series is norm convergent on $[-1,1]$.

Proof Set $x := t^2 - 1$ for $t \in [-1,1]$. Then

$$|t| = \sqrt{t^2} = \sqrt{1+t^2-1} = \sqrt{1+x} ,$$

and so the claim follows from Theorem 3.10. ■

4.6 Lemma *Let X be a compact metric space and A a closed subalgebra of $C(X,\mathbb{R})$ containing $\mathbf{1}$. If f and g are in A, then so are $|f|$, $f \vee g$ and $f \wedge g$.*

Proof Let $f, g \in A$. From Exercise I.8.11 we have

$$f \vee g = \frac{1}{2}(f+g+|f-g|) , \quad f \wedge g = \frac{1}{2}(f+g-|f-g|) .$$

Hence it suffices to prove that, if f is in A, then so is $|f|$. In addition, we need only consider the case $f \neq 0$. From Lemma 4.5 we have

$$\Bigl| |t| - \sum_{k=0}^{m} \binom{1/2}{k} (t^2-1)^k \Bigr| \le \sum_{k=m+1}^{\infty} \Bigl| \binom{1/2}{k} \Bigr| , \qquad t \in [-1,1] ,$$

where the right hand side converges to zero as $m \to \infty$. Thus, for each $\varepsilon > 0$, there is some $P_\varepsilon \in \mathbb{R}[t]$ such that

$$\bigl| |t| - P_\varepsilon(t) \bigr| < \varepsilon / \|f\|_\infty , \qquad t \in [-1,1] .$$

Setting $t := f(x)/\|f\|_\infty$, we get

$$\|f\|_\infty \Bigl| |f(x)/\,\|f\|_\infty| - P_\varepsilon\bigl(f(x)/\|f\|_\infty\bigr) \Bigr| < \varepsilon , \qquad x \in X .$$

Define $g_\varepsilon := \|f\|_\infty P_\varepsilon(f/\|f\|_\infty)$. Since A is a subalgebra of $C(X,\mathbb{R})$ containing $\mathbf{1}$, g_ε is in A. We have therefore shown that, for each $\varepsilon > 0$, there is some $g \in A$ such that $\bigl\| |f| - g \bigr\|_\infty < \varepsilon$. Thus $|f|$ is in $\overline{A}$. By hypothesis, A is closed, and so the claim follows. ■

A subset M of $C(X,\mathbb{K})$ **separates the points of** X if, for each $(x,y) \in X \times X$ with $x \neq y$, there is some $m \in M$ such that $m(x) \neq m(y)$. The set M is called **self adjoint** if $m \in M$ implies $\overline{m} \in M$.[1]

After this preparation we can now prove the main theorem of this section.

4.7 Theorem (Stone-Weierstrass theorem) *Let X be a compact metric space and A a subalgebra of $C(X,\mathbb{K})$ containing $\mathbf{1}$. If A separates the points of X and is self adjoint, then A is dense in $C(X,\mathbb{K})$. That is, for each $f \in C(X,\mathbb{K})$ and $\varepsilon > 0$, there is some $a \in A$ such that $\|f - a\|_\infty < \varepsilon$.*

Proof We prove the cases $\mathbb{K} = \mathbb{R}$ and $\mathbb{K} = \mathbb{C}$ separately.

(a) Suppose that $f \in C(X,\mathbb{R})$ and $\varepsilon > 0$.

(i) We claim that, for each pair $y,z \in X$, there is some $h_{y,z} \in A$ such that

$$h_{y,z}(y) = f(y) \quad \text{and} \quad h_{y,z}(z) = f(z) . \tag{4.1}$$

Indeed, if $y = z$, then the constant function $h_{y,z} := f(y)\mathbf{1}$ satisfies (4.1). If $y \neq z$, then, since A separates the points of X, there is some $g \in A$ such that $g(y) \neq g(z)$. Now define

$$h_{y,z} := f(y)\mathbf{1} + \frac{f(z) - f(y)}{g(z) - g(y)}\big(g - g(y)\mathbf{1}\big) .$$

Since $h_{y,z}$ is in A with $h_{y,z}(y) = f(y)$ and $h_{y,z}(z) = f(z)$, (4.1) holds.

(ii) For $y,z \in X$, set

$$U_{y,z} := \{ x \in X \,;\, h_{y,z}(x) < f(x) + \varepsilon \} , \quad V_{y,z} := \{ x \in X \,;\, h_{y,z}(x) > f(x) - \varepsilon \} .$$

Since $h_{y,z} - f$ is continuous, we know from Example III.2.22(c) that $U_{y,z}$ and $V_{y,z}$ are open in X. By (4.1), y is in $U_{y,z}$ and z is in $V_{y,z}$. Now fix some $z \in X$. Then $\{ U_{y,z} \ , \ y \in X \}$ is an open cover of the compact space X, and there are $y_0,\ldots,y_m$ in X such that $\bigcup_{j=0}^m U_{y_j,z} = X$. Set

$$h_z := \min_{0 \le j \le m} h_{y_j,z} := h_{y_0,z} \wedge \cdots \wedge h_{y_m,z} .$$

By Lemma 4.6, h_z is in $\overline{A}$. In addition, we have

$$h_z(x) < f(x) + \varepsilon , \qquad x \in X , \tag{4.2}$$

since, for each $x \in X$, there is some $j \in \{0,\ldots,m\}$ such that $x \in U_{y_j,z}$.

(iii) For $z \in X$, let $V_z := \bigcap_{j=0}^m V_{y_j,z}$. Then we have

$$h_z(x) > f(x) - \varepsilon , \qquad x \in V_z . \tag{4.3}$$

[1] This condition is always true in the real case: Any subset of $C(X,\mathbb{R})$ is self adjoint.

By (4.1), $\{ V_z \ ; \ z \in X \}$ is an open cover of X. Since X is compact, there are $z_0, \ldots, z_n$ in X such that $X = \bigcup_{k=0}^{n} V_{z_k}$. Set

$$h := \max_{0 \le k \le n} h_{z_k} := h_{z_0} \vee \cdots \vee h_{z_n} \ .$$

Then Lemma 4.6 and Example 4.1(e) show that h is in $\overline{A}$. In addition, from (4.2) and (4.3) follow the inequalities

$$f(x) - \varepsilon < h(x) < f(x) + \varepsilon \ , \qquad x \in X \ .$$

Thus $\|f - h\|_\infty < \varepsilon$. Since h is in $\overline{A}$, there is some $a \in A$ such that $\|h - a\|_\infty < \varepsilon$, and hence $\|f - a\|_\infty < 2\varepsilon$. Since $\varepsilon > 0$ was arbitrary, the claim now follows from Proposition 4.4.

(b) Let $\mathbb{K} = \mathbb{C}$.

(i) Let $A_{\mathbb{R}}$ be the set of all real valued functions in A. Then $A_{\mathbb{R}}$ is an algebra over the field $\mathbb{R}$. Because A is self adjoint, for each $f \in A$, the functions $\operatorname{Re} f = (f + \overline{f})/2$ and $\operatorname{Im} f = (f - \overline{f})/2i$ are in $A_{\mathbb{R}}$. Hence $A \subseteq A_{\mathbb{R}} + i\, A_{\mathbb{R}}$. Since also $A_{\mathbb{R}} + i\, A_{\mathbb{R}} \subseteq A$, we have shown that $A = A_{\mathbb{R}} + i\, A_{\mathbb{R}}$.

(ii) Suppose that $y, z \in X$ are such that $y \neq z$. Because A separates the points of X, there is some $f \in A$ such that $f(y) \neq f(z)$, that is, either $\operatorname{Re} f(y) \neq \operatorname{Re} f(z)$ or $\operatorname{Im} f(y) \neq \operatorname{Im} f(z)$. Thus $A_{\mathbb{R}}$ also separates the points of X. Using the result proved in (a), we now have $C(X, \mathbb{R}) = \overline{A}_{\mathbb{R}}$, and consequently

$$\overline{A} \subseteq C(X, \mathbb{C}) = C(X, \mathbb{R}) + i\, C(X, \mathbb{R}) = \overline{A}_{\mathbb{R}} + i\, \overline{A}_{\mathbb{R}} \ . \tag{4.4}$$

(iii) Finally, let $f \in \overline{A}_{\mathbb{R}} + i\, \overline{A}_{\mathbb{R}}$. Then there are $g, h \in \overline{A}_{\mathbb{R}}$ such that $f = g + i\, h$, and hence sequences (g_k) and (h_k) in A such that $g_k \to g$ and $h_k \to h$ in $C(X, \mathbb{R})$. Since the sequence $(g_k + i\, h_k)$ converges in $C(X, \mathbb{C})$ to $g + i\, h = f$, this implies that f is in $\overline{A}$, and hence $C(X, \mathbb{C}) = \overline{A}_{\mathbb{R}} + i\, \overline{A}_{\mathbb{R}} \subseteq \overline{A}$. This, together with (4.4), completes the proof. ■

4.8 Corollary *Let $M \subseteq \mathbb{R}^n$ be compact.*

(a) *Any continuous $\mathbb{K}$-valued function on M can be uniformly approximated by a polynomial in n variables, that is, $\mathbb{K}[X_1, \ldots, X_n] \,|\, M$ is dense in $C(M, \mathbb{K})$.*

(b) *The Banach space $C(M, \mathbb{K})$ is separable.*

Proof (a) Set $A := \mathbb{K}[X_1, \ldots, X_n] \,|\, M$. Then A is clearly a subalgebra of $C(M, \mathbb{K})$ containing $\mathbf{1}$. In addition, A separates the points of M and is self adjoint (see Exercise 7). Thus the claim follows from the Stone-Weierstrass theorem.

(b) If $\mathbb{K} = \mathbb{R}$, then $\mathbb{Q}[X_1, \ldots, X_n] \,|\, M$ is a countable dense subset of $C(M, \mathbb{R})$. If $\mathbb{K} = \mathbb{C}$, then $(\mathbb{Q} + i\mathbb{Q})[X_1, \ldots, X_n] \,|\, M$ has the desired properties. ■

4.9 Corollary (Weierstrass approximation theorem) *Let $-\infty < a < b < \infty$. Then, for each $f \in C([a,b],\mathbb{K})$ and $\varepsilon > 0$, there is a polynomial p with coefficients in $\mathbb{K}$ such that $|f(x) - p(x)| < \varepsilon$ for all $x \in [a,b]$.*

Using the Stone-Weierstrass theorem we can easily construct an example of a normed vector space which is not complete.

4.10 Examples (a) Let I be a compact perfect interval and $\mathcal{P}$ the subalgebra of $C(I)$ consisting of all (restrictions of) polynomials on I. Then $\mathcal{P}$ is a normed vector space, but not a Banach space.

Proof By Corollary 4.9, $\mathcal{P}$ is dense in $C(I)$. Since $\exp|X$ is in $C(I)$, but not in $\mathcal{P}$, $\mathcal{P}$ is a proper subspace of $C(I)$. It follows from Proposition 4.4 that $\mathcal{P}$ is not closed, and hence not complete. ■

(b) Let I be a compact interval and $\epsilon := \exp|I$. Then

$$A := \left\{ \sum_{k=0}^{n} a_k \epsilon^k \ ; \ a_k \in \mathbb{K},\ n \in \mathbb{N} \right\}$$

is a dense subalgebra of $C(I,\mathbb{K})$. So any continuous function on I can be uniformly approximated by 'sums of exponential functions' of the form $t \mapsto \sum_{k=0}^{n} a_k e^{tk}$.

Proof Clearly A is a subalgebra of $C(I,\mathbb{K})$ and $\mathbf{1} \in A$. Since $\epsilon(s) \neq \epsilon(t)$ for $s \neq t$, A separates the points of I. Since A is self adjoint, the claim follows from Theorem 4.7. ■

(c) Let $S := S^1 := \{ z \in \mathbb{C} \ ; \ |z| = 1 \}$ and $\chi(z) := z$ for $z \in S$. Define

$$\mathcal{P}(S) := \mathcal{P}(S,\mathbb{C}) := \left\{ \sum_{k=-n}^{n} c_k \chi_k \ ; \ c_k \in \mathbb{C},\ n \in \mathbb{N} \right\}$$

where $\chi_k := \chi^k$ for all $k \in \mathbb{Z}$. Then $\mathcal{P}(S)$ is a dense subalgebra of $C(S) := C(S,\mathbb{C})$.

Proof Clearly, $\mathcal{P} := \mathcal{P}(S)$ is a subalgebra of $C(S)$ with $\mathbf{1} \in \mathcal{P}$. Because $\chi(z) \neq \chi(w)$ for $z \neq w$, $\mathcal{P}$ separates the points of S, and, since $\overline{\chi_k} = \chi_{-k}$, $\mathcal{P}$ is self adjoint. So the claim follows again from Theorem 4.7. ■

The great generality of the Stone-Weierstrass theorem is obtained at the cost of a nonconstructive proof. In the context of the classical Weierstrass approximation theorem, that is, uniform approximations of continuous functions by polynomials, an explicit procedure for the construction of the approximating polynomials is possible (see Exercises 11 and 12).

Trigonometric Polynomials

We consider again Example 4.10(c) with the substitution $z = e^{it}$ for $t \in \mathbb{R}$. Then, for all $k \in \mathbb{N}$ and $c_k, c_{-k} \in \mathbb{C}$, Euler's formula (III.6.1) implies that

$$c_k z^k + c_{-k} z^{-k} = (c_k + c_{-k}) \cos(kt) + i(c_k - c_{-k}) \sin(kt) \ .$$

Setting
$$a_k := c_k + c_{-k}\ , \quad b_k := i(c_k - c_{-k}) \tag{4.5}$$
we can write $p := \sum_{k=-n}^{n} c_k \chi_k \in \mathcal{P}(S)$ in the form
$$p(e^{it}) = \frac{a_0}{2} + \sum_{k=1}^{n} \bigl[a_k \cos(kt) + b_k \sin(kt)\bigr]\ . \tag{4.6}$$

This suggests the following definition: For $n \in \mathbb{N}$ and $a_k, b_k \in \mathbb{K}$, the function
$$T_n : \mathbb{R} \to \mathbb{K}\ , \quad t \mapsto \frac{a_0}{2} + \sum_{k=1}^{n} \bigl[a_k \cos(kt) + b_k \sin(kt)\bigr] \tag{4.7}$$

is called a ($\mathbb{K}$-valued) **trigonometric polynomial**. If $\mathbb{K} = \mathbb{R}$ (or $\mathbb{K} = \mathbb{C}$), then T_n is called **real** (or **complex**). If $(a_n, b_n) \neq (0,0)$, then T_n is a trigonometric polynomial of degree n.

4.11 Remarks (a) Let
$$\mathcal{P}(S,\mathbb{R}) := \bigl\{ p = \textstyle\sum_{k=-n}^{n} c_k \chi_k\ ;\ c_{-k} = \overline{c}_k,\ -n \leq k \leq n,\ n \in \mathbb{N} \bigr\}\ .$$

Then $\mathcal{P}(S,\mathbb{R}) = \mathcal{P}(S,\mathbb{C}) \cap C(S,\mathbb{R})$, and $\mathcal{P}(S,\mathbb{R})$ is a real subalgebra of $C(S,\mathbb{R})$.

Proof For $p \in \mathcal{P}(S,\mathbb{R})$ we have
$$\overline{p} = \sum_{k=-n}^{n} \overline{c}_k \overline{\chi}_k = \sum_{k=-n}^{n} c_{-k} \chi_{-k} = p\ .$$

This shows that $\mathcal{P}(S,\mathbb{R}) \subseteq \mathcal{P}(S) \cap C(S,\mathbb{R})$. If $p \in \mathcal{P}(S)$ is real valued, then it follows from $\overline{\chi}_k = \chi_{-k}$ that
$$\sum_{k=-n}^{n} \overline{c}_k \chi_{-k} = \overline{p} = p = \sum_{k=-n}^{n} c_k \chi_k = \sum_{k=-n}^{n} c_{-k} \chi_{-k}\ ,$$

that is,
$$\sum_{k=-n}^{n} (c_{-k} - \overline{c}_k) \chi_{-k} = 0\ . \tag{4.8}$$

Since χ_{-n} is nowhere zero, it follows from $\chi_{-k} = \chi_{-n}\chi_{n-k}$ that (4.8) is equivalent to
$$\varphi := \sum_{k=0}^{2n} a_k \chi_k = 0 \tag{4.9}$$

with $a_{n-k} := c_{-k} - \overline{c}_k$ for all $-n \leq k \leq n$. Since φ is the restriction of a polynomial to S, it follows from the identity theorem for polynomials (Remark I.8.19(c)) that $a_k = 0$ for all $0 \leq k \leq 2n$. Thus p is in $\mathcal{P}(S,\mathbb{R})$, which proves the first claim. The second claim is now clear. ■

(b) Let $\mathcal{TP}(\mathbb{R},\mathbb{K})$ be the set of all $\mathbb{K}$-valued trigonometric polynomials. Then $\mathcal{TP}(\mathbb{R},\mathbb{K})$ is a subalgebra of $BC(\mathbb{R},\mathbb{K})$ and

$$\mathrm{cis}^* : \mathcal{P}(S,\mathbb{K}) \to \mathcal{TP}(\mathbb{R},\mathbb{K}) , \quad p \mapsto p \circ \mathrm{cis}$$

is an algebra isomorphism.

Proof It follows easily from (4.5), (4.6) and (a) that the function cis^* is well defined. It is also clear that $\mathcal{TP}(\mathbb{R},\mathbb{K})$ is a subspace of $BC(\mathbb{R},\mathbb{K})$ and that cis^* is linear and injective. Let $T_n \in \mathcal{TP}(\mathbb{R},\mathbb{K})$ be as in (4.7) and set $p := \sum_{k=-n}^{n} c_k \chi_k$ with

$$c_0 := a_0/2 , \quad c_k := (a_k - ib_k)/2 , \quad c_{-k} := (a_k + ib_k)/2 , \qquad 1 \le k \le n . \tag{4.10}$$

Then it follows from (a) that p is in $\mathcal{P}(S,\mathbb{K})$, and (4.5) and (4.6) imply that $T_n = p \circ \mathrm{cis}$. Thus cis^* is surjective and hence also a vector space isomorphism. Moreover

$$\mathrm{cis}^*(pq) = (pq) \circ \mathrm{cis} = (p \circ \mathrm{cis})(q \circ \mathrm{cis}) = (\mathrm{cis}^* p)(\mathrm{cis}^* q) , \qquad p,q \in \mathcal{P}(S,\mathbb{K}) ,$$

and so $\mathrm{cis}^* : \mathcal{P}(S,\mathbb{K}) \to BC(\mathbb{R},\mathbb{K})$ is an algebra homomorphism. It follows from this that $\mathcal{TP}(\mathbb{R},\mathbb{K})$, the image of $\mathcal{P}(S,\mathbb{K})$ under cis^*, is a subalgebra of $BC(\mathbb{R},\mathbb{K})$ and that cis^* is an isomorphism from $\mathcal{P}(S,\mathbb{K})$ to $\mathcal{TP}(\mathbb{R},\mathbb{K})$. ■

(c) The subalgebra $\mathcal{TP}(\mathbb{R},\mathbb{K})$ is not dense in $BC(\mathbb{R},\mathbb{K})$.

Proof Define $f \in BC(\mathbb{R},\mathbb{K})$ by

$$f(t) := \begin{cases} -2\pi , & -\infty < t < -2\pi , \\ t , & -2\pi \le t \le 2\pi , \\ 2\pi , & 2\pi < t < \infty . \end{cases}$$

Suppose, contrary to the claim, that $\mathcal{TP}(\mathbb{R},\mathbb{K})$ is dense in $BC(\mathbb{R},\mathbb{K})$. Then there is some $T \in \mathcal{TP}(\mathbb{R},\mathbb{K})$ such that $\|f - T\|_\infty < 2\pi$. In particular, $|T(2\pi) - f(2\pi)| < 2\pi$ and so $T(2\pi) > 0$. Since $T(2\pi) = T(0) = T(-2\pi)$ and $f(-2\pi) = -2\pi$, this implies

$$|T(-2\pi) - f(-2\pi)| = |T(2\pi) + 2\pi| > 2\pi ,$$

which contradicts $\|f - T\|_\infty < 2\pi$. ■

By Example 4.1(e), the closure of $\mathcal{TP}(\mathbb{R},\mathbb{K})$ in $BC(\mathbb{R},\mathbb{K})$ is a Banach algebra. We next show that this Banach algebra is precisely the algebra of continuous 2π-periodic $\mathbb{K}$-valued functions on $\mathbb{R}$.

Periodic Functions

First we prove several general properties of periodic functions. Let M be a set and $p \ne 0$. Then $f : \mathbb{R} \to M$ is called **periodic**[2] with **period** p (or simply p-**periodic**) if $f(t + p) = f(t)$ for all $t \in \mathbb{R}$.

---

[2]This is a special case of the definition given in the footnote for Corollary III.6.14.

4.12 Remarks **(a)** A p-periodic function is completely determined by its restriction to any interval with length p.

(b) Let $f : \mathbb{R} \to M$ be p-periodic and $q > 0$. Then the function

$$\mathbb{R} \to M , \quad t \mapsto f(tp/q)$$

is q-periodic. Consequently, for the study of periodic functions with a fixed period p, it suffices to consider only the case $p = 2\pi$.

(c) Let $\mathrm{Funct}_{2\pi}(\mathbb{R}, M)$ be the set of 2π-periodic functions from $\mathbb{R}$ to M. Then

$$\mathrm{cis}^* : M^S \to \mathrm{Funct}_{2\pi}(\mathbb{R}, M) , \quad g \mapsto g \circ \mathrm{cis}$$

is bijective. Using this bijection, we can identify the 2π-periodic functions with the set of functions on the unit circle.

Proof Since $\mathrm{cis} : \mathbb{R} \to S$ is periodic with period 2π, for each $g \in M^S$, the function $g \circ \mathrm{cis}$ is also 2π-periodic. By Proposition III.6.15, $\varphi := \mathrm{cis} \,|\, [0, 2\pi)$ is a bijection from $[0, 2\pi)$ to S. Thus, for $f \in \mathrm{Funct}_{2\pi}(\mathbb{R}, M)$, $g := f \circ \varphi^{-1}$ is a well defined function from S to M such that $g \circ \mathrm{cis} = f$. Hence cis^* is bijective. ■

(d) Suppose that M is a metric space and $f \in C(\mathbb{R}, M)$ is periodic and nonconstant. Then f has a least positive period p, the **minimal period**, and $p\mathbb{Z}^\times$ is the set of all periods of f.

Proof For $t \in \mathbb{R}$, let $P_t := \{\, p \in \mathbb{R} \ ; \ f(t + p) = f(t) \,\}$ and $P := \bigcap_{t \in \mathbb{R}} P_t$. Then $P \backslash \{0\}$ is the set of all periods of f. Since f is continuous, the function $p \mapsto f(t + p)$ is also continuous on $\mathbb{R}$. Because P_t is the fiber of the function $p \mapsto f(t + p)$ at the point $f(t)$, it follows from Example III.2.22(a) that P_t is closed in $\mathbb{R}$. Thus P, being an intersection of closed sets, is itself closed. Moreover, $P \neq \{0\}$ since f is periodic, and $P \neq \mathbb{R}$ since f is not constant. For $p_1, p_2 \in P$, we have $f(t + p_1 - p_2) = f(t + p_1) = f(t)$ for all $t \in \mathbb{R}$, meaning that $p_1 - p_2$ is in P. Setting $p_1 = 0$ in this we see that, if p is in P, then so is $-p$. Replacing p_2 by $-p_2$, we see that $p_1 + p_2 \in P$. Thus P is a closed subgroup of $(\mathbb{R}, +)$.

Because $P \neq \mathbb{R}$, there must be a smallest positive element p_0 in P. Otherwise there would be, for each $\varepsilon > 0$, some $p \in P \cap (0, \varepsilon)$, and so, for each $s \in \mathbb{R}$, some $k \in \mathbb{Z}$ such that $|s - kp| < \varepsilon$. Consequently P would be dense in $\mathbb{R}$, which, by Proposition 4.4 would imply $P = \mathbb{R}$. Clearly $p_0\mathbb{Z}$ is a subgroup of P. Suppose that $q \in P \backslash p_0\mathbb{Z}$ and, without loss of generality, that $q > 0$. Then there are $r \in (0, p_0)$ and $k \in \mathbb{N}^\times$ such that $q = kp_0 + r$. From this it follows that $r = q - kp_0 \in P$, which contradicts the minimality of p_0. This shows that $P = p_0\mathbb{Z}$.[3] ■

Let M be a metric space and

$$C_{2\pi}(\mathbb{R}, M) := \{\, f \in C(\mathbb{R}, M) \ ; \ f \text{ is } 2\pi\text{-periodic} \,\} .$$

The following discussion shows that the function cis^* of Remark 4.11(b) has a continuous extension on $C(S, \mathbb{K})$. This result, which is a considerable strengthening

[3]This proof shows that, if G is a closed subgroup of $(\mathbb{R}, +)$, then either $G = \{0\}$, $G = (\mathbb{R}, +)$, or G is infinite **cyclic** (that is, G is an infinite group generated by a single element).

of Remark 4.12(c), implies that we can identify continuous 2π-periodic functions with continuous functions on S.

4.13 Proposition *If M is a metric space, then cis^* is a bijection from $C(S,M)$ to $C_{2\pi}(\mathbb{R},M)$.*

Proof From Remark 4.12(c) and the continuity of cis it follows that cis^* is an injective function from $C(S,M)$ to $C_{2\pi}(\mathbb{R},M)$. Since cis^* is bijective from M^S to $\mathrm{Funct}_{2\pi}(\mathbb{R},M)$, it suffices to show that, for all $f \in C_{2\pi}(\mathbb{R},M)$, the function $(\mathrm{cis}^*)^{-1}(f)$ is continuous on S. Note first that, for all $\varphi = \mathrm{cis} \mid [0,2\pi)$, we have $\varphi^{-1} = \arg \mid S$. It follows from Exercise III.6.9 that φ^{-1} maps the set $S^\bullet := S \backslash \{-1\}$ continuously into $(-\pi,\pi)$. Thus $g := (\mathrm{cis}^*)^{-1}(f) = f \circ \varphi^{-1}$ maps the set $S^\bullet$ continuously into M. As $t \in (-\pi,\pi)$ approaches $\pm\pi$, we have $\mathrm{cis}(t) \to -1$, so the 2π-periodicity of f implies that

$$\lim_{\substack{z \to -1 \\ z \in S^\bullet}} g(z) = f(\pi) = (\mathrm{cis}^*)^{-1}(f)(-1) .$$

Consequently $(\mathrm{cis}^*)^{-1}(f)$ is continuous on S. ■

4.14 Corollary *Let $E := (E, |\cdot|)$ be a Banach space. Then $C_{2\pi}(\mathbb{R},E)$ is a closed subspace of the Banach space $BC(\mathbb{R},E)$ and hence a Banach space with the maximum norm*

$$\|f\|_{C_{2\pi}} := \max_{-\pi \le t \le \pi} |f(t)| ,$$

and cis^* is an isometric isomorphism[4] *from $C(S,E)$ to $C_{2\pi}(\mathbb{R},E)$.*

Proof By Remark 4.12(a), it is clear that $C_{2\pi}(\mathbb{R},E)$ is a subspace of $BC(\mathbb{R},E)$, and that $\|\cdot\|_\infty$ induces the norm $\|\cdot\|_{C_{2\pi}}$. It is also clear that the pointwise limit (and hence, in particular, the uniform limit) of a sequence of 2π-periodic functions is also 2π-periodic. Thus $C_{2\pi}(\mathbb{R},E)$ is a closed subspace of the Banach space $BC(\mathbb{R},E)$, and so is itself a Banach space. By Proposition 4.13, cis^* is a bijection from $C(S,E)$ to $C_{2\pi}(S,E)$ which is trivially linear. Since cis, by Proposition III.6.15, is a bijection from $[-\pi,\pi)$ to S, it follows that

$$\|\mathrm{cis}^*(f)\|_{C_{2\pi}} = \max_{-\pi \le t \le \pi} \bigl|f\bigl(\mathrm{cis}(t)\bigr)\bigr| = \max_{z \in S} |f(z)| = \|f\|_{C(S,E)}$$

for all $f \in C(S,E)$. Hence cis^* is isometric. ■

4.15 Remark For each $a \in \mathbb{R}$, we have

$$\|f\|_{C_{2\pi}} = \max_{a \le t \le a+2\pi} |f(t)| .$$

Proof This follows directly from the periodicity of f. ■

[4] Naturally, in connection with vector spaces, 'isomorphism' means 'vector space isomorphism'.

The Trigonometric Approximation Theorem

After this discussion of periodic functions, we can now easily prove the trigonometric form of the Weierstrass approximation theorem.

4.16 Theorem *$C_{2\pi}(\mathbb{R},\mathbb{K})$ is a Banach algebra with unity element* $\mathbf{1}$*, and the subalgebra of trigonometric polynomials $\mathcal{TP}(\mathbb{R},\mathbb{K})$ is dense in $C_{2\pi}(\mathbb{R},\mathbb{K})$. In addition,* cis^* *is an isometric algebra isomorphism from $C(S,\mathbb{K})$ to $C_{2\pi}(\mathbb{R},\mathbb{K})$.*

Proof By Corollary 4.14, cis^* is an isometric vector space isomorphism from $C := C(S,\mathbb{K})$ to $C_{2\pi} := C_{2\pi}(\mathbb{R},\mathbb{K})$. Example 4.10(c) and Remark 4.11(a) imply that $\mathcal{P} := \mathcal{P}(S,\mathbb{K})$ is a dense subalgebra of C. Remark 4.11(b) says that $\mathrm{cis}^* \,|\mathcal{TP}$ is an algebra isomorphism from $\mathcal{P}$ to $\mathcal{TP} := \mathcal{TP}(\mathbb{R},\mathbb{K})$. Now let $f, g \in C$. Then there are sequences (f_n) and (g_n) in $\mathcal{P}$ such that $f_n \to f$ and $g_n \to g$ in C. By the continuity of cis^* and the continuity of multiplication it follows that

$$\mathrm{cis}^*(fg) = \lim \mathrm{cis}^*(f_n g_n) = \lim(\mathrm{cis}^* f_n)(\mathrm{cis}^* g_n) = (\mathrm{cis}^* f)(\mathrm{cis}^* g) \ .$$

Thus cis^* is an algebra isomorphism from C to $C_{2\pi}$. Since $\mathcal{P}$ is dense in C and cis^* is a homeomorphism from C to $C_{2\pi}$, the image $\mathcal{TP}$ of $\mathcal{P}$ under cis^* is dense in $C_{2\pi}$ (see Remark 4.2(d)). ■

4.17 Corollary (trigonometric form of the Weierstrass approximation theorem) *For $f \in C_{2\pi}(\mathbb{R},\mathbb{K})$ and $\varepsilon > 0$, there are $n \in \mathbb{N}$ and $a_k, b_k \in \mathbb{K}$ such that*

$$\Bigl| f(t) - \frac{a_0}{2} - \sum_{k=1}^{n} \bigl[a_k \cos(kt) + b_k \sin(kt)\bigr] \Bigr| < \varepsilon$$

for all $t \in \mathbb{R}$.

Theorem 4.16 says, in particular, that the Banach algebras $C(S,\mathbb{K})$ and $C_{2\pi}(\mathbb{R},\mathbb{K})$ are isomorphic and isometric. This means that, for applications, as well as for questions about continuity and limits, we can use whichever of these spaces is most convenient. For algebraic operations and abstract considerations, this is often the algebra $C(S,\mathbb{K})$, whereas, for the concrete representations of 2π-periodic functions, the space $C_{2\pi}(\mathbb{R},\mathbb{K})$ is usually preferred.

Corollary 4.17 suggests several questions:

• What conditions on the coefficients (a_k) and (b_k) ensure that the **trigonometric series**

$$\frac{a_0}{2} + \sum_k \bigl[a_k \cos(k\,\cdot\,) + b_k \sin(k\,\cdot\,)\bigr] \tag{4.11}$$

converges uniformly on $\mathbb{R}$? When this occurs, the series clearly represents a continuous periodic function with period 2π.

• In the case that $f \in C_{2\pi}(\mathbb{R}, \mathbb{K})$ can be represented by a trigonometric series, how can the coefficients a_k and b_k be calculated? Are they uniquely determined by f? Can every 2π-periodic continuous function be represented in this way?

For the first of these questions, the Weierstrass majorant criterion provides an easy sufficient condition. We will return to the second question in later chapters.

Exercises

1 Verify that the Banach space $BC^k(X, \mathbb{K})$ of Exercise 2.10 is an algebra with unity and that multiplication is continuous. For which k is $BC^k(X, \mathbb{K})$ a Banach algebra?

2 Let $x_0, \ldots, x_k \in \mathbb{K}^n$ be nonzero. Show that $\{\, x \in \mathbb{K}^n \;;\; \prod_{j=0}^k (x \,|\, x_j) \neq 0 \,\}$ is open and dense in $\mathbb{K}^n$.

3 Let M be a metric space. Prove or disprove the following:

(a) Finite intersections of dense subsets of M are dense in M.

(b) Finite intersections of open dense subsets of M are open and dense in M.

4 Let D_k, $k \in \mathbb{N}$, be open dense subsets of $\mathbb{K}^n$ and $D := \bigcap_k D_k$. Show the following:[5]

(a) D is dense in $\mathbb{K}^n$.

(b) D is uncountable.

(Hint: (a) Set $F_k := \bigcap_{\ell=0}^k D_k$. Then F_k is open and dense, and $F_0 \supseteq F_1 \supseteq \cdots$. Let $x \in \mathbb{K}^n$ and $r > 0$. Then there are $x_0 \in F_0$ and $r_0 > 0$ such that $\bar{\mathbb{B}}(x_0, r_0) \subseteq \mathbb{B}(x, r) \cap F_0$. Choose inductively $x_k \in F_k$ and $r_k > 0$ such that $\bar{\mathbb{B}}(x_{k+1}, r_{k+1}) \subseteq \mathbb{B}(x_k, r_k) \cap F_k$ for all $k \in \mathbb{N}$. Now use Exercise III.3.4. (b) If D were countable, there would be $x_m \in \mathbb{K}^n$ such that $D = \{\, x_m \;;\; m \in \mathbb{N} \,\}$. Consider $\bigcap_m \{x_m\}^c \cap \bigcap_k D_k$.)

5 Show that there is no function from $\mathbb{R}$ to $\mathbb{R}$ which is continuous at each rational point and discontinuous at each irrational point. (Hint: Let f be a such function. Consider $D_k := \{\, x \in \mathbb{R} \;;\; \omega_f(x) < 1/k \,\}$ for all $k \in \mathbb{N}^\times$, where ω_f is the modulus of continuity from Exercise III.1.17. By Exercise III.2.20, D_k is open. But then $\mathbb{Q} \subseteq D_k$ and $\bigcap_k D_k = \mathbb{Q}$, contradicting 4(b).)

6 Let V be a finite dimensional normed vector space with basis $\{b_1, \ldots, b_n\}$, and D a countable dense subset of $\mathbb{K}$. Show that $\{\, \sum_{k=1}^n \alpha_k b_k \;;\; \alpha_k \in D \,\}$ is countable and dense in V.

7 Let $M \subseteq \mathbb{R}^n$ and $A := \mathbb{K}[X_1, \ldots, X_n] \,|\, M$. Show that A separates the points of M and is self adjoint.

8 Suppose that $-\infty < a < b < \infty$ and $f \in C([a, b], \mathbb{K})$. Show that f has an antiderivative. (Hint: Let (p_n) be a sequence of polynomials which converges uniformly to f. Find $F_n \in C^1([a, b], \mathbb{K})$ such that $F_n' = p_n$ and $F_n(a) = 0$. Now apply Exercise 2.11 and Theorem 2.8.)

9 Let $f \in C_{2\pi}(\mathbb{R}, \mathbb{R})$ be differentiable. Show that f' has a zero in $(0, 2\pi)$.

[5](a) is a special case of the Baire category theorem.

10 Let $D_0(\mathbb{R},\mathbb{K})$ be the set of all absolutely convergent trigonometric series with $a_0 = 0$ (see (4.11)). Show the following:

(a) $D_0(\mathbb{R},\mathbb{K})$ is a subalgebra of $C_{2\pi}(\mathbb{R},\mathbb{K})$.

(b) Each $f \in D_0(\mathbb{R},\mathbb{K})$ has a 2π-periodic antiderivative.

(c) Each $f \in D_0(\mathbb{R},\mathbb{R})$ has a zero in $(0,2\pi)$.

(d) Claim (c) is false for functions in $D_0(\mathbb{R},\mathbb{C})$.

11 For $n \in \mathbb{N}$ and $0 \le k \le n$, the (**elementary**) **Bernstein polynomial** $B_{n,k}$ is defined by

$$B_{n,k} := \binom{n}{k} X^k (1-X)^{n-k} .$$

Show the following:

(a) For each $n \in \mathbb{N}$, the Bernstein polynomials form a **decomposition of unity**, that is, $\sum_{k=0}^n B_{n,k} = \mathbf{1}$.

(b) $\sum_{k=0}^n kB_{n,k} = nX$, $\sum_{k=0}^n k(k-1)B_{n,k} = n(n-1)X^2$.

(c) $\sum_{k=0}^n (k-nX)^2 B_{n,k} = nX(1-X)$.

(Hint: For $y \in \mathbb{R}$, let $p_{n,y} := (X+y)^n$. Consider $Xp'_{n,y}$ and $X^2p''_{n,y}$ and set $y := 1-X$.)

12 Let E be a Banach space and $f \in C([0,1],E)$. Show that the sequence $(B_n(f))$ of **Bernstein polynomials for** f,

$$B_n(f) := \sum_{k=0}^n f\Big(\frac{k}{n}\Big) B_{n,k} , \qquad n \in \mathbb{N} ,$$

converges in $C([0,1],E)$ (and hence uniformly on $[0,1]$) to f. (Hint: For suitable $\delta > 0$ consider $|x - n/k| \le \delta$ and $|x - n/k| > \delta$, and use Exercise 11.)

13 Let X be a topological space. A family $\mathcal{B}$ of open sets of X is called a **basis for the topology** of X if, for each $x \in X$ and neighborhood U of x, there is some $B \in \mathcal{B}$ such that $x \in B \subseteq U$. Prove the following:

(a) Any separable metric space has a countable basis of open sets.

(b) Any subset of a separable metric space is separable (that is, a separable metric space with the induced metric).

(c) Any subset of $\mathbb{R}^n$ is separable.

14 Let X be a compact separable metric space. Show that $C(X,\mathbb{K})$ is a separable Banach space. (Hint: Consider linear combinations with rational coefficients of 'monomials' $d_{B_1}^{m_1} \cdot \dots \cdot d_{B_k}^{m_k}$ with $k \in \mathbb{N}$, $m_j \in \mathbb{N}$, $B_j \in \mathcal{B}$, where $\mathcal{B}$ is a basis for the topology of X, and $d_B := d(\cdot, B^c)$ as in Example III.1.3(l) for all $B \in \mathcal{B}$.)

Remark We will show in Proposition IX.1.8 that any compact metric space is separable.

Appendix

Introduction to Mathematical Logic

1 Logic is about statements and proofs. Examples of *statements* are: *The equation* $x^2 + 1 = 0$ *has no solution* and *2 is greater than 3* and *Given a line and a point not on the line, there is exactly one line which passes through the point which is parallel to the given line* (the parallel postulate as formulated by Proklos).

Statements can be 'true', 'false' or 'unprovable'. Standing alone, a statement may have no truth value, but may become true or false in connection with other statements.

In logic, statements are usually written in a special formal language. Such a language is based on simple word formation rules and grammar, and so avoids the ambiguities present in usual languages. This can however lead to immense, hard to understand, sentences.

Since we wish to use conventional language in this discussion, a precise definition of the word 'statement' is not possible. Our statements are sentences in the English language. But that does not mean that sentences and statements are the same thing.

Firstly it is possible for different sentences to be the same statement. For example, *There is no number* x *such that* $x^2 = -1$ is the same statement as the first example. Secondly, many sentences are ambiguous because words can have multiple meanings or because part of the intended statement is missing if it is seen as self-evident. For example, in the first example we have not explicitly said that x must be real. Finally most sentences from daily life are not statements in the sense intended here. We do not try to put a sentence such as *Team Canada strikes gold again* into a logical and coherent system of statements. We limit ourselves here to statements about *terms*, that is, about mathematical objects such as numbers, points, functions, and variables.

2 Even though we do not have a definition of a statement, we can at least provide rules for constructing statements:

a) *Equality*: Terms can always be equated. Thus we can construct the 'true' statement *The solution set of the equation* $x^2 - 1 = 0$ *is equal to* $\{-1, 1\}$ and the 'false' statement '$2 = [0, 1]$'.

b) *Membership*: Sentences such as *The point* P *lies on the line* $\mathcal{G}$, P *belongs to the line* $\mathcal{G}$ or P *is an element of the line* $\mathcal{G}$ are all the same statement. This kind of statement is often expressed using the membership symbol $\in$: '$P \in \mathcal{G}$'.

New statements can be constructed from other statements as follows:

c) Each statement ϕ has a *negation* $\neg\phi$. Thus *The equation* $x^2 + 1 = 0$ *has no solution* is the negation of *The equation* $x^2 + 1 = 0$ *has a solution.* The negation of *2 is greater than 3* is *2 is not greater than 3* (which is not the same as *2 is smaller than 3*).

d) From the two statements ϕ and ψ, we can construct the statement $\phi \to \psi$ (*if* ϕ *then* ψ). For example, we have the 'true', but seemingly abstruse, statement *If 2 is greater than 3, then the equation* $x^2 + 1 = 0$ *has a solution.*

e) The constructions in c) and d) can be combined. For example, from ϕ and ψ, we get the statements $\phi \vee \psi = (\neg\phi) \to \psi$ (ϕ *or* ψ) and $\phi \wedge \psi = \neg(\phi \to \neg\psi)$ (ϕ *and* ψ).

f) *Existence statements*: The statement *There exist real numbers* x *and* y *such that* $x^2 + y^2 = 1$ is often formally expressed using the symbol $\exists$ (*existential quantifier*):

$$\exists x\, \exists y \Big(\big((x \in \mathbb{R}) \wedge (y \in \mathbb{R})\big) \wedge (x^2 + y^2 = 1) \Big).$$

Here $\mathbb{R}$ is the set of real numbers.
The expression $\big((x \in \mathbb{R}) \wedge (y \in \mathbb{R})\big) \wedge (x^2 + y^2 = 1)$ is not a statement because x and y are *variables.* It is instead a *formula* which becomes a statement if the variables are replaced by numbers or, as above, becomes an existence statement using existential quantifiers.

g) A statement such as *For all real* x *and all real* y, *we have* $x^2 + y^2 > 0$ is a 'double' negated existence statement:

$$\neg(\exists x)(\exists y)\Big(\neg\big((x \in \mathbb{R}) \wedge (y \in \mathbb{R}) \to (x^2 + y^2 > 0)\big)\Big).$$

In practice this statement is abbreviated using the symbol $\forall$ (*universal quantifier*):

$$(\forall x)(\forall y)\big((x \in \mathbb{R}) \wedge (y \in \mathbb{R}) \to (x^2 + y^2 > 0)\big).$$

3 Each set of statements Γ has a *logical closure* $\overline{\Gamma}$, which is the set of all statements which are *implied* by Γ. Of course, $\overline{\Gamma}$ contains the set Γ (assumption rule) as well as the logical closure $\overline{\Delta}$ of any subset Δ of $\overline{\Gamma}$ (chain rule). In the following we collect only the most important of the remaining rules of logic. The notation $\Gamma \vdash \phi$ means that Γ implies ϕ. Similarly $\Gamma, \psi \vdash \phi$ means that ϕ is implied by the statements in Γ together with the statement ψ.

a) $\Gamma \vdash (t = t)$ for each set of statements Γ and each constant term t (equality rule). In particular, $t = t$ is implied by the 'empty' set of statements $\emptyset$.

b) $\psi, \neg\psi \vdash \phi$ for all statements ϕ and ψ (contradiction rule).

c) $\Gamma, \psi \vdash \phi$ and $\Gamma, \neg\psi \vdash \phi$ imply $\Gamma \vdash \phi$ (cases rule).

d) $\Gamma, \phi \vdash \psi$ implies $\Gamma \vdash (\phi \to \psi)$ (implication rule).

e) $\phi, (\phi \to \psi) \vdash \psi$ (modus ponens).

f) If $a, b, \ldots, c$ are constant terms and $\phi(x, y, \ldots, z)$ is a formula with the free variables $x, y, \ldots, z$, then $\phi(a, b, \ldots, c) \vdash (\exists x)(\exists y) \ldots (\exists z)\phi(x, y, \ldots, z)$ (substitution rule).

4 By combining the rules in **3** we get additional constructions:

a) $\Gamma \vdash (\phi \to \psi)$ implies $\Gamma, \phi \vdash \psi$ (the converse of the implication rule):
From $\phi, (\phi \to \psi) \vdash \psi$ (modus ponens) we get $\Gamma, \phi, (\phi \to \psi) \vdash \psi$. Then $\Gamma, \phi \vdash \psi$ because $\phi \to \psi$ is in $\overline{\Gamma}$ (chain rule).

b) $(\phi \to \psi) \vdash (\neg\psi \to \neg\phi)$ (First contrapositive rule):
From $\phi, (\phi \to \psi) \vdash \psi$ (modus ponens) we get $\phi, (\phi \to \psi), \neg\psi \vdash \psi$.
Since $\phi, (\phi \to \psi), \neg\psi \vdash \neg\psi$ also holds, this implies $\phi, (\phi \to \psi), \neg\psi \vdash \neg\phi$ (contradiction rule).
From $\phi, (\phi \to \psi), \neg\psi \vdash \neg\phi$ and $\neg\phi, (\phi \to \psi), \neg\psi \vdash \neg\phi$ it follows that $(\phi \to \psi), \neg\psi \vdash \neg\phi$ (cases rule).
Finally, this gives us $(\phi \to \psi) \vdash (\neg\psi \to \neg\phi)$ (implication rule).
Similarly, one can prove the following:
$(\phi \to \neg\psi) \vdash (\psi \to \neg\phi)$ (second contrapositive rule).
$(\neg\phi \to \psi) \vdash (\neg\psi \to \phi)$ (third contrapositive rule).
$(\neg\phi \to \neg\psi) \vdash (\psi \to \phi)$ (fourth contrapositive rule).
For example, to prove the fourth rule one replaces ϕ, $\neg\phi$, ψ and $\neg\psi$ by $\neg\phi$, ϕ, $\neg\psi$ and ψ respectively in the proof of the first rule.
Of course, the four contrapositive rules coincide if the underlying language is such that the double negation $\neg\neg\phi$ is the same as ϕ. This may be so in everyday conversation where we consider the double negation *It is not true that the equation* $x^2 + 1 = 0$ *has no solution* as a reformulation of the statement *The equation* $x^2 + 1 = 0$ *has a solution*. In the usual formal language of logic, ϕ and $\neg\neg\phi$ are distinct statements which are equivalent in the sense of implication:

c) $\phi \vdash \neg\neg\phi$ and $\neg\neg\phi \vdash \phi$ (double negation rule):
From $\neg\phi \vdash \neg\phi$ (assumption rule) we get
$\emptyset \vdash (\neg\phi \to \neg\phi) \vdash (\phi \to \neg\neg\phi)$ (implication and second contrapositive rules).
It then follows from $\emptyset \vdash (\phi \to \neg\neg\phi)$ (chain rule) that $\phi \vdash \neg\neg\phi$ (converse of the implication rule).

d1) $\psi \vdash (\phi \to \psi)$:
From $\psi, \phi \vdash \psi$ (assumption rule) we get $\psi \vdash (\phi \to \psi)$ (implication rule).

d2) $\neg\phi \vdash (\phi \to \psi)$:
This follows from $\neg\phi \vdash (\neg\psi \to \neg\phi)$ and $(\neg\psi \to \neg\phi) \vdash (\phi \to \psi)$ (fourth contrapositive rule) using the chain rule.

d3) $\phi, \neg\psi \vdash \neg(\phi \to \psi)$:
From $\phi, (\phi \to \psi) \vdash \psi$ (modus ponens) we get $\phi \vdash \big((\phi \to \psi) \to \psi\big)$ (implication rule) as well as $\phi \vdash \big(\neg\psi \to \neg(\phi \to \psi)\big)$ (first contrapositive and chain rules). The claim then follows from the converse of the implication rule.

e1) $\phi, \psi \vdash \phi \wedge \psi$ (conjunction rule):
From $\phi, (\phi \to \neg\psi) \vdash \neg\psi$ (modus ponens) we get $\phi \vdash \big((\phi \to \neg\psi) \to \neg\psi\big)$ (implication rule). Then $\phi \vdash \big(\psi \to \neg(\phi \to \neg\psi)\big)$ follows from the second contrapositive and chain rules. The claim is then a consequence of the converse of the implication rule.

e2) $(\phi \wedge \psi \vdash \phi)$:
From $\neg\phi \vdash (\phi \to \neg\psi)$ (d2) we get $\emptyset \vdash \big(\neg\phi \to (\phi \to \neg\psi)\big) \vdash \big(\neg(\phi \to \neg\psi) \to \phi\big)$ (third contrapositive rule) and $\neg(\phi \to \neg\psi) \vdash \phi$ (converse of the implication rule).

e3) $(\phi \wedge \psi \vdash \psi)$:
From $\neg\psi \vdash (\phi \to \neg\psi)$ (d1) we get $\emptyset \vdash \big(\neg\psi \to (\phi \to \neg\psi)\big) \vdash \big(\neg(\phi \to \neg\psi) \to \psi\big)$ (third contrapositive rule) and $\neg(\phi \to \neg\psi) \vdash \psi$ (converse of the implication rule).

f1) $\psi \vdash (\phi \vee \psi) \vdash (\psi \vee \phi)$ (disjunction rule):
By definition we have $(\phi \vee \psi) = (\neg\phi \to \psi)$. So the first implication follows from d1), and the second from the third contrapositive rule.

f2) $(\phi \vee \psi), \neg\phi \vdash \psi$ (modus ponens).

5 Using these construction rules we can construct statements α such that $\emptyset \vdash \alpha$. For example, from $\phi \vdash \phi$ and the implication rule we get $\emptyset \vdash (\phi \to \phi)$ for any statement ϕ. In particular, we have $\emptyset \vdash (\psi \vee \neg\psi) = (\neg\psi \to \neg\psi)$ (law of the excluded middle)

Statements which are implied by the empty set can be thought of as *absolutely true.* For example, the statements, $t = t$, $\neg\phi \to (\phi \to \psi)$, $(\phi \vee \psi) \to (\psi \vee \phi)$, $\phi \to \neg\neg\phi$, and $(\psi \wedge \neg\psi) \to \phi$ are absolutely true.

Since mathematicians usually thirst for more than 'absolute truth', it is common to start with a set of statements Γ, called *axioms*, which arise in some particular mathematical context. Examples of such axioms are the parallel postulate in Euclidean geometry or the *extensionality axiom* of set theory (Sets x and y are equal if and only if any z in x is in y, and any z in y is in x):

$$\forall x\, \forall y \Big(\forall z \big((z \in x \to z \in y) \wedge (z \in y \to z \in x) \big) \to x = y \Big).$$

The goal of mathematics is then the exploration of the logical closure $\overline{\Gamma}$ of the given set of statements. We want to suppose that these axioms can be trusted, that is, Γ does not imply any contradictions of the form $(\neg\phi \wedge \phi) = \neg(\neg\phi \rightarrow \phi)$. If so, we say that a statement ϕ is *true* if it is in $\overline{\Gamma}$, and we say it is *false* if $\neg\phi$ is true.

The statement $\phi \vee \psi$ is true if one of the statements ϕ and ψ is true (disjunction rule), and it is false if both ϕ and ψ are false (**4**.f2). However, it is possible for $\phi \vee \psi$ to be true even if none of the statements ϕ, $\neg\phi$, ψ, $\neg\psi$ are in $\overline{\Gamma}$. For example, the statement $\psi \vee \neg\psi$ is absolutely true. So, in general, it is not true that ψ must be either true or false. It is entirely possibly that ψ is not *decidable*, that is, neither ψ nor $\neg\psi$ is implied by Γ.

If we consider only *decidable* statements, then there is a *truth function* that maps each decidable statement to one of the values T (= true) or F (= false). The following 'truth table' gives the truth values of combinations of decidable statements. The decidability of these combinations follows easily from **3** and **4**. For example, if ϕ is true and ψ is false, then $\neg\phi$, $\phi \rightarrow \psi$ and $\phi \wedge \psi$ are false, and $\phi \vee \psi$ is true.

ϕ	ψ	$\neg\phi$	$\phi \rightarrow \psi$	$\phi \vee \psi$	$\phi \wedge \psi$
T	T	F	T	T	T
T	F	F	F	T	F
F	T	T	T	T	F
F	F	T	T	F	F

6 For a more detailed discussion of logic, the reader is referred to the literature, for example, [EFT96]. Even though the grammar of the formal languages developed in the literature is completely simple, we prefer in this presentation to express our statements in English. After sufficient practice, it allows compact and precise formulations of mathematical statements. In English there is no sharp distinction between syntax and semantics: A *set* is a *collection of objects* — not just a sequence of symbols devoid of meaning. In formal languages, the interpretation is left to the reader. In English, the interpretation is usually built in.

Bibliography

[Art91] M. Artin. *Algebra.* Prentice Hall, Englewood Cliffs, N.J. 1991.

[Ded95] R. Dedekind. *What Are the Numbers and What Should They Be?* translated by H. Pogorzelski, W. Ryan, and W. Snyder. Research Institute for Mathematics (RIM), Monographs in Mathematics. 1995.

[Dug66] J. Dugundji. *Topology.* Allyn & Bacon, Boston, 1966.

[Ebb77] H.-D. Ebbinghaus. *Einführung in die Mengenlehre.* Wiss. Buchgesellschaft, Darmstadt, 1977.

[EFT96] H.-D. Ebbinghaus, J. Flum, W. Thomas. *Mathematical Logic, 2nd. Edition.* Springer Verlag, New York, 1996.

[FP85] U. Friedrichsdorf, A. Prestel. *Mengenlehre für den Mathematiker.* Vieweg & Sohn, Braunschweig/Wiesbaden, 1985.

[Gab96] P. Gabriel. *Matrizen, Geometrie, Lineare Algebra.* Birkhäuser, Basel, 1996.

[Hal74] P. Halmos. *Naive Set Theory.* Springer Verlag, New York, 1974.

[Hil23] D. Hilbert. *Grundlagen der Geometrie.* Anhang VI: Über den Zahlbegriff. Teubner, Leipzig, 1923.

[IK66] E. Isaacson, H.B. Keller. *Analysis of Numerical Methods.* Wiley, New York, 1966.

[Koe83] M. Koecher. *Lineare Algebra und analytische Geometrie.* Springer Verlag, Berlin, 1983.

[Lan30] E. Landau. *Grundlagen der Analysis* (4th ed., Chelsea, New York 1965). Leipzig, 1930.

[Wal82] R. Walter. *Einführung in die lineare Algebra.* Vieweg & Sohn, Braunschweig, 1982.

[Wal85] R. Walter. *Lineare Algebra und analytische Geometrie.* Vieweg & Sohn, Braunschweig, 1985.

[WS79] H. Werner, R. Schaback. *Praktische Mathematik II.* Springer Verlag, Berlin, 1979.

Index